The Computational Neurobiology of Reaching and Pointing

Computational Neuroscience

Terrence J. Sejnowski and Tomaso A. Poggio, editors

Neural Nets in Electric Fish, Walter Heiligenberg, 1991

The Computational Brain, Patricia S. Churchland and Terrence J. Sejnowski, 1992

Dynamic Biological Networks: The Stomatogastic Nervous System, edited by Ronald M. Harris-Warrick, Eve Marder, Allen I. Selverston, and Maurice Moulins, 1992

The Neurobiology of Neural Networks, edited by Daniel Gardner, 1993

Large-Scale Neuronal Theories of the Brain, edited by Christof Koch and Joel L. Davis, 1994

The Theoretical Foundations of Dendritic Function: Selected Papers of Wilfrid Rall with Commentaries, edited by Idan Segev, John Rinzel, and Gordon M. Shepherd, 1995

Models of Information Processing in the Basal Ganglia, edited by James C. Houk, Joel L. Davis, and David G. Beiser, 1995

Spikes: Exploring the Neural Code, Fred Rieke, David Warland, Rob de Ruyter van Steveninck, and William Bialek, 1997

Neurons, Networks, and Motor Behavior, edited by Paul S. Stein, Sten Grillner, Allen I. Selverston, and Douglas G. Stuart, 1997

Methods in Neuronal Modeling: From Ions to Networks, second edition, edited by Christof Koch and Idan Segev, 1998

Fundamentals of Neural Network Modeling: Neuropsychology and Cognitive Neuroscience, edited by Randolph W. Parks, Daniel S. Levine, and Debra L. Long, 1998

Neural Codes and Distributed Representations: Foundations of Neural Computation, edited by Laurence Abbott and Terrence J. Sejnowski, 1999

Unsupervised Learning: Foundations of Neural Computation, edited by Geoffrey Hinton and Terrence J. Sejnowski, 1999

Fast Oscillations in Cortical Circuits, Roger D. Traub, John G. R. Jefferys, and Miles A. Whittington, 1999

Computational Vision: Information Processing in Perception and Visual Behavior, Hanspeter A. Mallot, 2000

Graphical Models: Foundations of Neural Computation, edited by Michael I. Jordan and Terrence J. Sejnowski, 2001

Self-Organizing Map Formation: Foundation of Neural Computation, edited by Klaus Obermayer and Terrence J. Sejnowski, 2001

Neural Engineering: Computation, Representation, and Dynamics in Neurobiological Systems, Chris Eliasmith and Charles H. Anderson, 2003

The Computational Neurobiology of Reaching and Pointing: A Foundation for Motor Learning, Reza Shadmehr and Steven P. Wise, 2005

The Computational Neurobiology of Reaching and Pointing

A Foundation for Motor Learning

Reza Shadmehr and Steven P. Wise

A Bradford Book
The MIT Press
Cambridge, Massachusetts
London, England

© 2005 Massachusetts Institute of Technology

All Rights Reserved. No part of this book may be reproduced in any form by any electronic or mechanical means (including photocopying, recording, or information storage and retrieval) without permission in writing from the publisher.

MIT Press books may be purchased at special quantity discounts for business or sales promotional use. For information, please email special_sales@mitpress.mit.edu or write to Special Sales Department, The MIT Press, 5 Cambridge Center, Cambridge, MA 02142.

This book was set in Palatino on 3B2 by Asco Typesetters, Hong Kong.
Printed and bound in the United States of America.

Library of Congress Cataloging-in-Publication Data

Shadmehr, Reza.
  The computational neurobiology of reaching and pointing : a foundation for motor learning / Reza Shadmehr and Steven P. Wise.
     p.   cm. — (Computational neuroscience)
  "A Bradford book"
  Includes bibliographical references and index.
  ISBN 0-262-19508-9 (alk. paper)
  1. Motor ability. 2. Motor learning. 3. Motor learning—Mathematical models.
I. Wise, Steven P. II. Title. III. Series.
QP303.S487 2005
152.3—dc22   2004042610

The views expressed in this book and in any electronic version do not necessarily represent those of the NIMH, the NIH, or the U.S. government.

10 9 8 7 6 5 4 3 2 1

To the first generation—Emilio Bizzi, Mahlon DeLong, Peter Strick, Jun Tanji, and Tom Thach

Contents in Brief

Contents in Detail   ix
Preface   xv

    1   Introduction   1

I Evolution, Anatomy, and Physiology   7

    2   Our Moving History: The Evolution of the Vertebrate CNS   9

    3   Burdens of History: Control Problems That Reach from the Past   27

    4   What Motor Learning Is, What Motor Learning Does   39

    5   What Does the Motor Learning I: Spinal Cord and Brainstem   61

    6   What Does the Motor Learning II: Forebrain   75

    7   What Generates Force and Feedback   93

    8   What Maintains Limb Stability   119

II Computing Locations and Displacements   141

    9   Computing End-Effector Location I: Theory   143

    10   Computing End-Effector Location II: Experiment   159

    11   Computing Target Location   179

    12   Computing Difference Vectors I: Fixation-Centered Coordinates   205

    13   Computing Difference Vectors II: Parietal and Frontal Cortex   229

    14   Planning Displacements and Forces   245

III Skills, Adaptation, and Trajectories   271

15 Aligning Vision and Proprioception I: Adaptation and Context   273

16 Aligning Vision and Proprioception II: Mechanisms and Generalization   295

17 Remapping, Predictive Updating, and Autopilot Control   319

18 Planning to Reach or Point I: Smoothness in Visual Coordinates   341

19 Planning to Reach or Point II: A Next-State Planner   353

IV Predictions, Decisions, and Flexibility   377

20 Predicting Force I: Internal Models of Dynamics   379

21 Predicting Force II: Representation and Generalization   403

22 Predicting Force III: Consolidating a Motor Skill   435

23 Predicting Inputs and Correcting Errors I: Filtering and Teaching   447

24 Predicting Inputs and Correcting Errors II: Learning from Reflexes   473

25 Deciding Flexibly on Goals, Actions, and Sequences   495

V Glossary and Appendixes   525

Glossary   527

Appendix A. Biology Refresher   533

Appendix B. Anatomy Refresher   537

Appendix C. Mathematics Refresher   539

Appendix D. Physics Refresher   543

Appendix E. Neurophysiology Refresher   547

Index   549

Contents in Detail

Preface   xv

1 Introduction   1
  1.1 Why Motor Learning?   1
  1.2 Why Now?   2
  1.3 Why a Theoretical Study?   3
  1.4 Why a Computational Theory?   3
  1.5 Why Vertebrates, Why Primates, and Why a Two-Joint Arm?   4

I Evolution, Anatomy, and Physiology   7

2 Our Moving History: The Evolution of the Vertebrate CNS   9
  2.1 Birth of the Motor System   9
  2.2 Components of the Motor System   10
  2.3 A Brief History of the Motor System   12
  2.4 First Steps: Inventing the Vertebrate Brain   15
  2.5 More Recent Steps: Cerebellum and Motor Cortex   21
  2.6 Summary   23

3 Burdens of History: Control Problems That Reach from the Past   27
  3.1 Limbs   28
  3.2 Muscles   32
  3.3 Nerves   37

4 What Motor Learning Is, What Motor Learning Does   39
  4.1 Motor Learning Undefined   39
  4.2 Motor Learning over Generations: Links to Instincts and Reflexes   41
  4.3 Learning New Skills and Maintaining Performance   46
  4.4 Making Decisions Adaptively   51
  4.5 Summary   58

5 What Does the Motor Learning I: Spinal Cord and Brainstem   61
  5.1 Spinal Cord   61

5.2 Hindbrain   65
5.3 Cerebellum   68
5.4 Red Nucleus   71
5.5 Superior Colliculus   73

6   What Does the Motor Learning II: Forebrain   75
6.1 Basal Ganglia   75
6.2 Thalamus   80
6.3 Cortical Organization I: General Considerations   81
6.4 Cortical Organization II: Cortical Fields for Reaching and Pointing   85

7   What Generates Force and Feedback   93
7.1 Biological Versus Mechanical Actuators   93
7.2 Muscle Mechanisms   94
7.3 Motor Units   98
7.4 A Muscle Model   99
7.5 Converting Force to Torque   102
7.6 Muscle Afferents   108
7.7 Muscle Afferents in Action   112

8   What Maintains Limb Stability   119
8.1 Equilibrium Points from Antagonist Muscle Activity   120
8.2 Restoring Torques from Length–Tension Properties   121
8.3 Stiffness from Muscle Coactivation   123
8.4 Reaching Without Feedback in Monkeys   123
8.5 Equilibrium Points from Artificial Stimulation   126
8.6 Rapid Movements from Sequential Muscle Activation   127
8.7 Passive Properties Produce Stability   129
8.8 Reflexes Produce Stability   131
8.9 Reaching Without Feedback in Humans   135
8.10 Passive Properties and Reflexes Combined   136

II   Computing Locations and Displacements   141

9   Computing End-Effector Location I: Theory   143
9.1 Reaching and Pointing Require Sensory Feedback   143
9.2 Kinematics and Dynamics   144
9.3 Degrees of Freedom and Coordinate Frames   144
9.4 End Effectors and Adaptive Mapping   146
9.5 Predicting the Location of an End Effector in Visual Coordinates   147
9.6 Predicting End-Effector Location with Proprioception: Virtual Robotics   148
9.7 Predicting End-Effector Location with Proprioception: Computations   151

10   Computing End-Effector Location II: Experiment   159
10.1 Role of Proprioceptive Signals in End-Effector Localization   159
10.2 Introduction to Frontal and Parietal Neurophysiology   162

10.3 Encoding of Limb Configuration in the CNS   165
10.4 Errors in Reaching due to Lesions of the PPC   175

11 Computing Target Location   179
11.1 Computing Target and End-Effector Locations in a Common Frame   180
11.2 Computing Target Location in a Vision-Based Frame   183
11.3 Combining Retinal Location with Eye Orientation Through Gain Fields   188

12 Computing Difference Vectors I: Fixation-Centered Coordinates   205
12.1 Planning Reaching and Pointing with Difference Vectors   205
12.2 Shoulder-Centered Versus Fixation-Centered Coordinates   209
12.3 Planning in Fixation-Centered Coordinates: Experiment   212
12.4 Planning in Fixation-Centered Coordinates: Theory   216
12.5 Localizing an End Effector in Fixation-Centered Coordinates   221
12.6 Encoding End-Effector Location in Fixation-Centered Coordinates   222
12.7 Issues Concerning Fixation-Centered Coordinates   225

13 Computing Difference Vectors II: Parietal and Frontal Cortex   229
13.1 Computing a Movement Plan   229
13.2 Planning Potential Movements but Not Executing Them   237
13.3 Planning the Next Movement in a Sequence   241

14 Planning Displacements and Forces   245
14.1 Representing the Difference Vector in the Motor Areas of the Frontal Lobe   247
14.2 Population Vectors, Force Coding, and Coordinate Frames in M1   261

III Skills, Adaptation, and Trajectories   271

15 Aligning Vision and Proprioception I: Adaptation and Context   273
15.1 Newts Cannot Adapt to Rotation of Their Eyes   275
15.2 Primates Adapt to Rotation of the Visual Field   276
15.3 Prism Adaptation Requires Modification of Both Location and Displacement Maps   279
15.4 Long-term Memories and Learning to Switch on Context   280
15.5 Prism Adaptation in Virtual Robotics   284
15.6 Consequences of Planning in Vision-Based Coordinates   286
15.7 Moving an End Effector Attached to the Hand   288
15.8 Internal Models of Kinematics   289
15.9 Estimate of Limb Location Is Influenced by the Likelihood of the Sensed Variables   291

16 Aligning Vision and Proprioception II: Mechanisms and Generalization   295
16.1 Neural Systems Involved in Adapting Alignments Between Proprioception and Vision   295
16.2 Generalization of Adaptation to Altered Visual Feedback   303

17 Remapping, Predictive Updating, and Autopilot Control   319
17.1 Remapping Target Location   319
17.2 Predictive Remapping of Target and End-Effector Location with Efference Copy   325
17.3 Remapping End-Effector Location   331

18 Planning to Reach or Point I: Smoothness in Visual Coordinates   341
18.1 Regularity in Reaching and Pointing   343
18.2 Description of Trajectory Smoothness: Minimum Jerk   350

19 Planning to Reach or Point II: A Next-State Planner   353
19.1 The Problem of Planning   353
19.2 Transforming a Displacement Vector into a Trajectory   354
19.3 The Next-State Planner   357
19.4 Minimizing the Effects of Signal-Dependent Noise   364
19.5 Online Correction of Self-Generated and Imposed Errors in Huntington's Disease   366
19.6 Transforming Plans into Trajectories: The Problem of Redundancy   371

IV Predictions, Decisions, and Flexibility   377

20 Predicting Force I: Internal Models of Dynamics   379
20.1 Internal Models of Dynamics   380
20.2 Correlates of Adapting to Altered Dynamics   391

21 Predicting Force II: Representation and Generalization   403
21.1 The Coordinate System of the Internal Model of Dynamics   403
21.2 Computing an Internal Model with a Population Code   410
21.3 Estimating Generalization Functions from Trial-to-Trial Changes in Movement   416
21.4 A Not-So-Invariant Desired Trajectory   432

22 Predicting Force III: Consolidating a Motor Skill   435
22.1 Consolidation   435
22.2 A Role for Time and Sleep in Consolidation of Motor Memories   441

23 Predicting Inputs and Correcting Errors I: Filtering and Teaching   447
23.1 Cancellation of Predicted Signals by Adaptive Filtering   449
23.2 Predicting and Responding to a Stimulus   454
23.3 Similar Learning Mechanisms in Basal Ganglia and Cerebellum   462

xiii   Contents in Detail

       23.4   A Training Signal for the Basal Ganglia   464
       23.5   Why Does Huntington's Disease Result in Disorders in Reaching?   468

24   Predicting Inputs and Correcting Errors II: Learning from Reflexes   473
       24.1   Climbing Fibers Encode a Signal That Represents Motor Error   475
       24.2   Predictively Correcting Motor Commands   480

25   Deciding Flexibly on Goals, Actions, and Sequences   495
       25.1   Deciding on a Target   496
       25.2   Choosing Among Multiple Potential Targets of Movement   503
       25.3   Deciding on Multiple Movements   504
       25.4   Action Selection Based on Estimates of State   505
       25.5   Moving to Places Other Than a Stimulus: Standard Mapping vs. Nonstandard Mapping   513
       25.6   Summary   519

V   Glossary and Appendixes   525

     Glossary   527

     Appendix A   Biology Refresher   533

     Appendix B   Anatomy Refresher   537

     Appendix C   Mathematics Refresher   539

     Appendix D   Physics Refresher   543

     Appendix E   Neurophysiology Refresher   547

Index   549

Preface

Precisely one century before we finished this book, Orville and Wilbur Wright invented the airplane. You might wonder why a book about reaching and pointing begins with the Wright brothers, but we think that their experience offers several relevant lessons. One involves the concepts of reverse and forward engineering. For example, the earliest attempts at aircraft design emulated flying birds, an approach called **reverse engineering**. (Note: Words in boldface type appear in the glossary.) You can find many examples of this, but Leonardo da Vinci surely produced the most famous one. To practice reverse engineering, you take an existing system, try to understand how it works, and perhaps design something like it. If you are a neuroscientist or studying to be one, you should recognize reverse engineering; it is more or less what neuroscientists do. The Wright brothers, however, did not rely on reverse engineering—at least not at first. In fact, something more like the reverse of reverse engineering occurred. Instead of an analysis of bird flight leading to better aircraft design, aircraft design led to an improved understanding of bird flight. How? The accomplishments of the Wright brothers led to better theories of aerodynamics, and this body of theory led to the improved understanding of bird flight. So *lesson 1* from the Wright brothers is that sometimes engineering advances lead to a better understanding of biological systems.

Another lesson from the Wright brothers concerns the importance of models in understanding complex systems. The Wright brothers based their designs on mathematical models, in the form of equations. Their experience shows that models, even flawed ones, can sometimes lead to something important. For example, after more than two years of work with gliders, the Wright brothers realized that their aerodynamic models had problems. Years earlier, an aviation pioneer named Otto Lilienthal had developed those models to predict the amount of upward force, called lift, produced by wings of various designs. The Wright brothers used Lilienthal's equations, but they soon learned that wings designed in accord with his models produced only about a third of the predicted lift. Given the fact that Lilienthal had died years earlier in a glider crash, the Wrights might have been more skeptical of his theories, but—reasonably enough—they began with what they had. Through frustration and failure, they eventually realized that they needed to develop their own models, and—in an astounding leap into modernity—they built a wind tunnel to test their theories. As a result, the Wright brothers developed better mathematical and physical models of airplane wings. *Lesson 2* from their experience, then, is that models can help in understanding the behavior of complex systems, but they need to be both tested and improved.

After the Wright brothers had solved the lift problem, their glider experienced continued difficulties with stability and control. Thus, *lesson 3* regards stability and control, problems with which the Wrights struggled for four years. Stability and control are general problems for moving systems, and your motor system is no exception. Temporarily stymied, Wilbur Wright returned to an examination of bird flight. That is, in a tight spot he resorted to reverse engineering. Wright noticed that bird wings warp during flight and, fiddling around with a flexible box, he saw how wing surfaces could warp in a similar way. Through such observations, the Wright brothers developed wing-warping controls to promote aircraft stability. *Lesson 4*: Combining reverse and forward engineering seems like a good idea.

After that breakthrough, the Wrights' glider no longer behaved erratically. Stability had been achieved, but problems remained with control. When, as pilots, the Wright brothers made maneuvers that should have turned the glider to the left, it often slid to the right instead. The addition of vertical tail fins alleviated that problem but generated new problems with stability. At that point, the Wright brothers must have begun to wonder whether stability and control were mutually exclusive. It is tempting in such situations to break down a complex system into its components and, by better understanding how each part works, hope to comprehend the overall system. This approach, called reductionism, has its place in understanding complex systems. However, the Wrights' biggest breakthrough came from enlarging the problem, not from reducing it. They recognized that it made no sense to study turning left or right in isolation from rolling the plane along its long axis; *they saw that airplanes turn by rolling*. If you want to turn an airplane to the right, you roll it so that its right wing rotates downward. This clockwise roll changes the direction of lift to the right, and the plane moves in that direction. *Lesson 5*, then, is that complex systems can be understood only at the systems level; reductionist methods help, but only in a limited way.

And the Rest Is History

In late 1903, Orville Wright took off on the first powered flight. He flew about 35 m, moving at less than 10 km/hour. The era of aeronautics had begun. (The era of really small seats and screaming infants followed shortly thereafter.) It is difficult in hindsight to appreciate the magnitude of the Wright brothers' breakthrough, but a few facts might help: Just 5 years later, Wilbur Wright flew his plane for 2 hours; 8 years later, a pilot flew across North America; 24 years after the first flight, a pilot flew an airplane from New York to Paris; and just 65 years after the Wright brothers' plane first took off—within the lifetime of many people—the aerodynamic theories that they pioneered allowed three people to return safely from a voyage around the moon.

The Wrights' breakthrough resulted from three factors that can help you understand the neurobiology of reaching and pointing movements: combining reverse and forward engineering, joining theory in the form of mathematical modeling with empirical testing, and a systems-level approach. The Wright brothers' success depended on the combination of these factors. For example, most of their competitors had envisioned controlling aircraft through a system of rudders, by analogy with ships at sea. Unlike the Wright brothers, those engineers did not test their models in wind tunnels. If they had, they might have realized that a maritime analogy has little relevance to flying machines. Remember that airplanes turn by rolling. Rudders can control watercraft because a ship's buoyancy makes rolling largely irrelevant to turning (although too much roll can, of course, make for a bad day). *Lesson 6*: Do not go out to sea in bad weather (see Sebastian Junger, *A Perfect Storm: A True Story of Men Against the Sea*, Norton, New York, 1997).

How to Use this Book

This book draws on information from a broad range of academic disciplines. To help you follow the discussion of topics outside your field of study, five appendices provide brief "refreshers" on some fundamentals of biology, anatomy, mathematics, physics, and neurophysiology. (The best part is, no one will know if you consult the refreshers. We suggest that you deny—under oath, if necessary—consulting any of them.)

Because of the book's computational nature, we have made supplemental material available on the Internet. These "web documents" provide source code for most of the simulations, step-by-step derivations of certain mathematical formulations, and expanded explanations of particular concepts. The documents are currently available at the Universal Resource Locator (URL) www.bme.jhu.edu/~reza, the Reza Shadmehr home page. In the event that the URL changes, an Internet search for the text string "Reza Shadmehr Home Page" should lead you to these documents.

For cross references within the book, a link such as "see section 1.2.3" or (section 1.2.3) indicates that you might look for further explanation or background information under the third subheading of the second main heading in chapter 1. Figure 1.2 refers to the second figure in chapter 1. Box 1.2 corresponds to the second box in chapter 1. A glossary contains some brief definitions of technical terms and concepts, and you can find words that appear in the text in boldface type defined there.

Note that we do not aim to present a comprehensive summary of the motor system, motor learning, or even the scientific literature on reaching and pointing movements. Many of the topics taken up—and not a small number of those omitted—deserve book-length treatment. Nearly all of the major topics presented in this book are already, or someday will be, the subject of full-length books. So this book leaves a lot out: the fields of oculomotor control, locomotion, speech, and movement disorders receive scant attention. Plastic change in motor maps, a subject often discussed in the context of motor learning, gets barely a mention. Obviously, we could not write—and you would not read—an encyclopedic dissertation on motor control and motor learning. Instead, this book presents an introduction to the computational neurobiology of reaching and pointing—with emphasis on motor learning in primates—based on an eclectic selection of topics. Chapter 1 explains why we have made those and other choices. A brief, annotated reading list and a selected (not comprehensive) list of citations appears at the end of most chapters.

We also draw your attention to another book, *Theoretical Neuroscience* by Peter Dayan and Larry Abbott (MIT Press, Cambridge, Mass., 2001). It is a useful complement to this one. *Theoretical Neuroscience* focuses on sensory processing, whereas this book addresses motor control and motor learning. We have, by and large, avoided duplication of the topics presented in *Theoretical Neuroscience*, especially background material such as the operations of neurons, et cetera.

The following acronyms and abbreviations are used throughout the book:

- CNS, central nervous system
- CPG, central pattern generator
- EMG, electromyographic (i.e., muscle) activity
- GABA, γ-aminobutyric acid
- ION, inferior olivary nucleus/nuclei
- PPC, posterior parietal cortex
- Abbreviated names of several cortical areas, illustrated in figure 6.3 and defined in its legend, including the primary motor cortex (M1), dorsal premotor cortex (PMd), ventral premotor cortex (PMv), and supplementary motor area (SMA).

Acknowledgments

We thank Barbara Murphy, Sara Meirowitz, and Margy Avery of The MIT Press. George Nichols, supported by a grant from The MIT Press, drew some of the illustrations. We also thank colleagues who read and commented upon certain chapters of this book: Paul Cisek, Giuseppe di Pellegrino, Eb Fetz, Richard Ivry, Mitsuo Kawato, Betsy Murray, Sandro Mussa-Ivaldi, Todd Preuss, Matthew Rushworth, Sharleen Sakai, Jeff Schall, Veit Stuphorn, Emo Todorov, and Dan Willingham. We express our special thanks to Paul Cisek for his numerous contributions. Benjamin Feinberg helped obtain permission to reproduce figures, and Melissa Miles assisted with the page proofs. Kevin Blomstrom developed the web documents.

The Computational Neurobiology of Reaching and Pointing

1 Introduction

Overview: Understanding reaching and pointing movements depends on knowledge of physics, biology, mathematics, robotics, and computer science. Physics plays a fundamental role because reaching and pointing require your central nervous system (CNS) to solve difficult mechanical problems: It must learn to control a limb that consists of linked segments, which interact with each other as well as with external objects as they accelerate in a gravitational field.

Neuroscience involves the study of the nervous system, and its topics range from genetics to inferential reasoning. At its heart, however, lies a search for understanding how the environment affects the CNS and how the CNS, in turn, empowers you and other vertebrates to interact with and to alter the environment. This empowerment arises from the fact that your CNS allows you to move, acquire skills, and adapt those skills to a variety of contexts.

1.1 Why Motor Learning?

Our devotion to *motor learning* might strike you as peculiar, given your undoubted interest in consciousness, abstract reasoning, language, and so forth. However, in evolutionary history, long before your ancestors possessed the capacity for language or abstract reasoning, they moved in relation to objects and places in their environment. So when we contend that *all learning depends on motor learning*—and we do—we mean this in two senses. First, the vertebrate CNS evolved to learn how, when, and where to move (see chapter 2). Second, the basic neuronal and synaptic mechanisms that evolved to do that job also support other forms of learning, and therefore provide the basis for all knowledge.

Nevertheless, to many students the motor system amounts to little more than an instrument of torture to be endured until the really interesting material appears. In his textbook *Physiological Psychology*, Robert Graham[1] summed up the view of many when he wrote that among his colleagues "the motor system is regarded as somewhat less interesting than a 1949 report on farm futures." Perhaps the approach taken in this book will help you overcome this feeling, if you have it. We try to avoid material that you could easily find in medical textbooks. We pick and choose topics and develop broad themes stretching over several chapters. Although this book focuses on reaching and pointing movements made by people and monkeys, topics include other kinds of movements in other kinds of animals. We hope you find this eclectic approach more engaging than the traditional textbook treatment of the motor system.

Even if you have little interest in the motor system per se, nearly everyone has some interest in learning, and the motor system provides some advantages for studying it. Through the theories of Newtonian mechanics, physics provides a mathematical framework for describing and analyzing movements. When muscles receive commands from the CNS, they produce forces, and equations from Newtonian mechanics describe the relationship between forces and motion. Thus, compared to other forms of learning, simulations of motor learning are less abstract and more physical. From a neurophysiological perspective, the study of motor learning has at least one additional advantage: It has the potential, at least, to lead directly to a causal analysis. All fields have a skeleton in the closet, and neurophysiology has a particularly disagreeable one: Very little contemporary neurophysiology accounts for behavior in terms of causal relationships. In the motor system, however, some cells synapse upon motor neurons, which cause the forces that generate movement. This fact enables a measurement of those neurons' contribution to behavior in direct, causal terms.[2] This advantage has yet to be fully realized, but we think that even its theoretical possibility is important.

Finally, motor learning matters because it allows you to act while directing your attention and intellect toward other matters. Imagine that you needed to attend to all of the routine aspects of your reaching or pointing movements. Motor learning provides you with freedom from such a life. Although many experts in **learning** and **memory** regard **skill** acquisition as a lower form of learning—mainly because it occurs subconsciously—Alfred North Whitehead summed up its importance in his *Introduction to Mathematics*:

It is a profoundly erroneous truism ... that we should cultivate the habit of thinking of what we are doing. The precise opposite is the case. Civilization advances by extending the number of important operations which we can perform without thinking about them. (Whitehead,[3] p. 61)

1.2 Why Now?

Why now? Because since the 1980s, robotics engineers have built machines with limblike structures that move. In doing so, they have had to face the problems inherent in learning to control systems with noisy sensors, delays in transmitting information, multiple interacting limb segments, and a changing environment. Their experience has led to new ways of thinking about motor learning, as well as a renewed interest in some older ways. In short, they have developed theories of how to control machines that move.

Although this book relies on some theories developed for controlling robots, the use of robots as metaphors involves some risks, and you need to recognize them. Robots rely on motors that develop force far faster than muscles, and wires that carry information a lot faster than axons. Thus, unlike a typical robot, your CNS cannot analyze the sensory consequences of its **motor commands** until a fair amount of time has passed. Your CNS, therefore, needs to *predict* the sensory feedback caused by its motor commands to a much greater extent than a typical robot does. Later chapters present the idea that your CNS predicts the sensory consequences of your reaching and pointing movements in terms of visual coordinates. As a result—and because your eyes can move—your CNS needs to use its **oculomotor** commands to control reaching and pointing movements: a counterintuitive notion that would not necessarily follow from contemporary robotics.

1.3 Why a Theoretical Study?

Since the 1980s, there have been many serious attempts at a theoretical study of motor control and motor learning. Those theories serve as a useful bridge between neuroscientists, who study the CNS at the cell and network levels, and engineers, who design and construct mechanical devices that emulate the movements of vertebrates. Biology and engineering have had a long, synergistic relationship, and an approach at the level of theory makes this interaction possible.

To cite just one example, the theory of **feedback control**—so fundamental to physiology—arose from interactions between biology and engineering during the 19th century. The French physician Claude Bernard developed the concept of **homeostasis**, which describes the ability of biological systems to maintain their physiological state within acceptable limits. At about the same time, the development of governors on steam engines—mechanical devices that slow an engine in proportion to its output—demonstrated a conceptual relationship between the principles of feedback in biology and in engineering that enriched both disciplines. This linkage could have been achieved only through a *theory* of feedback control.

1.4 Why a Computational Theory?

Lord Kelvin, the famous British physicist, is reputed to have said, "When you cannot express it in numbers, your knowledge is of a meager and unsatisfactory kind." Many would disagree—the theory of evolution does not require numbers for a quite satisfactory expression of vitally important scientific knowledge—but numbers do enable computations and, therefore, computational theories. In his often-quoted work on vision, David Marr described three levels of understanding CNS functions: the level of a *computational theory*, which clarifies the problem to be solved as well as the constraints that physics imposes on the solution; the level of an *algorithm*, which describes a systematic procedure that solves the problem; and the level of *implementation*, which involves the physical realization of the algorithm by a neural network. A computational-level theory thus explains some of what a complex system does and how it *might* work. To understand how evolution has solved motor-control problems such as those faced by the vertebrate CNS, it helps to understand—in some detail—at least one way that a system *could* solve the same problem, regardless of whether evolutionary processes arrived at that particular one. Ellen Hildreth and John Hollerbach expressed this opinion in the following way:

It is often true that before we can understand how a biological system solves an information processing problem, we must understand in sufficient detail at least one way that the problem can be solved, whether or not it is a solution for the biological system. (Hildreth and Hollerbach,[4] p. 606)

For example, reaching and pointing movements require the production of forces imposed on linked limb segments, compensation for **inertia** and gravity, and stabilization in the face of perturbations, among other factors. To overcome inertia, your CNS *might* estimate the mass of the limb and account for it in the production of its motor commands. To make the system stable in the face of perturbations, it *might* incorporate **feedback** loops. To cope with the possibility that targets move and an object held in your hand might have an unexpected mass, your CNS *might* use a readout of its current state to update its plan for the next state. To deal with the fact that movements around one joint produce unwanted forces at other joints, the CNS *might* learn to produce motor commands that predictively cancel this kind of error. Parts II–IV of this book present a computational theory of how the CNS *might* solve these and other problems. As Albert Einstein said, "As far as the laws of mathematics refer to reality, they are not certain; and as far as they are certain, they do not refer to reality." This book presents one plausible, if incomplete, framework for understanding reaching and pointing movements.

1.5 Why Vertebrates, Why Primates, and Why a Two-Joint Arm?

This book focuses on the vertebrate motor system, with emphasis on reaching and pointing movements in primates. Its principal model system is a two-joint arm. The choice of vertebrates is not meant to disparage other kinds of animals: They have impressive motor capabilities. And although the text often employs convenient shorthand statements, such as vertebrates "have such and such a trait," or primates do, these statements do not imply that other animals lack the structure or function mentioned.

Similarly, the focus on reaching and pointing with a two-joint arm is somewhat arbitrary. Many other kinds of movements have equal importance and, of course, primate limbs have more than two joints. We focus on two-joint reaching because that model system, or something like it, has attracted researchers in robotics, motor **psychophysics**, and neurophysiology, among other disciplines. Of course, as chapter 2 points out, the vertebrate CNS did not evolve to control reaching or pointing; it evolved to control swimming and eating. In fact, the earliest vertebrates did not have **appendages** at all, let alone arms. Nevertheless, evolution adapted the mechanisms of swimming for other purposes, including the control of reaching and pointing. Reaching to a target with a two-joint arm seems like a useful compromise between simpler motor behaviors (such as eye movements) that lack the richness of reaching and pointing, and more complex behaviors, such as vocalization, locomotion, dance, and athletics.

Given that reaching and pointing in primates is only one among many kinds of movements in one among many groups of vertebrates, part I of this book attempts to place this topic in a broader context:

- If the vertebrate CNS did not evolve to control reaching and pointing, what *did* it evolve to do, and how did it reach its present state (chapter 2)?
- How did its evolutionary history contribute to the problems that the CNS faces in learning to control reaching and pointing movements (chapter 3)?
- How does learning to reach and point relate to other kinds of motor learning and to other kinds of learning (chapter 4)?
- What parts of the CNS underlie learning to reach and point (chapters 5 and 6)?
- How does the CNS generate the forces and feedback used to control reaching and pointing (chapter 7)?
- How does the CNS maintain the limb stability needed for reaching and pointing (chapter 8)?

References

1. Graham, RB (1990) Physiological Psychology (Wadsworth, Belmont, CA).
2. Cheney, PD, Fetz, EE, and Mewes, K (1991) Neural mechanisms underlying corticospinal and rubrospinal control of limb movements. Prog Brain Res 87, 213–252.
3. Whitehead, AN (1911) Introduction to Mathematics (Henry Holt, New York).
4. Hildreth, EC, and Hollerbach, JM (1987) Artificial intelligence: Computational approach to vision and motor control. In Handbook of Physiology: The Nervous System, V, ed Plum, F (American Physiological Society, Washington, DC), pp 605–642.

I Evolution, Anatomy, and Physiology

2 Our Moving History: The Evolution of the Vertebrate CNS

*Overview: The vertebrate CNS originated approximately 500–550 million years ago (mya), in surprisingly recognizable form. In large part, what it did for those animals then, it does for you and other vertebrates today. A major role of the early vertebrate CNS involved the guidance of swimming based on receptors that accumulated information from a relatively long distance, mainly those for vision and **olfaction**. The original vertebrate motor system later adapted into the one that controls your reaching and pointing movements.*

2.1 Birth of the Motor System

According to fossil evidence,[1–3] the immediate ancestors of modern vertebrates evolved approximately 500–550 mya. These ancient animals had a conspicuous brain that enabled them to move themselves—with facility. They also probably had a pair of eyes, one on each side of their head. Although some of these conclusions remain controversial,[4,5] it seems likely that the earliest vertebrates made decisions about where and when to move with that brain, based on inputs from those eyes. Thus, the ability to move in a rapid, coordinated, and goal-directed manner probably represents *the* fundamental breakthrough in vertebrate evolution, and vision was important from the start. As one paleontologist[6] described the early vertebrates, their "most conspicuous structural features are associated with active swimming." Visually guided movement, then, represents one of the most deep-seated characteristics of vertebrates.

To put a time interval such as "500–550 mya" in perspective, note that the evolutionary **lineages** leading to humans and chimpanzees diverged approximately 6 mya.[7] New World and Old World primates—you number among the latter—went their separate ways 30 mya or so. And the vertebrate lineage appeared somewhat nearer to the origin of animals (perhaps 900 mya, or less) than to the present. (The footnote to table 2.1 presents more numbers like these.)

Fossil evidence also indicates that soft structures such as the brain and eyes evolved before hard structures such as bones and teeth. This sequence of events implies that the brain—and the visuomotor system—play a more fundamental role in the life of vertebrates than the bones that gave rise to our name. Partly for this reason, many biologists prefer the name *craniates* to *vertebrates*. In light of the primacy of the brain and eyes in vertebrate history, however, the name *visuomotors* might apply as well as any other.

Vertebrate motor learning began in those distant ancestors; their CNS solved many of the fundamental problems posed by an active, mobile life. How much of their CNS is reflected in yours? The answer is, More than you might think. But before that idea can be pursued in any more detail, you need to understand a little about the organization of the vertebrate CNS. (If you know neuroanatomy, skip to section 2.3, and if you know something about evolutionary biology as well, skip to section 2.3.2.)

2.2 Components of the Motor System

The CNS that vertebrates evolved has six major components: the spinal cord, medulla, pons, midbrain, diencephalon, and telencephalon (figure 2.1), the last five of which compose the brain. Looked at a little differently, the hindbrain (medulla and pons), the midbrain, and the forebrain (telencephalon and diencephalon) make up the brain. Taken together, the midbrain and hindbrain make up the brainstem, although aspects of brainstem architecture continue into the diencephalon. Regardless of how you choose to group its various components, all levels of the CNS participate in motor control and motor learning, and, in primates, all levels contribute to visually guided reaching and pointing. The remainder of this section presents a very brief introduction to the vertebrate CNS. A somewhat more detailed description appears in chapters 5 and 6.

The spinal cord contains neurons with numerous motor functions, including motor neurons. Motor neurons send command signals to

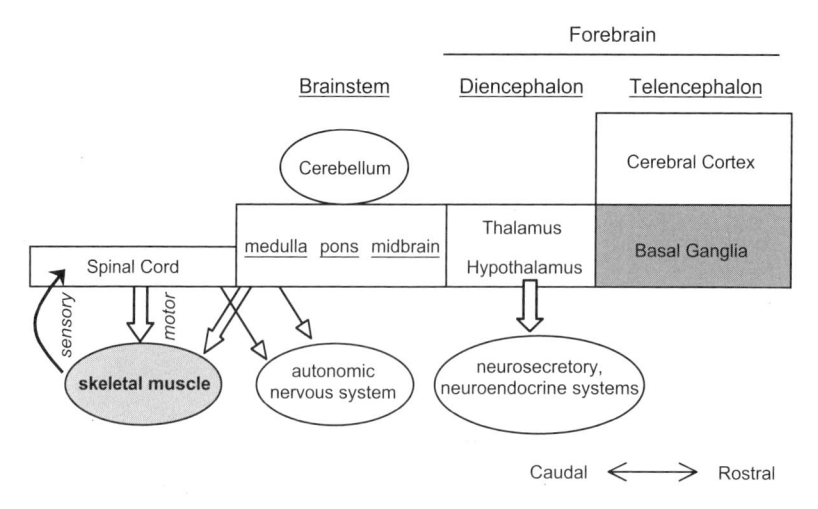

Figure 2.1
Layout of the vertebrate CNS.

muscles via motor nerves, and sensory nerves transmit information to the CNS from sensory receptors in the skin, muscles, and other parts of the body. Sensory nerve fibers terminate in the spinal cord or pass through it on the way to the brain. In addition, neural networks in the spinal cord called **central pattern generators (CPGs)** produce motor-command signals that underlie rhythmic behaviors such as swimming and walking.

The brainstem also has motor neurons, sensory nerves, and CPGs, much like those in the spinal cord. In addition to CPGs per se, the brainstem and diencephalon contain comparable neural networks that initiate and modulate the activity of CPGs in both the brainstem and the spinal cord. Throughout the core of the brainstem—and extending into the diencephalon as well—a diverse set of brain regions collectively called the *reticular formation* mediate a broad range of **reflexes** and other functions. In the midbrain (also known as the mesencephalon), both the superior colliculus and the red nucleus contribute to motor control. The superior colliculus of mammals is known more generally in vertebrates as the optic tectum; it receives input from visual receptors in the retina and controls the orientation of the retina through eye and head movements (see box 11.1, which discusses the ancestral function of eye movements). Other parts of the brainstem, including the cerebellum and the reticular formation, contribute to motor outputs both directly and indirectly.

Within the telencephalon, parts of both the cerebral cortex and the basal ganglia play a significant role in the motor system. In the cerebral cortex of primates, a dozen or so areas in the frontal lobe and another dozen or so parietal areas contribute relatively directly to motor control. All of these parts of the cerebral cortex are made up of "six-layered" cortex, called *neocortex*. Other mammals have fewer neocortical areas involved in motor control, and other vertebrates have no neocortex at all. A large number of nuclei make up the basal ganglia, which contributes to the motor system via interactions with the cerebral cortex, brainstem, and hypothalamus.

In the diencephalon, both the thalamus and the hypothalamus play important roles in motor control. The thalamus contributes to the motor system by providing inputs to the cerebral cortex and basal ganglia. It relays sensory inputs, such as those arising from visual, auditory, and somatosensory receptors, as well as inputs from the superior colliculus, cerebellum, basal ganglia and other structures. The hypothalamus consists of a diverse group of structures that contribute to the output of the CNS in a variety of ways. One of these outputs involves **neuroendocrine** functions. Such functions do not appear, at least at first glance, to be motor in nature. Like the **skeletomuscular** system, however, the neuroendocrine system provides a means by which the CNS controls the body. More obviously motor is the role played by parts of the hypothalamus in the **autonomic nervous system**. The motor functions of the hypothalamus extend far beyond neuroendocrine and autonomic outputs, however, and include the control of many instinctive behaviors.

2.3 A Brief History of the Motor System

2.3.1 Terms and Concepts

Now that you know some parts of the vertebrate CNS, consider the following question: How much of the motor system sketched in section 2.2 have you inherited from your most distant vertebrate ancestors? The answer is, Quite a lot, but to understand that idea, you need to know a few terms and concepts from evolutionary biology (see also appendix A).

The term *stem group* refers to the species (and their close relatives) that gave rise to a group of related, descendant species, known as a *crown group*. Members of a crown group receive many of their traits from the stem species through inheritance, usually in modified form. (Structures and functions inherited from a common ancestor are said to be *homologues* of each other.) An ancient member of one particular stem group, called *stem cephalochordates*, gave rise to *stem vertebrates*. Their descendants include the various fishes, amphibians, reptiles, birds, and mammals that compose the vertebrate crown group. The *chordate* part of the name "cephalochordate" comes from a larger group of animals that includes the cephalochordates, past and present. *Chordate*s are named for their *notochord*, a stiff, rodlike structure that runs along their body axis. You can assume that living cephalochordates have retained many key characteristics of stem cephalochordates, and although you should view such assumptions skeptically, a comparison of living cephalochordates with living and fossil vertebrates can tell you a lot about what made stem vertebrates successful.

You need to understand just a few more terms to follow the history of the vertebrate CNS. The first vertebrates to evolve are called *agnathans*, meaning "jawless" (fish). Their modern descendants include the lamprey (figure 2.2A) and the hagfish (figure 2.2B). The term *gnathostome* refers to fish and other vertebrates with jaws. Both terms appear forbidding, but they come from a word related to "gnaw." You can gnaw, so you are a gnathostome (meaning "gnawing mouth" or, more disturbingly, "gnawing hole").

All of this terminology can be difficult to follow and even harder to remember. Accordingly, refer to table 2.1 and figures 2.3 and 2.4 to help keep the terms (and the historical record) straight. Bear in mind that some of the names in table 2.1 indicate ancestor–descendant relationships, and some do not. For example, the term *vertebrate* refers to a group of animals that descended from a common ancestor, the stem vertebrate. Some group names may not imply that kind of evolutionary relationship. Experts in evolutionary biology wage brutal battles over such matters—and those "discussions" continue. Ideas about evolutionary relations change, not because history changes but because new information and ideas change current thinking.

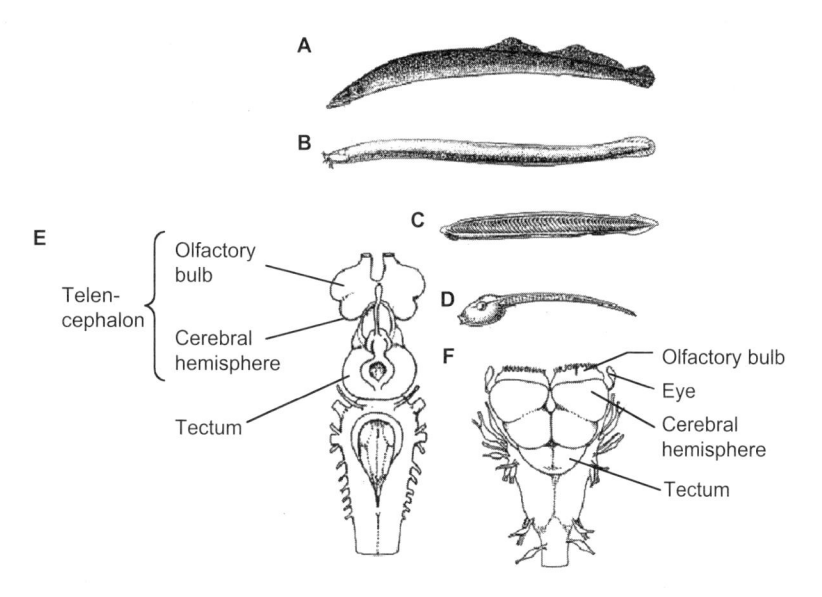

Figure 2.2
Chordates and vertebrates. (*A*) A lamprey, one kind of jawless fish (an agnathan). (*B*) A hagfish, another kind of jawless fish. (*C*) An amphioxus, a cephalochordate. (*D*) A larval tunicate, an immature chordate (see figure 2.3). (*E*) A lamprey brain. (*F*) A hagfish brain. A–D have rostral to the left; E and F have rostral at the top. (From Nieuwenhuys[10] with permission.)

2.3.2 The Origin of Animals

By the time of their appearance, the stem vertebrates already had a long evolutionary history. Animals first appeared approximately 700–900 mya,[8] during a time of increasing oxygen pressure in the atmosphere. Higher oxygen pressure allowed organisms to use a greater amount of energy, permitting the evolution of small, multicellular animals from single-cell ancestors. Because these animals relied on diffusion for gas exchange and nutrients, they adopted either tubular or planar shapes, and those body plans have persisted. With the later evolution of active circulation, respiration, and feeding, animals increased in size, and one such animal gave rise to the stem vertebrates.

The motor systems of vertebrates and other living animals have evolved separately for a long time, but have many common ancestors. Animals with planar architectures and a bilateral symmetry evolved into flatworms (*bilateria*). One branch of this evolutionary tree evolved into tubular animals, which form their tube from a cavity that appears very early in development (the *coelem*). The cavity resembles a collapsed, deflated basketball with an aperture that connects the outside with the hollow of the "ball." This cavity develops into a tube in one of two ways. The descendants of the animals in which the aperture of the coelem became the mouth (*protostomes*) are what you generally think of when you

Table 2.1
Vertebrates

					550 mya	Cephalochordates (amphioxus)
gnathostomes (jawed vertebrates)					530	Vertebrates
		fish		agnathans		Jawless fish of one sort (hagfish)
						Jawless fish of another sort (lamprey)
	bony vertebrates					Fish with jaws
						Fish with spines
						Fish with armor
						Fish with cartilage (sharks, rays)
						Fish with bones (goldfish, tuna)
		tetrapods & other fleshy-limbed vertebrates			370	Amphibians
			amniotes (vertebrates with 3-membrane eggs)		300	Reptiles
					200	Birds
					220	Mammals*
					70	Primates*
					3–6	Humans*

Note: The second column from the right gives an approximate time that each group appeared in terms of millions of yrs ago (mya). The cells to the left show groups of vertebrates, with those that descended from a single ancestor in shaded cells.
*Placental mammals appeared 100 mya; Old and New World monkeys diverged 33 mya; the monkey and humanlike lineages diverged 25 mya; and the human lineage diverged from that of (other) apes 3–6 mya; evidence for bipedal, upright gait occurs in fossil beds dating to about 4 mya; the first pebble tools date to 2.5 mya; evidence of fire begins at 1 mya; cave art and other artistic artifacts date to 0.04 mya; and the airplane (see the preface) was invented 0.0001 mya. All dates are approximate, to say the least.

hear the term "invertebrates." These are animals such as insects, mollusks, and segmented worms. The descendants of the animals in which the aperture of the coelem became the anus and a second opening became the mouth (*deuterostomes*) include additional groups of "invertebrates" as well as all vertebrates.

A number of these animals evolved visuomotor control systems, but the stem vertebrates did so in a particularly "brainy" way. If you read about such matters, you will find references to the "brain" of insects and other invertebrates. However, the neural **ganglia** called the insect brain evolved independently from that in vertebrates. No common ancestor of insects and vertebrates, the most recent of which lived a minimum of 600 mya, had anything resembling the vertebrate brain.[9]

2.3.3 The Origin of Vertebrates

The stem cephalochordates have a famous living descendant known by any of three names: amphioxus (which means "sharp on both sides," front and back), the lancelet (another reference to its sharp shape), or *Branchiostoma* (meaning "gill mouth"). Figure 2.2C illustrates this animal, originally identified in 1774 as a kind of slug. Although biologists corrected that mistake in the nineteenth century, the fact that the creature could be mistaken for a slug says quite a lot about it. Of course, amphioxus may have changed since its divergence from the stem cephalochordates, but it serves as a basis for comparison with modern vertebrates.

2.4 First Steps: Inventing the Vertebrate Brain

The stem vertebrates and their immediate ancestors appear to have evolved from animals that lived a *sessile* life.[10–13] Sessile animals do not move, but instead capture food that comes their way. Although some distant ancestors of vertebrates, such as flatworms, moved and hunted, somewhere along the line leading to vertebrates this ability was lost. Later, in an imperfectly understood series of events, a mobile, hunting life evolved again.

Many experts think that an ancestor of modern vertebrates went though a stage in which its larval forms moved but its adult forms did not. Somehow, through a dramatic change in this animal's maturation programs, the larval form transmogrified into a sexually mature adult. The term *neoteny* applies to this process, and it is a common one in evolution. Newts provide the classic example; they are sexually mature descendants of immature salamanders. (If your response is "I newt," then you do not need to refer to appendix A.)

The ancestral plan of the vertebrate motor system evolved in these animals, probably a chordate (figure 2.3). They had a genetic program for constructing a body axis consisting of three major parts: a dorsal neural tube derived from the outer layer of the embryo (*ectoderm*), a dorsal

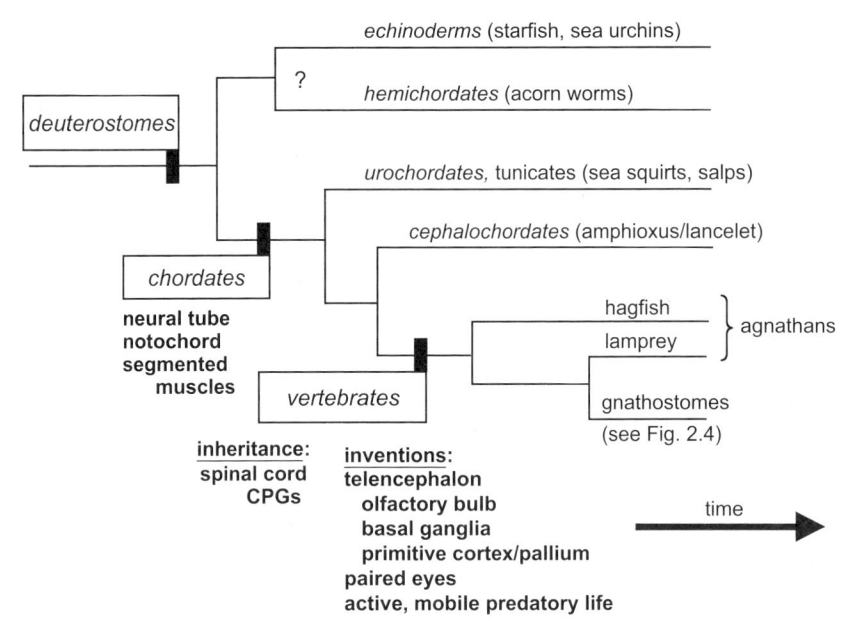

Figure 2.3
Evolutionary tree of the deuterostomes. CPGs are central pattern generators. Vertical black bars indicate the group that developed the innovations or inherited the traits shown, in bold type, below that group's name. (From a variety of sources, but especially Jeffries.[14])

notochord to stiffen the body axis, and a series of bilateral, segmentally controlled muscles to generate the forces needed for swimming.[10] Some larval chordates (called *tunicates*) have such a pattern, and so do all vertebrates, at least at some point during their development.

Regardless of how it happened, the ability of stem vertebrates to swim in a coordinated manner—and to direct their actions with a brain and sense organs concentrated in the head—was a dramatic development that almost certainly fostered further brain evolution. Your distant ancestors' predatory lifestyle was based, in large part, on inputs from distance sensors such as those for vision and olfaction, and the use of those inputs to guide their movements.

2.4.1 Inventing the Telencephalon

One current idea about vertebrate brain evolution is that the stem cephalochordates had brains with a diencephalon and brainstem, but probably no telencephalon (or olfactory bulb). (This point is not established beyond doubt, however, and it is possible that living cephalochordates lack a telencephalon because it was lost during evolution. This possibility is unlikely, however, because cases of such *degeneracy* in such an advantageous structure are usually accompanied by some conspicuous specialization that renders the "breakthrough" superfluous.) According to

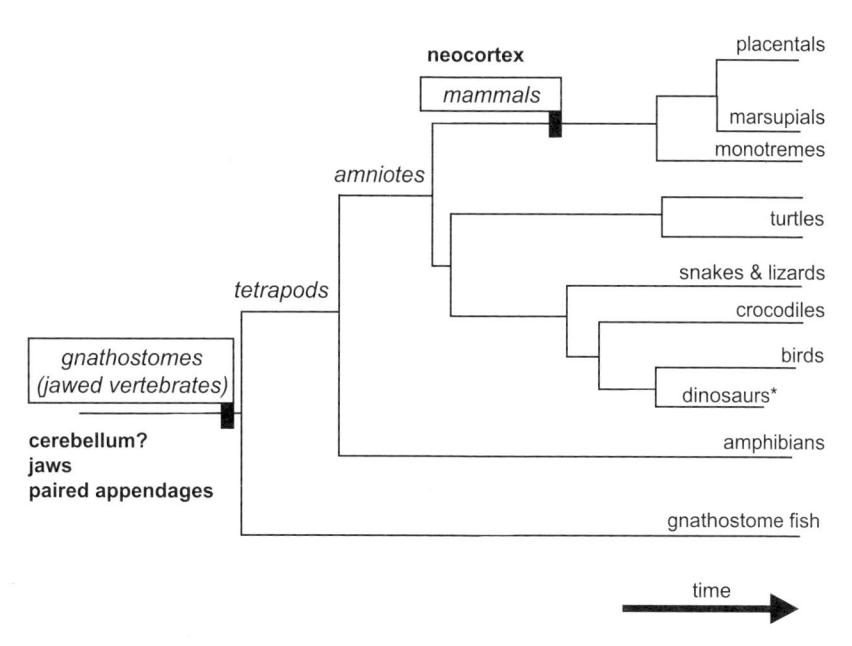

Figure 2.4
Evolutionary tree of gnathostomes. In the format of figure 2.3, except that an important innovation of mammals appears above the group's name. * indicates extinct group.

current thinking, the telencephalon (including the olfactory bulb) appears to be a vertebrate innovation.[15,16] As shown in figure 2.2, both of the living agnathans, lamprey and hagfish, have a telencephalon, as do all gnathostomes (see figures 2.3 and 2.4), and these structures appear to be absent in all other chordates. Such observations provide evidence that those structures evolved in stem vertebrates.

Along with the olfactory bulb and telencephalon, stem vertebrates used their eyes and brain to guide movements in relation to objects and places. Such behavior certainly involved the transformation of sensory inputs into the motor commands necessary to orient toward, approach, or avoid those inputs.

2.4.2 Inventing the Vertebrate Head

Along with the telencephalon, stem vertebrates developed sensory receptors concentrated on their heads. These receptors **transduced** information coming from a long distance, as typified by vision and olfaction. Although some experts have suggested that the vertebrate head was essentially added onto an existing body, this notion has not gained wide acceptance. Instead, it seems most likely that the development of certain key tissues augmented a head that already existed in chordates. The development of the **neural crest** and **placodes** (skin thickenings that develop into specialized sensory organs) appears to be a key step in the evolution of the vertebrate nervous system. Neural crest derivatives give rise to the **autonomic nervous system**, sensory nerves and cell bodies, and the cells that **myelinate** peripheral nerves, as well as skeletal and other tissues on the head, including muscle. Placodes form on the head and develop into olfactory, **vestibular**, and visual receptors, among other tissues.

Glenn Northcutt[17] has argued that the neural crest and sensory placodes give rise to most of the special characteristics of vertebrates. These structures include skull bones, **endocrine** organs, arches of cartilage that support the feeding and respiratory apparatus, and smooth muscle (common in the vasculature). According to his view, these new vertebrate tissues first provided an improved pumping mechanism for moving water near the animal. The older mechanism for obtaining nutrients involved small, hairlike cilia on the skin surface, which moved nutrient-containing water by agitation. The new solution to the problem of nutrient procurement involved the active "respiration" of fluids. A pump, called a *pharangeal pump*, made up of both muscle and stiffer skeletal elements (initially cartilage), provided improved movement of fluids. This new respiratory mechanism provided an important advantage for both oxygenation and nutrition. It also served as a **preadaptation**, a term used for an evolutionary development that enables some future advance. Perhaps elaborations of this pump mechanism led to further development of the motor system, which enabled hunting and the ingestion of more nutrients. Later, it also served as a precursor to the development of jaws, which allowed the ingestion of yet larger foods (see section 2.5). In animals with such motor capabilities, selection would favor those with better sensory receptors, brains to process the information provided by those sensors, and a motor system to make use of that information.

To sum up the presentation so far: The key evolutionary breakthrough accomplished by the stem vertebrates involved inventing the telencephalon and much of the head, including the development of distance receptors, such as those for vision and olfaction, to control swimming and eating.

All that having been said about early vertebrate evolution, notice some statements that have not been made. A traditional textbook view of vertebrate evolution as a steady progression from brainless filter feeders to brainy hunters probably exaggerates the facts. The entire brain probably did not appear de novo during the transition from stem cephalochordates to stem vertebrates (see table 2.1 and figure 2.3). Some experts now think that cephalochordates possess a rudimentary diencephalon, for example. Further, some yet more distant ancestors of vertebrates show some aspects of vertebrate brain organization,[18,19] which suggests that a common ancestor—among the stem chordates—might have had them as well. An acoustic **placode**[20] and **neural crest** cells[21] may have arisen more remotely than originally thought. None of these contentions lacks controversy, but it appears that at least some aspects of the vertebrate nervous system may have appeared quite early in chordate evolution.

Nevertheless, the CNS of the modern cephalochordate, amphioxus, displays more similarity to the vertebrate CNS than is the case for other chordates. One expert, Thurston Lacalli,[22] has identified as many as six homologies between the brain of larval amphioxus and that of modern vertebrates, and the expression patterns of certain regulatory genes back up some of his conclusions. One set of homologies involves neurons resembling tectal cells in vertebrates, which receive inputs from photoreceptors and project to motor output structures, much like neurons in the **oculomotor** system of vertebrates. Notwithstanding the existence of some putative homologies between cephalochordate and vertebrate brains, they remain very limited. As Rudolf Nieuwenhuys has characterized them, they

> pale into insignificance beside the scores of evident homologies that can be established between the brains of the various groups of [vertebrates]. The enormous increase in complexity that has occurred during the evolution from an Amphioxus-like ancestor to their [vertebrate] descendants is correlated with ... the development of a new muscular pharangeal pump, the genesis of new embryonic tissues including neural crest and ... placodes and the transition from filter-feeding to a more active, predatory lifestyle. (From Nieuwenhuys,[10] p. 268)

Modern cephalochordates have no obvious head, nor do they have the conspicuous brain and paired eyes found in the most ancient vertebrate fossils,[1,2] let alone a telencephalon that includes olfactory bulbs. Existing vertebrates have all of those features, and more. Accordingly, these structures probably number among the most important "inventions" of the stem vertebrates.

2.4.3 Inventing a Periphery Under Central Control

Another dramatic difference between the earliest vertebrates and their immediate ancestors involves muscle, an equally fundamental aspect of the motor system.[23] The ancestors of vertebrates had muscle cells with single nuclei, but, as described in section 3.2, vertebrate muscle has a very different structure. In vertebrates, precursor cells called **myoblasts** fuse to form cells with many nuclei, which grow longer and larger than can most cells having only a single nucleus.

In addition to their CNS, distance receptors, and multinucleate muscles, stem vertebrates evolved advances in the digestive and cardiovascular systems and in tissues for regulating tissue osmolarity (ultimately, the kidney and related organs).[24] These additional advances have a close functional relationship with those in the CNS: innervation of the heart allows modulation of that important organ. The development of two semicircular canals for gravitational sense and detection of body and head movements is important for a moving animal, and the development of a sensory lateral-line system, which subserves pressure sense, probably plays a related role. The appearance of an anterior pituitary can also be viewed as an aspect of centralization in the control of the body: It responds to hypothalamic secretions by producing hormones that regulate growth, metabolism, and reproduction. Indeed, a little-known fact about the development of the hypothalamus is that certain of its cells—notably some of those involved in reproduction—migrate into it from outside the brain. These "migrant workers" come into the brain from the nasal placode in all vertebrates studied so far, and thus appear to have developed very early in vertebrate history.

Although the brain can control head and eye movements, to control the body beyond the head, information needs to be transmitted to the spinal cord. The pattern of axonal projections descending from the brain to the spinal cord varies across vertebrate species, but many common features exist. All vertebrate species have descending projections from the brain to the spinal cord. These pathways include descending spinal projections from vestibular nuclei for gravity- and movement-related reflexes and from brainstem neurons that release either noradrenaline or serotonin at their terminals. In addition, many groups of cells in the reticular formation, extending from the medulla to the hypothalamus, project directly to the spinal cord. These pathways include spinal projections from the interstitial nucleus of Cajal (see box 2.1); the spinal nucleus of the trigeminal nerve, which receives sensory inputs from the head; and the nucleus of the solitary tract, which receives input from gustatory (taste) and visceral (organ) receptors.

Box 2.1
Neural integrators and brainstem motor control

> One particularly interesting source of brain input to the spinal cord lies near the junction of the midbrain and diencephalon. Brainstem projections to the spinal cord arise from cells in the *fields of Forel* and in the *interstitial nucleus of Cajal*, both of which are situated in this junction zone. Recent research[25] has demonstrated that the latter contains a *neural integrator* for transforming head velocity signals into a location signal, allowing the maintenance of a constant head **posture**. It is not surprising, given the importance of the head in vertebrate evolution, that an important descending projection to the spinal cord controls head orientation. The projection from the interstitial nucleus to the spinal cord is highly conserved during vertebrate evolution, and a similar mechanism controls eye orientation. Its neural integrator lies in a brainstem nucleus called the *prepositus hypoglossi* (in Latin, *hypoglossus* means tongue, and this nucleus is positioned in front of the motor neurons that control tongue movement; hence the term *prepositus*). The eye-orientation integrator receives motor commands that control the velocity of eye movements and computes an orientation signal, the integral of velocity. The output of this integrator holds the eyes at a desired orientation. Inactivation of this nucleus leads to an inability to maintain stable eye orientation, and inactivation of the interstitial nucleus of Cajal has a similar effect on head orientation.

2.5 More Recent Steps: Cerebellum and Motor Cortex

Since the revolutionary advances described above, which occurred about 500–550 mya, surprisingly little of a fundamental nature has changed. Vertebrates developed elaborate motor capabilities based mainly on the original system; the motor system with which you reach and point derives from the one that originally evolved to control swimming, eating, orientation of the head, and—early on, at least—movements of the eyes. You can be certain that the motor system did not evolve to control reaching and pointing, because the first vertebrates not only lacked jaws (hence their name, agnathans) but also lacked paired appendages of any kind. Those structures evolved later, as paired fins that appeared in the early gnathostomes and later evolved into limbs (see section 3.1).

For the following discussion, you need to know one additional biological term: *tetrapod*, for "four-footed" vertebrates (see table 2.1). Your status as a human entitles you to membership in several other groups. Primate and mammal, of course. You learned from the discussion above, if you did not already know, that you also belong to the group of jaw-mouthed vertebrates (gnathostomes). You also have a lifelong membership in a group called "fleshy-limbed" vertebrates (*sarcopterygians*). That is, you have descended from the group of vertebrates that innovated muscular limbs. These animals first exploited the opportunities of living on land by adapting their fleshy fins into the four limbs of land vertebrates and the four feet that give them the name tetrapods. All of the descendants of stem tetrapods remain tetrapods even if, like snakes and whales, some **lineages** have lost some feet or limbs during evolution. The original tetrapods and their descendants—at least for the first 60 million years or so of their history—would be considered amphibians, and some of these animals were the ancestors of modern amphibians such as frogs, newts, and salamanders. Tetrapods include all modern amphibians and the other descendants of stem tetrapods: reptiles, birds, and mammals.

Two sets of dramatic enhancements of the motor system appeared after the innovations of stem vertebrates: the *gnathostome–tetrapod augmentation* and the *mammalian augmentation*.

2.5.1 The Gnathostome–Tetrapod Augmentation

Gnathostomes or their immediate ancestors (see table 2.1, figures 2.3 and 2.4) appear to have innovated three highly prominent structures: the jaws for which they are named; the paired appendages (fins) just mentioned;[26] and another structure centrally important to the learning that underlies reaching (and other) movements, the cerebellum. The time of appearance of the cerebellum remains uncertain, but as a separate structure, such as it appears in most vertebrates, it seems to exist only in gnathostomes, not in agnathans. Some experts believe that they can recognize the beginnings of something like the cerebellum in agnathans, so it remains uncertain whether this structure was entirely an innovation of the stem gnathostomes, but that seems most likely. It has often been noted that the cerebellum seems to reach a most impressive state of development in tetrapods compared to other vertebrates.

The development of a cerebellum provided additional sources of descending spinal inputs: the deep nuclei of the cerebellum (see section 5.3.3) project directly to the spinal cord in tetrapods, but probably not in other vertebrates. The history of the red nucleus and its projection to the spinal cord is somewhat less clear, but it seems also to be especially prominent in tetrapods. (Although the direct projections from the red nucleus to the spinal cord are said by some experts to be dramatically diminished in humans, the evidence for this assertion remains unconvincing.)

Thus, of the fairly long list of structures in the CNS set forth in section 2.2, all but one appeared early in the history of vertebrates. By the time that jawed fish came on the scene, probably 450 mya or so, the motor system had motor neurons, striated multicellular muscles, CPGs, and descending control of the spinal cord from a brain that included a reticulospinal system, cerebellum, superior colliculus (optic tectum), hypothalamus, and telencephalon (containing, among other structures, the basal ganglia). All those elements were in place by the time that the first land vertebrates, tetrapods, evolved 370 mya or so. The brains of those animals received information from sensory receptors that accumulated information from a distance, most notably systems for vision and olfaction. They also made decisions about where to move and when to do so. Compared to the motor system that you use to reach, only one major component—a set of components, really—had yet to appear. Randy Nudo[27] has called those innovations the *mammalian augmentation*.

2.5.2 The Mammalian Augmentation

The most recent addition to the motor system is the *neocortex*. This "new" cortex contrasts with two more **primitive** types of cortex called paleocortex (old cortex) and archicortex (ancient cortex), also known as piriform (or olfactory) cortex and hippocampus, respectively. These older kinds of cortex appeared early in vertebrate evolution, certainly by the time of stem reptiles, in which they are called lateral cortex and medial cortex, respectively[28] (see figure 2.4 and table 2.1). In stem reptiles and their descendants, these two structures take the form of three-layered cortex known as *allocortex*. There is reasonable evidence that homologues of allocortex exist in amphibians, although their structure does not necessarily have the layered appearance of reptilian and mammalian cortex.[29] In fish, the anatomical data remain somewhat uncertain, but most experts think that something like a piriform cortex and a hippocampal cortex exist in all gnathostomes, and a homologue to piriform cortex seems to exist in all vertebrates, including agnathans.[30]

In contrast to the allocortex, neocortex occurs only in mammals (box 2.2). It seems nearly certain, therefore, that this cortex evolved much more

Box 2.2
Bird brains

> Birds merit a brief mention because comparisons between birds and mammals generate some confusion, especially regarding the telencephalon. The similarities in the telencephalon of mammals, birds, and reptiles represent mainly their inheritance from stem reptiles.[28] Despite some limited evidence for something like a corticospinal projection in certain birds, the ancestors of birds and extant reptiles diverged from the lineage leading to mammals very early in reptilian evolution (figure 2.4). In accord with this fact, recent analyses based on the expression of regulatory genes indicate that the motor systems of mammals and birds have taken very different paths. For example, the most prominent parts of the motor system in birds, including the well-studied birdsong system, appear to derive from an enigmatic structure called the claustrum in mammals.[31,32] Birds have many intriguing motor capabilities, but the lessons they teach may be limited by the distance of their divergence and subsequent specializations.

recently than allocortex, and that it did so in stem mammals. Given that parts of mammalian neocortex play a relatively direct role in the selection and control of reaching and pointing movements, the neocortex constitutes the most recent addition to the vertebrate motor system.

Many subsequent chapters address the functions of certain parts of the neocortex, but keep in mind two facts: The vertebrate motor system evolved long before the development of neocortex, and the ancestors of mammals got along very successfully without it. Regardless, this fact does not imply that you can get along without yours. Once the neocortex became fully integrated into the operations of your ancestors' CNS, it became indispensable. In mammals, the **corticospinal** projection arises from the frontal and parietal neocortex and joins the ancestral descending projections enumerated above.

2.6 Summary

Early vertebrates evolved a CNS that enabled them to succeed as mobile predators. Its features included improved mechanisms for

- Transducing information at a distance (better photoreceptors, for example)
- Processing and storing sensory information (the telencephalon, including the olfactory bulb)
- Orienting the head and its sense organs (the superior colliculus and brainstem)
- Regulating metabolism, maintaining **homeostasis**, and controlling the motor programs involved in reproduction, ingestion, exploration, and defense (the hypothalamus)

- Acquiring and storing information about a vertebrate's place in the world, and evaluating the biological value of objects, places, and events (the telencephalon)
- Predicting the effects of an animal's own **motor commands** on sensory inputs (see chapter 23).

References

1. Chen, J-Y, Huang, D-Y, and Li, C-W (1999) An early Cambrian craniate-like chordate. Nature 402, 518–522.
2. Shu, DG, Morris, SC, Han, J, et al. (2003) Head and backbone of the Early Cambrian vertebrate Haikouichthys. Nature 421, 526–529.
3. Mallatt, J, Chen, J, and Holland, ND (2003) Comment on "A new species of *yunnanozoan* with implications for deuterostome evolution." Science 300, 1372.
4. Shu, DG, and Morris, SC (2003) Response to Comment on "A new species of *yunnanozoan* with implications for deuterostome evolution." Science 300, 1372.
5. Shu, DG, Morris, SC, Zhang, ZF, et al. (2003) A new species of *yunnanozoan* with implications for deuterostome evolution. Science 299, 1380–1384.
6. Carroll, RL (1988) Vertebrate Paleontology and Evolution (Freeman, New York).
7. Haile-Selassie, Y (2001) Late miocene hominids from the middle Awash, Ethiopia. Nature 412, 178–181.
8. Vermeij, GJ (1996) Animal origins. Science 274, 525–526.
9. Lowe, CJ, Wu, M, Salic, A, et al. (2003) Anteroposterior patterning in hemichordates and the origins of the chordate nervous system. Cell 113, 853–865.
10. Nieuwenhuys, R (2002) Deuterostome brains: Synopsis and commentary. Brain Res Bull 57, 257–270.
11. Gans, C, and Northcutt, RG (1983) Neural crest and the origin of vertebrates: A new head. Science 220, 268–273.
12. Gans, C (1989) Stages in the origin of vertebrates: Analysis by means of scenarios. Biol Rev 64, 221–268.
13. Forey, P, and Janvier, P (1994) Evolution of early vertebrates. Am Scientist 82, 554–565.
14. Jeffries, RPS (2001). In Major Events in Early Vertebrate Evolution, ed Ahlberg, PE (Taylor and Francis, London), pp 40–66.
15. Holland, LZ, and Holland, ND (1999) Chordate origins of the vertebrate central nervous system. Curr Opin Neurobiol 9, 596–602.
16. Butler, AB (2000) Sensory system evolution at the origin of craniates. Phil Trans Roy Soc London B 355, 1309–1313.
17. Northcutt, RG (2001) Changing views of brain evolution. Brain Res Bull 55, 663–674.
18. Lacalli, TC, and Holland, LZ (1998) The developing dorsal ganglion of the salp *Thalia democratica*, and the nature of the ancestral chordate brain. Phil Trans Roy Soc London B 353, 1943–1967.
19. Wada, H, and Satoh, N (2001) Patterning the protochordate neural tube. Curr Opin Neurobiol 11, 16–21.

20. Wada, H, Saiga, H, Satoh, N, and Holland, PWH (1998) Tripartite organization of the ancestral chordate brain and the antiquity of placodes: Insights from ascidian *Pax*-2/5/8, *Hox* and *Otx* genes. Development 125, 1113–1122.

21. Holland, LZ, and Holland, ND (2001). In Major Events in Early Vertebrate Evolution, ed Ahlberg, PE (Taylor and Francis, London), pp 15–32.

22. Lacalli, TC, Holland, ND, and West, JE (1994) Landmarks in the anterior central nervous system of amphioxus larvae. Phil Trans Roy Soc London B 344, 165–185.

23. Carroll, RL (1997) Patterns and Processes of Vertebrate Evolution (Cambridge University Press, Cambridge).

24. Janvier, P (1996) Early Vertebrates (Oxford University Press, Oxford).

25. Klier, EM, Wang, H, Constantin, AG, and Crawford, JD (2002) Midbrain control of three-dimensional head orientation. Science 295, 1314–1316.

26. Northcutt, RG (1995) The forebrain of gnathostomes: In search of a morphotype. Brain Behav Evol 46, 275–318.

27. Nudo, RJ and Masterton, RB (1988) Descending pathways to the spinal cord: A comparative study of 22 mammals. J Comp Neurol 277, 53–79.

28. Ulinski, PS (1990) in Cerebral Cortex: Comparative Structure and Evolution of Cerebral Cortex, part I, eds Jones, EG and Peters, A (Plenum, New York), pp 139–215.

29. Neary, TJ (1990) in Cerebral Cortex: Comparative Structure and Evolution of Cerebral Cortex, part I, eds Jones, EG and Peters, A (Plenum, New York), pp 107–138.

30. Northcutt, RG (1996) The agnathan ark: The origin of craniate brains. Brain Behav Evol 48, 237–247.

31. Farries, MA (2001) The oscine song system considered in the context of the avian brain: Lessons learned from comparative neurobiology. Brain Behav Evol 58, 80–100.

32. Puelles, L, Kuwana, E, Puelles, E, et al. (2000) Pallial and subpallial derivatives in the embryonic chick and mouse telencephalon, traced by the expression of the genes *dlx-2, Emx-1, Nkx-2.1, Pax-6*, and *Tbr-1*. J Comp Neurol 424, 409–438.

Reading List

Comparative Vertebrate Neuroanatomy: Evolution and Adaptation, by Ann Butler and William Hodos (Wiley-Liss, New York, 1996), presents a general overview of the vertebrate brain and its evolutionary history. Its discussion of the motor system is not among the book's strengths, however. An excellent source on mammalian brain evolution is *Evolving Brains*, by John Allman (W. H. Freeman, New York, 2000).

3 Burdens of History: Control Problems That Reach from the Past

*Overview: The evolutionary history of the CNS accounts for many of the problems that the motor system must overcome in order to control reaching and pointing movements. In learning to control such movements, the CNS must generate force slowly with springlike **actuators** (muscles) that act against a skeleton. It also must analyze inputs from sensory **transducers** that provide **feedback**, but only after a relatively long delay.*

Many features of the vertebrate motor system seem inefficient, and if that is so, much of that inefficiency results from its evolutionary history. Indeed, the very fact that animals move, poses problems: Movement can damage the motor system, especially during early development. In animals that begin moving in early larval stages, a special class of molecule provides protection against damage that might otherwise be caused by motive forces.[1]

As noted in chapter 2, the motor system did not evolve to control reaching or pointing; it evolved to control swimming and eating. When the capability for reaching and pointing did evolve, the vertebrate motor system used building blocks that were, at least in certain fundamental ways, immutable.

Vertebrate motility depends on an **endoskeleton**, which means that the elements that stiffen the body lie within it. An alternative arrangement, called an **exoskeleton**, has the more rigid elements on the surface. Arthropods, such as insects and lobsters, have this kind of anatomy. In lobsters, for example, muscles inside the claw exert forces on the relatively hard covering of the shell to close it. In vertebrates, forces must work on structures within the body. The vertebrate endoskeleton did not begin as bone. In the earliest vertebrates, cartilage stiffened the body axis. Later, when jaws and **appendages** first evolved, cartilage stiffened them, as well. Eventually, these stiffening structures became bone, which increased stiffness further (at the expense of toughness; see box 3.1). Bones, cartilage, and muscles compose the major elements of vertebrate limbs.

In the terminology of robotics, the motor system you use for reaching and pointing includes springlike actuators (muscles), complex multijoint levers (bones), and sluggish communication lines (nerves). All of these characteristics, and others, present problems to the motor-control

Box 3.1
Stiffness versus toughness

> In vernacular usage, stiffness and toughness mean more or less the same thing. In material science, these terms refer to two different concepts. Stiffness refers to the amount of "give" in a material, the amount of deformation that will occur in response to a given applied force. Toughness pertains to the resistance to breakage and other damage. For many materials, including bones, the increase in stiffness comes at a cost in terms of toughness. The stiffer something gets, the more likely it is to break.

system. The CNS—the software—must overcome those limitations, and do so in the face of changes in those components due to growth, strengthening or weakening of muscles, and other factors. A significant amount of motor learning deals with solving these problems.

3.1 Limbs

Because this book focuses on reaching and pointing movements, you need to know some of the anatomical terms that apply to the limbs and their muscles. If you are unfamiliar with anatomical nomenclature, you should consult appendix B. Unfortunately, if you do not already know the anatomy of the arm, you must memorize many of the terms used in this section in order to follow the subsequent chapters.

3.1.1 The Skeletal Architecture of the Limb

The term *radial* refers to the side of your arm allied with your thumb and containing the *radius*, one of the two bones connecting your wrist and upper arm. The term *ulnar* (meaning "of the elbow") refers to the side with your small finger. The other bone in your forearm, the territory between your elbow and wrist, is the *ulna*, which attaches to the knob of your elbow. Now imagine that your thumb is "out" or lateral (away from your body), as it falls limply by your side. That posture is called **supinated**, from the term for lying on your back (supine). The opposite rotation, **pronation**, derives from the term for lying on your stomach, facedown (prone). In general, *flexion* refers to decreasing the angle subtended by two articulated limb segments; *extension* refers to increasing that angle. These two dimensions of movement around the elbow—flexion/extension and supination/pronation—compose the two **degrees of freedom** of movement around that joint. Chapter 9 uses these terms extensively, and figure 9.1 illustrates these movement dimensions.

The upper arm contains the *humerus* (from the Latin for shoulder). Movements of the upper arm go by names similar to those used for movements of the forearm. The upper arm also supinates (also known as outward rotation) and pronates (also known as inward rotation), in addition to flexing and extending (again, see figure 9.1). In addition, the upper arm can move along a third dimension. Movements that position the humerus closer to the trunk are called *adductions*. A useful mnemonic aid refers to the fact that to "add" means to put things together. Movements of the humerus away from the trunk are termed *abductions*. The three dimensions of movements of the upper arm correspond to the three degrees of freedom available for movements around the shoulder joint.

In the hindlimb, the same basic skeletal patterns occur, but the names differ. You might be tempted to ignore the hindlimb because the model system used in this book involves reaching and pointing by a two-joint *arm*. Some of the key experiments, however, have been done on the hindlimbs of humans and other animals, so you will need to know hindlimb anatomy as well. The concept of *serial homology* contributes to understanding the relationship between the forelimb and the hindlimb. This principle arises from the segmental nature of many animals. If an element at one segment has some important properties that appear in another segment, and has these properties because of a comparable developmental program, they are said to be **serial homologues**. In that sense, the forelimb and hindlimb are serial homologues. The serial homologue of the upper arm bone, the humerus, is the *femur* of the leg. The term *femur* derives, naturally enough, from the Latin word for thigh. The serial homologue of the radius is the *tibia* (Latin for shin), and that for the ulna is the *fibula*.

For both the forelimb and the hindlimb, the parts nearest the trunk are called *proximal* and those farthest from the trunk are called *distal*. The more proximal parts of the arm and the leg, those containing the humerus and the femur, respectively, are often called the upper arm and upper leg, whereas the more distal parts are often called the lower arm and lower leg. Limbs, in general, are called *appendages*, and their components, *appendicular*.

3.1.2 A Brief History of Limb Architecture: From Axis to Appendage

In stem vertebrates, locomotion was driven mainly, if not exclusively, by the axial (trunk) musculature. Neither appendages nor appendicular muscles existed. Not until the evolution of jawed fish,[2] or their immediate ancestors, did appendages appear.[3] Once they evolved, however, the structure of their more proximal parts remained more or less the same. The distal limbs, specifically hands and feet and their homologues, have undergone more modification. The fins of many fishes, alive and extinct, have clear skeletal homologues to limbs in land vertebrates, both for the fins nearest the head, called *pectoral fins* (from the Latin for breast), and for those nearest the tail, called *pelvic fins* (referring to the pelvis, or haunches of an animal). Once paired fins evolved, they provided advantages for swimming in terms of movement guidance, stopping body motion, and generating lift in the water. Paired lateral fins also served as a

preadaptation for limbs. The pectoral fins were the ancestors of forelimbs; the pelvic fins evolved into hindlimbs.

The architecture of the limb exemplifies some of the most conservative features of evolution, as well as some of its most revolutionary ones. The revolutionary aspect involves the transition from a structure that originally evolved to assist swimming to one supporting locomotion on land. Despite this revolutionary change, vertebrate appendages show a striking evolutionary conservatism. They all have the following features:

- A single proximal bone (the humerus or femur), with a bifurcation at the next segment to two and only two bones (the radius and ulna in the forelimb or pectoral fin, and the tibia and fibula in the hindlimb or pelvic fin)
- The derivation of most wrist, ankle, and digit bones from the *postaxial* part (i.e., small-finger, small-toe side) of the limb
- The ancestral pattern of having five digits on each hand and foot.

These principles of appendicular organization are preserved in bony fishes, amphibians, reptiles, mammals, and birds. The pattern of having five digits on your hands and feet reflects the conservation of a developmental process that dates back to the stem reptiles some 300–350 million years ago (mya).[4–8] Of course, many descendants of those five-toed animals evolved with reductions in the number of digits. For example, horses have lost two digits on each limb and retain only vestiges of two others. Some animals, such as whales and dolphins, have lost their hindlimbs entirely. On the other hand, no animal has managed to generate a sixth digit through the developmental mechanisms that produce the ancestral set of five. In an exception that proves this rule, the giant panda has a sixth digit on the forelimb, but this structure results from a somewhat heroic, secondary modification of a wrist bone.

Neglecting the wrist and fingers, the joints of the forelimb have five degrees of freedom, two at the elbow and three at the shoulder. You need only four degrees of freedom to navigate the limb through three-dimensional space. This is the sense in which the vertebrate limb, including human arms, has a property called **overcompleteness**.

How did this overcomplete limb come about? As noted in chapter 2, the stem vertebrates—all jawless fish (agnathans)—lacked paired lateral fins. Some jawless fish did develop appendages, however, and one among these species was presumably the ancestor of jawed fish (the stem gnathostomes; see table 2.1). They developed paired pectoral and pelvic fins. Fossil evidence indicates that paired fins of the kind found in jawed fish occurred relatively late in the evolution of agnathans, and that the particular organization that you and all other gnathostomes inherited represents only one of many "experiments" with paired appendages among agnathans.[9] Overcomplete limbs, then, appear to be a mainly "accidental" property.

Overcompleteness has both advantages and disadvantages. It allows the goals of reaching movement to be achieved in many different ways, from different starting points, and in the face of obstacles. But overcompleteness also creates computational burdens for controlling simpler movements. Section 19.6 returns to this topic by suggesting that your CNS learns to leave some degrees of freedom uncontrolled, if they are not important to achieving some goal. Thus, your CNS learns both to exploit the advantages of overcompleteness (also called *redundancy*) and to cope with the extra computational complexity it creates.

3.1.3 A Brief History of Limb Architecture: From Fins to Limbs

The transition from paired fins to limbs occurred in stem tetrapods. The shoulder, elbow, and wrist of the forelimb and the hip, knee, and ankle of the hindlimb appeared in unmistakable form in these animals. With the advent of limbs and adaptation to a terrestrial environment, axial muscles became more important for postural support of the body and for creating alternating negative and positive pressures in the lungs to support breathing. The ancestral function of the axial muscles, which was to control swimming with trunk undulations, became less important as tetrapods adapted to life on land.

Locomotion on land, however, differs dramatically from that in water, so this transition probably occurred slowly. Swimming involves movements propelled by axial muscles with an undulating, S-shaped movement of the body axis. Land locomotion, in contrast, involves transmitting force from the limbs to the body axis mainly through the axial endoskeleton. Thus, early tetrapods developed a way to use the ancient movement patterns, even when walking on land. They learned to locomote with motor patterns that resembled those used in swimming. This use of S-shaped trunk movements in terrestrial locomotion has sometimes been termed "swimming on land." When the first "fleshy-limbed" vertebrates adopted life on land, locomotion remained the function of axial muscles. According to Jennifer Clack,[10] the first land vertebrates were probably belly crawlers that used their limbs "for purchase [i.e., leverage] against the ground and for steering and braking, much as fish fins are used." They probably evolved from fish that were habitual bottom dwellers.

In birds and most mammals, especially placental mammals, the **CPGs** underlying such movements have been adapted to other functions, and they do not "swim on land." In many other land vertebrates, however, the older movement patterns have been retained, at least to an extent. As with the overcompleteness of the limb, you can see that the vertebrate motor system had to adapt existing mechanisms to new applications, and many of these mechanisms differ from those an engineer would have chosen.

Given the many developments required in adapting to the land, the "changes to the limbs and locomotory patterns may have been among those that occurred last in the suite of adaptations seen in the transition from fish to tetrapod," according to Clack. Only later, in other descendants, could reaching and pointing evolve. Of course, limb movements

3.2 Muscles

Muscles are the actuators for movement; they generate the force, **work**, and power needed for movement (see appendix D for definitions of these terms). In swimming vertebrates, muscles generate power that overcomes the drag of the water; in flying vertebrates, muscles generate lift. But these actuators do more than generate movement. During sound production, muscles transform work into acoustic energy, and modified muscle cells produce the currents generated by electric fish (see chapter 23). This section presents some of the basics of muscle architecture; chapter 7 picks up a description of its force-generating mechanisms.

3.2.1 The Muscular Architecture of the Limb

Because this book focuses on one particular model system, a two-joint arm, you do not need to know the muscles of the entire body. Familiarity with a selection of the muscles of the forelimb (figure 3.1) and their actions around the shoulder and elbow joints (figure 3.2) should suffice for following most of the presentation. Unfortunately, some of the key experimental results have been obtained from the hindlimb (figure 3.3), so

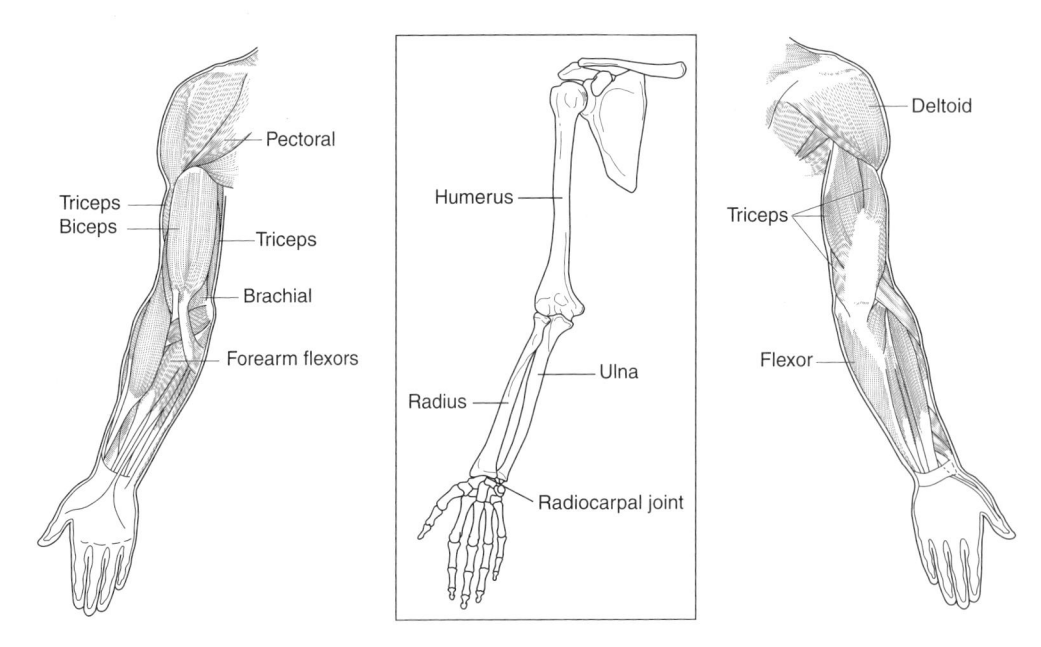

Figure 3.1
Selected forelimb muscles. (Drawn by George Nichols.)

3.2. Muscles

Figure 3.2
The actions of selected forelimb muscles on a two-joint arm.

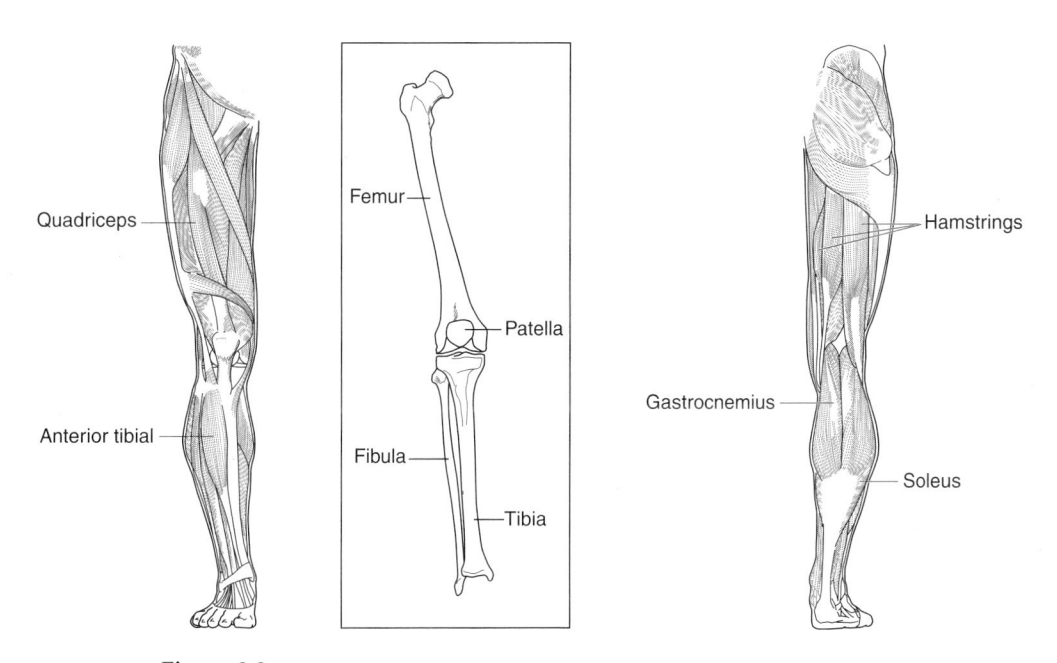

Figure 3.3
Selected hindlimb muscles. (Drawn by George Nichols.)

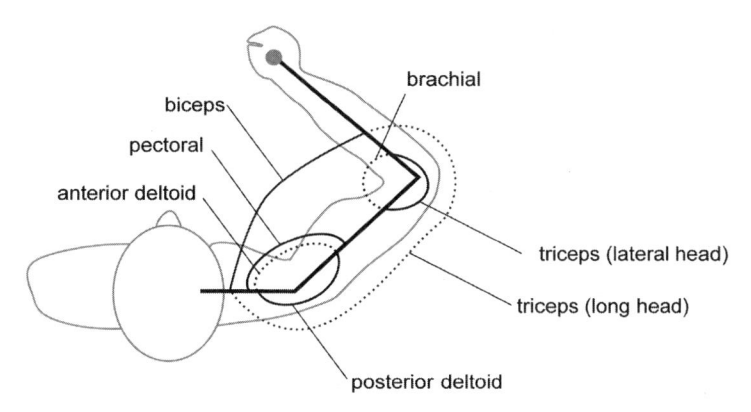

you occasionally need to know the names and locations of some of those muscles, as well.

3.2.2 A Brief History of Muscle Architecture

Section 2.4 discussed the pattern of the vertebrate motor system as one consisting of a CNS, a notocord, and segmented muscles. In stem vertebrates these muscles, called **myotomes**, are attached to the notochord. Because of its architecture, the notocord is stiffer along its dorsoventral dimension than along its mediolateral one, promoting more bending from side to side than from dorsal to ventral. Myotomes run for short distances along the long axis of the animal and are controlled by the output of a spinal segment. Each segment of the spinal cord has motor neurons that send axons to the muscles of the ipsilateral side of the body. When myotomes contract, they cause the body to bend in their direction.

Muscles have been elaborated into a diverse array of architectures derived from the myotomal arrangement of primitive vertebrates. Nevertheless, the pattern of muscles has remained remarkably stable during evolution. According to Robert Carroll:

Although the girdles and limbs are much altered among terrestrial vertebrates, the number, general location, and innervation of the muscles controlling the limbs have remained conservative throughout the evolution of terrestrial locomotion and flight. Nearly all of the muscles can be homologized from amphibians through reptiles, birds, and mammals. Some have been lost, others split or fused, but their number has remained generally similar. On the other hand, the way in which they function, such as adduction, abduction, supination, pronation, or rotation of the limb, may change in relationship to the geometry of the bony elements of the limbs. (Carroll,[4] p. 179)

3.2.3 Muscles as Actuators

The skeletal muscles that vertebrates evolved, however, have some interesting and important properties. Because they were adapted from mechanisms that originally evolved to move things within a cell (see section 3.2.6), they have properties that are somewhat unusual for mechanical actuators. Unlike the actuators used in robots and other mechanical devices, muscles are springy and generate force slowly.

3.2.4 The Problem of Springy Actuators

In mechanical engineering, actuators are usually rigid. For example, early in the industrial revolution, when running water powered mills and factories, the actuators were stiff leather belts connecting the **prime mover** to a set of subsidiary pulleys to run the equipment involved in manufacturing. The term "prime mover" is now used for the muscle doing the most work during a movement, but then it was used for the waterwheel that brought energy into the factory, and later for engines that did the work. Those early industrial engineers would not have replaced stiff belts with springy ones because the force generated by the prime mover would have been dissipated by stretching the belts rather than in moving the mills.

Evolutionary mechanisms, however, did not have the luxury of choosing the materials that made up their actuators. As one legacy of the motor system's history, your CNS must learn to control springlike actuators, in which force varies as a function of muscle length and depending upon recent usage, growth, and other factors. Unlike robots and other mechanical devices, therefore, biological controllers cannot simply send a command (e.g., a voltage to an amplifier) and expect the same force output every time. The force produced by the muscle depends not only on the command but also on the configuration of the limb (and therefore the length of the muscles) and on the limb's recent history. Chapter 8 considers some of the advantages of springy actuators: They promote stability with virtually instantaneous responses to imposed forces and, by being less stiff but tougher (see box 3.1), they can do so with minimal risk of rupturing. Notwithstanding these advantages, to control reaching and pointing, your CNS must overcome problems imposed by springy actuators.

The overcompleteness of the limb and the development of muscles that span more than one joint create additional computational problems. You might imagine better materials and architectures, but the CNS had to adapt to the mechanisms that evolution devised.

3.2.5 The Architecture of Individual Muscles

Skeletal muscles develop from **myoblasts** that fuse to form fibers 10–100 μm in diameter, extending on the order of a meter in length in large animals. As illustrated in figure 7.1A, each muscle fiber comprises bundles of *myofibrils* that house a collection of *filaments*, which in turn consist of *sarcomeres*. Within these sarcomeres, proteins called *actin* and *myosin* form strands, and interactions among these strands generate force.

A muscle comprises bundles of muscle fibers, with each bundle separated from the others by connective tissue. For example, in humans, the calf muscle *gastrocnemius* (from the Greek for belly of the calf) contains approximately 10^6 muscle fibers, each of which consists of about 10^3 myofibrils, arranged in parallel. Each myofibril consists of approximately 10^4 sarcomeres, the force-generating engines, arranged in series. Thus, a large muscle like gastrocnemius in humans consists of more than 10^{13} sarcomeres.

There are two types of muscle fibers: **extrafusal fibers**, which attach to tendons, which in turn attach to the endoskeleton, and **intrafusal fibers**, which attach to the extrafusal fibers. Extrafusal fibers produce the force that acts on bones and other structures. Intrafusal fibers also produce force, but they are much smaller than extrafusal fibers, and they produce negligible forces in comparison. Instead, intrafusal fibers play a sensory role. They contain muscle spindles which, innervated by **muscle spindle afferents**, provide the CNS with information about muscle length and changes in it.

3.2.6 Why Do Muscles Contract So Slowly?

Because of your evolutionary heritage, springlike muscles produce force slowly, taking relatively long periods (tens of milliseconds) after activation to begin producing force. This section addresses why muscles contract so slowly.

Muscles produce force and shorten when myosin molecules attach to actin molecules and change their shape in a ratchetlike manner (see section 7.2.1 for more detail about this mechanism). One reason that muscles contract relatively slowly is that this myosin–actin mechanism has evolved from a system that moves organelles within a cell and mediates cell division, among other functions. And it does not seem to put a particular premium on speed.

A similar system exists in single-cell organisms, called *protists*, some kind of which were ancestors of multicellular animals. Protists have a nucleus bound by an internal membrane, and thus are termed *eukaryotes* (meaning that they have true nuclei). Such cells move by a protrusion-and-stabilization mechanism, which unfolds at a glacial pace. First, these organisms temporarily break down the structural support mechanism in one part of their outer membrane. Then that part of the cell extrudes in the direction of movement by forcing the membrane outward. This process depends on the growth of actin filaments and other proteins that span the membrane to anchor the protrusion. Adhesions between the membrane and the external environment anchor the membrane and serve as a footing for moving the cell, a process that involves myosin motors like those that generate force in muscles.[12] Thus, the mechanisms that operate in muscles are not particular to muscles or even to animals. Instead, they evolved from an ancient mechanism for moving cells and moving organelles within cells. It is therefore not surprising that they generate force rather slowly.

Box 3.2
Fast movement with slow muscles

> If muscles produce force so slowly, then how can insects and some small birds, such as hummingbirds, flap their wings at nearly 100 beats per second? They can do so because the mechanics of the wing in these animals acts something like a spring-mass-damper system in which an initial force sets up an oscillation. The motor commands during flight consist of occasional pulses to the muscles that act something like someone giving you an occasional push when you are on a swing.

3.3 Nerves

3.3.1 Why Are Communication Lines So Slow?

Similar considerations apply to nerves. Grossly speaking, information traveling down an axon moves at something like the speed of sound, whereas the same information would move at the speed of light along a copper wire. The relatively slow conduction of neural signals to the actuators and the equally slow feedback about what is actually happening create serious problems for a central controller. The reasons for sluggish neural transmission go back to the origin of animals.

All cellular organisms, including single-cell organisms, have proteins integrated into their membranes to allow water-soluble (*hydrophilic*) substances, such as ions, to transverse the water-insoluble (*hydrophobic*) membrane that encloses all cells (see box 11.2). A cell membrane is hydrophobic because much of its interior is made up of fats, with the exterior elements being more water-soluble. Embedded in these membranes are a wide variety of proteins, some of which traverse the entire membrane to form pores or channels that allow ions to pass through the membrane. Yeast, protozoa, and bacteria have ion channels, suggesting that the existence of such channels goes back to near the beginning of life on Earth.

Some of these channels probably evolved to regulate cell osmolarity. Whenever too much or too little water entered a cell, stretching or shrinking its membrane, ions would move through membrane channels into or out of the cell to regulate its internal osmolarity, a crucial variable in a single-cell organism. It is easy to imagine how this kind of channel could evolve into a mechanical transducer for forces impinging on or within multicellular organisms, including sound or pressure waves and gravity fields. Single-cell organisms also have other kinds of channels. Some interact with receptors to transduce a concentration of molecules or a number of photons into an ionic or a voltage signal. These kinds of channels probably were preadaptations for neural transmission, hormonal responses, and sensation in multicellular animals.

As in multicellular organisms, in single-cell organisms these channels respond to mechanical forces, membrane voltages, and molecules that induce effects such as opening or closing the channel. An important example of a channel affected by membrane voltage is the voltage-activated Na^+ channel, which supports action potentials. As in multicellular animals, Ca^{++} ions enter protists during the rising phase of an overshooting action potential and serve to couple membrane excitation to the cell's contraction mechanism. In addition, as in multicellular animals, this coupling is based, in part, on the interaction of actin and myosin. Action potentials in single-cell organisms might serve to send a signal virtually instantaneously to the entire organism. However, what is nearly instantaneous for a single-cell organism, no more than 10 μm or so in diameter, is very sluggish for a large animal. It takes many tens or hundreds of milliseconds to send a signal along an axon, especially in motor nerves, which can be several meters long.

Why does neural transmission take so long? In general, the slowness of neural transmission results from the relatively long time that it takes for the proteins in channels to change configuration. These conformational shifts cause the channels to open and close in support of action potentials. An additional factor involves the passive properties of membranes. Thin axons have high electrical resistances that retard the progress of action potentials. In part to overcome these limitations, a **myelination** system (another of the **neural crest** derivatives mentioned in chapter 2) has evolved in which the axons receive "insulation" from dense wrappings of the membranes of other cells. These insulating membranes are separated from one another by uninsulated gaps called *nodes*, and they promote the progress of action potentials by *saltatory conduction*, in which the signal jumps from one patch of uninsulated membrane to another. This mechanism speeds the transmission of information, but cannot solve the fundamental problem faced by the central controllers in vertebrates: Transmission rates remain slow in relation to the events that unfold during active reaching and pointing movements.

References

1. Vogel, S (2001) Prime Mover (WW Norton, New York).
2. Coates, MI (1994) The origin of vertebrate limbs. Development 1994 Suppl, 169–180.
3. Carroll, RL (1988) Vertebrate Paleontology and Evolution (Freeman, New York).
4. Carroll, RL (1997) Patterns and Processes of Vertebrate Evolution (Cambridge University Press, Cambridge).
5. Wagner, GP, and Chiu, C-H (2001) The tetrapod limb: A hypothesis on its origin. J Exp Zool (Mol Dev Evol) 291, 226–240.
6. Ruvinski, I, and Gibson-Brown, JJ (2000) Genetic and developmental bases of serial homology in vertebrate limb evolution. Development 127, 5233–5244.
7. Capdevila, J, and Izpisúa Belmonte, JC (2000) Perspectives on the evolutionary origin of tetrapod limbs. J Exp Zool (Mol Dev Evol) 288, 287–303.
8. McGonnell, IM (2001) The evolution of the pectoral girdle. J Anat 199, 189–194.
9. Forey, P, and Janvier, P (1994) Evolution of early vertebrates. Am Scientist 82, 554–565.
10. Clack, JA (2002) Gaining Ground: The Origin and Evolution of Tetrapods (Indiana University Press, Bloomington).
11. Houk, JC, Fagg, AH, and Barto, AG (2002) Fractional power damping model of joint motion. Hum Kinet 2, 147–178.
12. Chicurel, M (2002) Cell migration research on the move. Science 295, 606–609.

Reading List

Gaining Ground: The Origin and Evolution of Tetrapods, by Jennifer A. Clack (Indiana University Press, Bloomington, 2002), presents a highly readable account of the origin of limbs and land animals.

4 What Motor Learning Is, What Motor Learning Does

*Overview: Your evolutionary history has given you a motor system that learns, and motor learning plays a fundamental role in reaching and pointing movements. Motor learning takes many forms, including (1) the learning, over generations, of reflexes and innate motor programs which become encoded in the **genome** and form the basis for learned (conditioned) reflexes; (2) learning new skills to augment your inherited motor repertoire, and adapting those skills to maintain performance at a given level; and (3) learning what movements to make and when to make them. Motor learning allows you and other animals to achieve beneficial goals and avoid harm. Reaching extends the range of goals available, and pointing has special importance for communication.*

Chapters 8–22 of this book deal primarily with reaching and pointing movements in monkeys and humans. This chapter attempts to place the learning that underlies those behaviors into a broader context.

Chapter 3 explained some of the burdens of your evolutionary history. This chapter spells out some of the benefits, which stem, in large part, from a motor system that learns. **Appendages** began as fins to guide swimming, later became limbs for locomotion, and, yet later, came to function in reaching and pointing. The motor system needed to adapt existing materials and control systems as those changes occurred. It did so, in part, through motor learning. To compensate for delays in receiving feedback and producing force, for example, your CNS learns to predict the feedback that should occur, based on what it has experienced (chapter 17). It also estimates the forces it needs to produce in order to receive a desired feedback signal (chapters 20–22). All of this requires motor learning. But motor learning is more than that. This presentation takes a very broad view of motor learning, and you should know how we think of it.

4.1 Motor Learning Undefined

We think of motor learning as the acquisition of information about movements (and other motor outputs), including what output to produce as well as how and when to produce it. Motor learning results in the formation of motor memories. Note that we say that we "think of" motor learning and memory that way, not that we *define* it that way. We think that a typical definition of motor learning would not really help you understand very much. For example, according to Richard Schmidt,[1] "Motor learning is a set of processes associated with practice or experience leading to relatively permanent changes in the capability for responding." Our reluctance to adopt that or any other definition of motor learning is not that it is wrong, but rather that it does not seem particularly helpful.

The reason that this book avoids defining motor learning is that in biological systems, rigid, formal definitions generally have less value than fuzzy ones. We agree with the Medawars, who, in writing about the definition of life and other terms in biology, expressed the following sentiment:

> In certain formal contexts—mathematical logic, for example, in which a definition is a rule for substituting one symbol for one or more others—definitions are crucially important, but in ... sciences such as biology their importance is highly exaggerated. It is simply not true that no discourse is possible unless all technical terms are precisely defined; if that were so, there would be no biology. A principal purpose of definition is to bring peace of mind. Sometimes, though, it is too dearly bought: a "definition," as the word itself connotes, has a quality of finality that is often unjustified and misleading and may have the effect of confining the mind instead of liberating it. (Medawar and Medawar,[2] p. 66)

The idea of fuzzy categories and the related idea of conceptual prototypes comes from the psychology of categorization and classification.[3] One type of category, often called a natural category, has a clear definition, unambiguous inclusion criteria, and definite boundaries. Four-minute milers—such as the world's most famous neurologist, Roger Bannister—exemplify a natural category. You either are a four-minute miler or you are not. (He was the first.) The concept of a crook, on the other hand, exemplifies a fuzzy category. Membership in a fuzzy category depends on the degree of correspondence to some prototypical or ideal member of that category. The prototypical crook might look like Fagin and mug people on the street, but malefactors such as Richard Nixon and researchers who fabricate data fit the prototype closely enough that some people classify them as crooks. Many might disagree. The concept of motor learning used in this book depends on fuzzy categorization in this sense. Its fuzziness involves primarily the distinction between motor learning and other kinds of learning.

Our view of motor learning extends the concept further than usual. Many discussions of motor learning concentrate exclusively on one prototypical form of motor learning—namely, skill acquisition, as in Schmidt's definition. This view excludes too much. Skill acquisition involves gaining a new level of performance or a new capability. Another form of motor learning, called motor adaptation, involves regaining (or retaining) a given level of motor performance. Reaching and pointing involve both of these kinds of motor learning, and more. Our concept of motor learning not only includes both skill acquisition and motor adaptation, but goes further in two main ways, which many experts exclude from the concept of motor learning:

It includes the learning of instinctive behaviors, a kind of evolutionary learning that accrues over generations and becomes encoded genetically. This genetic, cross-generational form of motor learning often contributes directly to biological **fitness**—the ability of a biological system to pass its genes to future generations—in a variety of ways. The existence of these mechanisms permits animals to associate new stimuli with those that trigger instinctive reactions, including relatively simple reflexes (such as Pavlovian conditioned reflexes) and more complicated actions (such as Pavlovian approach behavior). Section 4.2.4 explains these terms and concepts.

Our expansive view of motor learning also includes decision making (chapter 25): specifically, learning what to do in a given context, as well as when (and if) to do it. At first thought, this topic seems to belong to the **cognitive** realm and involve conscious intent. But behaviors that appear identical to those guided by conscious intent—including reaching and pointing movements—occur without awareness and, being sensorially guided movements, fall squarely within the domain of motor learning.

Figure 4.1 illustrates one view of motor learning. A great deal of learning—and not just motor learning—depends on one or more of several key mechanisms: prediction, error signals that reflect the accuracy of those predictions, and error correction. Note that the concept of "error" here is a broad one, encompassing anything that allows the CNS to evaluate its performance at any level, including both predictions and actions. The scheme depicted in the figure suggests three kinds of predictions, which correspond to what, in our view, are the three main contributions of motor learning: (1) maintaining fitness, (2) maintaining stability and control of the motor apparatus, and (3) deciding on targets and goals.

4.2 Motor Learning over Generations: Links to Instincts and Reflexes

The concept of fitness encompasses all of the properties of an animal (or any other biological system) that promote the transmission of its genes to future generations. Discussions of motor learning rarely consider this topic because maintaining fitness involves many **autonomic** and **neuroendocrine** functions, such as the regulation of heart and growth rates. These systems do not seem very "motor." However, like the prototypical motor system, autonomic and neuroendocrine functions involve the control of the body by the CNS. Other aspects of maintaining fitness—including many facets of procreation, defense, foraging, exploration, and ingestion—are more obviously motor: They all involve the **skeletomuscular system** in some way. These movements enhance an animal's fitness, clearly enough, but they might seem irrelevant to motor learning because animals know how to do these things instinctively.

But the behaviors listed above, and others, *are* learned, although through a different mechanism than the learning based on an individual's experience. Animals accrue this kind of motor learning over generations and encode it into their genes. Although contemporary psychology

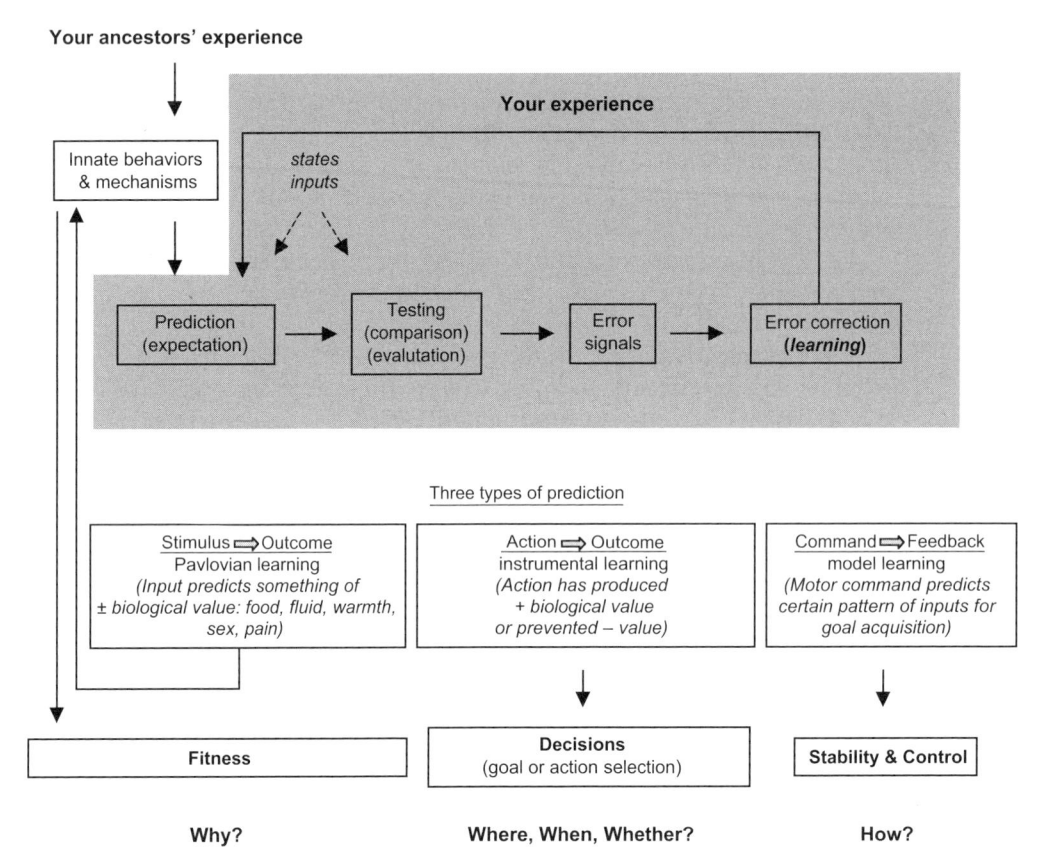

Figure 4.1
Three kinds of motor learning and some of their features.

typically underestimates the importance of instincts, they play an important role in motor control. For example, consider the precise sequence of movements made in the instinctive, procreative behavior of the three-spined stickleback fish. Its movements closely resemble the results of a highly sophisticated, learned skill. Indeed, Niko Tinbergen[4] called them a "dance."

Each male, if ready to receive a female, reacts to them by performing a curious dance towards and all around them. Each dance consists of a series of leaps, during which the male first turns as if going to swim away from the females, then abruptly turns toward them with its mouth wide open.... The male now immediately turns round and swims hurriedly towards the nest. The female follows it. Arrived at the nest, the male thrusts its snout into the entrance, turning along its body axis, so that it lies on its side, its back towards the female, which now tries to wriggle into the nest. (Tinbergen,[4] p. 11)

Those precise moves and timing, repeated by individuals for millennia, have been genetically encoded and reproduced over generations.

4.2.1 Instinctive Behaviors

Motor learning and memory thus include a variety of inherited, genetically encoded behaviors that are modifiable to various degrees by an individual's experience. These behaviors include instinctive behaviors, various reflexes, autonomic motor outputs, and both **homeostatic** control and neuroendocrine systems.

Genes encode many motor memories. Birds know how to build a nest even though they need not ever have seen another bird make one. Like the mating dance of the stickleback fish, nest building makes an obvious contribution to reproductive success and requires knowing how to put together a sequence of actions to produce something that meets a goal or criterion. This kind of motor memory must be acquired over generations, and similar behaviors include aspects of locomotion, orientation, and reproduction. These behaviors fulfill obvious essentials, such as the need for nutrients, maintaining cells and tissues within a certain range of temperature and osmolarity, and elimination of metabolic waste products. They also include less obvious requirements, such as a broad range of behaviors that give an individual a better chance to transmit his or her genes to the next generation.

In a word, these behaviors are instinctive, and as the Medawars put it:

Instinct is a concept more easily described, explained, and exemplified than made the subject of a formal definition. For the benefit of those insecure people who feel that no discourse is possible unless words are defined—a counsel of despair nullified by the fact that biologists have discussed instincts for many years without attempting a formal definition—the term "instinct" is almost invariably used to describe not a single action, but a functionally connected train of performances having what ordinary people describe as a purpose (for example, a newborn marsupial's spectacularly precarious climb from womb to pouch).... Biology has no greater triumph to look forward to than a solution of the problem of how a program of instinctual behavior is genetically stored and **epigenetically** retrieved. (Medawar and Medawar,[2] pp. 160–161)

Instinctive behaviors need not be entirely genetically encoded. In the much-studied example of birdsong, for example, a great deal of the behavior is instinctual, but a great deal also must be learned. Instinctive behaviors range from those that are highly modifiable during an individual's lifetime to those that have virtually no flexibility at all.

4.2.2 Taxis and Kinesis

The **primitive** forms of behavior known as taxis and kinesis involve innate responses to a stimulus, such as light. Phototaxis, for example, refers to movements in relation to light. Positive phototaxis involves movement toward light; negative phototaxis, movement away from light. Kinesis refers to movements in the presence of a stimulus. Photokinesis, for example, involves movements made in the light, but those movements are not necessarily directed toward or away from light. The contribution of these primitive behaviors to fitness seems obvious, but they have only limited relevance to reaching or other vertebrate behaviors. Nevertheless, vertebrates do show such behaviors, such as *thigmotaxis*, the tendency to maintain contact with a border or barrier sensed by somatosensory receptors (often called wall-hugging behavior; see figure 15.2C for an example in humans).

4.2.3 Habituation and Sensitization

Habituation involves a decrease in a response to some stimulus upon repeated presentations. The reverse, sensitization, consists of an increasing response over successive presentations. In habituation and sensitization, the behavior of the animal has no relevance to what the CNS learns. Repeated presentation of a stimulus, alone, causes subsequent responses to decrease or increase.

An example of habituation involves the response to an unexpectedly loud noise. Reflexively, people and other animals find loud noises startling. The startle response consists of a generalized tensing of muscles, among other components. If the noise occurs sufficiently frequently, however, the CNS learns not to respond to such inputs. Habituation contributes to an animal's fitness, in part, by allowing it to perform other behaviors in the presence of stimuli that it has learned will not cause harm.

4.2.4 Linking Reflexes and Instincts to Neutral Stimuli: Conditioned Reflexes

One major mechanism by which motor learning enhances fitness involves the prediction of sensory inputs associated with potential harm or benefit. This learning mechanism typically involves the triggering of a reflex. This kind of motor learning depends on an individual's experience, as well as on the innate learning mechanisms that subserve these behaviors. Collectively these learned responses are called conditioned reflexes.

Pavlovian conditioning, also known as classical conditioning, requires the association of an initially neutral stimulus, called the conditioned stimulus, with a different stimulus, one genetically programmed to trigger a reflex response. This reflex-triggering stimulus is generally called the unconditioned stimulus and the reflex is called the unconditioned response. The choice of conditioned stimulus is arbitrary; any neutral input that an animal's sensory receptors can **transduce** and transmit to the CNS can be conditioned to produce a reflex response. (Note, however, that some stimuli have a special status that can lead to very rapid conditioning.)

Pairing of the conditioned stimulus with the unconditioned stimulus in time causes the induction of a conditioned reflex. The timing must be fairly precise, and the mere occurrence of one stimulus after another does not suffice for learning, a point taken up in section 23.2.1. In this form of Pavlovian conditioning, the animal responds as if the stimulus has acquired a new attribute. Consider, for example, conditioning a reflex response to a tone when an electric shock serves as the unconditioned stimulus. After learning, the animal responds to the tone in a manner appropriate for a shock. The response resembles the unconditioned reflex, but with different timing. The animal need not wait for the innately programmed (unconditioned) stimulus to arrive if some preceding event predicts that stimulus with high reliability. The animal learns to make a response, called the conditioned reflex or conditioned response, that has an effect similar to the unconditioned reflex, although these two kinds of action need not be identical in all particulars.

Pavlovian conditioning therefore represents a form of motor learning that involves learning when to make a reflexive movement. This form of motor learning deviates, however, from the prototype of skill acquisition enough that some experts would dispute its inclusion as a form of motor learning.

Pavlovian conditioning comes in more than one type. In one, as described above, an initially neutral stimulus predicts an unconditioned stimulus, which triggers a somatic reflex such as limb flexion. In another form, the CNS learns a similarly predictive relationship between an initially neutral stimulus and the appearance or availability of substances (e.g., water or food) that satisfy an innate drive. Unlike the reflexes involved in the former variety, the latter kind of Pavlovian conditioning involves movements of the whole animal. For example, animals will move to a stimulus associated with water or fruit juice if that stimulus has been associated with the availability of that fluid. This kind of behavior is sometimes called *Pavlovian approach behavior*. Often, autonomic reflexes accompany the animal's approach, as in the prototype for Pavlovian conditioning: the salivation of Pavlov's dog.

Another form of conditioned reflexes, essentially the opposite of Pavlovian approach behavior, is called fear conditioning. Fear conditioning leads to "escape" behavior based on initially neutral stimuli, and it, too, is associated with autonomic reflexes. The first kind of Pavlovian conditioning mentioned above relies on the fact that one stimulus predicts another stimulus, often a threatening or damaging one, and it involves movements of part of the body to withdraw from or mitigate the effects of those stimuli. Those stimuli are often known as *aversive*. The second type mentioned above, Pavlovian approach behavior, may involve movements of the whole animal toward the source of a biological drive reducer. Those stimuli are often known as *appetitive*. The third type mentioned here, fear conditioning, may involve movements of the whole animal away from a potentially harmful situation or other aversive input.

Stated more generally, conditioned reflexes allow a mechanism for implementing innate behaviors, of either the whole animal or a part of it, based on initially neutral stimuli. The CNS learns to predict the occurrence of inputs genetically programmed to trigger those movements and other aspects of behavior (such as autonomic outputs). These behaviors underlie aspects of the famous "four Fs" of Paul MacLean, which are fundamental to the fitness of any vertebrate: feeding, fleeing, fighting, and mating behavior.

Conditioned reflexes probably play only a limited role in learning to reach and point, but the mechanisms of this kind of learning have become reasonably well understood and serve as a model for how parts of the brain such as the cerebellum and basal ganglia work and, especially, how they learn. Chapters 23 and 24 take up this topic in some detail. At that point in the book, it will be important to remember the distinctions among various forms of Pavlovian learning (see section 23.4.3). Of course, you cannot assume that these structures function only in Pavlovian conditioning. In the cerebellum, for example, it is likely that the same regions that control withdrawal and eyeblink reflexes in Pavlovian conditioning also play a role in the control of limb movements during both locomotion and reaching tasks.[5]

4.3 Learning New Skills and Maintaining Performance

Motor programs and reflexes learned across generations do not suffice for biological success. To behave adaptively under changing circumstances, animals must extend their motor repertoire during their lifetime and continually adapt those new and existing motor programs to changing circumstances. Expansion of the motor repertoire is called skill learning or skill acquisition. Modification of existing elements of that repertoire is called motor adaptation, calibration, or, simply, adaptation. Adaptation applies to reflexes as well as to learned movements such as reaching and pointing. This book uses the terms *skill acquisition* and *motor adaptation* to distinguish these two kinds of motor learning. Together, they contribute to the stability and control of the limb.

4.3.1 Skill Acquisition

Once the CNS selects the targets (or goals) of reach, discussed in section 4.4, it must eventually compute a motor plan and generate the coordinated forces needed to achieve the goal, even if this computation evolves during the movement. The ability to achieve such goals typically requires a motor skill. In goal-directed movement, the CNS generates a sequence of motor commands indicating the forces involved in the movement (often called the **dynamics** of movement) and/or the changes in joint angles needed to reach the target (often called the **kinematics** of movement). A consideration of movement kinematics takes up most of chapters 9–19, and chapters 20–22 deal extensively with dynamics.

Skill acquisition serves as the prototype for the concept of motor learning in the minds of most experts. As noted above, according to some definitions skill acquisition is all there is to motor learning. Learning a new skill involves extending the motor system's performance beyond its prior limits.[6] On day one, the system—for instance, a child—cannot ride a bicycle. After a certain amount of practice and an uncertain number of bandages, the child can ride a bicycle without too much damage to itself or the environment.

Various principles have been advanced for skill acquisition. For example, some experts have proposed that motor learning consists of constraining the redundant degrees of freedom in the limbs[7] (see section 3.1.1). In reaching, for example, the arm can adopt several different final configurations: Sometimes the elbow can be high (when the upper arm is abducted), at other times low (when the upper arm is adducted). According to this view of skill acquisition, motor learning involves selecting one of these configurations in reaching a goal. Other views emphasize notions such as minimizing endpoint error, independent of the redundant degrees of freedom.[8] Other authorities have suggested that skill learning involves changing the **coactivation** of **antagonist** muscles.[9]

Early in motor learning, rigidity can be established through coactivation to help stabilize the limb, more or less regardless of the activity being performed (see chapter 8). According to this idea of skill acquisition, then, the stability attained by coactivation is replaced, over time, with a more specific sequence of motor commands that accomplishes a goal while maintaining stability. There have been several other general ideas about skill acquisition like those mentioned here, and although they have their place, each reflects only an aspect of motor learning. Skill acquisition appears to be highly dependent on the specific requirements of a particular activity, and the strategies adopted by people and other animals in learning those skills vary widely, depending on the precise circumstances.

4.3.2 Motor Adaptation

Adaptation involves changes in motor performance that allow the motor system to regain its former capabilities in altered circumstances. In general, motor adaptation does not enhance the motor system's capabilities overall, but returns the system to a former state of performance. Unlike skill acquisition, no new capability emerges after motor adaptation. For example, if you go swimming underwater with a mask, the air–water interface causes a deflection of visual images falling on your retina. To work underwater, you need to adapt your reaching movement to account for this distortion. In the laboratory, experimenters have used prism glasses to study this kind of adaptation because prisms shift a visual image by a known amount. It takes several attempts to correct the reaching errors induced by such prisms (see chapter 15).

To understand something about this kind of adaptation, imagine reaching or pointing to a target directly in front of you, at eye level. For the present purpose, call that a 0° movement. Now imagine that you put on eyeglasses that distort the visual world so that a target directly in front of you appears to be located 17° to your right. In reaching or pointing to that target, you will, of course, reach about 17° to your right at first. You will notice, however, that your hand does not move toward the target, which is, in reality, directly in front of you at 0°. Reach by reach, you will begin your movement in a direction closer to 0°, until you reach or point directly toward the target. It takes approximately half a dozen movements to correct for a 17° displacement. After adaptation, your motor system has regained the ability to reach or point accurately, but no more so than before adaptation. Almost everyone accepts motor adaptation as a form of motor learning, but the concept must be clearly distinguished from skill acquisition.

The key difference between skill learning and motor adaptation is that the latter adjusts the system's operations for only one context. For example, when you remove the prisms mentioned above, your motor system no longer performs as well as it did earlier. You first reach about 17° to the left of the visual target, and it takes about as long to correct this error as it did to adapt in the first place. An **aftereffect** like that is the telltale sign of adaptation: you have to "de-adapt" as if the system can perform only one computation at a time. In skill learning, by contrast, no "de-adaptation" occurs. With sufficient experience, you can learn to reach accurately with the prisms as soon as you put them on and *switch* back to "normal" reaching almost immediately upon taking them off. Such switching signifies a motor skill, as taken up in more detail in chapters 15 and 16.

Much of the motor learning discussed in this book involves motor adaptation. Later chapters emphasize three forms of adaptation: adaptation to distortions of sensory inputs (exemplified by prism adaptation), adaptation to distortions in the trajectory generated by a given motor command (often called "rotation" experiments), and adaptation to imposed forces. Although motor adaptation differs from skill learning, it has no less importance. Adaptation represents a crucial capability of the motor system as well as a prerequisite for skill acquisition. In reaching, for example, the movements of one part of the limb generate forces on other parts. These forces, termed **interaction torques**, change during life and vary for different objects held in the hand. Without adaptation to interaction torques, skill learning would become very difficult, if not impossible. For skill acquisition and motor adaptation, every movement teaches the CNS something about its motor apparatus and its interaction with the world.

Skill acquisition and motor adaptation produce stable and controlled movements. For a two-joint arm, this means stability and control of reaching and pointing.

4.3.3 Stability

One of the fundamental requirements of the vertebrate motor system involves maintaining stability, including constant position of the body and limbs (often known as **posture**), as well as providing a sound "platform" for the control of movements. Limb stability is achieved, in part, through an antagonist architecture (see chapter 8). You can appreciate the reason that an antagonist architecture promotes stability by considering flexion and extension around your elbow. Imagine holding your arm steady, so that your elbow does not flex or extend. The decision to hold your arm steady requires the motor system to send commands that activate both biceps and triceps. A motor command to two antagonist muscles, called coactivation, results in both flexion and extension torques exerted around the elbow. If the net torque equals zero, then the forearm remains still. A net torque of zero can also be attained by simply shutting down both muscles, which would conserve energy. As explained later in detail (see section 7.2.3), however, muscles become stiffer as their activation level increases. Thus, active muscles promote limb stability. Why? Imagine that something exerts a small force on your hand in the flexion direction. During coactivation, the muscle that gets stretched because of an applied force (triceps; see figures 3.1 and 3.2) will resist more when active than when inactive. Without coactivation (i.e., with an inactive triceps), the external force would result in a much larger flexion of the elbow.

This peripheral contribution to stability partially overcomes some of the limitations of slowly conducting nerves, one of the burdens of history emphasized in section 3.3. Information travels from place to place in the motor system at speeds restricted by membrane properties. One advantage of intrinsic muscle stiffness, therefore, involves the speed of reaction to perturbations that threaten stability, taking into account the long times required to transfer sensory information to the CNS and back to the periphery. The CNS can adjust the level of coactivation in advance of perturbing inputs and thereby provide almost immediate resistance to imposed forces. Most discussions of coactivation emphasize its role in maintaining a given posture, but in addition it helps stabilize the rate of movement. Coactivation has important limits, however, as a mechanism for stability. During large movements, the strategy will cause lengthening of antagonist muscles because they are active, which wastes energy and might, under some circumstances, cause muscle damage.

In addition to intrinsic muscle stiffness, stretch reflexes contribute to limb stability. For example, as a force presses against your hand in the flexion direction, the triceps muscle not only resists lengthening based on the level of coactivation and its intrinsic stiffness, but a stretch of the muscle activates receptors that excite motor neurons supplying input to the triceps. Chapter 8 examines the nature and mechanism of these reflexes.

Stretch reflexes, intrinsic muscle properties, and antagonistic architecture illustrate some of the ways the motor system achieves stability, a prerequisite for control.

4.3.4 Control

Maintaining a given arm configuration can serve as a goal in its own right, but when you think of reaching and pointing, you probably think of movements. Reaching and pointing movements require a control mechanism, and your control system requires learning.

To understand why your control mechanism requires learning, consider the movement of your hand from a resting location to a cup of coffee. To achieve the goal, your CNS must generate patterns of muscle activation, which causes forces to act on the bones in your arm. A computation that estimates the motion that will result from an applied force is called **forward dynamics**. The opposite computation, called **inverse dynamics**, allows the motor system to transform the desired movement (to the handle of the coffee cup) into a pattern of muscle activation that produces the required forces.

Subsequent chapters return to these concepts and topics many times, but, as an introduction, consider why the computation of forward dynamics requires learning. Recall that forward dynamics involves estimating the movements that will result from a given set of forces produced by the limb's muscles. The joint rotations that result from a given force depend on the interaction of the limb with objects in the environment. For example, if a command reaches your arm muscles after you grasp the cup, the increased **inertia** of the limb will cause a more slowly accelerating motion than without the cup's weight. Under many circumstances, this inertial effect might matter little, and some theories of motor control stress that the hand will eventually reach its target (see chapter 8).

If, however, your goal includes control of the initial speed and direction of hand movement, then waiting until your limb slowly overcomes the inertia of the cup's weight will not do. It also will not do to await a feedback signal from your arm to compensate for this slowed acceleration; the signals do not arrive fast enough (see section 3.3). According to the model presented in this book, the CNS learns to estimate effects of holding the cup's weight and what sensory feedback might result from a given motor command. The computations underlying those estimates are called **internal models** (see section 17.3.3 and chapters 20–22), and a great deal of motor learning involves learning new internal models and adapting existing ones.

4.3.5 Benefits of Skill Acquisition, Motor Adaptation, Reaching, and Pointing

Accurate reaching and pointing, then, depend crucially on both skill acquisition and motor adaptation, which help maintain stability and control. The motor system evolved this kind of plasticity because the requisite information could not all be encoded into the genes. That is, more is needed for biological success than the kinds of motor learning described in section 4.2. Part of the reason involves the changes in the peripheral motor system as an animal grows, and part comes from variability in the world.

In addition to the general benefits gained from skill acquisition and motor adaptation, reaching and pointing confer specific advantages. Many kinds of animals have developed the ability to reach, especially with the forelimbs. Raccoons, for example, are excellent reachers. Primates, however, appear to have emerged as specialized graspers.

The most recent fossil evidence indicates that stem primates evolved to exploit a particular niche: feeding on flowers and nectar, perhaps along with young, tender leaves, at the end of relatively thin tree branches. Jonathan Bloch and Doug Boyer[10] have concluded that the first primates appeared about 55 million years ago as specialized, tree-dwelling "graspers." According to their analysis, a specialized grasping ability apparently preceded the capacity for advanced leaping and frontally directed eyes. Bloch and Boyer believe that primates "evolved grasping first and convergent orbits later," which argues for visually guided reaching as a driving force for early primate evolution.

This topic comes up again in chapter 18, and figure 18.5 shows what they think this animal looked like and how it might have used its forelimbs and hindlimbs. Again according to Bloch and Boyer, these developments occurred at about the same time as an increased diversity of flowers, fruits, kinds of leaf buds and nectars, so it makes sense that some animals would exploit this new resource. An animal able to grasp distal tree limbs with its hindlimb and foot—and to use these anchors for stability while using forelimb and mouth movements to acquire food—would have a useful advantage. This manner of feeding not only allows the exploitation of an unexploited niche, but also opens the way to shifting the eyes to a forward binocular configuration, with growing emphasis on the visual cortex and foveal scrutiny of objects. The accurate control of reaching, then, may have occurred early in primate evolution. (This account, like many in primate paleontology, has vociferous critics.[11,12])

Our primate ancestors, then, appear to have put together a reaching mechanism, a grasping hand, eyes that provided forward, foveal, and binocular vision, and the ability to back up those sensory inputs by acquiring and adapting the skills necessary for accurate reaching. Pointing must have evolved after reaching; it can provide benefits only in a social context. It does no good for an animal to point at an object too distant to reach. This kind of behavior makes sense only as a communicative gesture.

4.4 Making Decisions Adaptively

Section 4.3 discussed why animals need more for biological success than genetically encoded behaviors, the kind of behaviors listed in section 4.2. Motor control and stability require skill acquisition and motor adaptation during an individual's lifetime. More is needed than the kinds of motor learning described in section 4.2 for yet another reason. As mentioned in that section, Pavlovian learning allows animals to respond to an initially neutral stimulus in an adaptive way: The CNS picks up on the regularities in the environment to predict the occurrence of a decidedly nonneutral stimulus and generate some appropriate response. Accordingly, Pavlovian conditioning is a form of *predictive learning*, but one based only on innate mechanisms involving the "nonneutral" stimulus and the reflex response it evokes. The reason that **advanced** animals require additional flexibility for success has been summarized by Bernard Balliene:

It is rarely acknowledged that both prediction and [response] control are required for successful adaptation in a changing environment. Although predictive [Pavlovian] learning provides animals with the capacity to elicit anticipatory responses as a result of learning about associations between events, the ... responses [are] clearly determined by evolutionary processes rather than individual learning. As a consequence, an animal that can engage only in predictive learning is at the mercy of the stability of the causal consequences of these responses. For its responses to remain adaptive in an unstable environment, therefore, the animal must be capable of modifying its behavioral repertoire in the face of changing environmental contingencies; that is, it must be capable of learning to control responses instrumental to gaining access to sources of benefit and avoiding events that can maim or kill. (Balliene,[13] p. 311)

To put it somewhat differently, an advanced animal cannot be successful if its behavioral repertoire consists only of the movements—responses, in the terminology of Balliene—generated reflexively or instinctively. Such an animal would be "at the mercy" of the environment because, if the environment changed so that a given action began to cause a different outcome, the animal would be lost. Even predictive learning, which extends the innate behavioral repertoire by linking neutral sensory stimuli to instinctive and reflex responses, will not do. The behavioral repertoire needs to be modified (through skill acquisition), and the animal needs to be able to learn to select its actions on the basis of the potential consequences of those actions. This chapter concludes with a consideration of that process, often called *decision making*.

Decision making does not depend on reflexes or innate behaviors, as do the kinds of motor learning mentioned in section 4.2, or upon the physics of the body or the world, as do the kinds of the motor learning discussed in section 4.3. Decisions involve learning what action to select—usually based on some context—as well as whether and when to execute or withhold that action (figure 4.1). The decision-making process is not always considered to be a part of motor learning, but it is essential to reaching and pointing, among other movements. Before either kind of movement begins, some goal must be selected, usually from among several potential ones. Sometimes the motor system selects a sequence of such goals.

Table 4.1 shows one view of how decision making fits within our concepts of motor learning and of learning more generally. It is bound to be somewhat controversial, so we present it as a "personal" view. Note the relative importance of rows and columns in the table. We think that people come in two types: row people and column people. Traditional psychology emphasizes the rows in table 4.1. We, however, are column people. Note especially the fuzzy distinction between **implicit** and **explicit** tasks (concepts taken up next), as captured by the misalignment of the row divider in the left column compared to the other columns of the table. This device reflects the idea that the kinds of behaviors included in the bottom row come in both explicit and implicit forms. Table 4.2 presents a more conventional view for comparison.

Table 4.1
Types of motor learning (shaded cells) in relation to other forms of learning

		Long-term memory					
		decision making		motor learning			
				maintaining stability and control		maintaining fitness		
		arbitrary motor learning: context ⇔ action (S–R)		skill acquisition	motor adaptation	conditioned reflexes: CS ⇔ US ⇔ reflex	taxis, kinesis, habituation, sensitization	instinct, reflex
implicit (motor learning)		skeletomotor: voluntary S–R (arbitrary mapping)	skeletomotor: predictive S–R (instrumental learning), nonpredictive S–R (habits)	skeletomotor	skeletomotor	skeletomotor, autonomic	skeletomotor, autonomic	skeletomotor, autonomic, neuroendocrine, neurosecretory
implicit (nonmotor learning)	priming, perceptual learning	mental models, implicit grammar						
	guess-guided S–R*, object recognition, event memory, fact memory							
explicit		mental logic, language, explicit grammar						

*See Lebedev and Wise[20] for an explanation of this phrase. Abbreviation: S–R, stimulus–response association.

Table 4.2
Types of long-term memory according to a noted authority on learning and memory, Larry Squire (see Reading List)

Long-term memory					
explicit (declarative)		implicit (nondeclarative)			
facts	events	procedural	priming	classical conditioning	non-associative learning
		skills	habits		emotional responses	skeletal musculature	reflex pathways

Note: The rows show the hierarchical relationship among various types of memory. The shaded blocks show the types of learning considered to be "motor learning," according to this view.

4.4.1 What Is Decision Making?

What movement should you make and when should you make it? These are the questions that have plagued the vertebrate "mind" from time immemorial. The answers are called decisions and choices.

The discussion of decision making is problematic, however, because of some vexing philosophical and conceptual problems. The proverbial "elephant in the room," which no one quite wants to discuss but is difficult to ignore, is consciousness. The concept of consciousness is related loosely to *awareness* and *volition*; also to *intention*, *attention*, and *planning*; and yet more loosely to *cognition* (the Latin word for knowledge) and *knowledge* (the English word for knowledge). Many treatments of motor learning, and even some discussions of learning in general, keep the problem of consciousness and its large family of querulous relatives securely in the closet. According to the accepted wisdom, memories with access to consciousness are called **declarative** (or, referring to tasks rather than "memory systems," explicit), whereas all other kinds of memory are called **procedural** (or implicit). Some authorities use the term **habit** as a rough synonym for procedural memory. Indeed, the use of evasive terminology to mask the fact that the concept under discussion is consciousness, with all of its tenacious problems, has attractions that readers appreciate almost as much as authors do. In order to avoid these problems, this presentation takes a different approach: We simply admit that we do not know. We can summarize what *we* think, as neuroscientists:

- no one understands the neural correlates of consciousness;
- much behavior in humans and other animals—and even robots—resembles that following conscious decisions in people, including reaching and pointing movements;
- through introspection, we believe that many of our movements, especially reaching and pointing movements, follow conscious decisions on our part and, for this reason, we call them *voluntary movements*; and

- we could be wrong about that some of the time, much of the time, or, in the view of a minority of experts, all of the time.[14]

The biggest problem is that nonhuman animals may or may not have consciousness: Expert opinion remains divided. You can easily find books—there are many examples—in which the author of one chapter gnashes his or her teeth, wrings his or her hands, and generally obsesses about whether chimpanzees have even the faintest glimmer of conscious awareness, once one examines the issue sufficiently rigorously. In another chapter of the same book, a different expert describes with all the confidence in the world his or her detailed studies of conscious memory in laboratory rats, as if no one would doubt it.

It might seem easy to tell the difference between voluntary movements, made on the basis of conscious decisions, and involuntary movements, made on the basis of subconscious ones. The following thought experiment, however, shows how difficult it can be to do that. Imagine that you observe some "thing" reach or point to an object, but at such a large distance that you cannot tell whether the "thing" is a person or a robot. If you assume that "thing" is a person, you might suppose that he or she made a conscious decision to reach or point. If, however, you think that it is a robot, you would not reach the same conclusion, even if you are a big science-fiction fan. You would instead suppose that a robot's movement did not require conscious mediation, that it was executed as a "simple" sensorimotor transform. To many experts, a sensorimotor transform in this sense represents an instance of motor learning, sometimes called a habit or an S–R (stimulus–response) association. To these experts,[15–17] having classed a movement in that way is tantamount to saying that it is subconsciously controlled (although they use the terms *habit*, *implicit*, or *procedural* to convey this message.)

When you see a robot reaching, you might imagine that it guided those movements with conditional logic implemented by its computers or a complex assessment of probabilities, et cetera, but you would not think that the robot had made a conscious decision. This line of thought applies to the person in the thought experiment as well. According to this view, then, the reaching or pointing movement you see at a distance could result from conscious control or subconscious processes that many experts consider to result from motor learning (a form of procedural knowledge for implicitly controlled tasks). It turns out that it is very difficult to tell whether a system that has conscious awareness is using that awareness to guide its actions.

To make a long argument short, a lot of what appears to be voluntary movement—implying that it is consciously controlled and therefore not *merely* the subconsciously controlled result of motor learning—may not be consciously controlled, and therefore is not voluntary in that sense. Some experts maintain that your sense of volition is entirely illusory, that all your actions result from sensorimotor transforms. They suppose that you make up the "reasons" for your actions in retrospect.[14] We do not go that far, and leave it as one of the many open questions in this book. But you should not assume that a *seemingly* "voluntary" movement necessarily implies conscious awareness.

4.4.2 Instrumental Learning

One process that leads to decisions and choices is called **instrumental learning**. In this form of learning, choices among goals or actions are based on two factors: arbitrary *contexts*, usually sensory contexts, and the expected outcome of making a particular movement in that context. (Often, the focus is on the movement per se, without much emphasis on the sensory context.) In the laboratory, a typical example of this kind of decision making involves an experiment in which one color cue instructs one motor response and a different color cue instructs a different response. Often, these responses are reaching and pointing movements to alternative targets in space. For example, a red stimulus at the center of a video screen might instruct a participant to reach to the leftmost of two potential targets. A blue stimulus at the same location might instruct reaching for or pointing to the rightmost target. Motor learning of this kind enables you to associate any feature of any stimulus with any action in your motor repertoire. The stimulus provides the context for the decision to take some action, and these actions have some expected value. You can call this kind of learning *arbitrary motor learning*, but it usually goes by other names, including conditional motor learning, voluntary action,[18] arbitrary sensorimotor mapping, and stimulus–response association. (You need to know these synonyms if you want to pursue further reading on this topic, but not for the purposes of this book.)

When people and other animals make a movement in the context of a given stimulus, they are more likely to repeat that response in the future if the movement is followed by a "reward." Reward (also known as a reinforcer) is something of biological value that people and other animals have the motivation to obtain. Rewards might consist of food in the state of hunger, water in the state of thirst, a particular kind of food or, these days, even a particular kind of water. (For example, a hotel in New York City recently hired a water sommelier to help customers decide what kind of water they should drink with their meal!) As in Pavlovian learning, in instrumental learning the context for an action is sometimes called a conditioned stimulus. In the example experiment described in the previous paragraph, reaching or pointing to the left target in the context of a red *conditioned stimulus* increases if a reinforcer follows *that* movement in *that* context. Instrumental learning thus contributes to decisions and choices. Instrumental learning may also contribute to lower-order control and stability, such as caused by reinforcement for faster or slower movements, faster or slower response latencies, and other features of an action.

4.4.3 Habits as a Form of Decision

In traditional psychology, the end result of instrumental learning is called a habit. Used loosely, this term has come to mean any behavior that depends upon subconscious guidance, including habits of thought as well as of action. According to a more rigorous concept, however, habits develop only after persistent experience with making a certain response to a given stimulus, and can be recognized as habits only when the expected outcome of the action no longer influences decisions and choices (see chapter 25). In psychological terms, an action is a habit when the animal continues to perform it regardless of whether the predicted outcome materializes. Thus, habits are context-dependent movements that are so **overlearned** that they become difficult to stop, even when the outcome no longer produces the expected results. Commonly, a habit is an action that is repeated in a certain context without any awareness of a decision or choice having been made. Nevertheless, a decision and a choice *have* been made, and thus habits are a form of what we term *arbitrary motor learning*.

Note, however, that a great deal of arbitrary motor learning has nothing to do with habits. Other forms of arbitrary motor learning do not depend on any experience with particular stimuli, responses, and outcomes. They instead are based on inferences, rules, and strategies that can be applied abstractly to problems with substantial generalization.

4.4.4 Flexible Decision Making

Decisions thus range from those based on learned habits to those based on highly abstract knowledge. They vary widely in their degree of flexibility. The relationship between the location of a visual stimulus and the goal or target of an action serves to illustrate the point. In its least flexible form, visuomotor behavior involves the transformation of visual input into the motor commands necessary to orient, locomote, or reach toward or away from that input, called *standard sensorimotor mapping*.[19] In standard mapping, the input directly guides action, either as a target to be approached or as one to be avoided (table 4.3). In behaviors guided by standard mapping, the only decisions required are whether to approach or avoid the stimulus, and when to do so. It is possible that some animals can behave only in this way. However, your motor system has much more flexibility. Decisions about where you reach and point need not be limited to the

Table 4.3
Types of sensorimotor mapping

standard mapping	approach	stimulus as target of action	vision-for-action (procedural knowledge)
	avoidance	stimulus as anti-target	
nonstandard mapping	transformational	stimulus as input to spatial algorithm	
	arbitrary	stimulus as symbolic instruction	transitional
	abstract, guess-guided	stimulus as subject of report	vision-for-perception (declarative knowledge)

locations of visible stimuli, but those stimuli can guide your movement to other places as well.

In contrast to standard sensorimotor mapping, in *nonstandard mapping* the visual input is neither approached, acquired, nor avoided. Table 4.3 lists several different kinds of nonstandard mapping. In one kind, the CNS somehow applies a spatial algorithm or transform to a stimulus to guide movements. Termed *transformational mapping*, this kind of movement might involve choosing a target that is rotated 90° from a starting location, so that in seeing a target at 3 o'clock (to the right of the center of the clock), you reach to 12 o'clock (top). In another kind of nonstandard mapping, the location of a stimulus does not matter, but features such as its color and shape guide movement. Termed *arbitrary sensorimotor mapping*, a common example involves stopping in response to a red traffic light while driving a car. For both transformational and arbitrary mapping, a remembered target location can be substituted for the location of a current stimulus.

Another form of nonstandard mapping, termed *abstract mapping*, was alluded to at the end of section 4.4.3. This kind of mapping refers to the ability to use abstract rules, strategies, and inferences to guide movement. It can be viewed as a distinction between using available and remembered sensory information to guide movement versus using hypotheses about the world and yourself. You might also think of this form of sensorimotor mapping as "guess-guided," as opposed to sensorially or mnemonically guided. As Mikhail Lebedev and Wise wrote:

Some species might use guesses often, some rarely, some never; the guesses must vary widely in sophistication. The precise moment when, during evolution, some animal started using guesses to guide behavior was not occasioned by fanfare or bolts of lightning. More likely, a step of this monumental significance began humbly, not really as a step at all, but more like a gently sloping ramp from using visual inputs and stored visual information, alone, to using novel guesses about that information. (Lebedev and Wise,[20] p. 126)

According to this view, motor learning includes the ability to make decisions with a flexibility that extends far beyond using stimuli as objects to be approached or avoided, and beyond your specific history with stimuli, responses to those stimuli, and their outcome (i.e., instrumental learning).

4.5 Summary

This chapter has presented the idea that motor learning takes several forms: (1) learning, over generations, the reflexes and instincts that serve to maintain fitness, and linking initially neutral stimuli with them; (2) learning, within an individual's lifetime, of new motor skills to expand the motor repertoire and adapting motor programs in order to maintain the stability and control of movements; and (3) learning to make decisions about movements, targets, and goals likely to achieve desirable outcomes.

4.5. Summary

There are many reasons for motor learning during your lifetime, and several apply to reaching and pointing movements. You must adapt to changes in the mechanics of your limbs and variations in the environment. The slowness of information flow to and from your periphery and the interaction among body and limb segments complicates this problem for your central **controller**. According to the theory presented in this book, to compensate, your CNS predicts feedback and coordinates outputs for the body as a whole. Further, you need to do more than predict damaging or beneficial inputs and react with genetically programmed responses. You must expand your motor repertoire and decide which of several potential actions is most likely to yield a desirable outcome in a given context.

References

1. Schmidt, R (1988) Motor Control and Learning: A Behavioral Emphasis, (Human Kinetics, Champaign, IL).
2. Medawar, PB, and Medawar, JS (1983) Aristotle to Zoos: A Philosophical Dictionary of Biology (Harvard University Press, Cambridge, MA).
3. Lakoff, G (1987) Women, Fire, and Dangerous Things (University of Chicago Press, Chicago).
4. Tinbergen, N (1990) Social Behavior in Animals (Chapman and Hall, London).
5. Bracha, V, Kolb, FP, Irwin, KB, and Bloedel, JR (1999) Inactivation of interposed nuclei in the cat: Classically conditioned withdrawal reflexes, voluntary limb movements and the action primitive hypothesis. Exp Brain Res 126, 77–92.
6. Hallett, M, Pascual-Leone, A, and Topka, H (1996) in The Acquisition of Motor Behavior in Vertebrates, eds Bloedel, JR, Ebner, TJ, and Wise, SP (MIT Press, Cambridge, MA), pp 289–301.
7. Bernstein, NA (1967) The Coordination and Regulation of Movements (Pergamon, New York).
8. Todorov, E, and Jordan, MI (2002) Optimal feedback control as a theory of motor coordination. Nat Neurosci 5, 1226–1235.
9. Moore, SP, and Marteniuk, RG (1986) Kinematic and electromyographic changes that occur as a function of learning a time-constrained aiming task. J Mot Behav 18, 397–426.
10. Bloch, JI, and Boyer, DM (2002) Grasping primate origins. Science 298, 1606–1610.
11. Kirk, EC, Cartmill, M, Kay, RF, and Lemelin, P (2003) Comment on "Grasping primate origins." Science 300, 741.
12. Ni, X, Wang, Y, Hu, Y, and Li, C (2004) A euprimate skull from the early Eocene of China. Nature 427, 65–68.
13. Balleine, BW (2001) in Handbook of Contemporary Learning Theories, eds Mower, RR and Klein, SB (Lawrence Erlbaum, Mahwah, NJ), pp 307–366.
14. Libet, B (1985) Unconscious cerebral initiative and the role of conscious will in voluntary action. Behav Brain Sci 8, 529–566.
15. White, NM, and McDonald, RJ (2002) Multiple parallel memory systems in the brain of the rat. Neurobiol Learn Memory 77, 125–184.

16. Fernandez-Ruiz, J, Wang, J, Aigner, TG, and Mishkin, M (2001) Visual habit formation in monkeys with neurotoxic lesions of the ventrocaudal neostriatum. Proc Natl Acad Sci USA 98, 4196–4201.
17. Mishkin, M, and Appenzeller, T (1987) The anatomy of memory. Sci Am 256, 80–89.
18. Passingham, RE (1993) The Frontal Lobes and Voluntary Action (Oxford University Press, Oxford).
19. Wise, SP, di Pellegrino, G, and Boussaoud, D (1996) The premotor cortex and nonstandard sensorimotor mapping. Can J Physiol Pharmacol 74, 469–482.
20. Lebedev, MA, and Wise, SP (2002) Insights into seeing and grasping: Distinguishing the neural correlates of perception and action. Behav Cogn Neurosci Rev 1, 108–129.

Reading List

Memory: From Mind to Molecules, by Larry R. Squire and Eric R. Kandel (Scientific American Library, New York, 1999) presents the prevailing views on learning and memory mechanisms. Lebedev and Wise[20] review the literature on relating perception and action.

5 What Does the Motor Learning I: Spinal Cord and Brainstem

Overview: All levels of your CNS contribute to motor learning, including those lowest in its hierarchy: the spinal cord and brainstem. One highly specialized part of the brainstem, the cerebellum, plays a particularly important role in learning to reach and point, among other aspects of motor learning.

Figure 2.1 depicts the major subdivisions of the CNS: spinal cord, medulla, pons, midbrain, diencephalon, and telencephalon. All of those structures contribute importantly to the control of reaching and pointing movements. This chapter and the next briefly describe selected aspects of motor-system anatomy, more or less what you will need to follow the presentation in the remainder of the book. They begin by describing some components of the motor system, one at a time. Bear in mind, however, that no part of the motor system works in isolation. The motor system functions as a distributed neural network, performing computations collectively rather than as a set of motor centers. Section 6.2.1 takes up this point explicitly.

5.1 Spinal Cord

Note that *rostral* and *caudal* mean toward the beak and toward the tail, respectively. This nomenclature works well for many vertebrates, but for reference to humans, the spinal cord segments go from superior (rostral) to inferior (caudal).

5.1.1 Organization

Your spinal cord has four major divisions, from rostral to caudal: cervical, thoracic, lumbar, and sacral. Some of these names might be easier to remember if you know what they mean. *Cervix* is the Latin word for neck, thorax means chest, lumbar refers to the loins, and sacral refers to some imagined similarity to a religious symbol. In mammals, the cervical spinal cord comprises eight segments, the thoracic twelve, and the lumbar and sacral five each (see box 5.1). The spinal segments that control the muscles of the arms and legs have larger cross-sectional diameters than other segments and thus are called "enlargements." The cervical enlargement

Box 5.1
Why are there eight cervical segments?

> The number of cervical segments varies greatly among vertebrates, but has become fixed in mammals. Despite their impressive length, many snakes have no cervical vertebrae, but some dinosaurs had more than 70.

controls your arms, and the lumbar enlargement controls your legs. Each segment has a number. For example, C_1 stands for the first cervical segment and the notation C_{1-8} refers to all eight cervical segments.

In any slice through the spinal cord, you see a ring of white matter surrounding a central core of gray matter. A high concentration of **myelin** in the fiber pathways makes white matter lighter in appearance than regions with many neuronal cell bodies. The spinal gray matter bulges to form the *dorsal horn* and *ventral horn*.

Each spinal segment has motor neurons that send axons to the muscles of the ipsilateral side of the body. These motor neurons lie in the ventral horn of the spinal cord, and each segment has a pair of motor nerves exiting from it. Motor neurons collect into ***motor pools***, each of which provides input to specific muscles. The axons that emanate from the motor pools collect in bundles, called motor nerves, that exit the spinal cord in the *ventral roots*.

In addition to motor neurons, the spinal motor system includes sensory pathways, the proprioceptive system, and central pattern generators (CPGs). Sensory afferents provide information to the CNS from the skin, joints, and muscles. In contrast to the motor nerves and their ventral roots, the sensory nerves enter the spinal cord through the *dorsal roots*. Collections of cell bodies of these sensory nerves are called *dorsal root **ganglia***, and they provide the *primary* sensory afferents. Some sensory axons terminate in the dorsal horn, but others terminate elsewhere in the spinal gray matter, and many ascend to the brain to terminate on cells of the *dorsal column nuclei* in the medulla and elsewhere. For example, cells that project directly from the spinal gray matter to the thalamus are called *spinothalamic* neurons, and those that project to the cerebellum are called *spinocerebellar* neurons. Spinocerebellar neurons contribute to the spinocerebellar tracts, including the dorsal spinocerebellar tract (DSCT). Axons from cells in the dorsal column nuclei project to the thalamus through a fiber pathway called the *medial lemniscus*.

Most of these terms figure prominently in the discussions that appear later in this book. For example, chapter 10 discusses one aspect of the spinocerebellar tract in some detail, and spinothalamic neurons play an important role in a variety of movements. This part of the book may seem somewhat dry, but it might boost your spirits to know that spinothalamic neurons figure prominently in sexual behavior, as described below. Before you get to the sexy parts, though, you should memorize a few facts about spinal anatomy.

5.1. Spinal Cord

Box 5.2
A controversy

> Currently, one part of the proprioceptive systems is involved in some controversy. Originally described in domestic cats, the C_3–C_4 proprioceptive system consists of interneurons in the third and fourth cervical spinal segments. An extensive series of experiments has shown that cats need this system for reaching movements, but not for grasping movements (provided the cat's arm is adequately supported). Cats have few, if any, direct projections from the motor cortex to motor neurons, and so their C_3–C_4 proprioceptive system conveys cortical signals to motor neurons. The controversy involves whether this system has the same functional importance in primates, including humans. All experts agree that this proprioceptive relay system conveys a weaker signal in primates than in cats. Experiments show the signal remains almost undetectable without suppression of inhibition in macaque monkeys. This difference may be due to the fact that certain primates have extensive projections from motor cortex to motor neurons. Interestingly, in another kind of monkey, the squirrel monkey, a robust C_3–C_4 proprioceptive system exists. This finding correlates with the fact that squirrel monkeys, like domestic cats, have relatively few direct cortical projections to motor neurons, and those that do exist cause less excitation. It seems that in the absence of a strong direct connection from the motor cortex to motor neurons, the CNS instead uses a robust C_3–C_4 proprioceptive system to control reaching. Humans, who have a very large projection from cortex to motor neurons, would, on these grounds, be expected to have a weak proprioceptive system. This issue remains unresolved, however, because several papers report indirect evidence for a reasonably strong C_3–C_4 proprioceptive system in humans.

In addition to *primary sensory afferents*, the spinal cord contains a group of interneurons serving a number of functions collectively termed proprioceptive. Technically, proprioceptors are sensory transducers in muscles, tendons, and other deep tissues. However, the concept of a proprioceptive or **propriospinal** system extends beyond the kind of information provided by those receptors to include a broad network of interneurons that relay somatosensory and motor signals within and among segments of the spinal cord. (See box 5.2.)

CPGs are neural networks that control rhythmic movements such as those involved in swimming and other modes of locomotion. Probably the best-understood CPGs are those of lamprey, introduced briefly in chapter 2 (figure 2.2A). Each spinal segment controls a compact set of muscles called a **myotome**. In lampreys, as in many other fish, swimming arises from waves of contraction caused by myotomal contractions, beginning from the head and continuing toward the tail through the myotomes on one side of the body. These waves alternate with comparable waves of muscle contraction on the opposite side of the body, resulting in propulsion through a fluid medium by means of S-shaped swimming movements. The CPG contains the circuitry needed to generate and modulate these patterned rhythmic movements.

5.1.2 The Spinal Cord and Control of Ejaculation

CPGs contribute to many behaviors besides swimming and other forms of locomotion. As section 4.2.1 describes, motor learning functions, in part, to maintain **fitness** and transmit an animal's genes to future generations. This section details one CPG that plays an important role in this function. After spinal cord injuries, including those that completely disconnect the brain from the lumbar and sacral spinal cord, the ejaculation reflex remains intact. This indicates that the CPG for the reflex resides in the spinal cord, mainly if not exclusively at lumbar and sacral levels (collectively, the *lumbosacral* spinal cord). You should not, however, think of the ejaculation reflex as being much like a simple knee-jerk reflex. Ejaculation involves intricately timed rhythms of emission (the movement of seminal fluid into the urethra) and forceful expulsion of the seminal fluid. The output—to use the term advisedly—depends on a highly synchronized and coordinated motor pattern generator. Recent experiments by William Truitt and Lique Coolen[1,2] in rats discovered that a certain class of spinothalamic cells in the lumbar spinal cord make up part of the CPG that controls ejaculation. These cells, located at the L_3 and L_4 levels, project to the thalamus. Taking advantage of the fact that these particular spinothalamic cells express a receptor called neurokinin–1, Truitt and Coolen killed those cells by attaching a toxin to the **ligand** for that receptor. After confirming the selectivity of the lesion and that the animals had a normal pain threshold, they examined the rats' sexual behavior. Male rats without these neurons, unlike a control group that did not have the toxin attached to the ligand, lost the ability to ejaculate. Other aspects of their sexual behavior remained normal, and it did not matter whether the rats were sexually experienced.

Occasional appearances to the contrary notwithstanding, the brain controls behavior, not the penis. Once intromission behavior begins, ejaculation continues as a reflex that involves neurons in the lumbosacral spinal cord, but this reflex remains under descending control from the brain. The spinothalamic cells within the CPG for ejaculation transmit information to the brain (specifically to the cerebral cortex and basal ganglia) to mediate that control.

5.1.3 Proprioceptive Pathways to the Brain

To review some aspects of spinal anatomy that you need to know for the presentation that follows, recall that primary afferent fibers enter the spinal cord through the dorsal roots and terminate on spinal motor neurons, spinal interneurons, and nuclei in the brain. Among the interneurons targeted by primary sensory afferents, many also relay somatosensory information to the brain. Some of these cells receive inputs from primary afferents and project to the cerebellum through the spinocerebellar tract, including the DSCT. These neurons provide proprioceptive information to the cerebellar cortex and the output nuclei of the cerebellum.

5.2 Hindbrain

5.2.1 Organization

Like the spinal cord, the brainstem contains motor neurons. Brainstem motor neurons send their axons through cranial nerves, primarily to muscles of the tongue, face, and eyes. Like the spinal motor pools, many of these cranial motor nuclei receive direct input from sensory neurons and influences from proprioceptive interneurons. Also like the spinal cord, the brainstem contains CPGs. In addition, the brainstem has a diverse, broadly distributed system for control of the spinal cord. This "system" of projections, called the *reticulospinal system*, comprises neurons in the reticular core of the brainstem that send axons into the spinal cord.

Brainstem CPGs, like those described above for the spinal cord, generate rhythmic movements. In general, spinal CPGs subserve locomotion and the orientation of the head, whereas brainstem CPGs underlie breathing, chewing, eye movements, and other highly specialized forms of rhythmic motor behavior, such as the whisking of whiskers by rats and mice. Brainstem networks, partly through the reticulospinal system, also influence the activity of spinal CPGs. One brainstem region particularly involved in initiating the activity of CPGs in the spinal cord has been termed the *midbrain locomotion region*. In addition to CPGs and descending control mechanisms, the brainstem contains higher-order networks, akin to CPGs but having more complex output patterns. For example, aggressive posturing and vocalizations, such as crying, laughing, and shrieking with fear, depend on neural networks in and near a midbrain structure called the *periaqueductal gray*.

CPGs are for more than ongoing, fixed rhythms, however. They also control adjustments to ongoing behaviors, such as turning during locomotion. The reticulospinal system appears to be especially important for these functions. It, along with other higher-order components of the motor system, controls movement by affecting CPGs in several ways:

- Modulating CPGs (i.e., speeding or slowing the rhythmic motor output)
- Suppressing CPGs (i.e., halting CPG output or preventing the CPG from beginning to produce an output)
- Initiating CPGs (i.e., releasing an inhibited or quiescent CPG or driving it to produce its output)
- **Fractionating** CPG output (i.e., modifying the output so that some, but not all, of the motor outputs of the CPG occur).

5.2.2 Reticulospinal Escape Behaviors

As an example of higher-order control of a CPG, imagine that a stimulus on one side of an animal induces it to turn toward the other side. An experiment on lampreys[3] (see table 2.1) showed that they move away from such a stimulus by increasing the duration and intensity of contractions on the side opposite the stimulus. The mechanism for turning thus depends on a slowing of the periodicity of the CPG's cycle. Not only do muscles on the opposite side contract longer and more forcefully in response to the stimulus, but their contractions occupy a larger proportion of the movement cycle. The pathways necessary for this turning behavior in lampreys include the reticulospinal system.[3] For mechanical inputs to the head, the information comes to the CNS through the trigeminal nerve and is transmitted through a pathway that connects the reticulospinal system of one side of the brain with the CPGs (and, to a lesser extent, the motor neurons) on the opposite side. Higher levels of the CNS play little role in this behavior.

The experimenters[3] recorded from a sample of cells in the lamprey's reticulospinal system and found that they respond to the stimulus in several ways, but most of them increase their activity during ipsilateral turns and decrease their activity during contralateral turns. Given that these cells send their axons to the opposite side of the spinal cord, this pattern of activity is consistent with a role in turning the body away from the mechanical stimulus: in other words, an avoidance reflex. Other reticulospinal cells show increased activity during turns in either direction or are inhibited for either direction. These cells could provide a general increase in the level of activation of the CPGs. Perhaps the ratio of bilaterally excited to bilaterally inhibited neurons determines the overall excitability of the system or, alternatively, the inhibited neurons may contribute to motor outputs that the system suppresses during avoidance turning. Similar mechanisms activate the CPGs when they are "off," control vertical as well as horizontal trajectories, adjust the speed of swimming, and make adjustments in body **posture** in response to local conditions.

5.2.3 Reticulospinal Adjustments to Support Reaching

The example of turning in lampreys illustrates the general principle of reticulospinal modulation of spinal pattern generators. The cells of origin for the reticulospinal system extend from medullary and pontine levels of the brain through the midbrain and into the diencephalon. The system's functions include the regulation of muscle tone and the control of posture and locomotion.

Parts of the reticulospinal system serve as a rapid transmission route to postural motor neurons that help maintain balance during movement. Imagine, for example, that you intend to lift a heavy object and that you will do so by activating your biceps. The weight of the object, which extends far beyond your body's center of gravity, will tend to pull you forward and off balance. This weight, if uncompensated for, will produce unwanted movements around your knees and ankles. To counter these forces, certain of your leg muscles need to increase their activity just before you pick up the weight.

In this situation, the activity in your leg muscles precedes the activity in your biceps. Your CNS recognizes that an upcoming arm movement will have consequences for the leg, and it anticipates those consequences. To compensate for them, it activates your leg muscles in advance of arm-muscle activation. The consequences of a given arm movement differ, depending on the configuration of your body. For example, imagine yourself lifting a weight while you stand on solid ground, as opposed to when you sit on a tall stool with your legs dangling. In the standing posture, it would make sense to activate your leg muscles slightly before you begin to lift the weight. However, when your legs dangle in the air, your back muscles become active to counter the force of the weight and the leg muscles can contribute nothing. Therefore, the activity of your leg muscles prior to lifting weight cannot result from a simple, stereotypical program that plays out every time you pick up a load. Rather, your CNS needs to take into account the *context* in which the lifting of weight occurs in computing the consequences for other body parts.

Paul Cordo and Lew Nashner[4] investigated this context dependency by asking participants in an experiment to pull on a fixed handle when they heard a tone. Because they attached the handle to a solid wall, the pulling action caused the participants to sway forward. This sway would produce a flexion in their ankles and knees. Cordo and Nashner observed that just before the CNS activated the biceps to pull on the handle, it activated two muscles: the hamstring, a knee flexor, and the gastrocnemius, an ankle extensor (see figure 3.3 for illustrations of those muscles). On average, activity in the leg muscles led that in the arm by 40 msec. To test whether this anticipatory postural adjustment depended on context, Cordo and Nashner examined a condition in which they supported the participant with a padded crossbrace at shoulder height. The brace prevented the act of pulling on the handle from causing a swaying motion of the body. The introduction of the brace thus changed the movement context. In this condition, when the participant's CNS activated his or her biceps, it did not significantly activate either the gastrocnemius or the hamstring. Therefore, the CNS activates the leg muscles only when it anticipates that arm movement will have a consequence for the leg.

The ability to predict consequences of a planned movement plays a fundamental role in motor control. You may be aware only of your plan to lift an object, but your CNS computes the consequences of this plan and activates your leg muscles before your arm muscles. This mechanism serves as an example of an internal model in the CNS: a system that computes the predicted consequences of a planned action. This internal model depends on context because a given action, such as pulling on a handle, has different consequences depending on what supports your body. Chapter 17 and chapters 20–22 discuss internal models in considerable detail. For the present, these findings serve to exemplify ways in which descending pathways, originating in the brainstem, mediate postural adjustments.

The brainstem motor pathways also play a key role in your body's activity during sleep and wakefulness. During wakefulness, the pontine reticulospinal system has a mainly facilitative influence on motor pools, which promotes maintenance of balance during execution of motor commands. During sleep, however, medullary reticulospinal neurons exert a strong inhibitory influence on motor neurons, which, among other effects, (usually) prevents the performance of imagined actions during dreams.

Important influences over the reticulospinal system arise from other systems, including vestibular afferents, which signal movements of the head and its orientation with respect to Earth's gravitational field. Through its direct projections to the spinal cord, called the *vestibulospinal* pathway, and by inputs to brainstem oculomotor nuclei, the vestibular system can contribute directly to various reflexes that adjust eye orientation, posture, and limb movements. Chapter 24 takes up some vestibular reflexes. Vestibular afferents also provide inputs to the reticulospinal system, and thus influence spinal-cord activity by this indirect route.

Finally, consider the role of your reticulospinal system as you run through a field of obstacles. The signals conveyed by the reticulospinal system to CPGs and spinal motor pools adjust your posture and movement primarily on the basis of vestibular and proprioceptive inputs. Other descending inputs supply information from the cerebral cortex, which the CNS uses for the dynamic adjustments required to step over and around visible obstacles, for example. Many of these higher-order inputs also affect the operations of your reticulospinal system.

5.3 Cerebellum

The cerebellum is the largest component of the brainstem motor system. It consists of two hemispheres, one on each side of the midline, and a medial region, spanning the midline and termed the vermis (meaning worm). The medial cerebellum mainly controls posture, whereas the lateral cerebellum participates more directly in reaching and pointing movements. Accordingly, vestibular and propriospinal inputs predominate in the medial cerebellum, and inputs from the cerebral cortex dominate the lateral cerebellum. Section 16.1.4 and chapters 23 and 24 discuss certain functions of the cerebellum, but one can make the general statement that the cerebellum contributes to motor skills by reducing variability in the timing and force of muscle contractions.

5.3.1 Intrinsic Architecture

Each piece of the cerebellar cortex consists of small folds called folia, each of which has the same architecture. The main output cells of the cerebellar cortex, the *Purkinje cells*, send their output to the *deep cerebellar nuclei*.

Among the locally projecting neurons in the cerebellar cortex, *basket cells* inhibit Purkinje cells, and *granule cells* excite them, mainly through *parallel fibers*. Purkinje cells have an elaborate and stereotyped dendritic field. Each Purkinje cell dendrite occupies a narrow plane perpendicular to the orientation of the parallel fibers and to the long axis of each folium. This architecture allows each Purkinje cell to receive a large number of parallel-fiber inputs. In some vertebrates, the entire cerebellum consists of a single folium. Advanced vertebrates, and especially birds and mammals, have hundreds of folia, organized into larger structures called lobules. One of these lobules is called the flocculus, and section 24.2.4 discusses the role of this part of the cerebellar cortex in oculomotor control. Other lobules and folia probably work in much the same way.

The cerebellum thus adheres to a common principle in brain evolution: a unit of organization, such as a folium, replicates during evolution, and those units diverge in some particulars while retaining their fundamental pattern of organization. They develop functional specializations not so much by changing their intrinsic computations or operations as by altering the inputs and outputs of each replicated unit.

5.3.2 Inputs

Inputs to the cerebellum come from both *mossy fibers* and *climbing fibers*. Climbing fibers arise from neurons in the inferior olivary nuclei (ION), so called because these nuclei form an oval bulge on the ventral surface of the medulla. Climbing fibers terminate on Purkinje cells and cells in the deep cerebellar nuclei, and their name comes from the impression that they climb up the Purkinje cell. Inputs from climbing fibers cause *complex spikes*, a pattern of discharge in Purkinje cells that, in the view of many experts, signals motor errors or loss of coordination. Thus, climbing fibers play a central role in the motor-learning functions of the cerebellum, a topic discussed in chapters 23 and 24. Each axon from the ION averages six branches, each of which distributes approximately 250 synaptic terminals per branch. Thus, each climbing fiber has about 1500 terminals, with each branch also distributing approximately 55 terminals within the deep cerebellar nuclei. With minor exceptions, each branch of a climbing fiber terminates mainly on a single Purkinje cell, mostly on its dendrite. The individual branches of a climbing fiber project to a discrete part of a longitudinal band or "stripe." Stripes are oriented in the parasagittal plane and span much of the rostrocaudal extent of the cerebellar cortex, extending across folia. The branches of one climbing fiber interdigitate with those of other climbing fibers to fill in (and, in large part, define) each stripe.

Mossy fibers terminate on deep cerebellar nuclei as well as on granule cells in the cerebellar cortex (figures 5.1 and 5.2A). Granule cells send parallel fibers to excite the Purkinje cells and cause *simple spikes* (figure 5.2B). In contrast to climbing fibers, which have their terminals confined to a single longitudinal stripe, mossy fibers distribute axons to many stripes, often bilaterally. They always have axon collaterals to the deep

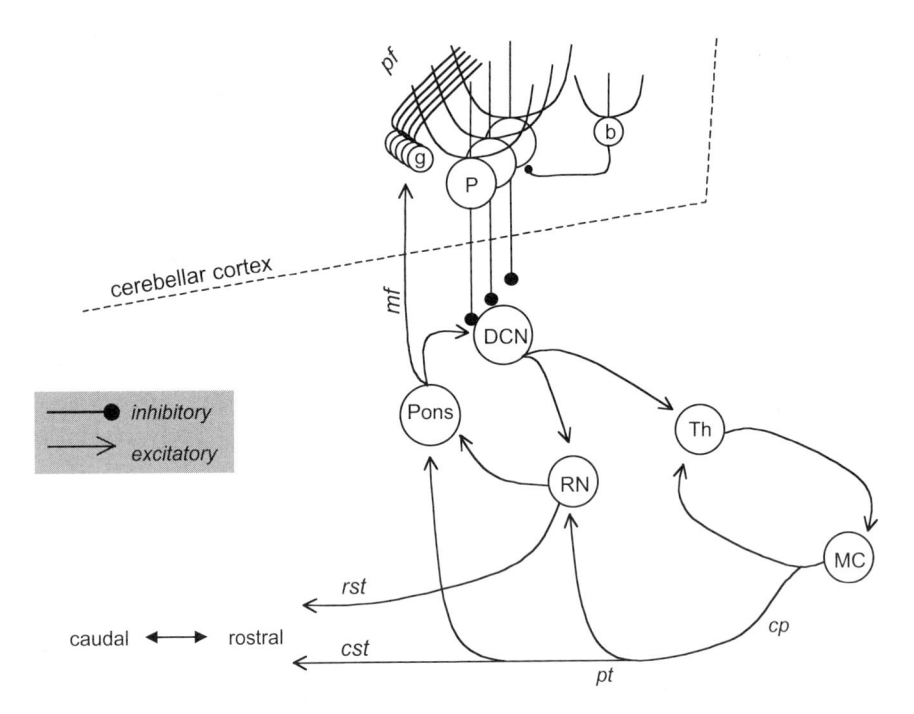

Figure 5.1
Sketch of cerebellar connectivity. P, Purkinje cell; b, basket cell; g, granule cell; Th, thalamus; DCN, deep cerebellar nucleus; MC, motor cortex; RN, red nucleus; mf, mossy fibers; pf, parallel fibers; rst, rubrospinal tract; cp, cerebral peduncle; pt, pyramidal tract; cst, corticospinal tract. The dashed line shows the limits of the cerebellar cortex. (Modified from Kiefer and Houk[5] with permission.)

cerebellar nuclear neurons, averaging about seven collaterals per mossy fiber. These collaterals form a mean of 155 terminals on the granule cells of the cerebellar cortex. These, in turn, give rise to the parallel fibers, which run along the cerebellar cortex.

5.3.3 Outputs

The output of the cerebellar cortex inhibits neurons in the deep cerebellar nuclei (and in one of the vestibular nuclei) by releasing the neurotransmitter γ-aminobutryic acid (**GABA**). As noted above, all of the outputs of the cerebellar cortex arise from Purkinje cells. The three deep cerebellar nuclei are called the fastigial, interpositus, and dentate nuclei. The fastigial nucleus lies nearest the midline. The dentate nucleus lies furthest from the midline, with the interpositus—as its name suggests—between the fastigial and dentate nuclei. In the human brain, the dentate nucleus has a crenulated appearance that reminded early anatomists of teeth; hence the name. In general, the fastigial nucleus receives inputs from the medial parts of the cerebellar cortex, the dentate nucleus from the lateral cerebellar cortex, and the interpositus nucleus from intermedi-

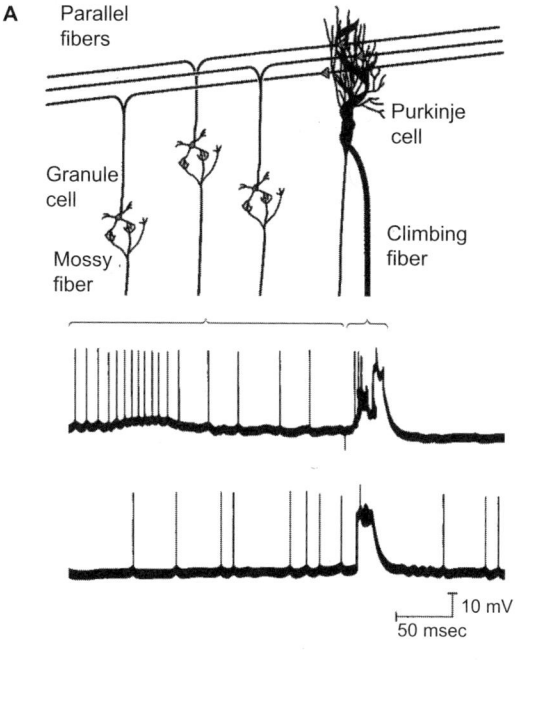

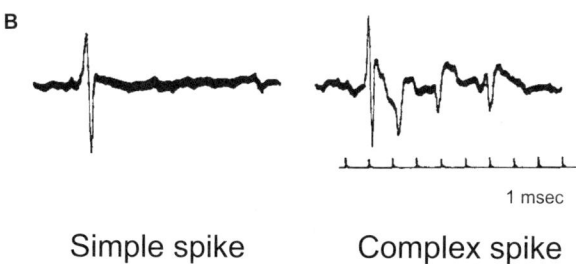

Figure 5.2
Parallel-fiber inputs to Purkinje cells generate simple spikes; complex spikes result from climbing-fiber inputs. (*A*) Cerebellar inputs (top) and neuronal responses (bottom). (*B*) Simple and complex spikes from a cerebellar Purkinje cell. (A from Ghez and Thach[6] and Martinez et al.[7]; B from Mano et al.[8] with permission.)

ate regions. The deep cerebellar nuclei send excitatory outputs to a variety of structures, including the thalamus (figure 5.3), reticulospinal system, red nucleus, ION, superior colliculus, and spinal cord. They also send inhibitory, GABAergic outputs to the ION.

5.4 Red Nucleus

The red nucleus receives a major projection from the deep cerebellar nuclei, as well as from the motor cortex (figure 5.1). The largest part of the red nucleus, its parvocellular (small-cell) component, projects

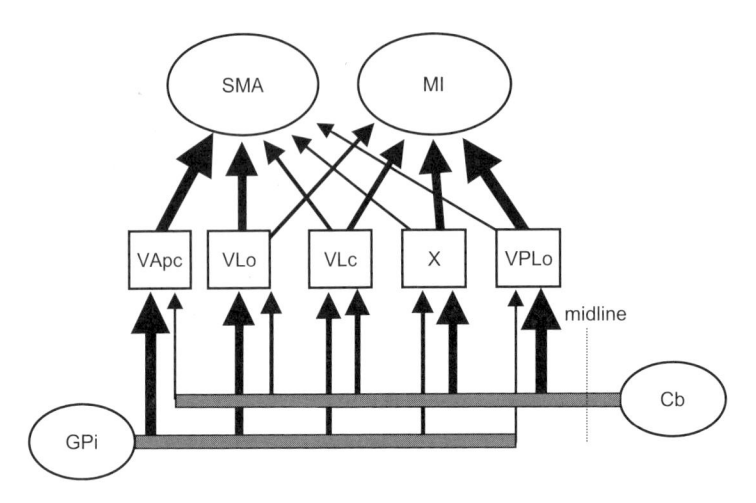

Figure 5.3
Output from cerebellum (Cb) crosses the midline and projects to the thalamus (squares). The thalamus relays this information to the primary motor cortex (M1), the supplementary motor area (SMA), and other parts of the frontal motor cortex (not shown). Cerebellar outputs have a bias toward M1 (shown by the thicker arrows). Output from one part of the basal ganglia (the internal segment of the globus pallidus, GPi) also targets several motor areas, but with a bias toward SMA. Abbreviations for thalamic nuclei: ventroanterior nucleus, parvocellular part (VApc); ventrolateral nucleus, oral (rostral) part (VLo); ventrolateral nucleus, caudal part (VLc); nucleus X (X); and ventroposterolateral nucleus, oral (rostral) part (VPLo). (From Sakai et al.[9] with permission.)

predominantly to the ION, the source of climbing fibers. The magnocellular (large-cell) component of the red nucleus sends its axons directly to the spinal cord through the *rubrospinal tract*, which some experts believe plays a particularly important role in stabilizing the limb by coactivating agonist and antagonist muscles. However, as you think about reaching and pointing movements in primates, bear in mind that some experts maintain that the magnocellular red nucleus is said to be relatively small in the human brain, which may reflect a dominant role of the cerebral cortex. (Convincing neuroanatomical evidence for this commonly held opinion has yet to appear in print, however.)

A traditional view of the red nucleus, one mainly referring to the rubrospinal system, holds that it specializes in the control of extension movements. In traditional experimental neurology, a great deal of effort went into understanding the function of various motor "centers" and how they might affect the balance between extensors and flexors in the limbs. In retrospect, you can understand what led these investigators to study the motor system in this way: When they damaged the red nucleus or the rubrospinal tract, monkeys adopted a highly flexed posture. It is a mistake, however, to consider the function of some structure to be the inverse of lesion effects. The fact that lesions of the magnocellular red nucleus or the rubrospinal tract lead to highly flexed postures does not imply that the red nucleus functions primarily in extensor movements or provides "balance" between flexion and extension. More likely, the extensor bias of the magnocellular red nucleus and its rubrospinal projection reflects some deeper control principle.

As sections 2.4 and 2.5 explain, the red nucleus (along with the cerebellum) became much more elaborate in tetrapods, the first vertebrates to use their limbs for navigation on land. Not only did these animals have to innovate the use of paired appendages to move on land, but they had to do so without the buoyancy provided by water. Gravity poses a much bigger problem on land than in the water. Perhaps the well-known bias of magnocellular red nucleus neurons toward extensor muscles (such as deltoid and triceps,[10,11] illustrated in figure 3.1) results from adaptation to gravity. This idea accords with the finding that electrical stimulation of the magnocellular red nucleus preferentially excites extensor muscles of the proximal and distal limbs.[10-12]

5.5 Superior Colliculus

The superior colliculus, although usually discussed in terms of eye-movement control, also has an important role in controlling head movements, and there is recent evidence that it may play a role in reaching movements, as well.[13] It functions to orient the retina and other receptors on the head toward stimuli. Known as the optic tectum in other vertebrates, the mammalian superior colliculus guides head and eye movements through its projections to the brainstem and spinal cord.

References

1. Truitt, WA, and Coolen, LM (2002) Identification of a potential ejaculation generator in the spinal cord. Science 297, 1566–1569.
2. Truitt, WA, Shipley, MT, Veening, JG, and Coolen, LM (2003) Activation of a subset of lumbar spinothalamic neurons after copulatory behavior in male but not female rats. J Neurosci 23, 325–331.
3. Fagerstedt, P, and Ullen, F (2001) Lateral turns in the Lamprey. I. Patterns of motoneuron activity. J Neurophysiol 86, 2246–2256.
4. Cordo, PJ, and Nashner, LM (1982) Properties of postural adjustments associated with rapid arm movements. J Neurophysiol 47, 287–302.
5. Keifer, J, and Houk, JC (1994) Motor function of the cerebellorubrospinal system. Physiol Rev 74, 509–542.
6. Ghez, C, and Thach, WT (2000) in Principles of Neural Science, eds Kandel, ER, Schwartz, JH, and Jessell, TM (McGraw-Hill, New York), pp 832–852.
7. Martinez, FE, Crill, WE, and Kennedy, TT (1971) Electrogenesis of cerebellar Purkinje cell responses in cats. J Neurophysiol 34, 348–356.
8. Mano, N, Kanazawa, I, and Yamamoto, K (1989) Voluntary movements and complex spike discharges of cerebellar Purkinje cells. In The Olivocerebellar System in Motor Control, ed Strata, P (Springer, Berlin), pp 265–280.

9. Sakai, ST, Inase, M, and Tanji, J (2002) The relationship between MI and SMA afferents and cerebellar and pallidal efferents in the macaque monkey. Somatosens Mot Res 19, 139–148.

10. Belhaj-Saif, A, Karrer, JH, and Cheney, PD (1998) Distribution and characteristics of poststimulus effects in proximal and distal forelimb muscles from red nucleus in the monkey. J Neurophysiol 79, 1777–1789.

11. Miller, LE, van Kan, PLE, Sinkjaer, T, et al. (1993) Correlation of primate red nucleus discharge with muscle activity during free-form arm movements. J Physiol (London) 469, 213–243.

12. Mewes, K, and Cheney, PD (1991) Facilitation and suppression of wrist and digit muscles from single rubromotoneuronal cells in the awake monkey. J Neurophysiol 66, 1965–1977.

13. Stuphorn, V, Bauswein, E, and Hoffmann, KP (2000) Neurons in the primate superior colliculus coding for arm movements in gaze-related coordinates. J Neurophysiol 83, 1283–1299.

Reading List

There are many neuroanatomy texts, most of which are designed for medical students. Larry W. Swanson takes a somewhat different approach in *Brain Architecture: Understanding the Basic Plan* (Oxford University Press, Oxford, 2003).

6 What Does the Motor Learning II: Forebrain

*Overview: The forebrain comprises the diencephalon and telencephalon. Among its major components, the hypothalamus functions as the body's chief controller for **homeostasis**, reproduction, and defense. The basal ganglia plays an important, but enigmatic, role in motor control and learning, including reaching and pointing movements. The thalamus acts as a key node in recurrent, **distributed modules**—often known as "loops"—which integrate the cerebral cortex into subcortical motor-control systems. As in other advanced mammals, your cerebral cortex makes up most of your CNS, and your neocortex makes up most of your brain. Two large parts of it, the motor cortex and the posterior parietal cortex (PPC), make important contributions to reaching and pointing.*

The forebrain contains several structures that play an important role in reaching and pointing movements, including the basal ganglia, thalamus, and cerebral cortex.

6.1 Basal Ganglia

6.1.1 Intrinsic Architecture

You can think of most components of the basal ganglia as either input or output structures. The basal ganglia's input structure, the striatum, consists of the *putamen* (a term from the Latin for husk or outer shell), the *caudate nucleus* (named for its tail), the *nucleus accumbens* (meaning lying down, because it is located near the base of the brain), and other nuclei and regions within the ventral forebrain, such as the olfactory tubercle (from the Latin for swelling). Together, the caudate nucleus and putamen compose most of the *dorsal striatum*; the nucleus accumbens, olfactory tubercle, and certain other structures compose the *ventral striatum*. The basal ganglia's output structure, the *pallidum*, includes not only the globus pallidus (Latin for pale globe), but also the reticular part of the *substantia nigra* and additional parts of the ventral forebrain, such as the *substantia innominata*. The pallidum sends **GABAergic**, inhibitory projections to the brainstem and thalamus. In addition, the subthalamic nucleus plays an important role in control of the basal ganglia's output.

6.1.2 Inputs

The major excitatory inputs to the striatum come from the cerebral cortex (including the hippocampus) and the thalamus. Cortical inputs to the striatum make excitatory synapses with medium spiny neurons, which are inhibitory, GABAergic neurons of roughly uniform size and dendritic organization. They, in turn, project from the striatum to the globus pallidus and the reticular part of the substantia nigra (often abbreviated SNr). In addition, the substantia nigra has another component, its compact part (SNc), which sends dopaminergic inputs to the striatum. Degeneration of its dopaminergic neurons causes Parkinson's disease, and degeneration of the medium spiny neurons causes Huntington's disease.

Nearly the whole of the cerebral cortex projects to the striatum. Some of the cortical input to the striatum branches off axons that continue on to the brainstem or the spinal cord. These branching corticostriatal neurons probably provide the striatum with an **efference copy** of descending motor commands, among other information.[1]

Small patches of cerebral cortex send projections that terminate fairly extensively in the striatum. That is, a small part of the cortex projects to a large part of the striatum. But only very sparsely. This property leads to a high level of convergence from cortical neurons onto individual striatal spiny neurons. As a general rule, cortical areas that are interconnected with each other typically provide overlapping inputs to the striatum. For example, in the somatosensory and motor cortex, regions representing the hand project to partially overlapping parts of the striatum (see section 6.4 for an explanation of somatotopic **representations**, such as hand representations). Similarly, partially overlapping striatal projections arise from two eye fields in the frontal cortex, the frontal eye field and the supplementary eye field, which also are reciprocally interconnected.

As noted above, the corticostriatal projection distributes widely but sparsely onto medium spiny neurons. In general, corticostriatal axons make few synaptic contacts with any individual striatal neuron, but synapse upon many. The best quantitative data on basal ganglia anatomy come from rats, but the general principles discerned from these studies almost certainly apply to primates as well. In rats, ~17 million corticostriatal neurons contact ~1.7 million medium spiny neurons. Each corticostriatal neuron makes, on average, one to four synapses on each neuron it contacts, and those neurons are spread out over ~4% of the striatum's volume. Within that 4%, each corticostriatal neuron makes, on average, 800–900 synapses. However, 68,000 cells or so have dendrites in that volume, so those 800–900 synapses affect only a very small minority of striatal cells, even within a corticostriatal neuron's terminal field. Thus, any given corticostriatal neuron contacts a large number of medium spiny neurons with a few synapses, often only one. (Contrast this pattern of input with climbing-fiber contacts on cerebellar Purkinje cells, as described in section 5.3.2. Each climbing fiber makes hundreds of synapses on a Purkinje cell, producing a powerful excitatory effect.)

The numbers presented above for the corticostriatal system imply that no single cortical cell influences a striatal neuron very strongly. Furthermore, they indicate that each striatal medium spiny neuron receives a virtually unique set of inputs, and no two striatal neurons receive more than a small proportion of excitatory inputs in common. This architecture implies that each medium spiny cell requires many convergent inputs—probably thousands of inputs discharging nearly simultaneously—to drive a striatal neuron to produce an output.

Other factors also limit the firing of action potentials by medium spiny neurons. The membranes of these neurons generally adopt one of two states. In one state, called the *down state*, the membrane potential rests at −80 mV or so. Given an action potential threshold of approximately −45 mV, down-state neurons almost never discharge. When a sufficient number of excitatory inputs arrive, the striatal cell makes a transition to its *up state*. In this condition, the membrane potential averages about −60 mV, much closer to the threshold for action potentials. Nevertheless, for a period after they attain their up state, striatal neurons still resist discharging because certain current channels take a relatively long time to close.

6.1.3 Outputs

A high level of convergence also exists at the next stage of the basal ganglia's output, the projection from the striatum onto pallidal output neurons. Two subtypes of striatal **projection neurons** have been identified: one gives rise to the *direct*, striatal output pathway to the pallidum; the other, to an *indirect pathway*. The indirect striatal-output pathway relays via the external segment of the globus pallidus and, in its next stage, the subthalamic nucleus. The indirect and direct pathways converge on the same output nuclei, and probably on the same neurons, but the projections via the external globus pallidus and the subthalamic nucleus invert the sign of the signals. Thus, the direct pathway *disinhibits* neurons in the thalamus, superior colliculus, and brainstem, whereas the indirect pathway *inhibits* them.

This sketch of the basal ganglia has heuristic value (figure 6.1), for example, in understanding Parkinson's disease and other consequences of basal ganglia dysfunction. However, you should recognize that it represents a simplification. The subthalamic nucleus, for example, excites not just the internal segment of the globus pallidus but also the external segment, which sends inhibitory inputs "back" to the striatum; and the motor cortex sends a direct, excitatory projection to the subthalamic nucleus. Further, in addition to dopaminergic inputs, serotonergic inputs come from the raphe nucleus, and the striatum contains a large variety of intrinsic neurons which use other neurotransmitters, including acetylcholine.

Corticostriatal convergence seems particularly well suited for a pattern-recognition network. The nearly unique input received by each

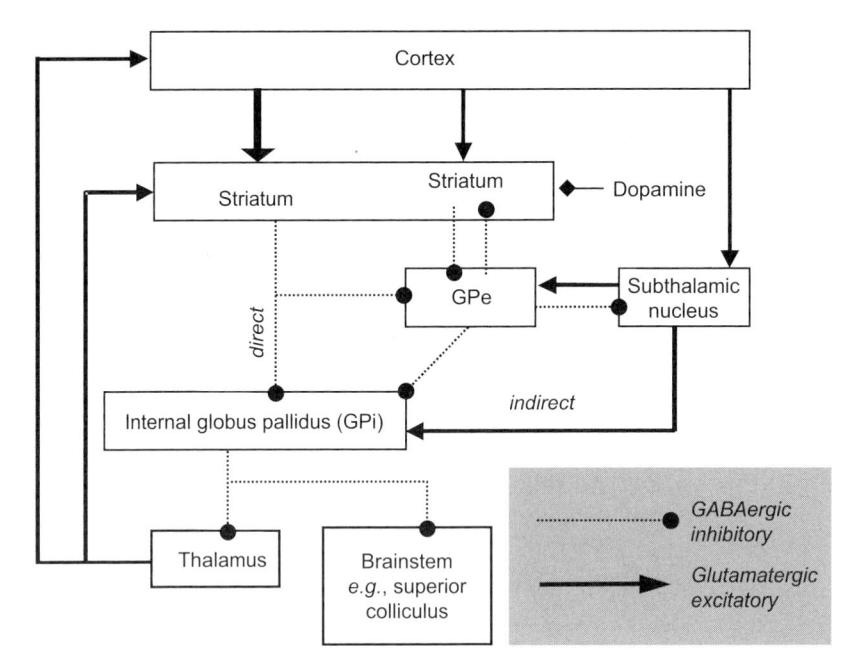

Figure 6.1
Basal ganglia circuitry. Lines ending in circles depict inhibitory connections; lines ending in arrowheads indicate excitatory ones. Lines ending in diamonds show projections that do not fit into either of those two categories. GPe is the external segment of the globus pallidus. (Drawing by Dietmar Plenz)

striatal neuron, the weak nature of each individual input, and the small number of inputs from any given corticostriatal or thalamostriatal cell suggest that the striatal medium spiny neurons detect some combination or pattern of incoming signals. Many inputs must be rejected because they do not arrive with sufficient coherence, with sufficient strength, or within a sufficient time frame. The high level of convergence suggests that many input signals will produce the same output, even though the causes of those outputs will vary. Given the dramatic convergence from cortex and thalamus onto striatal medium spiny neurons, plus another major "round" of convergence from medium spiny neurons to pallidal output neurons, the striatum and pallidum probably act as a kind of coincidence detector, one that classes several kinds of input patterns to produce a smaller number of output states. This kind of organization is consistent with the idea that the basal ganglia recognizes a number of contexts for a given behavioral state.

6.1.4 Context Switching and Basal Ganglia

In section 5.2.3, you saw that when your CNS plans for your hand to pull on an object, say a doorknob, it predicts and compensates for certain consequences of that act. For example, pulling on a doorknob might tilt your body forward and result in imbalance. So when you pull on a handle, your CNS activates extensor muscles in your leg, such as the gastrocnemius and the hamstrings (figure 3.3), before activating the arm flexor biceps. This anticipatory behavior prevents your legs from bending too much at the knees and ankles as your elbow pulls on the handle.

Because not every pull on a handle requires leg-muscle activation, this anticipatory activity appears to be based on an **internal model** of the state of your body. If you lean back against something that supports your shoulders, then the anticipatory EMG activity in your leg muscles no longer accompanies the handle pull. Therefore, the consequences of pulling on the handle depend on the state of your body. Your CNS recognizes the difference between these two states, and in one case activates leg muscles in anticipation of potential imbalance, while in the other case it leaves the leg muscles quiet.

These observations illustrate two important ideas about how your CNS controls movements. First, an act like lifting an object with your arm has consequences for other parts of your body. Your CNS predicts those consequences and minimizes them. Second, predicting the consequences of an action requires that your CNS know not only about the planned action but also about the current state of your body.

Fay Horak and her colleagues[2] observed that patients who had Parkinson's disease did not have a normal ability to use their body's state for predicting the consequences of their actions. The experimenters examined control participants and parkinsonian patients as they stood, unsupported, on a platform that suddenly shifted backward. This shift produced a forward tilt, resulting in the bending of the participant's hips, knees, and ankles. About 100 msec after the onset of the perturbation, the control group activated the muscles that extend the ankle, knee, and hip joints, thus opposing the bending effect. The experimenters then asked the participants to sit on a stool as they applied the same perturbation. In this *supported* condition, the perturbation caused an activation of the hip extensor muscles, but no responses occurred in the knee and ankle extensors. This posture-dependent change in the EMG response did not require practice, but occurred from the first trial onward.

Patients with Parkinson's disease behaved similarly in many respects. While they were standing, a perturbation caused an EMG response in the extensor muscles of the hips, knees, and ankles with a latency comparable to that of controls. However, when the parkinsonian patients sat down on a stool, the same perturbation produced the same responses in all three sets of extensor muscles, despite the fact that the change in the context led to little advantage in activating the knee and ankle extensors. Thus, the response in healthy people was based on *context*. When the context changed from standing to sitting, the response changed. Parkinson's patients showed inflexibility in their response; they did not take into account the change in context. These patients' trouble with balance, a common complaint in Parkinson's disease, may result from reflex responses inappropriate to a given context. These findings point to a role for basal ganglia in context-dependent action.

6.2 Thalamus

The thalamus and hypothalamus compose a large proportion of the diencephalon. Technically, the thalamus (a Greek term for an inner, secret chamber) consists of both a dorsal and a ventral thalamus. The dorsal thalamus sends its largest outputs to the cerebral cortex and basal ganglia; the ventral thalamus has a diverse pattern of connections, including direct projections to the spinal cord. Unmodified, the term "thalamus" usually refers to the dorsal thalamus. A major component of the thalamus, in this sense, receives projections from the cerebellum and basal ganglia (figure 5.3) and relays information to the cerebral cortex. The ventroanterior and ventrolateral nuclei receive most of these inputs. These parts of the thalamus send excitatory projections to the frontal cortex and receive excitatory projections from the same cortical areas. Rostral parts of these nuclei of the "motor thalamus" receive denser basal ganglia projections; caudal parts get denser cerebellar input, as illustrated in figure 5.3. Together, these thalamocortical projections directly influence most, if not all, of the frontal lobe, in addition to other parts of the cerebral cortex.

The thalamus serves as a key node in integrating brain systems that involve components distributed over several interacting structures, such as the cerebellum and the cerebral cortex, on the one hand, and the basal ganglia and the cerebral cortex, on the other. These systems have become known as "loops," a term that conveys the idea of recurrence. Jim Houk and his colleagues[3–5] have called these loops *distributed modules* to emphasize the fact that they involve several distinct brain structures but operate in an integrated network architecture. Figure 6.2 sketches the basic components of two kinds of modules.

6.2.1 The Principle of Distributed Modules

An overview of the peripheral and central motor system might look something like that shown in figure 2.1. Unfortunately, such sketches make it appear that the motor learning system operates through motor centers, each performing some particular function. Contemporary theory suggests that instead of isolated centers, the various components of the central motor system operate collectively in distributed neural networks. The concept of distributed modules captures the essence of that kind of architecture.

Cortical–basal ganglionic modules consist of cortical, striatal, pallidal, and thalamic elements that form, at least in principle, a recurrent excitatory pathway (figure 6.2). This circuit includes the direct pathway described in section 6.1. The GABAergic spiny neurons of the direct pathway project directly to the internal segment of the globus pallidus. These inhibitory GABAergic neurons project to the thalamus.

Cerebellar outputs to many targets have "return" projections through a variety of pathways. The motor areas of the cortex reciprocate their cerebellothalamocortical projections by a cortical projection to the

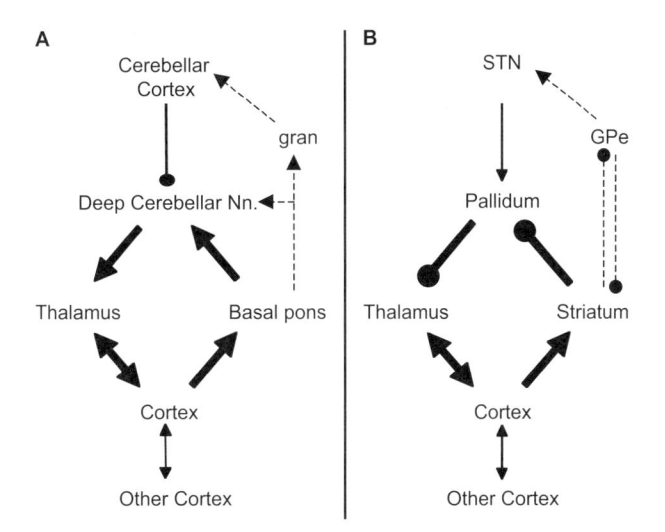

Figure 6.2
Recurrent modules (thicker arrows) and some of their inputs. (*A*) Cerebellar–cortical modules. (*B*) Cortical–basal ganglionic modules. Note the embedded, recurrent connections between motor cortex and thalamus (cortical–thalamic modules) and between different parts of cortex (cortical–cortical modules), as indicted by the two-headed arrows. Lines ending in circles indicate inhibitory projections; those ending in arrowheads indicate excitatory projections. Abbreviations: GPe, external segment of the globus pallidus; Nn, nuclei; STN, subthalamic nucleus; gran, granule cells of the cerebellar cortex.

cerebellum via the basilar pontine nuclei (figure 6.2). Houk and his colleagues have called recurrent, excitatory circuits of this kind cortical–cerebellar modules. Similar recurrent modules involve the cerebellum, red nucleus, and precerebellar nuclei (figure 5.1).[6]

Additional recurrent, excitatory modules link the various cortical areas to each other, and others interconnect cortical areas with the thalamus. For example, many frontal motor areas and the posterior parietal cortex (PPC) have reciprocal projections with each other and with their main-relay thalamic nuclei. These reciprocal cortical–cortical and cortical–thalamic modules have some of the characteristics of recurrent cortical–basal ganglionic and cortical–cerebellar modules.[4]

6.3 Cortical Organization I: General Considerations

The cerebral cortex dominates the external appearance of the human brain. Indeed, the two cerebral hemispheres make up more than 80% of the brain's volume in humans and only a little less, 70–75%, in many other primates, such as Old World monkeys. The term *cerebrum* comes from the Latin for brain, and cortex means an outer covering. As one might imagine for a structure that so dominates the human brain, the cerebral cortex has long fascinated researchers, and no detail has been considered too inconsequential for investigation.

6.3.1 Intrinsic Architecture

Even expert neuroanatomists make the common mistake of thinking about the cerebral cortex as if it comprised only neocortical areas. However, cerebral cortex comes in two types, *allocortex* and *neocortex*, with many transitional and intermediate forms. Allocortex means "other cortex" in Greek, and it evolved much earlier than neocortex. Neuroanatomists often refer to allocortex as three-layered cortex. The principal output cells of the cerebral cortex, pyramidal cells, collect in sheets parallel to the surface of the cortex, with one long dendrite extending toward the surface, called an apical dendrite. These cells also have a set of shorter dendrites arranged roughly in parallel with the surface, called basilar dendrites. In allocortex, inputs arrive in and terminate in layer 1, and outputs arise from the pyramidal cells of layer 2. You have two large regions of allocortex in your cerebral cortex, called the hippocampus and the piriform cortex, along with several smaller allocortical areas near and part of the amygdala. Allocortex, in general—and probably both the hippocampus and piriform cortex—had evolved by the time of the stem reptiles (see table 2.1), and most likely much earlier.

In addition to allocortex, mammalian brains—and only mammalian brains—have varying amounts of neocortex (also called isocortex). Neocortex, of course, means new cortex, and it indeed evolved significantly later than most other parts of the telencephalon (including the basal ganglia, as an entity). **Primitive** mammals have a relatively small amount of neocortex, and the volume of allocortex approaches or exceeds the volume of neocortex. In **advanced** mammals such as humans, neocortex makes up the vast majority of the cerebral cortex (and, indeed, the vast majority of the brain). Although neuroanatomists often refer to neocortex as six-layered cortex, they recognize that neocortex comes in many variations. In the simplest description, the layers are numbered from outermost (nearest the surface) to innermost, 1 to 6. Inputs arrive in layers 1 and 4, for the most part, interconnections among cortical areas arise from layers 2 and 3 and terminate in the same layers of other cortical areas, outputs to most subcortical structures come from layer 5, and projections to the thalamus arise from layers 5 and 6.

The four lobes of the neocortex have names that come from the skull bones overlying them: frontal, parietal, occipital, and temporal—at the front, top, back, and sides of the skull, respectively. Most pertinent to motor control are the frontal lobe, which lies in front of the central sulcus, and the parietal lobe, behind the central sulcus.

You can divide the neocortex into cortical fields, also known as "areas," and each lobe contains a large number of them. You need to accept the existence of these cortical fields on faith—as a surprising number of experts in the field do. Note, however, that only a small number of cortical-field definitions have received universal acceptance; most serve as battlefields scarcely less contested than European boundaries of the twentieth century.

Experts often cite one source for cortical-field definitions, Korbinian Brodmann, a German neuroanatomist who worked in the early twentieth century. Accordingly, you will often see reference to "Brodmann's areas." He numbered areas from dorsal to ventral as he examined sections through the cortex. He called one famous area in the occipital cortex, which contains a map of visual space, area 17. Also known as the striate cortex and the primary visual cortex (V1), area 17 serves as the prototype of a cortical area. Other areas are more difficult to identify. Cortical field definitions range from those established with the reliability of V1, to those established with *fairly* convincing data, to pure invention. Agreement comes, unfortunately, from two very different sources: one legitimate, the other from the copying of one authority's maps by others, usually accompanied by what politicians call plausible deniability.

Despite disagreement about their specific details and boundaries, experts agree that cortical fields differ from each other in their inputs and outputs and also, to an extent, in their intrinsic architecture. *Cytoarchitectonics* involves the recognition of cortical fields by the pattern and constituents of the cortical layers, as revealed by stains that demonstrate the distribution of neuronal cell bodies in the cortex. Other methods approach the same problem, some of them involving the staining of myelin, the distribution of neurotransmitters or their receptors, and the visualization certain proteins. All of these methods have some value, but none has a high level of reliability. In effect, cortical architectonics depends on a high-dimensional pattern-recognition skill that remains largely **implicit**. The **explicit** features that neuroanatomists describe in demarcating cortical fields represent only the most prominent among the features actually used.

Because of the terms that experts use to refer to these areas, you will need to keep in mind several different ways of naming them. This book limits the names to those arising from three "systems" of nomenclature. One relies on a modification of Brodmann's numerical names, usually through subdivision of areas he recognized. For example, the region he called area 5 comprises many functionally distinct areas, including one called area 5d, for its dorsal part, and another called (for the first time in this book) area 5ip for a part within the intraparietal (ip) sulcus of monkeys. The second naming system relies on functional designations, such as V1 for primary visual cortex and M1 for primary motor cortex. The third involves descriptors such as "medial intraparietal area," which indicates where you might find the area in question, at least in monkeys.

This book has excluded a fourth system, one devised by a contemporary of Brodmann, Constantin von Economo, who named areas by lobe (using the first letter of each) and then (using letters) sequentially by area. Accordingly, you will not see cortical names such as PE, which stands for area 5 of parietal cortex. To go from four naming systems to "only" three, we introduce a few new names, such as "area 5ip." To fully follow the discussion, you will also have to keep in mind the rough synonyms for caudal and rostral—posterior and anterior, respectively—as well as the even rougher synonyms for dorsal and ventral (superior and inferior, respectively). To follow many of the later chapters in this book, you must—to put it simply—memorize this diverse, conflicting, and confusing neuroanatomical nomenclature. Monkeys and humans have dozens of cortical areas that contribute to reaching and pointing, and section 6.4 takes up that topic.

Much of the neocortex processes visual, auditory, somatosensory, and motor information. (The somatosensory system includes proprioception.) You can think of the organization of the mammalian neocortex as comprising four domains—for visual, auditory, somatosensory, and motor information processing—arranged roughly from caudal to rostral. Visual cortex consists of a large number of areas in the occipital, posterior parietal, and inferior temporal cortex; auditory areas are concentrated mainly in the superior temporal cortex; somatosensory areas, in the anterior parietal cortex; and motor cortex in the posterior frontal lobe. Many areas receive information from combinations of visual, auditory, somatosensory, and motor input.

6.3.2 Inputs

Inputs to the cerebral cortex come from several sources, including nonspecific modulatory inputs from the locus coeruleus, the raphe nucleus, and the dopaminergic cells of the midbrain. The main excitatory inputs to cortex come from the thalamus and are reciprocated in the form of corticothalamic projections, as outlined in section 6.2.

Cortical areas interconnect with each other extensively, both within a hemisphere and between hemispheres. This fact becomes important to the ideas developed in later chapters: Many of the models of reaching and pointing depend on reciprocal connections between cortical areas (see section 12.4). Although exceptions exist, you can usually assume that if cortical area x sends projections to area y, then area y has projections to area x. However, these projections may terminate in different layers. In addition to such *corticocortical* projections within a hemisphere, the two hemispheres project extensively to each other. In placental mammals, most of the connections travel via the *corpus callosum*. In section 16.1.2, you will read about experiments in which this fiber pathway was cut, a procedure often known as a *commissurotomy*. Additional interhemispheric projections run through the anterior commissure, and a complete commissurotomy would include that pathway.

6.3.3 Outputs

Cortical outputs come from the *corticofugal* system. Corticofugal projections to the brainstem and spinal cord arise exclusively from neurons with cell bodies in layer 5. Projections to the basal ganglia also arise from layer 5 for the most part. As noted above, projections from cortex to thalamus come from layer 6 in abundance, as well as more sparingly from layer 5.

6.4 Cortical Organization II: Cortical Fields for Reaching and Pointing

Because more is known about the brains of monkeys than those of other animals, this presentation focuses on their neuroanatomy.

6.4.1 Frontal Cortex Anatomy and Organization

Everything rostral to the central sulcus is frontal cortex. It has three parts, from caudal to rostral: (1) the primary motor cortex (M1), (2) the *nonprimary motor cortex*—a group of areas collectively called premotor areas, although only some of its members are specifically called "premotor cortex"—and (3) the prefrontal cortex.

6.4.2 Motor Cortex

M1 corresponds approximately to Brodmann's area 4. In most primates, it lies in the rostral bank of the central sulcus and contains a **somatotopic representation**. Despite decades of research indicating otherwise (box 6.1), some textbooks continue to depict its somatotopy in the form of a **homunculus** (i.e., a projection of the body onto the cortical surface). Those pictures do not reflect the actual organization of M1, but do reflect the fact

Box 6.1
Does the M1 cortex contain a homunculus?

> No. The homunculus presents a highly inaccurate image of M1 organization. Other ideas, such as a nested-ring organization,[7] also have little validity.[8] Indeed, in his classic paper on the organization of the motor cortex in monkeys, Clinton Woolsey[9] warned against simplistic interpretations of his drawing. It did not work. The following example shows how misleading it is to imagine a homunculus in M1. A recent and particularly careful neurophysiological study in monkeys failed to reveal any clustering of neurons encoding the activity of any given finger,[10] and evidence for such clustering in humans is also unconvincing.[11] Thus, M1 lacks an "index-finger center." Instead of a homunculus-like organization, in which you would expect an index-finger representation, M1 consists of a mosaic of broadly overlapping muscle representations, with each part of the body represented repeatedly. M1 distributes information concerning the control of index-finger movements throughout much of its hand representation. The lack of an index-finger center does not, however, prevent you from making relatively independent movements of your index finger. Such control arises from the operations of the distributed network.

that the medial part of M1 contains the leg and foot representations, a more lateral part contains the arm and hand representations, and the most lateral part has the face, tongue, and mouth representations.

The projection from M1 to the spinal cord (the **corticospinal** tract) makes **monosynaptic** excitatory connections with α-**motor neurons**. However, its synapses on spinal interneurons vastly outnumber those on motor neurons. In macaque monkeys and in humans, excitatory postsynaptic potentials on motor neurons have amplitudes of 4–8 mV, and direct cortical inputs to wrist extensor muscles could provide up to 60% of the excitation needed to maintain steady discharge of motor neurons.[12]

The organization of the corticospinal projection from M1 respects its somatotopic organization. With few exceptions, its hindlimb (foot and leg) representation projects to lumbar spinal segments, which innervate the muscles of the leg. Similarly, its forelimb representation projects to cervical segments of the spinal cord, which have the motor neurons for arm and hand muscles, and its mouth and face representation sends axons to brainstem nuclei that control mouth and face muscles.

Corticospinal axons accumulate in large bundles traversing the internal capsule, cerebral peduncle, pyramidal tract, and corticospinal tract on the way from the telencephalon to the spinal cord (see figure 2.1). Because corticospinal axons run through the *pyramidal tract*, experts sometimes call these neurons *pyramidal tract neurons*. Additional outputs from the motor cortex, including from M1, terminate in the basal ganglia (see section 6.1), the red nucleus, the cerebellum (via the basilar pontine nuclei), and the reticulospinal system.

The nonprimary motor cortex comprises about a dozen areas by current estimates. Many of these fields occupy parts of Brodmann's area 6, but parts of areas 8 and 24 also contain nonprimary motor areas. Figure 6.3 shows the approximate location of several nonprimary motor areas in monkeys, including the dorsal premotor cortex (PMd), the ventral premotor cortex (PMv), the supplementary motor area (SMA), and some other areas referred to from time to time in the following chapters.

PMd, PMv, and SMA all play a fairly direct role in reaching and pointing movements and, like M1, all project directly to the spinal cord. Some SMA neurons terminate monosynaptically on spinal motor neurons. Thus, the idea that M1 serves as the final, common path for motor command signals to the spinal cord receives little support from the anatomy of the corticospinal system. Instead, it appears that the output from the cerebral cortex to the spinal cord has a parallel organization.

PMd and PMv might be further subdivided into rostral and caudal components, as shown in figure 6.3, and both are considered *lateral* premotor areas. Among the *medial* premotor areas, the SMA, the pre-supplementary motor area (pSMA), and a group of motor areas within the cingulate sulcus, called cingulate motor areas (CMAs) also contribute to reaching and pointing. In addition, figure 6.3 shows the rough locations of some areas important in **oculomotor** control, including the frontal eye field (FEF) and the supplementary eye field (SEF). The distinction between

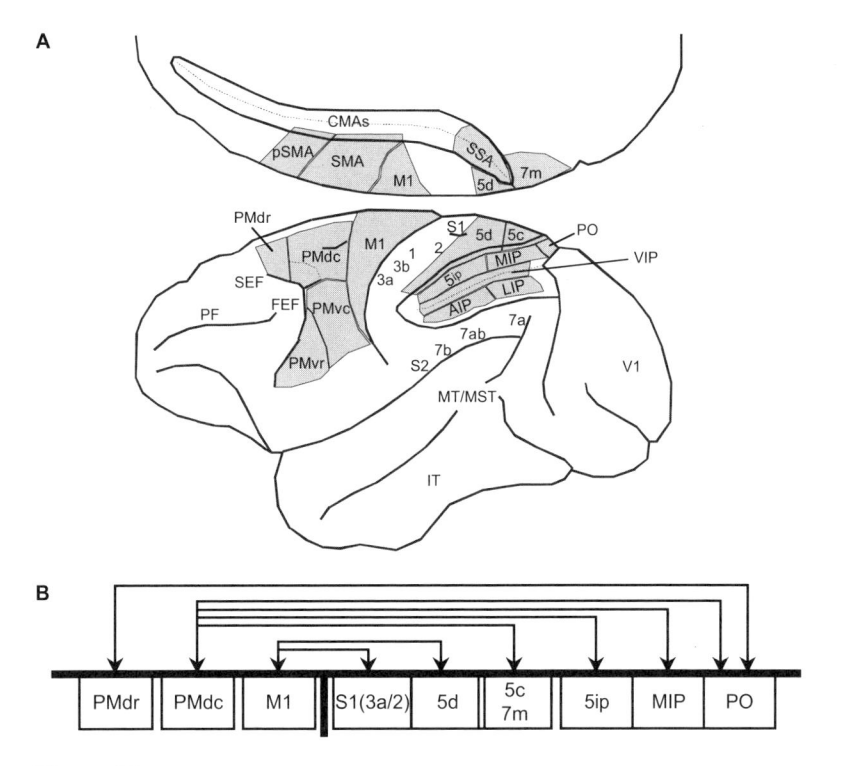

6.4. Cortical Organization II: Cortical Fields for Reaching and Pointing

Figure 6.3
Some subdivisions of the posterior parietal cortex (PPC), frontal cortex, and other areas. (*A*) Cortical fields in the PPC and frontal cortex. Abbreviations, in addition to Brodmann's areas, which are indicated by numbers 1–7: AIP, anterior intraparietal area; c, caudal; CMAs, cingulate motor areas; d, dorsal; FEF, frontal eye field; ip, intraparietal; IT, inferotemporal cortex; LIP, lateral intraparietal area; m, medial; M1, primary motor cortex; MIP, medial intraparietal area (also known as the parietal reach region, PRR); MT/MST, middle temporal area and middle superior temporal area; PF, prefrontal cortex; PMdc, caudal part of the dorsal premotor cortex (PMd); PMdr, rostral part of the dorsal premotor cortex; PMvc, caudal part of the ventral premotor cortex (PMv); PMvr, rostral part of the ventral premotor cortex; PO, parieto-occipital area; pSMA, presupplementary motor area; r, rostral; S1, primary somatosensory areas; S2, secondary somatosensory areas; SEF, supplementary eye field; SMA, supplementary motor area; SSA, supplementary sensory area; v, ventral; V1, primary visual area (striate cortex); VIP, ventral intraparietal area. (*B*) Interconnections among the PPC and certain frontal areas. (Modified from Kalaska et al.[16] with permission.)

oculomotor and skeletomotor areas has become less clear in the recent past, however. PMd, PMv, and SMA each have an oculomotor representation in addition to a skeletomotor one, and the SEF, originally considered an oculomotor field, also plays a role in skeletal movements.

Both the primary and the nonprimary motor areas receive visual and proprioceptive inputs, but do so very indirectly. Visual areas in the occipital cortex do not project directly to frontal cortex, at least not in primates.[13,14] Accordingly, both visual and proprioceptive information comes to the motor cortex via the parietal cortex and the prefrontal cortex.

6.4.3 Parietal Cortex Anatomy and Organization

Figure 6.3 also shows some of the parietal areas important for reaching and pointing movements. Brodmann named the cortical fields immediately caudal to the central sulcus areas 3, 1, and 2 (from rostral to caudal). He called everything superior to the intraparietal sulcus area 5, and everything inferior to it area 7. Together, areas 5 and 7 compose the posterior parietal cortex (PPC), and areas 3, 1, and 2 contribute to the somatosensory cortex.

In the somatosensory cortex, some areas receive relatively direct input from the *medial lemniscus*, which transmits sensory signals via the thalamus. One such area is called the primary somatosensory cortex (S1), which includes at least part of area 3. Brodmann's area 3 comprises two separate cortical fields: area 3a, which receives prominent input from muscle spindle afferents (see section 7.6.2), and area 3b, which receives most of its input from receptors in the skin. Some experts[15] apply the term S1 exclusively to area 3b, but others use it to refer to areas 3, 1, and 2, collectively.

Similar to the division of Brodmann's area 3 into 3a and 3b, experts have recognized several finer subdivisions of areas 5 and 7 of the PPC. Although the exact number of areas in the PPC and their lines of division remain open issues, figure 6.3 shows the rough locations of many of them.

The most posterior parietal areas, called the parieto–occipital areas (PO), are often considered part of the visual cortex. They have an unusual feature compared to other parts of the visual system. Almost all other cortical and subcortical visual structures show a strong **foveal** magnification factor: a disproportionate amount of each nucleus or cortical area represents the part of visual space near the fovea. PO lacks this characteristic, which suggests that it plays an important role in target localization and detection.[17] The response of some neurons in PO to a visual target also depends on the eye's orientation within the orbit, although most cells have response fields that move with the eyes.[18] Therefore, PO's neural signals seem suitable for computing target location (chapter 11) and, in line with this idea, damage to PO leads to deficits in visually guided reaching.[19]

Parts of the PPC on and near the medial surface of the hemisphere, such as area 7m (not illustrated) and area 5c (often called PEc; see figure 6.3), appear to have similar properties: current hand position, saccadic eye movements, reaching movements, visual inputs, and planning for movements in both the light and the dark all affect cell activity in those areas.[20,21]

Another important part of the PPC, the medial intraparietal area (MIP), corresponds—at least to an approximation—to what some experts call the parietal reach region (PRR). This notion becomes important in chapters 11 and 12. Neurons in MIP respond to both visual and proprioceptive stimuli, and their activity reflects both limb movement and hand location.[22] Similarly, the ventral intraparietal area (VIP) receives convergent input from the somatosensory and visual systems,[23] and the lateral intraparietal area (LIP) receives both visual and auditory signals.[24] In these multimodal areas, a given neuron may respond to stimuli of either modality or both.[25,26] Chapters 11 and 12 present the idea that, for reaching and pointing movements, the primate CNS uses extrinsic, vision-based coordinates to compute the difference between hand and target locations.[27-29] MIP and LIP, along with area 5d, play an important role in this function.[30-32] Cells in LIP and MIP encode information important for saccades and reaching movements, respectively. LIP also reflects the perception of the targets of movement[33] and the allocation of attentional resources.[34-36]

Other parts of the PPC appear to be more concerned with the manipulation of objects, as opposed to reaching for them. Inactivation of one of these areas, termed the anterior intraparietal cortex (AIP) or the part of frontal cortex connected with AIP (PMv), leads to deficits in object manipulation, but not in reaching.[37-39]

Several sources contribute information to PPC, but a principal input comes from the superior colliculus. As chapter 17 explains in detail, reaching and eye movements both depend on copies of motor commands, known as **efference copy** or **corollary discharge**. These signals update the relative locations of potential movement targets as the eyes move. The projection from the superior colliculus to the thalamus, known as the *tectothalamic projection*, provides one source of efference copy to the cortex.

References

1. Zheng, T, and Wilson, CJ (2002) The implications of corticostriatal axonal arborizations. J Neurophysiol 87, 1007–1017.
2. Horak, FB, Nutt, JG, and Nashner, LM (1992) Postural inflexibility in parkinsonian subjects. J Neurol Sci 111, 46–58.
3. Beiser, DG, Hua, SE, and Houk, JC (1997) Network models of the basal ganglia. Curr Opin Neurobiol 7, 185–190.
4. Houk, JC, and Wise, SP (1995) Distributed modular architectures linking basal ganglia, cerebellum, and cerebral cortex: Their role in planning and controlling action. Cerebral Cortex 5, 95–110.
5. Barto, AG, Fagg, AH, Sitkoff, N, and Houk, JC (1999) A cerebellar model of timing and prediction in the control of reaching. Neural Comput 11, 565–594.
6. Keifer, J, and Houk, JC (1994) Motor function of the cerebellorubrospinal system. Physiol Rev 74, 509–542.

7. Wong, YC, Kwan, HC, MacKay, WA, and Murphy, JT (1977) Topographic organization of afferent inputs in monkey precentral cortex. Brain Res 138, 166–168.
8. Donoghue, JP, and Sanes, JN (1994) Motor areas of the cerebral cortex. J Clin Neurophysiol 11, 382–396.
9. Woolsey, CN, Settlage, P, Meyer, DR, et al. (1952) Patterns of localization in precentral and "supplementary" motor areas and their relation to the concept of a premotor area. Res Publ Assoc Res Nerv Ment Dis 30, 238–264.
10. Poliakov, AV, and Schieber, MH (1999) Limited functional grouping of neurons in the motor cortex hand area during individuated finger movements: A cluster analysis. J Neurophysiol 82, 3488–3505.
11. Sanes, JN, and Schieber, MH (2001) Orderly somatotopy in primary motor cortex: Does it exist? Neuroimage 13, 968–974.
12. Kirkwood, PA, Maier, MA, and Lemon, RN (2002) Interspecies comparisons for the C3–C4 propriospinal system: Unresolved issues. Adv Exp Med Biol 508, 299–308.
13. Pandya, DN, and Kuypers, HGJM (1969) Cortico-cortical connections in the rhesus monkey. Brain Res 13, 13–36.
14. Jones, EG, and Powell, TPS (1970) An anatomical study of converging sensory pathways within the cerebral cortex of the monkey. Brain 93, 793–820.
15. Krubitzer, L, Manger, P, Pettigrew, J, and Calford, M (1995) Organization of somatosensory cortex in monotremes: In search of the prototypical plan. J Comp Neurol 351, 261–306.
16. Kalaska, JF, Cisek, P, and Gosselin-Kessiby, N (2003) In Advances in Neurology: The Parietal Lobe, eds Siegel, AM, Andersen, RA, Freund, H-J, and Spencer, DD (Lippincott Williams and Wilkins, Philadelphia), pp 97–119.
17. Colby, CL, Gattass, R, Olson, CR, and Gross, CG (1988) Topographical organization of cortical afferents to extrastriate visual area PO in the macaque: A dual tracer study. J Comp Neurol 269, 392–413.
18. Galletti, C, Battaglini, PP, and Fattori, P (1995) Eye position influence on the parieto-occipital area PO (V6) of the macaque monkey. Eur J Neurosci 7, 2486–2501.
19. Battaglini, PP, Muzur, A, Galletti, C, et al. (2002) Effects of lesions to area V6A in monkeys. Exp Brain Res 144, 419–422.
20. Battaglia-Mayer, A, Ferraina, S, Genovesio, A, et al. (2001) Eye-hand coordination during reaching. II. An analysis of the relationships between visuomanual signals in parietal cortex and parieto-frontal association projections. Cerebral Cortex 11, 528–544.
21. Marconi, B, Genovesio, A, Battaglia-Mayer, A, et al. (2001) Eye-hand coordination during reaching. I. Anatomical relationships between parietal and frontal cortex. Cerebral Cortex 11, 513–527.
22. Johnson, PB, Ferraina, S, Bianchi, L, and Caminiti, R (1996) Cortical networks for visual reaching: Physiological and anatomical organization of frontal and parietal lobe arm regions. Cerebral Cortex 6, 102–119.
23. Colby, CL, Duhamel, J-R, and Goldberg, ME (1993) Ventral intraparietal area of the macaque: Anatomic location and visual response properties. J Neurophysiol 69, 902–914.

24. Stricanne, B, Andersen, RA, and Mazzoni, P (1996) Eye-centered, head-centered, and intermediate coding of remembered sound locations in area LIP. J Neurophysiol 76, 2071–2076.

25. Linden, JF, Grunewald, A, and Andersen, RA (1999) Responses to auditory stimuli in macaque lateral intraparietal area. II. Behavioral modulation. J Neurophysiol 82, 343–358.

26. Grunewald, A, Linden, JF, and Andersen, RA (1999) Responses to auditory stimuli in macaque lateral intraparietal area. I. Effects of training. J Neurophysiol 82, 330–342.

27. Andersen, RA, Snyder, LH, Batista, AP, et al. (1998) Posterior parietal areas specialized for eye movements (LIP) and reach (PRR) using a common coordinate frame. Novartis Found Symp 218, 109–122.

28. Andersen, RA, and Buneo, CA (2002) Intentional maps in posterior parietal cortex. Annu Rev Neurosci 25, 189–220.

29. Buneo, CA, Jarvis, MR, Batista, AP, and Andersen, RA (2002) Direct visuomotor transformations for reaching. Nature 416, 632–636.

30. Snyder, LH, Batista, AP, and Andersen, RA (1998) Change in motor plan, without a change in the spatial locus of attention, modulates activity in posterior parietal cortex. J Neurophysiol 79, 2814–2819.

31. Snyder, LH, Batista, AP, and Andersen, RA (2000) Saccade-related activity in the parietal reach region. J Neurophysiol 83, 1099–1102.

32. Snyder, LH, Batista, AP, and Andersen, RA (1997) Coding of intention in the posterior parietal cortex. Nature 386, 167–170.

33. Shadlen, MN, and Newsome, WT (2001) Neural basis of a perceptual decision in the parietal cortex (area LIP) of the rhesus monkey. J Neurophysiol 86, 1916–1936.

34. Bisley, JW, and Goldberg, ME (2003) Neural activity in the lateral intraparietal area and spatial attention. Science 299, 81–86.

35. Kusunoki, M, Gottlieb, J, and Goldberg, ME (2000) The lateral intraparietal area as a salience map: The representation of abrupt onset, stimulus motion, and task relevance. Vision Res 40, 1459–1468.

36. Gottlieb, JP, Kusunoki, M, and Goldberg, ME (1998) The representation of visual salience in monkey parietal cortex. Nature 391, 481–484.

37. Jeannerod, M, Arbib, MA, Rizzolatti, G, and Sakata, H (1995) Grasping objects: The cortical mechanisms of visuomotor transformation. Trends Neurosci 18, 314–321.

38. Rizzolatti, G, Fogassi, L, and Gallese, V (1997) Parietal cortex: From sight to action. Curr Opin Neurobiol 7, 562–567. Curr Opin Neurobiol 7, 562–567.

39. Fogassi, L, Gallese, V, Buccino, G, et al. (2001) Cortical mechanism for the visual guidance of hand grasping movements in the monkey—A reversible inactivation study. Brain 124, 571–586.

40. Gerfen, CR (2004) In The Rat Nervous System, 3rd ed, ed Paxinos, G (Elsevier Science, Amsterdam), pp 455–508.

Reading List

Charles Gerfen provides a thorough overview on the anatomy of the basal ganglia.[40]

7 What Generates Force and Feedback

*Overview: Muscles convert chemical energy into force and act like an integrated system of springs, dampers, and force generators. This chapter describes the relationship between linear forces, as produced by muscles, and **torques** generated in a two-joint arm. Muscle fibers not only generate force but also give rise to feedback signals that convey information about forces and muscle lengths to the CNS.*

7.1 Biological versus Mechanical Actuators

Engineers have devised many mechanisms for moving things. Take, for example, a typical torque motor in a robot's arm. For most of these motors, the amount of current received from a **controller** determines the amount of torque it produces. With a given input, the motors produce the desired torque regardless of the configuration of the robot's arm or the rate at which that configuration changes. Accordingly, the amount of force generated by a given motor command signal does not depend on the angle of the joint. Your muscles work differently. As section 7.2.3 details, when you change joint angles, you also change muscle lengths, and muscle length affects the amount of force generated. This fact, among many others, presents problems to the CNS controllers.

A robot's torque motor also produces forces much more quickly than a typical muscle. It takes approximately 30 msec from the time that a motor command arrives at a cat's leg muscle to produce force—and cats are very quick animals. A torque motor generates force in one-tenth that amount of time.

Further, by changing the polarity of the current that drives a torque motor, engineers can reverse the torque the motor produces. For example, if positive current flexes a joint, negative current extends it. However, your muscles can only pull; they cannot push. Limb muscles produce forces that either flex or extend the limb, not both. Accordingly, limb movements (and stable postures) depend on the fact that muscles occur in antagonistic pairs. Contraction of one muscle both moves the joint and lengthens the antagonist muscle.

Finally, in robots, information flows to motors and from sensors at the speed of light, but your axons conduct action potentials to and from your muscles only at approximately the speed of sound (see section 3.3). Delays in communication not only slow the speed of muscle contraction but also make muscles difficult to control. As explored in subsequent chapters, delays in information transmission present perhaps the most fundamental challenge that the CNS faces in controlling movement. These delays make it extremely difficult to respond to a perturbation, and provoke instability during both reaching movements and maintenance of steady postures.

7.2 Muscle Mechanisms

7.2.1 Active Mechanisms: The Motor

The generation of force by muscle fibers depends mostly on two proteins: myosin and actin (figure 7.1C). Myosin forms *thick filaments*, and actin along with two other proteins, troponin and tropomyosin, makes up *thin filaments*. Muscles function principally by converting chemical and kinetic energy into mechanical force and **work** (see appendix D).

In an ideal muscle fiber at rest, the globular head of each myosin molecule (figure 7.1C) rests in a "cocked" position: mechanically stretched, but detached from actin. The kind of myosin in muscles, called myosin II, consists of molecules with a head connected through a necklike structure to a tail. The angle of the neck changes by approximately 60° from the cocked to the released configuration, which results in a head movement of about 5–10 nm. In its cocked position, the myosin head binds to a relatively low-energy molecule, adenosine diphosphate (ADP), and phosphate. In its released position, it binds to adenosine triphosphate (ATP), a relatively high-energy molecule. This association results in recocking the myosin along with breaking down ATP into ADP and phosphate. Thus, chemical energy gained by breaking down a phosphate bond in ATP becomes stored as kinetic energy in the configuration of myosin. (See box 7.1.) When the muscle generates action potentials, a cascade of chemical events results in the attachment of the myosin head to actin and the release of the kinetic energy stored in the myosin. The release of this energy is something like the release of a stretched spring. As a result, actin and myosin move relative to each other. How?

The cascade of events that generates force usually begins with the release of acetylcholine by motor neuron synapses, which concentrate at *neuromuscular junctions*. This excitatory neurotransmitter binds temporarily with the muscle's cholinergic receptors, which leads to depolarization of the muscle fiber's postsynaptic membrane. Depolarization, caused by an influx of sodium ions, affects internal structures called transverse tubules (figure 7.1A), also called *T-tubules*. Once the action potential affects these tubules, intracellular calcium ions "mobilize" (i.e., leave the tubules) to enter the cytoplasm of the muscle (also known as *sarcoplasm*). When present in sufficient concentration, intracellular calcium ions expose a site on the actin filaments that binds myosin. Recall that the myosin head is, at this point in the process, in its cocked position.

7.2. Muscle Mechanisms

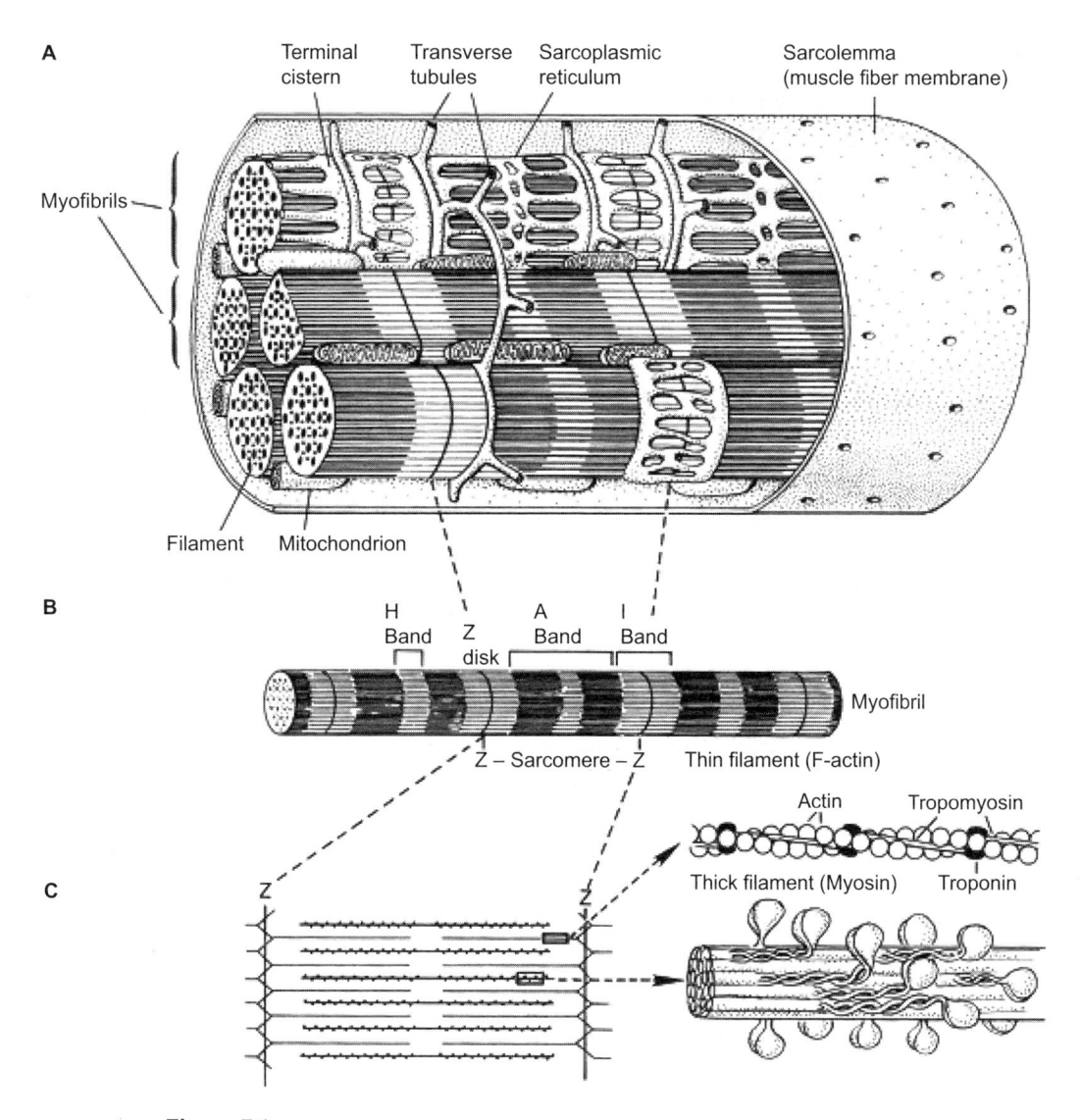

Figure 7.1
Structure of skeletal muscle at three different levels. (*A*) The muscle-fiber level. (*B*) The myofibril level. (*C*) The filament level. (From Ghez,[1] as adapted from Bloom and Fawcett[2] and Loeb and Gans[3] with permission.)

Box 7.1
Sources of chemical energy

> As a practical matter, ATP acts more like a final common path in the conversion of chemical into mechanical energy than as the principal energy source. Cellular stores of ATP last only a few seconds. Through interactions with ATP-related mechanisms, a different compound, creatine phosphate, provides the energy and phosphate to support several seconds of work. Adipose and glycogen metabolism within muscle fibers provides support for several minutes; for longer periods, the body must mobilize energy resources in the liver and elsewhere in the body.

Once attached to the actin-based thin filaments, the myosin molecules undergo a change in structure known as a *conformational shift*. This reconfiguration of myosin's internal structure, in which the head rotates around the neck, releases its stored kinetic energy and forces the actin and myosin filaments to slide relative to each other. This key event either shortens the muscle or generates force, in either case creating mechanical energy.

Whether the relative sliding of actin and myosin filaments causes a movement or, alternatively, a force without any movement depends on the interaction of muscles with their tendons, the skeleton, the environment, and other muscles. If the force generated does not suffice to overcome the **inertia** of the skeleton and attached tissue, there is no movement. Objects linked to the skeleton, such as a weight grasped in your hand, have a similar effect. Furthermore, if the skeleton pushes against some immovable object, such as a wall, the limb cannot move. Muscles produce their maximum force in this situation, termed **isometric** contraction.

7.2.2 Passive Mechanisms: The Spring

The active properties of a muscle explain how it behaves when it receives inputs. Passive properties account for a muscle's behavior at rest. Stretching a resting muscle results in that muscle producing a force, which arises because elements within the muscle act like springs.

Sarcomeres contain both thick and thin filaments, which provide the muscle's active, contractile elements (figure 7.1B). At the boundary between sarcomeres, the thin filaments attach directly to a structure called a *Z disk* (figure 7.1B). The ends of the thick filaments also attach to Z disks, but only through noncontracting filaments containing *connectin*.

The connectin filaments behave like springs and confer this property on the muscle as a whole. To understand why, consider a sarcomere of a given length, with its connectin filaments at their resting length. As the thick filaments pull on the thin filaments to shorten the muscle, the sarcomere shortens. Accordingly, the connectin filaments also shorten. When the muscle stops receiving inputs, myosin is released from actin and the muscle relaxes. During relaxation, the length of the sarcomere increases until it reaches the resting length of the connectin filaments. If you now pull the muscle beyond this length, the connectins will resist much as when you pull a spring beyond its resting length.

Like metal springs, muscles vary in stiffness. For example, when you pull on a spring with a given amount of force, a thick spring increases its length much less than a thin one. This property is called stiffness (K), which is defined as the ratio of change in force (F) to change in length (L). Muscles with high levels of stiffness require large increments in force to produce a given amount of lengthening, as captured by equation 7.1:

$$K = \frac{\Delta F}{\Delta L}. \tag{7.1}$$

In chapter 3, you saw that evolution led to certain features of muscles that seem immutable. Similarly, across vertebrate species, the active and passive mechanisms of muscles vary remarkably little. Vertebrates have sarcomeres of a standard 2.5 μm length. (In contrast, invertebrate sarcomeres can range from ~2 to 20 μm.) The force generated by skeletal muscles remains within a fairly narrow range, 1–8 kg/cm^2 of cross-sectional area.[4] Vertebrate muscles contract by no more than 40% of their fully stretched length, and typically by 20% or less. Although no vertebrate has evolved a muscle that operates outside these parameters, muscles have evolved in two major ways to alter their function. Serial architectures stack sarcomeres to lengthen muscles and thereby increase the amount and speed of contraction; parallel architectures thicken muscles to increase the number of myofibrils and thereby multiply force.

7.2.3 Active and Passive Mechanisms: Length–Tension Properties

Because of the properties of actin and myosin, and the way they interact to produce force, the length of a muscle affects the force that it generates, a property known as the force–length or the length–tension relation. In part, the length–tension relation results from the mechanical advantage gained by a stretched muscle. In that state, the myosin molecules can reconfigure to a greater extent than in shortened muscles, which generates more force. The length–tension relation contributes to limb stability, as described in section 8.2. Stated very generally, shortened muscles produce relatively small forces, and lengthened muscles produce relatively large forces.

In shortened muscles (and therefore shortened sarcomeres), the connectin and Z-disk formations compress the thick filaments, crumpling their ends, where the myosin heads concentrate. Because the binding of myosin to actin occurs only when the myosin heads rest near a binding site on the thin filament, this crumpling reduces the number of myosin heads that can bind with actin. With fewer myosin molecules available for binding, the sarcomere (and therefore the muscle) produces less force.

Box 7.2
Variants

> In an interesting variation from the vertebrate pattern, in the cephalo-chordate amphioxus (see section 2.3.1), muscle cells have "tails" that extend toward the CNS to form neuromuscular junctions.[5] In essence, the muscles go to the CNS rather than vice versa.

As muscle length increases, the thick filaments stretch and the amount of overlap between eligible myosin in the thick filaments and actin in the thin filaments increases. The mechanical advantage and binding availability thereby both increase, resulting in a greater ability to produce force.

As muscle length increases further, at some point a maximum number of myosin heads can bind to actin. Beyond this length, the thin and thick filaments overlap less than the optimal amount, and the amount of force decreases. The length at which maximum contractile force can be produced often corresponds approximately to the resting length of the muscle. Also at approximately this length, the passive properties of the muscle begin to resist stretching. Therefore, when muscle length increases to the point that contractile forces begin to decline, the passive forces begin to increase. These two components sum to produce the total force in the muscle. Taken together, these properties of muscle tissue account for the length–tension relationship.

7.3 Motor Units

Motor neurons are segregated into **motor pools**, each of which innervates a particular muscle. (See box 7.2.)

There are two types of motor neurons: α-motor neurons, which send their axons to **extrafusal** fibers, and γ-motor neurons, which innervate **intrafusal** fibers. Motor pools extend over two to four spinal segments, with medially situated motor pools innervating axial muscles (e.g., those of the neck and trunk), and laterally situated motor pools projecting to limb muscles. Motor neurons contacting the most distal muscles occur in the most lateral parts of the ventral horn.

The term *motor unit* applies to a motor neuron and the muscle fibers that it innervates. Except for periods early in development and in some diseases, each muscle fiber receives input from only one motor neuron. Each motor neuron, however, branches to contact several muscle fibers, a number that varies according to function. In muscles controlling fine movements, a single motor neuron controls only a small number of muscle fibers. For example, each motor neuron projecting to the eye muscles terminates on only three to six muscle fibers. In contrast, in muscles that function in coarser movements, such as the gastrocnemius (calf) muscle of the leg (see figure 3.3), motor neurons innervate 2000–3000 fibers. As a general rule, motor neurons with larger cell bodies typically innervate

7.4 A Muscle Model

As noted above, a muscle produces two kinds of force, active and passive, which compose a muscle's total force. A muscle's contractile elements provide its active force through the actin and myosin "ratcheting" mechanism. Noncontractile elements contribute its passive force and have properties resembling a spring. Technically, a muscle's force should be called *elastic* (or viscoelastic) rather than springlike (see appendix D), although a spring will do as an approximation. Because this springlike element attaches in series with the contractile one, you can think of it as a *series elastic element*.

In the 1920s, A. V. Hill[6] first noted that activated muscles produce more force when held isometrically (i.e., at a fixed length) than when they shorten. When muscles shorten, they appear to waste some of their active force in overcoming an inherent resistance. This resistance could not result from the series elastic element because it resists lengthening, not shortening. Hill thought of this resistance as another kind of passive force in the muscle. He found that the faster a muscle shortens, the less total force it produces. Assuming a constant active force, Hill concluded that the faster shortening leads to a larger resistive force.

Hill drew an analogy between the resistive force and a shock absorber. A piston in a viscous fluid is a simple shock absorber, also known as a *damper*. If you push on its piston, a shock absorber will resist by a **tension** T (equivalent to a force) that depends on the viscosity b of the fluid in its cavity. The faster you try to push the piston, the more strongly the fluid resists. For a given speed $\dot{x}$, the force that you need to move the piston is $T = b\dot{x}$ (see appendix C for an explanation of this notation). To account for the fact that muscle produces less force when it shortens, Hill proposed that a viscous element lies in parallel with the contractile element. Accordingly, it can be called a *parallel elastic element*. Together, the series and parallel elastic "elements" make up a muscle's passive force.

To investigate the properties of this viscous element, Hill and his colleagues performed a simple experiment. They attached a muscle to a bar that pivoted around a point (figure 7.2A). One end of the bar had a catch mechanism that they could release at any time. A basket held a weight on the other end of the bar. When Hill released the catch, this weight would pull on the muscle by a force T. The experiment began with the catch in place and the muscle stimulated maximally. The stimulation resulted in the production of force T_0 in the muscle. Because the muscle pulled on a bar that could not move, this force did not change the muscle's length.

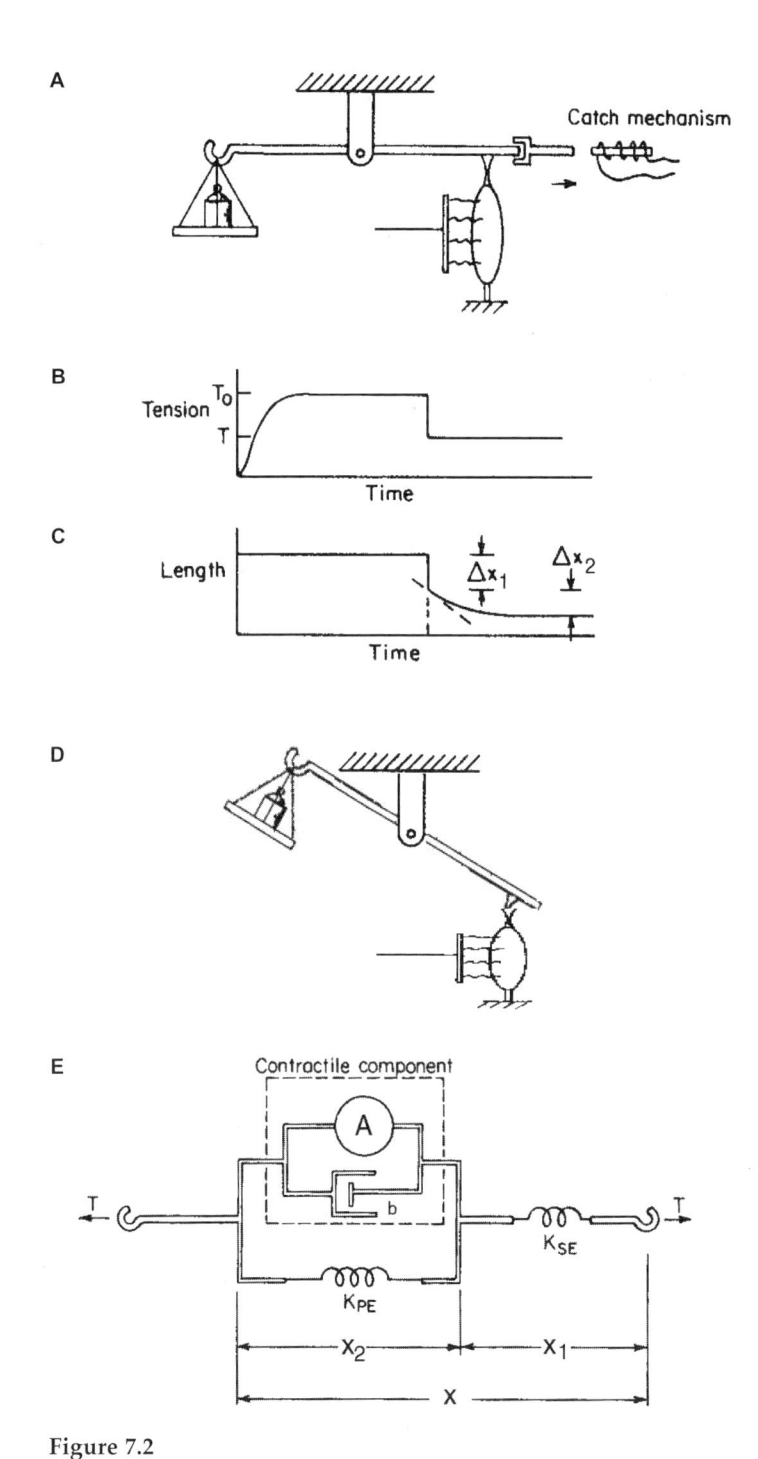

Figure 7.2
Development of a mathematical muscle model. In this experiment, a muscle receives stimulation (A) prior to release of the catch (D). Because the scale (left in A and D) has less weight on it than the force (B) that the muscle produces, the muscle shortens, first rapidly and then gradually (C). A muscle model (E) results from the changes in the muscle's length and tension. Abbreviations: A, active force component; b, coefficient of viscosity; K_{PE}, stiffness of the parallel element; K_{SE}, stiffness of the series element; T, tension (equivalent to force); x, length; Δx, change in length. (From McMahon[7] with permission.)

7.4. A Muscle Model

At this point, the experimenters released the catch. Figure 7.2C shows that the length of the muscle suddenly shortened, and figure 7.2B shows, that the force dropped from T_0 to T. The fact that the muscle quickly shortened by amount Δx_1 and reduced its force from T_0 to T suggests that something in the muscle acted like a spring. If you put tension on a spring by pulling it, then suddenly release it, the spring will rapidly shorten and its tension will decrease. This springlike element in the muscle is the series elastic (SE) element, and figure 7.2E depicts its stiffness as K_{SE}. Recall that stiffness relates changes in force (or tension) to changes in length (equation 7.1).

After the immediate change in muscle length and force, a slow, gradual change in length developed (figure 7.2C), without any change in force (figure 7B). Whereas a part of the muscle's mechanism changed length rapidly in response to the force change, another part did not change as quickly—as if a "shock absorber" acted on the "spring," slowing its response to the force change. The parallel elastic (PE) element, referred to above, represents this second passive element in the muscle, and figure 7.2E depicts its stiffness as K_{PE}.

The muscle's viscosity, the parallel elastic element, and the series elastic element compose the passive components of an elementary model muscle. In figure 7.2, the length of the series elastic element is x_1 and the length of the parallel elastic element is x_2. Note that this is a *model* of a muscle: It does not imply that the various components have this physical arrangement within a muscle. Further, the viscous component of muscle tension results from mechanisms very different from the mechanical damper. Mathematically, however, the characteristics of the muscle accord reasonably well with this depiction. You might think of the model as a simile. In the model, as in a muscle, when the tension in the system suddenly decreases, the series elastic element responds immediately, but the parallel elastic element responds gradually because of its viscous component.

The muscle's active component contributes the final piece of the muscle model. This active force acts against the passive components to produce the final force that acts on the bar in figure 7.2D. Function A indicates the active component in figure 7.2E. The model can now describe how the total force produced by the muscle depends on its passive and active components. Assume that the series elastic element—its most springlike component—has a resting length x_1^* and the parallel elastic element has a resting length x_2^*. The same force T develops in both of these elements because a muscle can have only one force at any given time. Thus,

$$T = K_{SE}(x_1 - x_1^*) \quad \text{and}$$

$$T = K_{PE}(x_2 - x_2^*) + b\dot{x}_2 + A,$$

where A represents the active force; x, the muscle length; $\dot{x}$, its rate of change; and T, its force (or tension). With substitutions (described in the web document *musclemodel*; see the Preface), you can show that tension changes in the muscle are described by the following expression:

$$\dot{T} = \frac{K_{SE}}{b}\left(K_{PE}\Delta x + b\dot{x} - \left(1 + \frac{K_{PE}}{K_{SE}}\right)T + A\right).$$

The web document *musclemodel* shows how the parameters of this equation can be found from experimental data and provides a simple description of active force A as a function of muscle-stimulation rate.

7.5 Converting Force to Torque

Thus far, the presentation has dealt with linear mechanics, mainly changes in muscle length and force. In a limb, however, angular mechanics become important, specifically, changes in joint angles and torque.

7.5.1 Torque and Angular Velocity as Vectors

Imagine that you are holding a ball on a countertop. If you give the ball a quick twist, it will rotate. You have imposed a torque on the ball, and that torque has caused motion. However, the ball does not move from point to point (a *translational movement* or *displacement*); it rotates in a given location (i.e., it moves around a pivot). The ball rotates at a given speed, termed **angular velocity**, which, like translational velocity, corresponds to a vector composed of speed and direction. For angular velocity, this vector points along the axis of rotation of the ball, with a magnitude often expressed in terms of degrees per second or radians per second. Because the vector could orient in either direction along the axis of rotation, a convention—the right-hand-rule—describes the vector's direction unambiguously. When the fingers of your right hand curve around the ball in its direction of rotation—for example, clockwise as you look down on the spinning ball—your thumb points down, which shows the direction of the angular-velocity vector. The torque used to spin the ball has the same direction.

7.5.2 Relating Force to Torque Through Virtual Work

When a muscle such as the triceps (figures 3.1 and 3.2) contracts, it produces a force that results in a torque on the elbow joint (see figure 7.3). The force f that the muscle produces, imposes a torque τ on the joint, resulting in an extension of the joint angle by an amount $\Delta\theta$. The angle θ decreases when the joint extends, which means that a negative angular velocity results in extension. Note that the coordinate system used to represent the angle of the joint also specifies the coordinate system for torque. This convention leads to extension being a negative angular velocity.

If you imagine the muscle acting along a line and label the length of the muscle as λ, as in figure 7.3, the contraction may cause the muscle to shorten by amount $\Delta\lambda$. When the muscle shortens, the work that it performs is the force (f) that it produces times the displacement that

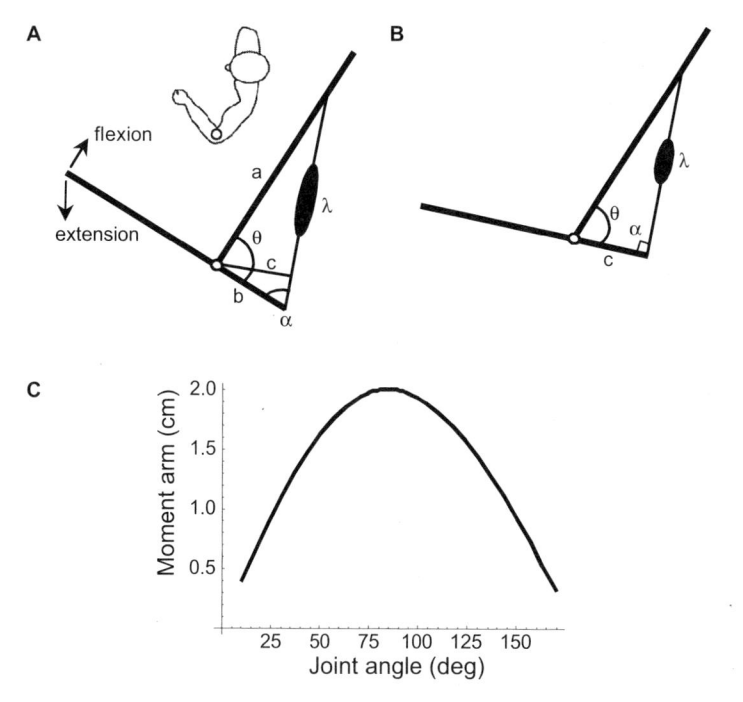

7.5. Converting Force to Torque

Figure 7.3
Moment arm of a triceps-like muscle. (A) One arm configuration, where a and b equal the length of skeletal segments from joint to insertion; c is the length of the moment arm; λ is muscle length; and α and θ indicate joint angles. (B) A more extended arm configuration. (C) Torque as a function of muscle length. This plot is the value of the moment arm, equation 7.5, as a function of joint angle, where a = 20 cm and b = 2 cm. The value of the moment arm is related to joint angle, whereas the value of the joint angle is a function of muscle length.

it undergoes (see appendix D). The muscle shortens when it contracts; therefore $\Delta\lambda$ is a negative number. When the muscle changes length by amount $\Delta\lambda$, the joint rotates by amount $\Delta\theta$. The work that the muscle performs in shortening its length, $-f\,\Delta\lambda$ (force times distance), must equal the work it performs in rotating the joint, $\tau\,\Delta\theta$; thus

$$\tau\,\Delta\theta = -f\,\Delta\lambda.$$

Solving for τ you get:

$$\tau = -\frac{\Delta\lambda}{\Delta\theta}f. \qquad (7.2)$$

If you allow the length and angular changes in equation 7.2 to approach 0 (i.e., you make each change step minuscule), you can represent them as a derivative. This function relates changes in length and force, which are linear measures, to changes in joint angle and torque, which are angular measures.

$$\tau = -\frac{d\lambda}{d\theta} f. \tag{7.3}$$

Equation 7.3 expresses an important idea: The torque that a muscle produces on a joint depends on how the muscle's length changes with respect to the angle of the joint. This notion also explains the concept of **virtual work**: It is virtual because the amount is infinitesimal as the steps approach 0. Because of the reference frame and the muscle chosen in this example, an increase in the angle θ results in an increase in the length of the muscle. To express the derivative in equation 7.3, you need to consider the geometry of the limb. Consider the triangle with muscle length λ and lengths a and b in figure 7.3. You can express how the length of the muscle depends on the angle of the joint through the law of cosines:

$$\lambda = \sqrt{a^2 + b^2 - 2ab\cos(\theta)}. \tag{7.4}$$

The derivative of this function is

$$\frac{d\lambda}{d\theta} = \frac{ab\sin(\theta)}{\sqrt{a^2 + b^2 - 2ab\cos(\theta)}}. \tag{7.5}$$

Interestingly, equations 7.5 and 7.3 both specify how muscle forces relate to joint torques. The larger the value of the function in equation 7.5, the larger the magnitude of torque that a given force in the muscle produces. Figure 7.3C plots how equation 7.5 varies with respect to joint angle θ for given values of a and b. In this example, the muscle exerts its greatest torque on the joint when the angle is approximately 85°. At very flexed or very extended positions, the function becomes small, which means that at these joint angles the muscle's force results in very little joint torque. This relationship occurs, in part, because for a force to generate torque, only perpendicular force counts. For example, imagine a hinged door. If you apply force directly along the door from the edge toward the hinge, it will not rotate at all. In contrast, however, when you apply force perpendicular to the door, you need very little force to open and close the door. Applying force at intermediate angles will "waste" proportionate amounts of force. The same kind of thing happens as muscles shorten and joints rotate. The muscles' **insertions** adopt systematically varying angles with respect to the skeleton. Figure 7.3A shows the insertion as a smaller angle; figure 7.3B, as a right angle. The change in this *angle of insertion* causes some of the variation in how much torque the muscle generates for a given force, a concept related to the **moment arm**.

7.5.3 The Moment Arm

Figure 7.3A illustrates the moment arm of the muscle: the length c that connects the center of rotation of the joint with a line perpendicular to the line of action of the muscle. As noted above, only forces perpendicular to the skeleton generate torque. Consider that the length of the moment arm is

$$c = b \sin(\alpha).$$

From the law of sines, you have

$$\frac{\sin(\alpha)}{a} = \frac{\sin(\theta)}{\lambda},$$

which, using equation 7.4, gives you an expression for the length of the moment arm:

$$c = \frac{ab \sin(\theta)}{\lambda} = \frac{ab \sin(\theta)}{\sqrt{a^2 + b^2 - 2ab \cos(\theta)}}.$$

The right side of this expression is identical to equation 7.5 from the principle of virtual work. That is, the change in muscle length λ with respect to joint angle θ is the moment arm of this mechanical system:

$$\frac{d\lambda}{d\theta} = c. \tag{7.6}$$

Both the virtual-work and moment-arm methods produce a function that relates muscle force to torque. However, the principle of virtual work provides some power worth exploiting because it helps describe muscles with complex geometry, as taken up in the next section.

7.5.4 The Jacobian

Contraction of a muscle can result in torques on multiple joints. For example, consider the muscle in figure 7.4. How does force in this muscle relate to torques on the joints? You can use the principle of virtual work to describe this relationship.

Work equals force times distance, and therefore, in terms of vector algebra, work is a scalar value that equals the dot product of a force vector (the amount of force in a given direction) and a difference vector (the difference, for example, between hand location at the beginning and end of a movement). Thus, $W = \vec{f} \cdot \Delta \vec{x}$. As appendix C explains, the dot product is the "projection" of one vector onto another in a given dimension, which depends on the angle between the vectors. In this case, this projection is force in the direction perpendicular to the skeleton, the moment arm of the muscle. You can also represent that same work in terms of torque and joint-angle changes. Work in this coordinate frame equals the dot product of torque and the joint's angular change. However, because the muscle shortens when it produces its force, the dot product of force and length change is negative. To take care of this relationship, you can write equation 7.2 as

$$\tau \Delta \theta = -f \Delta \lambda,$$

which says that the work described in terms of torques and joint-angle changes has to match work in terms of forces and muscle-length changes. Alternatively, for use below, you can transpose the torque and force vectors:

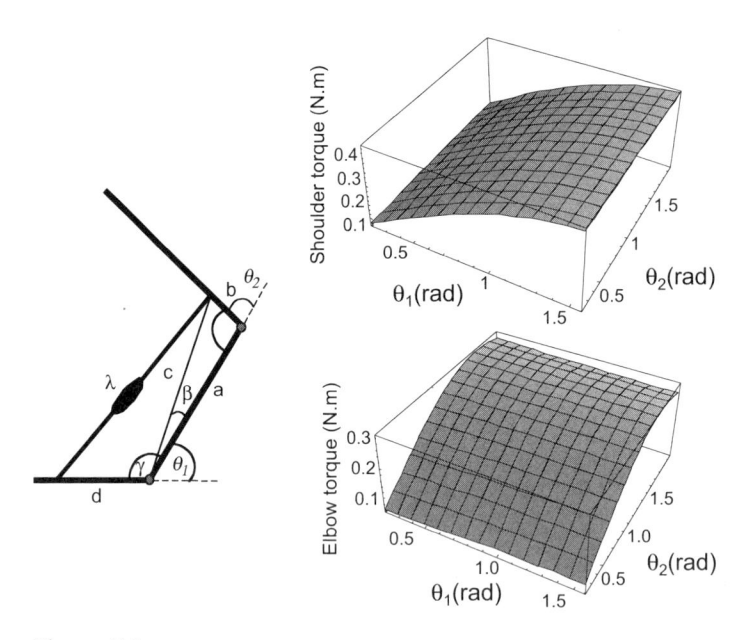

Figure 7.4
A two-joint muscle. In the format of figure 7.3, with d and γ for the additional link length and joint angle, respectively. The torque on the shoulder (θ_1) and elbow (θ_2) is plotted as a function of the angle of each joint for a constant muscle force of 10 N. Link lengths: a = 33 cm, d = 4 cm, b = 3 cm. N.m, newton-meters (N-m).

$$\tau^T \Delta\theta = -f^T \Delta\lambda. \tag{7.7}$$

In this equation, torque and force may be multidimensional vectors. If the displacements become infinitesimal, as in virtual work, you can define a Jacobian matrix (see appendix C). You can think of a Jacobian, in general, as a matrix of equations that relates one coordinate system to another. Another useful idea about Jacobians is that they serve as local linear approximations of nonlinear functions. For example, one Jacobian might convert the changes in joint angle into the displacements of the hand that occur when the limb segments rotate around a joint:

$$\Delta x = J_x(\theta) \Delta\theta. \tag{7.8}$$

However, Jacobian matrices come in many forms. Another Jacobian matrix is often called the moment-arm matrix. This Jacobian consists of a matrix of equations that relates changes in muscle length to changes in joint angle:

$$\Delta\lambda = J_\lambda(\theta) \Delta\theta$$
$$J_\lambda = \frac{d\lambda}{d\theta}. \tag{7.9}$$

7.5. Converting Force to Torque

Note that in the equation $\Delta\lambda = J_\lambda(\theta)\,\Delta\theta$ (and $\Delta x = J_x(\theta)\,\Delta\theta$), the dependence of the Jacobian on joint angle θ is explicitly denoted. In the remainder of this book, this notation is dropped in favor of the simpler and more conventional $\Delta\lambda = J\,\Delta\theta$.

Equation 7.9 together with equation 7.6 yields $J = c$ in the case of the system in figure 7.3: This Jacobian *is* the moment arm. To understand this equality intuitively, consider that the moment arm for a muscle involves the component of force perpendicular to the skeleton upon which it acts, at the point of insertion. Recall from figure 7.3A and 7.3B that the angle of insertion changes with joint angle. A nearly perpendicular insertion angle generates the most torque. Thus, the moment arm for a muscle changes as the joint angle changes because a change in joint angle also changes the angle of insertion.

Further, substituting equation 7.9, $\Delta\lambda = J \cdot \Delta\theta$, into equation 7.7, $\tau^T \Delta\theta = -f^T \Delta\lambda$, yields

$$\tau^T \Delta\theta = -f^T J \Delta\theta,$$

which must hold true for all $\Delta\theta$; therefore, simplifying the equation, you have

$$\tau^T = -f^T J.$$

Transposing both sides of that equation yields the final result:

$$\tau = -J^T f. \tag{7.10}$$

Equation 7.10 represents an interesting relationship because it allows you to convert a force in muscle coordinates into a torque in **joint coordinates**. The relationship depends on how the length of the muscle changes with respect to the joint angle. If the length of the muscle depends on multiple joint angles, then the Jacobian will be a multidimensional vector.

Consider the muscle illustrated in figure 7.4. Here, the length of the muscle λ depends on angles θ_1 and θ_2. To find the Jacobian, you first need to express this dependence and then find its derivative. Begin by writing the length c in terms of the angle θ_2:

$$c = \sqrt{a^2 + b^2 + 2ab\,\cos(\theta_2)}. \tag{7.11}$$

Next use the law of sines to express angle β and then write length λ in terms of it:

$$\beta = \arcsin\left(\frac{b\,\sin(\theta_2)}{c}\right)$$

$$\lambda = \sqrt{d^2 + c^2 + 2dc\,\cos(\beta + \theta_1)}.$$

Using the identity $\cos(a+b) = \cos a \cos b - \sin a \sin b$, you get

$$\lambda = \sqrt{d^2 + c^2 + 2dc[\cos(\beta)\cos(\theta_1) - \sin(\beta)\sin(\theta_1)]}.$$

Next, insert the expression for β and use the identity $\cos(\arcsin(a)) = \sqrt{1-a^2}$ to get

$$\lambda = \sqrt{d^2 + c^2 + 2dc\left(\sqrt{1 - \frac{b^2 \sin^2(\theta_2)}{c^2}} \cos(\theta_1) - \frac{b \sin(\theta_2)}{c} \sin(\theta_1)\right)}. \quad (7.12)$$

After you insert the expression for c from equation 7.11 into the above relation, you are left with an expression for λ that depends on lengths a, b, and d, and angles θ_1 and θ_2. To find the relationship between force in the muscle and the torques on each joint, find the Jacobian, which in this case is a 1×2 vector:

$$J = \frac{d\lambda}{d\theta} = \begin{bmatrix} \frac{d\lambda}{d\theta_1} & \frac{d\lambda}{d\theta_2} \end{bmatrix}.$$

If you insert this vector into equation 7.10, $\tau = -J^T f$, you can compute the torques on each joint for a given muscle force. The product of the transposed Jacobian, a 2×1 vector and a 1×1 force vector, is a 2×1 torque vector.

Figure 7.4 plots how a force of 10 newtons in the muscle is distributed to each joint. Note that the maximum torque is produced when both joints are flexed (large values for θ_1 and θ_2). When those joints are extended, the muscle cannot produce a significant torque on either joint. These equations allow you to relate linear muscle force to torque and to assign the torques to both joints of a two-joint arm.

The Jacobian describes how length changes in one coordinate system depend on changes in another coordinate system. If you change the coordinate system in which you represent the angles of the limb, you should expect that the Jacobian would change. Because of this, the way in which torques are described will also change. This leads to the counterintuitive conclusion that the force that your muscle produces will generate different torques depending on how you choose to measure the angle of that joint. This idea is explored in the web document *forceandtorque*.

7.6 Muscle Afferents

In addition to generating force and torque, muscles provide **feedback** about the state of the arm and its movements. Why? Consider the **actuators** that control the movement of a two-joint robotic arm. Its torque motors produce a given torque regardless of joint angle. Engineers need only to provide a command (typically a voltage), and the motor will produce a torque proportional to that command. As you have seen, however, muscles produce different amounts of force for a given input, depending on their length. Therefore, the CNS needs to know the length of the muscle in order to control forces and movements. Muscle afferents provide a relatively fast way for your CNS to acquire information about the state of the limb.

7.6.1 Types of Muscle Afferents

Muscle afferents come in two types: Golgi tendon organs and muscle-spindle afferents. Golgi tendon organs transduce force and occupy a re-

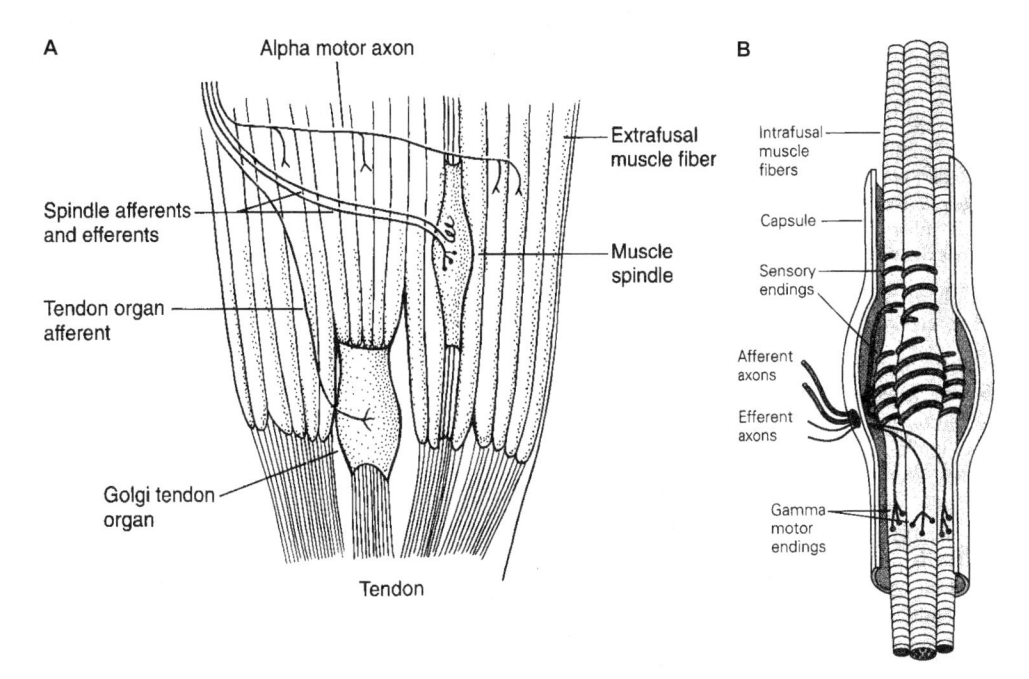

Figure 7.5
Sensory systems in muscles. (A) Golgi tendon organs lie between extrafusal muscle fibers and tendons. Thus, these force transducers lie in series with the extrafusal fibers. Muscle-spindle afferents innervate the intrafusal muscle fibers, which lie in parallel to the extrafusal fibers. (B) Muscle spindles, which are intrafusal muscle fibers, receive innervation from both afferent neurons and motor neurons. Axons from γ-motor neurons control the muscle spindle's activity. Afferent neurons, in this illustration primary muscle-spindle afferents, wrap their axons around the intrafusal muscle and respond to length changes. (A from Houk et al.,[8] B from Hulliger[9] with permission.)

gion between the muscle fibers and the tendon. Section 7.6.3 deals with these receptors, also known as group Ib afferents. Muscle-spindle afferents also come in two types: group Ia afferents, also known as *primary* muscle-spindle afferents, and group II afferents, also known as *secondary* muscle-spindle afferents (figure 7.5). (Note that *secondary* muscle-spindle afferents remain *primary* sensory afferents in that they project from the periphery to the CNS.) The "group" names come from the diameter of their axons. Group I afferents, subdivided into Ia and Ib, have larger fiber diameters (12–20 μm) than group II fibers (6–12 μm), and therefore have faster transmission rates.

7.6.2 Muscle-Spindle Afferents

The term "muscle spindle" refers to the structure of fine intrafusal muscle fibers that taper at their ends, called *poles*, and contain a fluid-filled capsule at their center. The sensory fibers wrap around the intrafusal fibers within the capsule and serve as transducers of muscle length (figure 7.5).

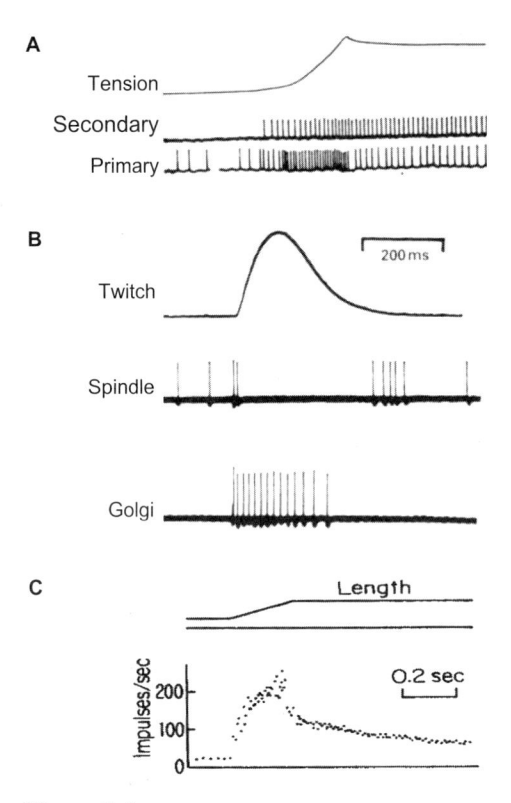

Figure 7.6
Properties of muscle-spindle afferents and the afferents of Golgi tendon organs. (A) Response of two muscle-spindle afferents to the lengthening of a muscle. With the lengthening caused by increased tension on the muscle, the activity of primary and secondary muscle-spindle afferents increases. (B) Response of a primary muscle-spindle afferent and a Golgi-tendon-organ afferent to a twitch in the muscle. (C) Activity of a primary muscle-spindle afferent during lengthening of a cat's soleus muscle. (A and B from Matthews,[10] C from Bessou et al.[11] with permission.)

In the intrafusal fiber the contractile region is limited to its poles, which receive innervation from γ-motor neurons. Primary muscle-spindle afferents innervate the central region (figure 7.5); secondary afferents innervate the poles. Because the intrafusal muscle lies in parallel to the extrafusal muscle, they change length together. However, the two kinds of muscle-spindle afferents encode this length change differently, in part because they innervate different parts of the spindle. As shown in figure 7.6A and 7.6C, primary muscle-spindle afferents respond strongly to a stretch but then decrease activity as the stretching period ends. On the other hand, secondary muscle-spindle afferents respond to the stretch with increased activity and, for the most part, maintain that increase after the stretching period ends. To a first approximation, it appears that the primary muscle-spindle afferents encode the velocity of lengthening, whereas the secondary muscle-spindle afferents encode the length.

Box 7.3
The discovery of reflexes

> Thomas Willis coined the term "reflex" in the seventeenth century to describe how "spirits" in sensory nerves could be "reflected" to muscles by the CNS. Reflexes seemed to him like light bouncing off a mirror. Wilhelm Erb and Carl Westphal first described the knee-jerk reflex in the 1870s. Erb saw the quick extension of the knee as something that involved transfer of information to the spinal cord and back to the muscles. Westphal saw it as a mere mechanical twitch of the tightened muscle. In the 1890s, Charles Sherrington demonstrated that sensory neurons mediate the reflex.

Some reflex pathways involve muscle-spindle afferents. The **monosynaptic** action of the primary muscle-spindle afferents on α-motor neurons allows the nervous system to detect errors in muscle length and respond quickly. The result is the stretch reflex, also known as the monosynaptic stretch reflex, the myotactic reflex, and the knee-jerk response. For example, tapping the patellar tendon, the tendon just beneath the knee, stretches the quadriceps muscle (see figure 3.3). That increase in length also stretches the muscle spindles in the quadriceps, which causes increased activity of the primary and secondary muscle-spindle afferents. That sensory input excites α-motor neurons, causing a knee jerk. (See box 7.3.) Chapter 8 takes up stretch reflexes in more detail.

7.6.3 Golgi Tendon Organs

Golgi tendon organs are located between extrafusal muscle fibers and tendons, at a transition area surrounded by a capsule. The muscle fibers insert into the proximal end of this capsule, and the distal end of the capsule attaches to the tendon. In the capsule, the endings of the group Ib afferents intertwine with collagen fibers. When the capsule stretches, its collagen squeezes the endings of the Ib afferents, producing action potentials and thereby encoding force.

The contrast between muscle-spindle afferents and Golgi-tendon afferents has important consequences for the signals they send to the CNS. As shown in figure 7.6B, when a muscle produces a single twitch, the spindle afferents reduce or stop their activity, whereas the Golgi-tendon afferents increase theirs. Why? The twitch causes an increase in the force of the muscle, pulling on the tendon region and exciting the group Ib afferents. The twitch also results in shortening of the muscle, including the intrafusal elements, and "unloading" of the muscle spindles. This unloading reduces the activity in the primary and secondary muscle-spindle afferents.

Figure 7.7A shows the discharge of two group Ib afferents during an experiment. As tension in the muscle increased, the discharge in the Golgi tendon organs increased, reaching a peak immediately after the tension

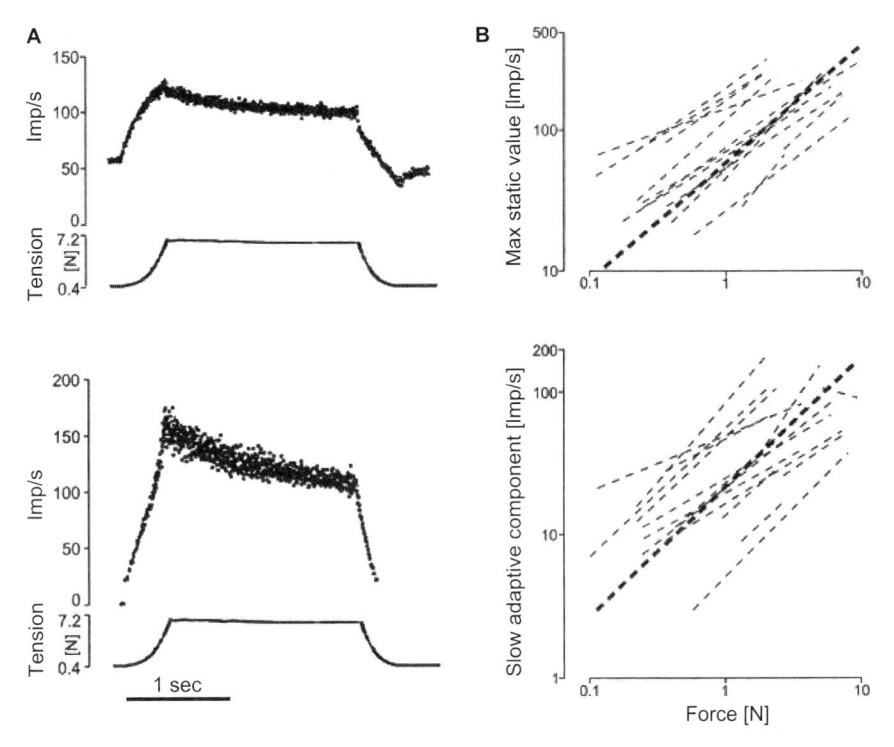

Figure 7.7
Properties of Golgi tendon organs. (*A*) Response of two Golgi-tendon-organ afferents to a stretch of the anterior tibial muscle (see figure 3.3) in an anesthetized cat. Abbreviations: imp, impulses; N, newtons. (*B*) The relationship between peak and steady-state discharge and muscle force for various Golgi tendon organs in the same muscle (linear regressions). The heavy dashed line shows the average of the individual linear regressions. (From Schafer[12] with permission.)

attained its plateau phase. After the experimenters released the tension, the discharge returned to baseline levels. Both the peak discharge and the steady-state discharge (during the plateau phase of the tension) increased linearly with the magnitude of the tension in the muscle (figure 7.7B).

Unlike muscle spindles, Golgi tendon organs receive no motor innervation. Their group Ib fibers project to the spinal cord, where they synapse on interneurons. These interneurons inhibit the α-motor neurons that innervate the same muscle.

7.7 Muscle Afferents in Action

To review the structure of muscle-spindle afferents, the contractile elements of the muscle spindle lie in the *pole* region, which receives inputs from γ-motor neurons and innervation from secondary muscle-spindle afferents. The central region—called the *nuclear bag* region—lacks contractile properties and receives innervation from primary muscle-spindle afferents. Forces that stretch the muscle spindle result in length changes in both the nuclear bag and pole regions, and the muscle-spindle afferents transduce this length change into firing rate.

Figure 7.8A represents the muscle spindle as a system with two elastic elements and one viscous element, much like the muscle model elaborated above for force generation. The length of the series elastic (SE) element represents the length of the nuclear bag region. The primary muscle-spindle afferent discharges at a rate proportional to this length. Here x signifies how much the entire intrafusal muscle extends beyond its resting length, and in the equations below, x_2 signifies how much the series elastic element extends beyond its resting length. The contractile component lies in the pole region and corresponds to the length of the parallel elastic (PE) element: x_1 signifies how much it extends beyond its resting length. The web document *musclemodel* derives the dynamics of the system, as described in the following equation:

$$\dot{T} = \frac{K_{SE}}{b}(b\dot{x} + K_{PE}x + g) - \left(\frac{K_{SE} + K_{PE}}{b}\right)T.$$

If you assume that the primary muscle-spindle afferent responds linearly to length changes in the SE element, then you can write its discharge rate as $S_{Ia} = ax_2 = a(x - x_1)$. Similarly, the discharge at time t of the secondary afferent may be written as $S_{II}(t) = ax_1$. Although the responses of these neurons do not correspond exactly to these assumptions of linearity, they serve as useful approximations. Finally, when the intrafusal muscle contracts via stimulation of the γ-motor neuron, the contractile component produces force $g(t)$.

Consider what the spindle afferents in the model should do if you suddenly stretch the muscle spindle and maintain it at a given increased length. Remember that the parallel elastic element cannot change length immediately, due to its viscous component, but the series elastic element can. Accordingly, you should see a large initial increase in the discharge of the primary muscle-spindle afferent. Gradually, as the parallel elastic element overcomes the effect of its viscosity, it will increase in length. Because the sum of the lengths of the parallel and series springlike elements equals the total length of the muscle spindle, and the spindle length remains constant after the stretch, the length of the series elastic element should gradually decrease after the stretching stops. As a result of this relaxation in the polar region of the muscle spindle, the nuclear-bag regions should gradually shorten, which should result in a reduction in the discharge of the primary muscle-spindle afferents. In fact, this predicted pattern of activity—a rapid increase followed by a gradual decrease—resembles the discharge dynamics of primary muscle-spindle afferents during a sudden stretch (figures 7.8B and 7.6A). These properties create the appearance of a velocity signal. The faster the muscle lengthens, the higher the peak of the transient discharge in the primary spindle afferents.[13]

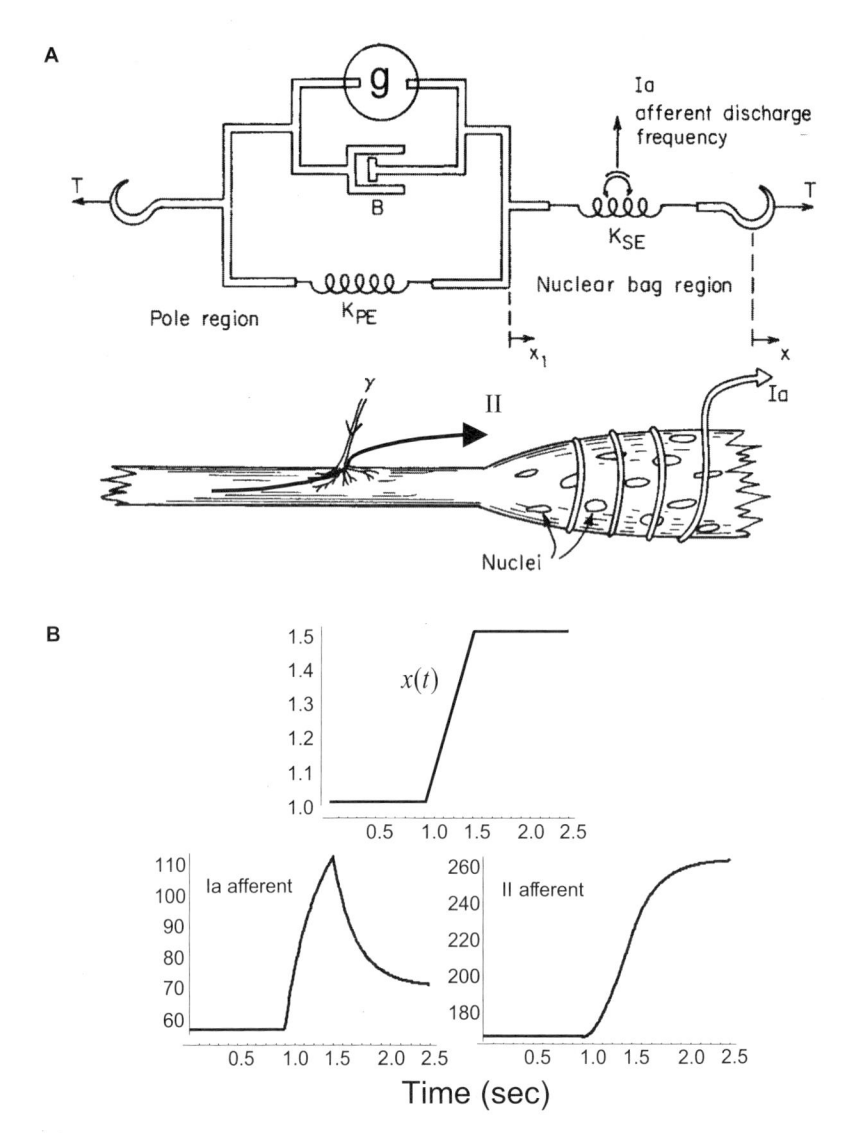

Figure 7.8

Model of a muscle spindle. (*A*) A muscle spindle (also known as an intrafusal muscle fiber) comprises a nuclear bag region where group Ia afferents (primary muscle-spindle afferents) wrap around the intrafusal fiber, and a pole region that group II afferents (secondary muscle-spindle afferents) and γ-motor neurons innervate. In this schematic drawing, the bag region is represented by the series elastic element (K_{SE}) and the pole region by the contractile element in parallel with the parallel elastic (K_{PE}) and viscous (B) elements. Format is as in figure 7.2E. Abbreviation: g, active force component, driven by γ-motor neurons. (*B*) Simulation of the dynamics of a muscle spindle. The muscle undergoes a stretch, which causes changes in the discharge of the primary and secondary muscle-spindle afferents. (*A* from McMahon[7] with permission.)

7.7. Muscle Afferents in Action

7.7.1 Role of the γ-Motor Neuron

The activity of γ-motor neurons contracts the spindle, which affects the discharge of the muscle-spindle afferents. If your muscle spindles did not have γ-motor neurons, their length changes would simply reflect the length changes in the extrafusal muscles. Activation of γ-motor neurons allows the CNS to bias the sensitivity of the primary muscle-spindle afferents and turns them into sensors that measure movement errors.

Suppose that you decide to flex your elbow, as illustrated in figure 7.8B. Your CNS activates α-motor neurons that act on the biceps, producing force in the extrafusal muscles and shortening the muscle along trajectory $x(t)$. In this case, the term "trajectory" applies to the change in muscle length as a function of time. Now suppose that your CNS programs the γ-motor neurons on the basis of its *expectation* of how the biceps should shorten during this movement. For the present purpose, you can think of this conceptually as the desired trajectory of muscle length $x_d(t)$, as a function of time. As you activate your α-motor neurons, you also activate the γ-motor neurons—a coupling termed α–γ coactivation—and if the muscle changes length according to your expectation, force in the muscle spindle will not change (i.e., $\dot{T}(t) = 0$). The level of intrafusal-fiber force consistent with this expectation is

$$g(t) = g(0) - b\dot{x}_d(t) - K_{PE}(x_d(t) - x(0)); \quad t > 0.$$

The above equation tells you that as your biceps shortens, your CNS must activate the γ-motor neurons by the amount $g(t)$ to maintain tension in the spindle. Figure 7.9 shows the way α–γ coactivation works, in principle. As a muscle shortens by 5 mm, for example, the desired trajectory programs activity in the γ-motor neuron, which produces force $g(t)$ in the muscle spindle (figure 7.9A). If this increased force precisely cancels the decrease in force due to shortening of the muscle, the series elastic element (in the nuclear bag region) will not change length and, accordingly, the primary muscle-spindle afferents will not discharge. However, because of activating the contractile element in the spindle, the parallel elastic element (in the pole region) will shorten, which causes reduced discharge in the secondary muscle-spindle afferents (figure 7.9B).

Now suppose that as the biceps shortens, an unexpected perturbation prevents it from flexing the elbow as much or as fast as desired. The γ-motor neurons have activity that corresponds to the desired trajectory, but—because of the external, perturbing force—the actual trajectory falls short. In principle, then, the primary muscle-spindle afferents should increase their activity (figure 7.10A). By modulating the activity of the γ-motor neurons, the CNS has gained a mechanism through which it can detect the error in a movement. Less or slower shortening than expected results in the activation of muscle-spindle afferents (figure 7.10A), which, through spinal circuitry, further activates the extrafusal muscles to enhance muscle shortening. Conversely, more or faster shortening than expected depresses primary muscle-spindle activity (figure

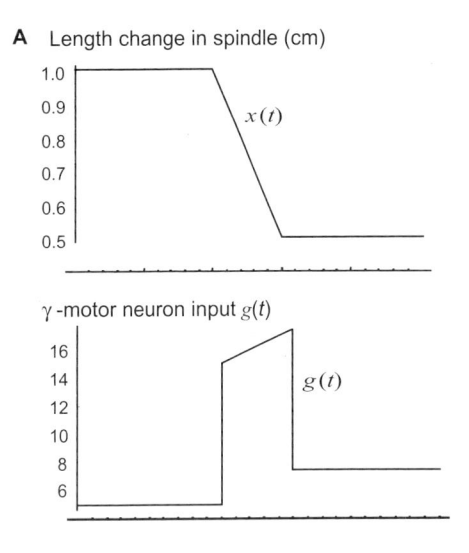

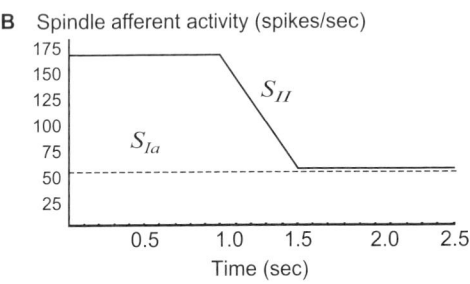

Figure 7.9
Simulation of a muscle spindle shortening by 5 mm. (A) The γ-motor neuron activity is programmed on the basis of the desired change in muscle length, with a time course of activity modulation that cancels the change in tension in the muscle spindle. (B) As the muscle spindle shortens, primary muscle-spindle afferent discharge does not change (S_{Ia}, dashed line), but secondary muscle-spindle afferents continue to respond to the decrease in muscle length (S_{II}, solid line).

7.10B), decreases extrafusal activity, and diminishes muscle shortening. In order for the primary spindle afferents to act as an error detector, the γ-motor neurons must coactivate with α-motor neurons to take the "slack" out of the shortening muscle spindles.

In contrast, activity in secondary muscle-spindle afferents does not reflect an error signal, but instead sends the CNS a fairly faithful measure of actual muscle length—with brief deviations from a pure length signal due to the viscous component of the parallel elastic element. These neurons lack monosynaptic input to the α-motor neurons, but instead act on interneurons in the spinal cord. Section 8.8 takes up the issue of **feedback control** and reflexes initiated by signals from muscle afferents in more detail.

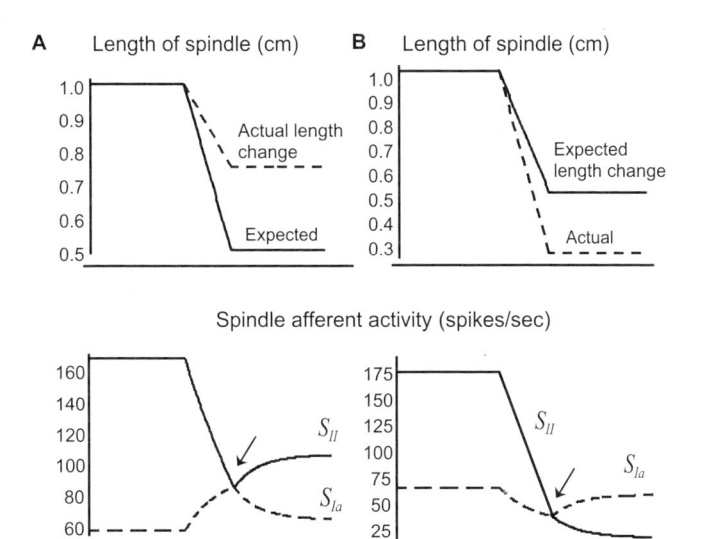

Figure 7.10

Simulation of a muscle spindle as it shortens by less or more than a desired amount. (*A*) A simulated perturbation causes less muscle shortening than desired. (*B*) A perturbation causes more muscle shortening than desired. Because of the mismatch between the tension produced by the γ-motor neurons and the change in tension due to the shortening of the muscle, primary muscle-spindle afferents (S_{Ia}, dashed line) increase their activity in *A* and decrease their activity in *B*. This activity thus reflects an error signal, which indicates that the muscle did not shorten as much as expected (increased firing, as in *A*) or that the muscle shortened more than expected (decreased firing, as in *B*). Secondary muscle-spindle afferents (S_{II}, solid line) behave differently. Their activity more faithfully reflects muscle length, although the viscous component of the parallel elastic element in the pole of the muscle spindle leads to transient deviations from a pure length signal (arrows).

References

1. Ghez, C (1991) In Principles of Neural Science, eds Kandel, ER, Schwartz, JH, and Jessell, TM (Appleton & Lange, East Norwalk, CT), pp 548–563.
2. Bloom, W, and Fawcett, DW (1975) A Textbook of Histology (WB Saunders, Philadelphia).
3. Loeb, GE, and Gans, C (1986) Electromyography for Experimentalists (University of Chicago Press, Chicago).
4. Hildebrand, M (1995) Analysis of Vertebrate Structure (Wiley, New York).
5. Yasui, K, Tabata, S, Ueki, T, et al. (1998) Early development of the peripheral nervous system in a lancelet species. J Comp Neurol 393, 415–425.
6. Hill, AV (1970) First and Last Experiments in Muscle Mechanics (Cambridge University Press, Cambridge).

7. McMahon, TA (1984) Muscles, Reflexes, and Locomotion (Princeton University Press, Princeton, NJ).
8. Houk, JC, Crago, PE, and Rymer, WZ (1980) In Spinal and Supraspinal Mechanisms of Voluntary Motor Control and Locomotion, ed Desmedt, JE (Karger, Basel), pp 33–43.
9. Hulliger, M (1984) The mammalian muscle spindle and its central control. Rev Physiol Biochem Pharmacol 101, 1–110.
10. Matthews, PBC (1972) Mammalian Muscle Receptors and their Central Actions. (Arnold, London).
11. Bessou, P, Emonet-Denand, F, and Laporte, Y (1965) Motor fibres innervating extrafusal and intrafusal muscle fibres in the cat. J Physiol 180, 649–672.
12. Schafer, SS, Berkelmann, B, and Schuppan, K (1999) The contribution of muscle afferents to kinaesthesia shown by vibration induced illusions of movement and by the effects of paralysing joint afferents. Brain Res 846, 210–218.
13. Goodwin, GM, McCloskey, DI, and Matthews, PBC (1972) Two groups of Golgi tendon organs in cat tibial anterior muscle identified from the discharge frequency recorded under a ramp-and-hold stretch. Brain 95, 705–748.

Reading List

Prime Mover: A Natural History of Muscle, by Steven Vogel (W.W. Norton, New York, 2001) presents a compendium of interesting and useful information about muscles.

8 What Maintains Limb Stability

*Overview: Pairs of muscles act against each other. This **antagonist** architecture produces an equilibrium point—a balance of forces—that helps stabilize the limb. The passive, springlike properties of your limb promote its stability, but your CNS also uses reflexes to stabilize the limb. These mechanisms maintain your hand at a reach target or in a given direction of pointing.*

When activated, your muscles produce forces that tendons transmit to your skeleton, resulting in a torque on the joint or joints that the muscle crosses (see section 7.5). For a robotics engineer, a single torque motor can produce both positive and negative torques. Muscles, however, do not work like that: they can only pull. Often, the torque produced by flexor muscles counters (and cancels) the torque generated by extensor muscles. When both kinds of muscles generate force at the same time, the net torque acting on the joint may amount to nothing. This approach to motor control might seem wasteful and inefficient, but this apparent extravagance provides important benefits for control of the limb.

Because a muscle can only pull, moving a simple joint such as that shown in figure 8.1 requires a pair of muscles. Recall that a Jacobian matrix converts force into a torque (equation 7.10). The Jacobian for muscle 1 has the same form as in equation 7.5:

$$\frac{d\lambda_1}{d\theta} = \frac{ab\sin(\theta)}{\sqrt{a^2 + b^2 - 2ab\cos(\theta)}}.$$

This quantity represents the **moment arm** of muscle 1 (length c in figure 7.3A). It always has a positive value in the convention adopted here. From equation 7.10, $\tau = -J^T f$, you can see that because the Jacobian for muscle 1 is always positive, the force produced by muscle 1, f_1, results in a negative torque τ_1. On the other hand, the Jacobian for muscle 2 is always negative:

$$\frac{d\lambda_2}{d\theta} = \frac{-ab\sin(\theta)}{\sqrt{a^2 + b^2 + 2ab\cos(\theta)}}.$$

Therefore, muscle 2's force, f_2, produces a positive torque, τ_2. Their forces sum to produce a net torque acting around the joint illustrated in figure 8.1A:

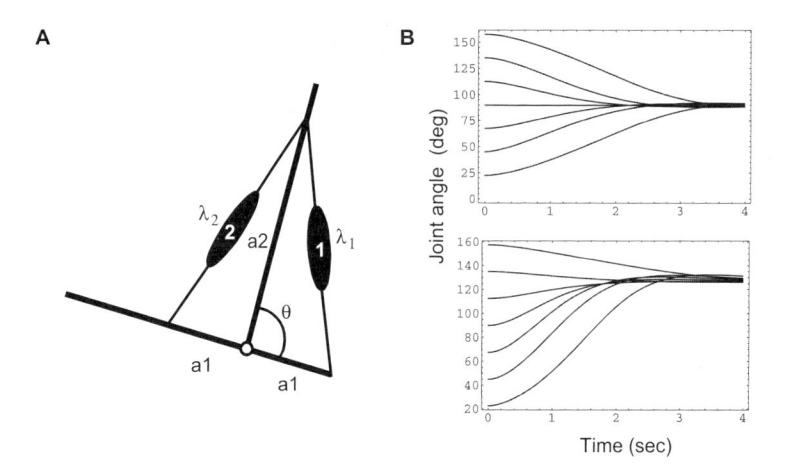

Figure 8.1
A pair of muscles acting on a single joint. (*A*) Antagonist architecture in a two-joint limb. (*B*) Top: A simulation of the joint-angle changes for stimulation of both muscles at 15 Hz. Each curve represents the joint-angle trajectory for a given initial angle. Note that the limb adopts a configuration with the joint at a 90° angle regardless of the initial configuration of the limb. This angle corresponds to the equilibrium point of the system. Bottom: Same as top except that stimulation of muscle 2 is 20 Hz. The limb adopts a more flexed final state.

$\tau = \tau_1 + \tau_2.$

Because the torques produced by antagonistic muscles oppose each other, when one of the two muscles produces more force, it pulls the limb into a new configuration, known as an *equilibrium point*, and stretches the antagonist muscles, in part due to the springlike properties of muscles (see section 7.2.2).

8.1 Equilibrium Points from Antagonist Muscle Activity

To appreciate how your limb behaves when you activate your arm muscles, consider a simple thought experiment. Recall that a moment arm converts force to torque. These torques sum to produce a net torque on the joint, which causes a motion that depends on the limb's moment of **inertia** I and the joint's viscosity v. The angular acceleration $\ddot{\theta}$ that the torque will cause can be written as

$$\ddot{\theta} = \frac{1}{I}(\tau - v\dot{\theta}).$$

Now imagine that you stimulate each of your muscles at 15 Hz, and observe how your limb moves and where it ends up, as if the stimulation came from an external agent (figure 8.1B, top). You begin with a configuration of the limb θ at $t = 0$, which corresponds to a given joint angle. Assume that your hand is stationary in this configuration (i.e., $\dot{\theta} = 0$). Based on the stimulation going to your muscles, you can calculate the torque τ acting on the joint at $t = 0$ and calculate $\ddot{\theta}$. To know where your limb will go, you need to know θ and $\dot{\theta}$ at some slightly later time, $t + \varepsilon$. If you can calculate those values, you can solve the problem because you could repeat this calculation for a series of time steps. The famous physicist Richard Feynman devised a clever approach to this calculation. He noted that if you know the velocity now and also know that it is changing, you can compute the configuration for a future time step based on information about velocity halfway through it:

$$\dot{\theta}(t + \varepsilon/2) = \dot{\theta}(t - \varepsilon/2) + \varepsilon\ddot{\theta}(t)$$
$$\theta(t + \varepsilon) = \theta(t) + \varepsilon\dot{\theta}(t + \varepsilon/2).$$

To start the calculation, you need to know $\dot{\theta}(\varepsilon/2)$. To calculate this value, use $\dot{\theta}(\varepsilon/2) = \dot{\theta}(0) + (\varepsilon/2)\ddot{\theta}(0)$. Feynman's approach results in a surprisingly accurate numerical solution. Figure 8.1B (top) shows the result of 15-Hz stimulation of the two muscles illustrated in figure 8.1A. Regardless of its initial configuration, the joint angle rotates to 90°. When $\theta < 90°$, the force in muscle 2 exceeds that in muscle 1 because it has greater length. As the joint angle approaches 90°, the forces in the two equally stimulated muscles balance, and zero net torque results, stopping the movement. When you stimulate muscle 1 at 15 Hz, but increase the activation of muscle 2 to 20 Hz (figure 8.1B, bottom), the joint now settles at a larger angle (~130°). In this condition, with the joint angle at 90°, the greater activity in muscle 2 results in a positive net torque on the joint. As the joint angle increases, the active force in muscle 2 declines because of its shortening. Meanwhile, the active force in muscle 1 increases because of its lengthening. At 130°, the effect of the longer length of muscle 1 matches the effect of larger stimulation on muscle 2. The system has found an equilibrium point.

You might be surprised to observe that a relatively small change in the activation of muscle 2 (from 15 to 20 Hz) resulted in a joint-angle change of 40°. The moment arms of the muscles account for this large effect. Recall that each muscle's Jacobian depends on the joint angle. For constant inputs, as in the present example, the torque produced grows slower than linearly as the joint angle increases. However, for muscle 2, an increase in joint angle produces an increase in the moment arm. Therefore, as the system finds its equilibrium at 130°, the torques imposed by the two muscles balance because the longer length of muscle 1 produces a force large enough to compensate for the larger stimulation rates and larger moment arm of muscle 2.

8.2 Restoring Torques from Length–Tension Properties

The passive properties of muscles not only play a crucial role in moving the limb, but they also contribute to its stability. Section 7.2.3 discussed

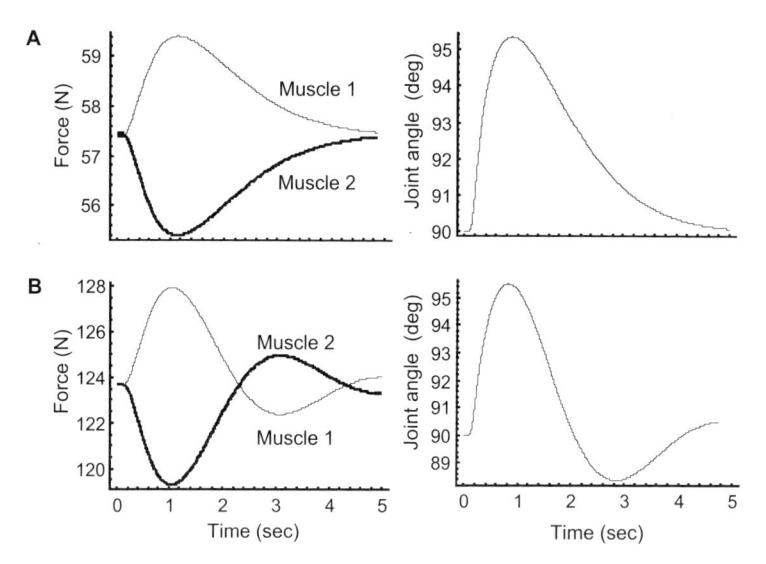

Figure 8.2
Effect of perturbations on a simulated limb. (*A*) A 100 msec, 1-N-m torque perturbs the limb depicted in figure 8.1A, with the activity of both muscles set at 10 Hz. The plots show the force produced in each muscle (left) and the resulting angle of the elbow (right). (*B*) Same as A, but with a 30-Hz stimulation rate for both muscles.

the molecular basis of this contribution to stability. But how does the limb as a whole respond to a perturbation? To understand this feature of limb stability, imagine that you impose a perturbation on the system illustrated in figure 8.1A.

Begin with each muscle receiving a 10-Hz stimulation and the system in equilibrium at $\theta = 90°$. Then you impose a 100 msec torque disturbance of 1 N-m on the limb (see appendix D for an explanation of N-m). For this positive (flexor) torque, the joint angle increases, in this case by about 5° (figure 8.2A, right). However, as the joint rotates, muscle 1 lengthens and muscle 2 shortens. These length changes produce an increased force in muscle 1 and a decreased force in muscle 2 (figure 8.2A, left). Accordingly, the muscles resist the perturbation and gradually restore the limb to its equilibrium point at 90° (figure 8.2A, right).

Now consider how the system responds to the same perturbation with a higher stimulation level for both muscles. Figure 8.2B plots the response for muscles stimulated at 30 Hz. With this heightened activity, the perturbation displaces the limb by about the same amount (~5°), but the joint angle returns much faster, overshooting its equilibrium point and eventually settling at 90° (figure 8.2B, right). The higher stimulation rate made the muscles respond more vigorously to the perturbation, but at a cost: the limb overshoots its initial configuration and settles only after a period of oscillation.

Figure 8.3

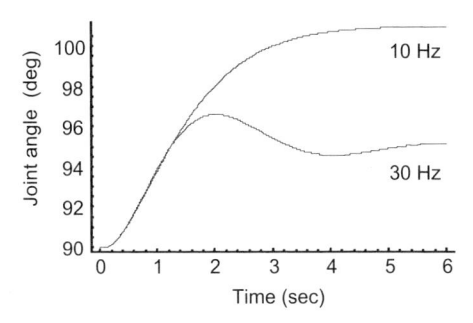

Effect of coactivation. In this simulation, the limbs receive a constant 0.1 N-m torque, with stimulation rates to the muscles of either 10 Hz or 30 Hz.

8.3 Stiffness from Muscle Coactivation

The increased activation of the muscles did not change the system's equilibrium point, but it did change how fast the limb responded to the perturbation. Increasing the amount of muscle coactivation tended to make the limb return to its equilibrium point faster because of an increase in limb stiffness. To understand this property more fully, imagine that you disturb the system with a *step* (i.e., constant) torque of 0.1 N-m, rather than the *pulse* perturbation described in the previous section. Then compare two activity levels: 10 Hz and 30 Hz, as before. Figure 8.3 shows the limb's response to the perturbation. At the lower level of coactivation, the imposed torque of 0.1 N-m displaces the limb from 90° to approximately 100°. At the high level of coactivation, the same imposed torque displaces the limb by only 5° or so. The system's stiffness has increased. According to equation 7.1, at 10 Hz the limb had a stiffness of ~0.01 N-m/deg; at 30 Hz, its stiffness doubled to ~0.02 N-m/deg.

8.4 Reaching Without Feedback in Monkeys

The presentation to this point deals in theory. To examine whether these theories reflect the properties of the biological system, Andreas Polit and Emilio Bizzi[1] examined pointing movements in **deafferented** monkeys. They cut the dorsal roots of the spinal cord at the levels C_2–T_3 to eliminate afferent input from all the muscles of the arms and the hands, on both sides of the body. The absence of stretch reflexes confirmed the completeness of the lesion. The monkey sat in a chair with its right arm in a splint. The experimenters connected the splint to a **manipulandum**, which allowed rotation of the forearm around the elbow in the horizontal plane (figure 8.4). An opaque plastic sheet covered the monkey's arm. Thus, given the deafferentation, the monkey could neither see nor feel its arm. Before the surgery, Polit and Bizzi had **instrumentally** conditioned the monkey to make movements in response to visual cues arrayed in a

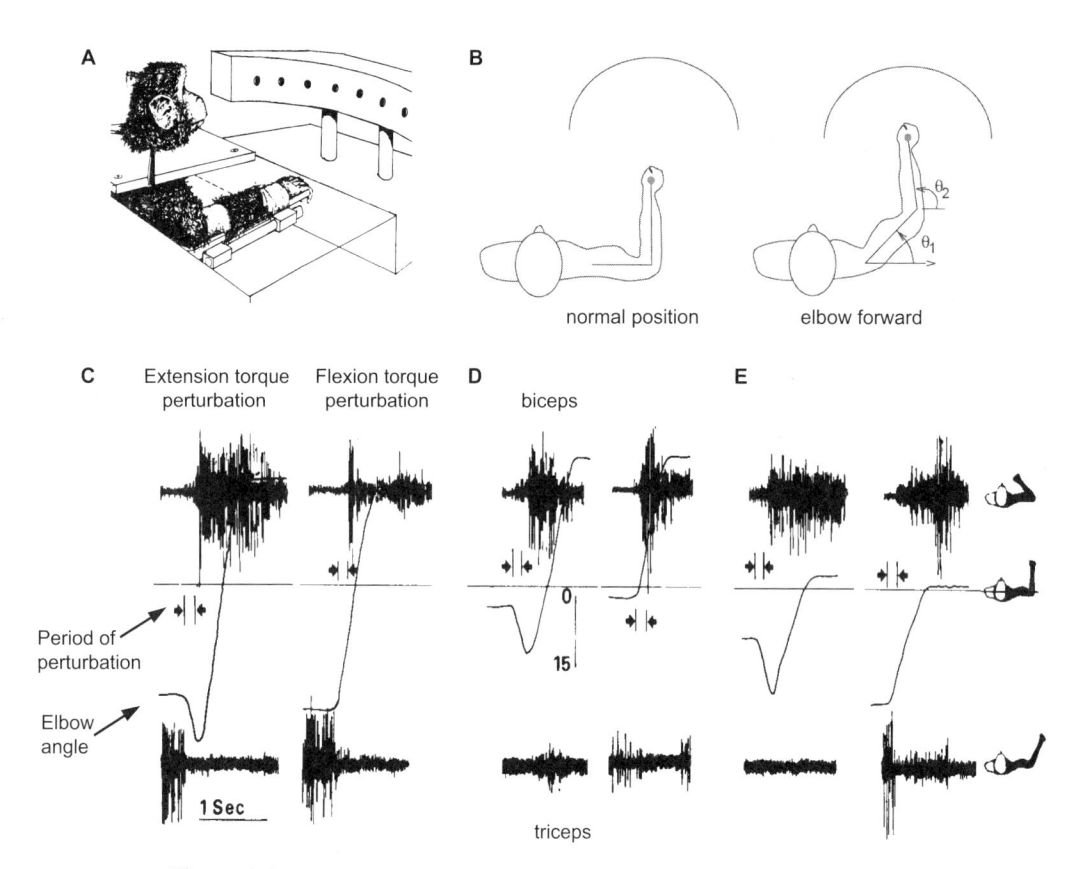

Figure 8.4

Perturbations to the elbow joint during movements in a deafferented monkey. (A) After Polit and Bizzi trained a monkey to make elbow movements to align the forearm with the visual cues, they cut the sensory nerves to both forelimbs. The monkey could not see its arm. (B) Overhead view of a right arm, showing joint angles for the elbow (θ_2) and shoulder (θ_1). In the experiment, the monkey's shoulder was set at an angle of 0° throughout training. (C–E) Activity from biceps and triceps. On randomly selected trials, a motor perturbed the monkey's arm movements, always near the beginning of movement. The angle of the elbow joint θ_2 is plotted for each movement. Scale bar in (D) indicates a 15° elbow rotation. In all of the movements shown, the monkey flexed the elbow in order to point toward a visual cue. In (C), the monkey flexed the elbow from an extended limb configuration. In the left subplot, the torque motor imposed an extension torque at the start of movement, displacing the elbow from the target. In the right subplot, the torque motor imposed a flexion torque, accelerating the elbow toward the target. In both cases, the elbow reached the target accurately despite the perturbed trajectories. D and E show the effects of perturbations as in C, but for movements from different initial arm configurations and for different visual-cue locations. (A and C–E from Polit and Bizzi[1] with permission.)

horizontal pattern. When a light cue appeared, the monkey rotated its arm so that its hand pointed toward the cue. The monkey became highly skilled in this task.

Two days after the surgery, the monkey again performed the task. Polit and Bizzi hypothesized that as it did so, the monkey's CNS activated its arm muscles to change the arm's equilibrium point. According to this idea, after a cue appeared at a given location, the equilibrium point changed to one for pointing at that target. Polit and Bizzi predicted that a transient perturbation of the limb should affect the trajectory of the movement but not the final direction of pointing. According to the equilibrium-point hypothesis, the passive properties of the muscles would get the limb into the proper configuration in the face of the perturbation.

To test their prediction, Polit and Bizzi perturbed the monkey's forearm on randomly selected trials. A motor imposed torques around the elbow through a connection with the manipulandum. Imagine that the monkey needed to flex its right forelimb in order to point to a visual cue. On some trials, the motor imposed a torque that pushed the forearm away from the target (figure 8.4C, extension torque), and on other trials, it pushed the forearm toward the target (figure 8.4C, flexion torque). Recall that the monkey could neither feel these perturbations nor see their effects. Despite that insentience, Polit and Bizzi observed that the monkey's forearm consistently pointed toward the target. Because the monkey lacked proprioceptive reflexes, restoring torques must have provided the response to the perturbation. The passive properties of antagonistically arranged muscles achieved an equilibrium point that accomplished the monkey's goal: pointing toward the visual cue. This finding suggested that the passive properties of muscles play a significant role in stabilizing a limb during a movement.

As predicted, the monkey could not perform the task when the experimenters replaced the transient torque pulse with a constant, step torque. In the presence of that additional torque, the sum of externally produced and muscle-produced torques led to inaccurate pointing.

Interestingly, the monkey also failed when the experimenters changed the initial configuration of the limb. In the monkey's original conditioning, the angle of the shoulder joint never varied from 0° (directly to the right for the right shoulder). A different angle will affect the elbow angle needed for the forearm to point toward the visual cue. For example, as illustrated in figure 8.4B, when θ_1 was 0°, pointing to the center target required a θ_2 of 90°. If θ_1 equaled 25°, pointing at the same target required an elbow angle significantly smaller than 90°, ~65°. This fact led to another prediction: Rotation of the upper arm around the shoulder joint should lead to inaccurate pointing. Polit and Bizzi confirmed this prediction in deafferented monkeys. In contrast, normal monkeys had no trouble adapting to this change in upper-arm configuration.

Polit and Bizzi concluded that the monkey learned a muscle-activation pattern for each visual-cue location, and that this motor command produced an equilibrium point for the muscles acting around the elbow. Because muscles act like springs, the forelimb pointed toward the target despite *transient* perturbations during its motion, but did not do so for *step* (constant) perturbations. When the orientation of the upper arm differed from the one expected from long and consistent experience, the same equilibrium point for the elbow no longer aligned the hand with the target.

The experiment of Polit and Bizzi on deafferented monkeys demonstrates that the antagonistic architecture of muscles, and their passive properties, can result in significant restoring forces in response to a perturbation. Keep in mind, however, that other mechanisms aid in stabilizing movements in intact animals. As discussed below, these mechanisms include reflexes. Passive, mechanical properties of muscles provide the first line of defense against perturbation, responding with almost no delay; reflexes rely on neural transfer of information, which takes longer.

8.5 Equilibrium Points from Artificial Stimulation

In the experiment of Polit and Bizzi, the animal activated its arm muscles to produce an equilibrium point for its arm. Recent experiments suggest that equilibrium points can be produced by directly stimulating the CNS.

8.5.1 Spinal Cord Stimulation

Recall that the **propriospinal** system consists of connections within the spinal cord, usually connecting segment to segment (section 5.1). The axons of the propriospinal system arise from neurons with cell bodies in the spinal gray matter. In a series of experiments, Simon Giszter, Sandro Mussa-Ivaldi, and Emilio Bizzi found that artificially stimulating the spinal gray matter of frogs produced an equilibrium point for the frog's leg.[2] In these experiments, an electrode produced currents that excited neurons and axons, some of which synapse on motor neurons. These influences occur both directly and indirectly, and through both excitatory and inhibitory pathways. As a result, each **motor pool** eventually attains a certain level of excitation, which in turn leads to a certain level of activity in the muscles of the frog's leg. As you have seen, activity levels in antagonistically arranged muscles lead to an equilibrium point. The final configuration of the limb does not depend very much on the initial one; a given amount of muscle activity will, in the absence of any imposed loads, always lead to a given configuration of joint angles.

The number of distinct limb configurations observed in such an experiment was not as large as you might guess. By slowly moving the stimulating electrode in the spinal gray matter, you might observe a gradual change in the configuration of the leg. However, Bizzi and his colleagues[2] found that only a handful of distinct postures resulted from this kind of stimulation. They invoked the concept of **motor primitives** to explain this finding. According to this idea, there may be only a small number of equilibrium points that are genetically coded into neural networks in the frog spinal cord. Each of these networks activates the leg muscles in a particular way and produces an equilibrium point. When two networks are simultaneously activated, they sum to an equilibrium point between the two points produced independently by each network.[3] Their hypothesis predicts that the CNS takes advantage of these spinal neural networks to move the equilibrium point of the limb through weighted coactivation of these networks. However, many experts object to this method of studying the motor system and any of the conclusions that follow from it. Repetitive stimulation at a site in the CNS will necessarily activate diverse circuits, with the net outcome that the limb reaches a configuration determined by the dominant neural elements and the mechanics of the muscles and limb segments. More elaborate conclusions that follow from these observations, such as the concept of motor primitives, remain controversial.

8.5.2 Motor Cortex Stimulation

Michael Graziano obtained a similar result by electrically stimulating the motor cortex of monkeys.[4–6] The stimulation results in prolonged excitation of corticospinal pathways (see section 6.3.3), eventually exciting the motor pools innervating the muscles of the forearm. As in the spinal-stimulation studies of Bizzi and his colleagues, the establishment of a given state of activity in the motor pools leads to the production of an equilibrium point for the arm.

According to Graziano, the way primates use their forelimbs results in the encoding of motor primitives in the motor cortex. A primate's arms and hands adopt many different postures, and by electrically stimulating the motor cortex for 500 msec—the approximate duration of many reaching movements—Graziano and his colleagues found that they evoked complex, coordinated movements. Surprisingly, these movements appeared much like those performed by monkeys in their daily activities. For example, stimulation of one region of cortex caused the shoulder and elbow joint angles to rotate so that the hand reached toward the mouth (and the mouth opened at about the same time). Stimulation of parts of motor cortex evoked other complex postures of the hand and arm, which appeared to combine to form a rough map of hand locations in the space around the monkey's body. If this idea has merit, then the CNS could recruit various combinations of such learned motor primitives to make novel movements. As noted above, this idea remains highly controversial.

8.6 Rapid Movements from Sequential Muscle Activation

A given stimulation pattern sent to the muscles produces an equilibrium point for the system. It seems reasonable to hypothesize, then, that in order to move the limb from one configuration to another, the CNS simply changes the activation patterns for your muscles in a steplike fashion. For example, equal stimulation of the muscles depicted in figure 8.1A produces an endpoint elbow angle of 90° (figure 8.1B, top). When the activation of muscle 2 increases from 15 Hz to 20 Hz, the limb's equilibrium point shifts to 130° (figure 8.1B, bottom). However, the change in limb configuration occurs very slowly: It took more than 2 sec to move ~40°. Clearly, you can move your limbs much faster. How?

One possible answer lies in the *pattern* of stimulation that your CNS sends to your limb muscles. Consider again the two-muscle system illustrated in figure 8.1A. You could move the limb simply by increasing the stimulation rate to one of the muscles. For example, if you wish to increase the angle of the joint, you could increase the stimulation rate for muscle 2, as in figure 8.1B (bottom). Muscle 2 is the agonist in this movement, also known as the prime mover. This enhanced level of stimulation results in a new equilibrium point for the system, and the limb slowly moves into that configuration. You could move the limb in other ways, however. For example, you could begin the movement by greatly increasing the stimulation rate to the agonist muscle while reducing the stimulation to the antagonist muscle. Because muscles produce force only slowly, the increased stimulation enhances the rate of force generation, accelerating the limb more quickly. After this period of acceleration, you could activate the antagonist muscle to act as a "brake" on the system, while keeping the agonist muscle minimally stimulated. Finally, you might reactivate the agonist muscle to the level appropriate for maintaining the final equilibrium point.

Figure 8.5 shows how this might work. In figure 8.5A, stimulation for muscle 2 increases from 10 Hz to 15 Hz while stimulation for muscle 1 remains at 10 Hz. The limb slowly moves toward the equilibrium point of the system at 146°. Figure 8.5B shows a "three-burst" stimulation pattern. The input to the agonist muscle (muscle 2) increases strongly while that to the antagonist decreases. After this event, at about the time when the speed of movement reaches its height, the antagonist receives strong stimulation while the agonist gets less. Finally, input to the agonist increases again, briefly, then assumes a level appropriate for maintaining the desired equilibrium point. This three-burst, sequential pattern of activity reduces the time for completion of this movement from more than 4 sec to 1 sec.

The CNS uses this three-burst pattern of activity for making rapid movements. Figure 8.6A shows an example of a rapid wrist flexion. The commands to the muscles begin with a burst of activity to the agonist, then a burst for the antagonist, and finally a small burst for the agonist. Accordingly, the joint angle slightly overshoots the target, and then returns to its equilibrium point. Brain damage sometimes disturbs this simple pattern, as occurs in a condition called *essential tremor*, thought to be associated with damage to the cerebellum. Figure 8.6B illustrates essential tremor for a rapid wrist flexion. In both a healthy person and an essential-tremor patient, this rapid movement produces a slight overshoot. In both people, the wrist returns toward the target—a *corrective movement*—mainly because of the activity in the wrist extensor (i.e., because of the antagonist burst). However, in the healthy person the movement terminates accurately because a second agonist burst partially

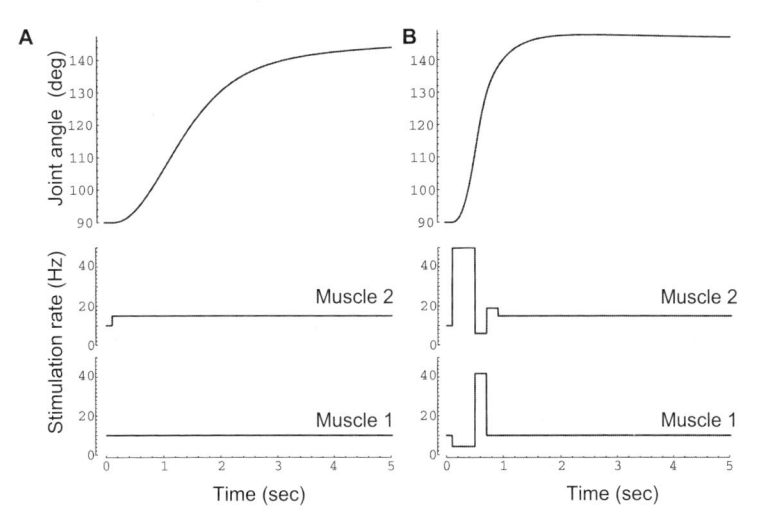

Figure 8.5
Simulation of a three-burst, agonist–antagonist activation pattern, using the two-muscle system of figure 8.1A, with two different patterns of activation. (*A*) Muscles 1 and 2 begin at 10 Hz, but at 100 msec muscle 2 receives 15 Hz of input and maintains this level (bottom) as the joint angle slowly changes (top). (*B*) As in A, but at 100 msec the muscles receive a three-burst activation pattern. In this pattern, the agonist muscle (muscle 2) receives strong stimulation, followed by stimulation of the antagonist muscle, and finally stimulation of the agonist muscle (bottom). The limb rotates rapidly (top).

overlaps the antagonist burst. In the patient, however, the second agonist burst comes too late; therefore, the action of the antagonist goes unopposed and the corrective movement overshoots the target. This takes the wrist past the target, a repeating pattern of overshooting movements that results in oscillation around the target.

8.7 Passive Properties Produce Stability

The experiment of Polit and Bizzi, described in section 8.4, suggests that when the CNS plans a movement, it does so in terms of transitions between equilibrium points for the limb: one for the initial limb configuration, another for a configuration that achieves some goal. This basic idea is often called the *equilibrium-point hypothesis*. If you think of the three-burst pattern of muscle activity in terms of transitions in the equilibrium point of the system, then the equilibrium point initially moves rapidly beyond the intended target of the movement and then returns to the target. Thus only the slowest movements can be thought of as gradual transitions of equilibrium points of the limb.

Now imagine holding a textbook as you intend to make a rapid movement of your forearm. The textbook increases the inertia of your limb, and if you compare a rapid elbow movement with the book versus

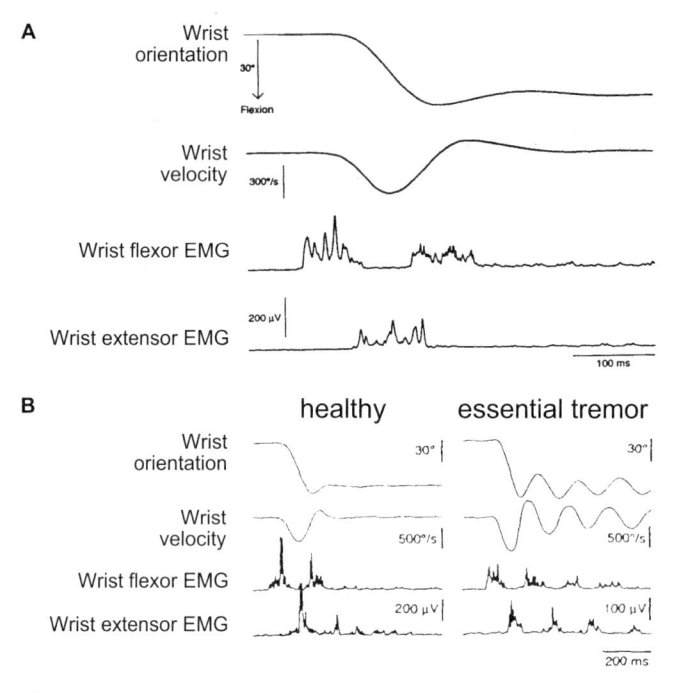

Figure 8.6
Three-burst, agonist–antagonist EMG activity. (*A*) The EMG activity that accompanies a rapid wrist movement in a healthy person, with the flexor as the agonist. (*B*) The three-burst, agonist–antagonist activation pattern in a healthy person (left) and a patient with essential tremor (right). (From Britton et al.[7] with permission.)

one without it, you can appreciate that the same movement requires larger torques on the elbow joint when you are holding the book. For example, to achieve the same initial speed, the agonist muscle needs to generate more force with the book in your hand. Rapid movements in which the inertial mass of the system changes from one movement to the next depend on much more than simply changing the equilibrium point of the limb.

Thus, although the activation of muscles does indeed produce an equilibrium point for the limb, making a movement involves much more than that. A particularly clear demonstration of this idea was provided by Jim Lackner and Paul Dizio.[8] They showed that external forces affect not only limb trajectory but also the final configuration of the arm. Coriolis forces occur in spinning bodies (see box 20.1). If a body rotates, that movement generates a force proportional to the velocity of rotation. However, when the velocity decreases to 0, no residual force remains. In this sense, a Coriolis force resembles the pulse force described above and, according to the equilibrium-point hypothesis, the limb should reach the same final configuration regardless of whether Coriolis forces impinge on the limb during a reaching movement. Lackner and Dizio

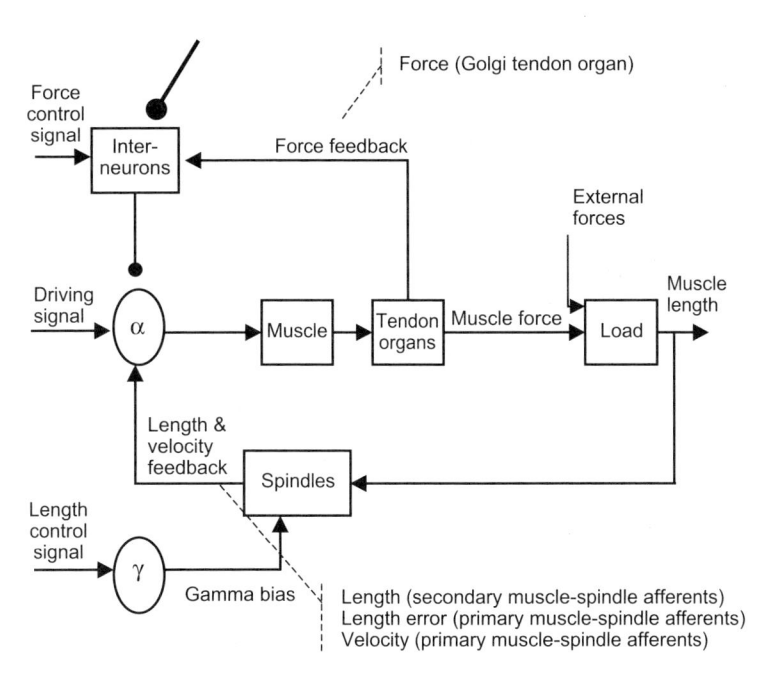

Figure 8.7
A model of the spinal controller. Schematic of the muscle-spindle and Golgi tendon organ feedback system. Symbols ending in arrowheads indicate excitatory connections; symbols ending in circles indicate inhibitory ones. The inhibitory input at the top left represents descending projections from the brain. Abbreviations: α, α-motor neurons; γ, γ-motor neurons.

showed, however, that Coriolis forces significantly affect the limb's final configuration.

Accordingly, instead of looking at the equilibrium point as the key to controlling movements, as originally thought, you might view it more as a stabilizing mechanism, one reflecting the springlike properties of the muscles and the antagonist architecture of the limb.

The springlike properties of muscles provide an important part of the mechanism by which the limb can respond to a perturbation, but cannot account for all such responses. Reflexes also enhance the limb's ability to respond to perturbations.

8.8 Reflexes Produce Stability

Passive, springlike properties of muscles provide one line of defense against perturbations, but the errors and instability provoked by external perturbations can be countered only by a combination of active and passive muscle properties. Reflexes that depend on primary muscle-spindle afferents appear to play the largest role in such corrections during rapid movements; those which depend on secondary muscle-spindle afferents do so during the maintenance of steady posture.

Figure 8.7 schematically summarizes the functions of the muscle-spindle (see section 7.6.2) and Golgi tendon organ (see section 7.6.3) afferents in terms of a feedback control system. Inputs to α-motor neurons cause muscles to produce forces that act on tendons, shorten muscles, and change joint angles. Muscle-spindle afferents sense length and changes in muscle length, as biased by activity in the γ-motor neurons. The Golgi tendon organs sense force changes, and their signal to inhibitory interneurons is modulated by the descending inputs from the CNS.

The feedback control system depicted in figure 8.7 rejects perturbations and helps ensure that the limb follows the desired trajectory. Consider a situation in which the load exceeds expectations. For example, imagine that you see a milk carton you assume to be empty, so you decide to dispose of it. When you begin to lift the carton after grasping it, its weight—the load on your muscles—exceeds your prediction. The prime-mover muscles will shorten, but because of the increased load, they will not shorten as much as expected for an empty milk carton.

You saw in section 7.7.1 that γ-motor neuron activity reflects the expected length change in the muscle. When the muscle shortens less than expected, primary muscle-spindle afferents increase their activity as if stretched (figure 7.10A). These afferents project to the spinal cord to further activate the α-motor neuron, producing more force and countering the effect of the load, at least to some extent. The force feedback system, mediated by the Golgi tendon organ, works similarly. As depicted in figure 8.7, inputs signaling force from the Golgi tendon organs inhibit activity on α-motor neurons via a spinal interneuron. This influence would have the effect of a negative feedback loop in which high levels of force decrease the excitatory drive on motor neurons, therefore moderating the force. By inhibiting the interneuron that the Golgi tendon organ acts upon (unfilled symbol at the top of figure 8.7), descending inputs can modulate how effectively force feedback inhibits the α-motor neurons. Strong inhibition from the descending inputs imposes a high threshold for the force feedback pathway. Only if the force exceeds this threshold will the feedback pathway inhibit the α-motor neuron.

Not all feedback pathways rely entirely on spinal mechanisms. Those that do, often called **short-loop reflexes**, serve as the fastest reflex pathway by which the CNS can respond to a perturbation. In primates—particularly for the muscles of the arm, wrist, and fingers—a second pathway can result in a compensatory activity in the motor neurons. A **long-loop reflex** pathway begins with the primary muscle-spindle afferents and travels through the dorsal column–medial lemniscal pathway to the thalamus, which projects to cortical neurons that contribute to the corticospinal tract. You can voluntarily alter your response to a stimulus that stretches your muscles; if you expect the stretch, you can choose either to ignore it or to respond more vigorously than some default value. Modulation of long-loop reflexes probably accounts for this ability.

These afferent pathways also provide a mechanism by which the CNS can monitor the progress of a movement and respond in case of errors. However, it takes time for these signals to reach the CNS, and this

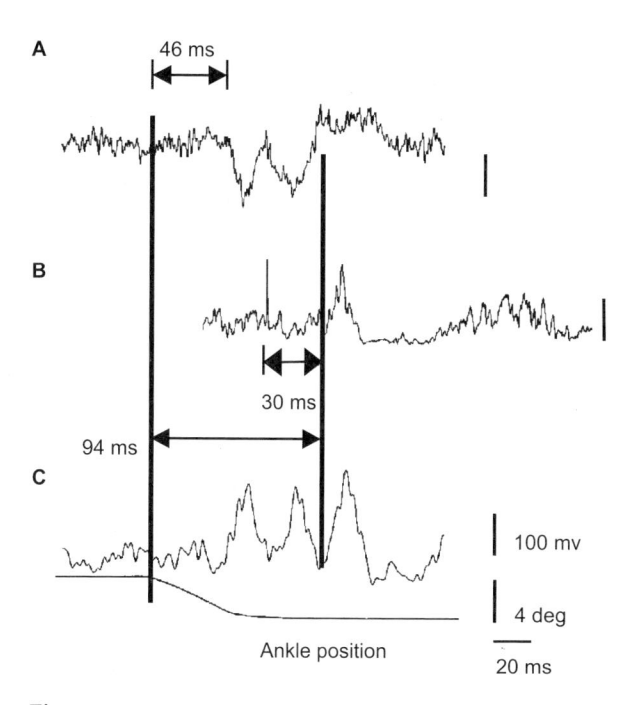

Figure 8.8
Time delays in neural transmission. The experimenters suddenly extended the ankle. (A) The plot shows somatosensory evoked potentials at a latency of 46 msec from the onset of ankle extension. (B) EMG recordings showed a "motor evoked potential" at 30 msec latency from cortical stimulation. (C) EMG reflex responses. (From Petersen et al.[11] with permission.)

delay represents one of the most fundamental problems of controlling reaching and pointing.

8.8.1 Time Delays in Reflex Pathways

David Marsden and his colleagues[9,10] examined some of the neural pathways involved in long-latency reflexes. Damage to the brainstem that prevented proprioceptive signals from reaching the thalamus—or damage to the somatosensory or motor cortex—blocked long-latency reflexes (50–100 msec) but not short-latency ones (<50 msec). Other experimenters[11] have examined these latencies by stretching ankle muscles. The study of the leg provides some advantages because of the longer distances involved in both central and peripheral pathways. One group of investigators recorded evoked potentials from the somatosensory cortex (figure 8.8A). Ankle extension produced a response in the somatosensory cortex at a latency of ~45 msec. To estimate the delay in the descending pathways, the investigators stimulated the foot representation of the motor cortex via **transcranial magnetic stimulation**. Stimulation produced a movement of the foot and potentials recorded from the muscle. Figure

8.8B shows that it took 30 msec for the signal to travel from the cortex to the ankle muscle. Figure 8.8C plots the muscle activity from the same muscle after ankle extension. The muscle stretch caused by that passive extension causes a complex pattern of activations at ∼45 msec, ∼70 msec, and ∼95 msec. The latencies of the second and third peaks appear to be consistent with a transcortical, long-loop reflex.

Are there any neural signals in the cerebral cortex that could account for these findings? Cells in the somatosensory cortex, specifically in area 3a, respond much like muscle-spindle afferents.[12] Cells in M1 respond similarly, although fewer have the strong bias toward the velocity of muscle stretch seen in area 3a.[12] Because many cells in M1 project to spinal motor neurons and respond to muscle stretch, they must contribute to long-loop reflexes in some way.

8.8.2 Contribution of Cerebellum and Basal Ganglia

Peter Strick[13] instrumentally conditioned monkeys to use a cue either to oppose or to assist a torque pulse to the elbow. He found that muscle activity in the two conditions did not differ until about 70 msec after the onset of the perturbation. Neurons in one deep cerebellar nucleus, the dentate nucleus, showed activity at 30–50 msec after the perturbation that depended on the instruction to the animal. (A different deep cerebellar nucleus, the interpositus, did not.) The timing and behavior of activity in the dentate nucleus points to a role for the cerebellum in the generation of a relatively late component of the response to a perturbation.

Evidence also points to a role for the basal ganglia in the modulation of long-loop reflexes. In Huntington's disease, which initially affects the basal ganglia (see section 19.5), the short-latency responses to perturbations of the hand and arm remain normal, but long-latency responses decrease in magnitude or disappear.[14,15] Cortical responses to peripheral nerve stimulation, as measured by evoked potentials from the somatosensory cortex, also decrease early in the course of the disease.[16,17]

In Parkinson's disease, which also affects the basal ganglia, reflexes exceed normal levels. In one experiment, extension of the wrist evoked reflex activity; figure 8.9 shows responses for the wrist flexor in two conditions. In the first condition, the participants had received an instruction to "actively" oppose the perturbation. In the other condition, they were told to "passively" allow the perturbation to displace their hand.[18] Healthy people had a much greater EMG response in the active than in the passive condition. They could "turn off" their reflexes. In the active condition, they responded at ∼50 msec and again at ∼100 msec. In parkinsonian patients, however, the response did not depend on the person's intent. Their EMG response occurred at ∼50 msec in both conditions.

Taken together, these findings suggest that basal ganglia dysfunction affects the ability of the CNS to modulate its response to perturbations. Huntington's patients appear to have deficient long-loop reflexes; Parkinson's patients have hyperactive reflexes that fail to habituate. Sec-

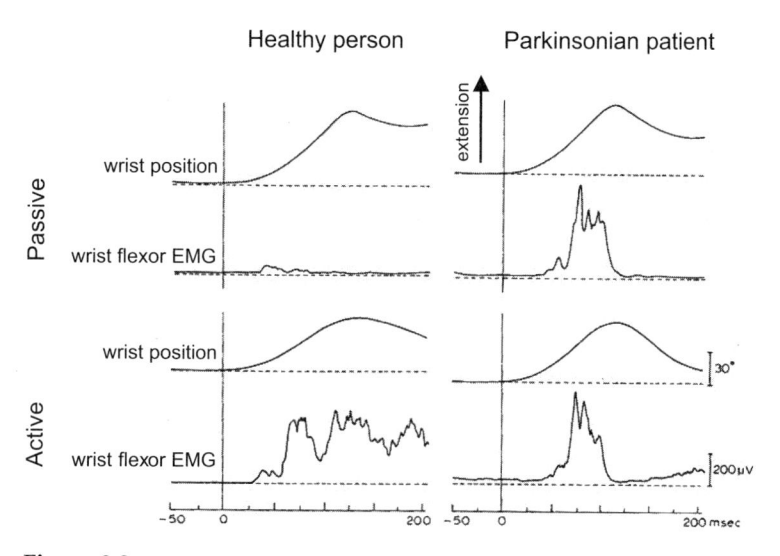

Figure 8.9
Reflex response in Parkinson's disease. Response to sudden extension of the wrist in a normal participant and a parkinsonian patient in two conditions. In one condition, the experimenters instructed the participants to remain "passive." In the second condition, they instructed the participants to be "active" in resisting the perturbation. (From Lee and Tatton[18] with permission.)

tion 19.5 presents the idea that inputs to basal ganglia play a crucial role in generating the next step toward a goal.

8.9 Reaching Without Feedback in Humans

The circuits that respond to perturbations rely on the ability of the peripheral nervous system to sense changes in the state of the limb. How would the system behave if the CNS could not rely on these inputs and reflexes? In a rare condition called *large-fiber sensory neuropathy*, certain primary afferent fibers degenerate. The ones most affected are primary muscle-spindle afferents and those from Golgi tendon organs. These patients cannot sense the location or motion of their hands without vision.

John Rothwell and his colleagues[19] examined a patient who suffered from sensory neuropathy. This patient could produce a wide range of movements, including rapid movements in which a three-burst pattern of EMG activity is typical. Thus, like the monkey illustrated in figure 8.4, deafferented people can make accurate forelimb movements. Although they can make relatively normal movements, their ability to maintain joint angles for more than a brief period is badly impaired. Without visual feedback, healthy people can maintain thumb position without difficulty. In marked contrast, deafferented patients move their thumbs in an apparently random manner. Stable joint configurations require sensory feedback.

8.10 Passive Properties and Reflexes Combined

To this point, the consideration of limb stability has dealt separately with active, proprioceptive reflexes and passive muscle properties. This section explores how those passive properties combine with reflexes to influence how the limb as a whole responds to a perturbation. Shadmehr, Sandro Mussa-Ivaldi and Emilio Bizzi[20] performed an experiment in which they asked participants to hold the handle of a robotic arm. Figure 8.10A illustrates the experimental setup. The experimenters made measurements at two different arm configurations, labeled "left" and "right" in the figure. They instructed the participants to close their eyes and "try not to intervene" as the robot displaced their hand from its resting location. The robot's handle recorded the forces that the muscles produced as the robot pulled the participant's arm from its resting location. The further the robot pulled, the larger the force the muscles produced. Eventually, the robot stopped pulling, and the hand returned to its resting location.

Figure 8.10B plots the forces for the left and right configurations. You can see that the pattern of forces differs for these two conditions. For the left resting location, pulling the hand along 135° produced the largest restoring forces. For the right, pulling toward 45° did so. Note also that the force vectors rarely point exactly toward the center (the origin). For example, with the hand to the left, when the robot pulled the hand toward 270° (straight down in the figure), the restoring forces pushed upward but also to the left.

To represent the limb's stiffness, and how it varied with respect to resting location, Sandro Mussa-Ivaldi, Neville Hogan, and Emilio Bizzi[21] suggested that the restoring forces measured at the hand approximated a linear function of hand displacement. If you represent force at the hand and the location of the hand as vectors,

$$\mathbf{f} \equiv \begin{bmatrix} f_x \\ f_y \end{bmatrix} \qquad \mathbf{x} \equiv \begin{bmatrix} x \\ y \end{bmatrix},$$

then the stiffness at the hand equals the change in force divided by the change in location:

$$K_x \equiv \frac{d\mathbf{f}}{d\mathbf{x}} = \begin{bmatrix} \frac{df_x}{dx} & \frac{df_x}{dy} \\ \frac{df_y}{dx} & \frac{df_y}{dy} \end{bmatrix} = \begin{bmatrix} k_{11} & k_{12} \\ k_{21} & k_{22} \end{bmatrix}.$$

K_x represents stiffness in Cartesian coordinates based on a given resting location of the hand. The web document *stiffness* describes how you can estimate the parameters of this matrix from measurements of force and hand displacement.

To represent their result, Mussa-Ivaldi and his colleagues took the K_x value that they had estimated for displacements of the hand and multiplied it by $d\mathbf{x}$, a vector that had a unit length and a direction that changed gradually from 0° to 360°. When they multiplied the "circle" described by

8.10. Passive Properties and Reflexes Combined

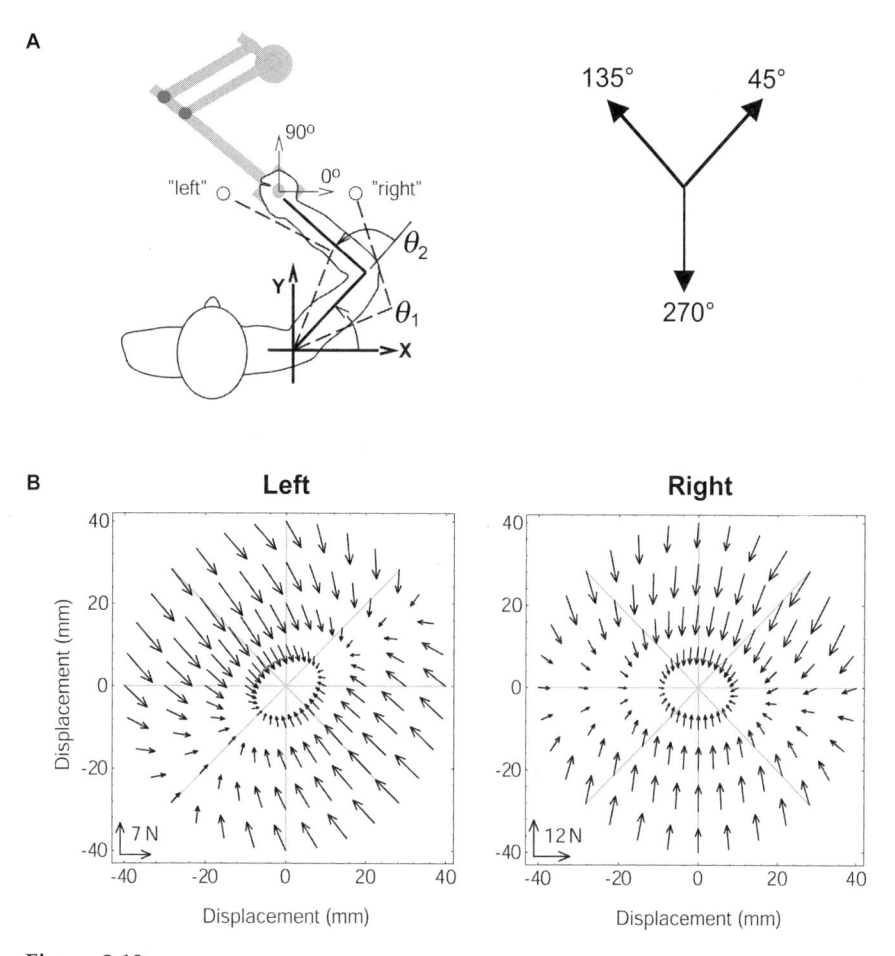

Figure 8.10
Measurement of arm stiffness at two different limb configurations. (*A*) Experimenters asked participants to hold the handle of a robotic arm. They then made measurements at two configurations of the arm, termed the left and the right. (*B*) Forces measured at the hand as a function of hand displacement. The left subfigure shows data for the "left" arm configuration; the right subfigure for the "right" configuration. (From Shadmehr et al.[20] with permission.)

displacements $d\mathbf{x}$ by the stiffness matrix K_x, the result was an ellipse. Figure 8.11A shows how this representation of hand stiffness changed with the hand's resting location, for five different locations. The long axis of each ellipse shows the direction of hand displacement associated with the largest restoring forces; the short axis shows the direction of displacement associated with the least restoring forces. The arm's greatest stiffness occurs along a line that connects the hand to the shoulder joint. As the shoulder's angle changes, so does the direction of maximum stiffness.

Mussa-Ivaldi and his colleagues observed that although the stiffness ellipses varied greatly by resting location, they varied little from person

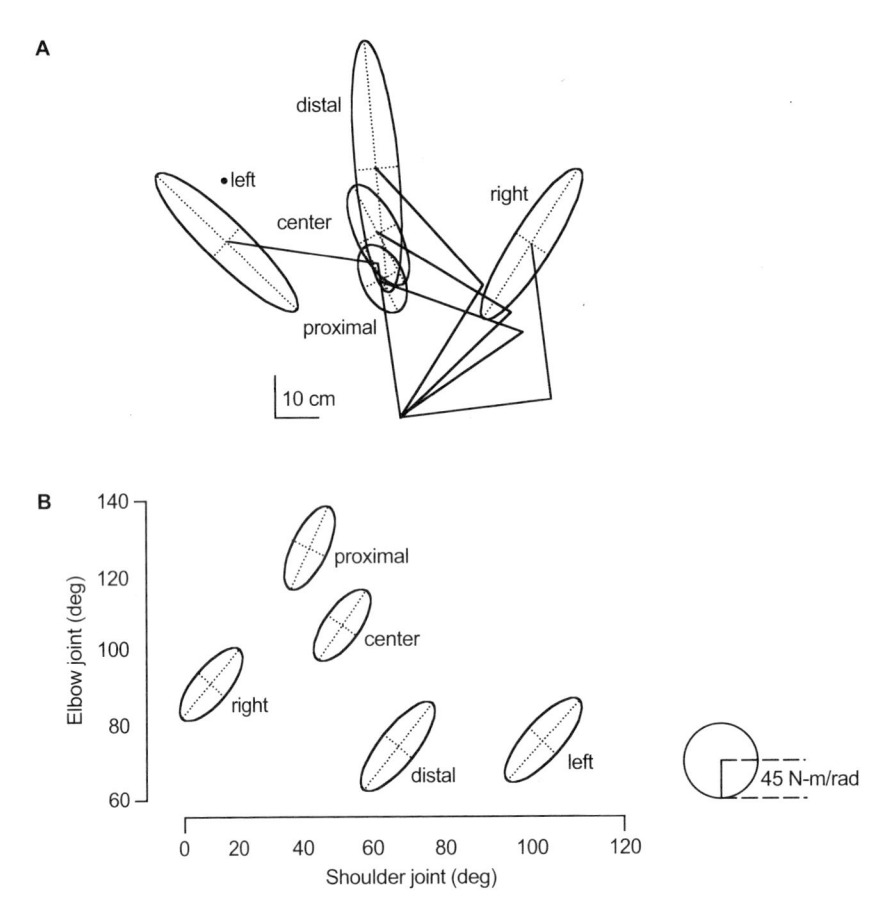

Figure 8.11

Representation of stiffness as a function of arm configuration. (*A*) An ellipse at each arm configuration plots the stiffness of the arm as measured at the hand. The major axis of each ellipse specifies the direction of hand displacement for which the limb has the greatest stiffness (at that configuration). The minor axis specifies the direction of displacement for which the limb has the least stiffness. (*B*) Joint stiffness at various configurations of the arm. Abbreviations: deg, degree; N-m/rad, newton-meters per radian (A from Mussa-Ivaldi et al.[21] with permission.)

to person.[21] To gain some insight into this finding, they transformed stiffness in terms of hand displacements into stiffness in terms of joint-angle changes, called *joint stiffness*. And, after some mathematical derivation (see the web document *stiffness*), when they plotted the joint-stiffness matrix as an ellipse, they observed that its major axis always had an angle of approximately 45° in joint-space coordinates (figure 8.11B). This means that the muscles produced their maximum restoring torque when forces rotated both the elbow and the shoulder in the same direction by approximately the same amount, a property that probably results from muscles that span both joints. These results suggest that the stiffness of the arm remains roughly constant when expressed in terms of **joint coordinates**. The seemingly complex changes seen at the level of the arm (figure 8.11A) result from something rather simple at the level of the muscles (figure 8.11B).

References

1. Polit, A, and Bizzi, E (1979) Characteristics of motor programs underlying arm movements in monkeys. J Neurophysiol 42, 183–194.
2. Giszter, SF, Mussa-Ivaldi, FA, and Bizzi, E (1993) Convergent force fields organized in the frog's spinal cord. J Neurosci 13, 467–491.
3. Mussa-Ivaldi, FA, and Bizzi, E (2000) Motor learning through the combination of primitives. Phil Trans Roy Soc London B 355, 1755–1769.
4. Graziano, MS, Taylor, CS, Moore, T, and Cooke, DF (2002) The cortical control of movement revisited. Neuron 36, 349–362.
5. Graziano, MS, Patel, KT and Taylor, CS (2004) Mapping from motor cortex to biceps and triceps altered by elbow angle, J Neurophysiol. 10.1152/jn.01241.2003.
6. Graziano, MS, Taylor, CS, and Moore, T (2002) Complex movements evoked by microstimulation of precentral cortex. Neuron 34, 841–851.
7. Britton, TC, Thompson, PD, Day, BL, et al. (1994) Rapid wrist movements in patients with essential tremor: The critical role of the second agonist burst. Brain 117, 39–47.
8. Lackner, JR, and Dizio, P (1994) Rapid adaptation to Coriolis force perturbations of arm trajectory. J Neurophysiol 72, 299–313.
9. Marsden, CD, Merton, PA, Morton, HB, et al. (1978) Automatic and volunary responses to muscle stretch in man. Prog Clin Neurophysiol 4, 167–177.
10. Marsden, CD, Merton, PA, Morton, HB, and Adam, J (1978). In Cerebral Motor Control in Man: Long Loop Mechanisms, ed Desmedt, JE (Karger, Basel) pp 334–341.
11. Petersen, N, Christensen, LOD, Morita, H, et al. (1998) Evidence that a transcortical pathway contributes to stretch reflexes in the tibialis anterior muscle in man. J Physiol (London) 512, 267–276.
12. Wise, SP, and Tanji, J (1981) Neuronal responses in sensorimotor cortex to ramp displacements and maintained positions imposed on hindlimb of the unanesthetized monkey. J Neurophysiol 45, 482–500.
13. Strick, PL (1978). In Cerebral Motor Control in Man: Long Loop Mechanisms, ed Desmedt, JE (Karger, Basel) pp 85–93.
14. Noth, J, Podoll, K, and Friedemann, HH (1985) Long-loop reflexes in small hand muscles studied in normal subjects and in patients with Huntington's disease. Brain 108, 65–80.
15. Thilmann, AF, Schwarz, M, Topper, R, et al. (1991) Different mechanisms underlie the long-latency stretch reflex response of active human muscle at different joints. J Physiol (London) 444, 631–643.
16. Meyer, BU, Noth, J, Lange, HW, et al. (1992) Motor responses evoked by magnetic brain stimulation in Huntington's disease. EEG Clin Neurophysiol 85, 197–208.
17. Topper, R, Schwarz, M, Podoll, K, et al. (1993) Absence of frontal somatosensory evoked potentials in Huntington's disease. Brain 116, 87–101.

18. Lee, RG, and Tatton, WG (1975) Motor responses to sudden limb displacements in primates with specific CNS lesions and in human patients with motor system disorders. Can J Neurol Sci 2, 285–293.
19. Rothwell, JC, Traub, MM, Day, BL, et al. (1982) Manual motor performance in a deafferented man. Brain 105, 515–542.
20. Shadmehr, R, Mussa-Ivaldi, FA, and Bizzi, E (1993) Postural force fields of the human arm and their role in generating multijoint movements. J Neurosci 13, 45–62.
21. Mussa-Ivaldi, FA, Hogan, N, and Bizzi, E (1985) Neural, mechanical and geometric factors subserving arm posture in humans. J Neurosci 5, 2732–2743.

Reading List

Readers interested in putting the discussion of short- and long-loop reflexes in the context of motor cortex structure and function are encouraged to consult *Corticospinal Function and Voluntary Movement*, by Robert Porter and Roger Lemon (Clarendon Press, Oxford, 1993).

II Computing Locations and Displacements

9   Computing End-Effector Location I: Theory

*Overview: Collectively, your hand and other things controlled by it are **end effectors**. According to the model presented in this book, in order to control a reaching movement, the CNS computes the difference between the location of a target and the current location of the end effector. This chapter considers the problem of computing end-effector location from sensors that measure muscle lengths or joint angles, a computation called **forward kinematics**.*

9.1   Reaching and Pointing Require Sensory Feedback

As discussed in chapter 8, simulation of **antagonist** muscles produces an equilibrium point. You can think of the equilibrium point as analogous to the bottom of an oddly shaped bowl. If you flick a steel ball at the bottom of the bowl—just a little, so that it does not fly out—the ball will eventually return to its initial location, or near it, if you allow for some "noise." By analogy, if something pushes the limb away from its current equilibrium point, passive forces tend to return it. At first glance, this model appears to be an excellent way to control reaching movements. If you want to reach for a coffee cup, you can just compute the joint angles needed to position your hand at the cup, then determine the muscle activations that will produce that equilibrium point for your limb. According to this idea, once you send those motor commands to the arm muscles, your hand will (eventually) get to the cup and remain there.

Although this approach to controlling reaching movements should work if implemented in a robot, an equilibrium-point control method is not very useful for controlling movements in biological systems. Reaching movements controlled in this way would be too slow (see section 8.6). It is not enough for your hand to reach a target "eventually"; sometimes you need to reach the target quickly. Further, convincing evidence shows that the CNS does not use the equilibrium-point control method: Normal reaching movements require sensory feedback, in contrast to the predictions of that model. People who cannot sense the configuration of their arm or the location of their hand with either proprioception or vision cannot make accurate movements (section 8.9). Similarly, **deafferented** monkeys cannot make accurate movements unless they can see their limbs (section 8.4). The CNS needs to estimate hand location at the beginning and throughout the movement to move it swiftly and accurately to a target, and this computation requires sensory feedback.

9.2 Kinematics and Dynamics

The problem of computing hand location from proprioceptors—transducers that measure joint angles or muscle lengths—is called forward kinematics. The term "forward" here reflects the causal nature of the relationship: You change joint angles to move your hand. Therefore, if you use joint angles to compute hand location, you solve the problem of forward kinematics. If you wanted to calculate your joint angles, given that your hand is at a particular location, you would perform the reverse computation, called **inverse kinematics**.

Kinematics contrasts with **dynamics**. If you need to consider forces, then the problem would belong to a class of computations called dynamics. The CNS must deal with dynamics because, ultimately, muscles produce force and those forces move the limb. It will be a long journey, however, from this point until the topic of dynamics reappears in chapter 20.

9.3 Degrees of Freedom and Coordinate Frames

To understand forward kinematics, you need to understand a concept called **degrees of freedom**. Chapter 3 introduced this topic. A two-joint arm has five degrees of freedom: three at the shoulder plus two at the elbow. Like the hip, the shoulder is a ball-and-socket joint. A structure like that provides three degrees of freedom. When your arm is hanging by your side (figure 9.1), you can rotate your shoulder inward or outward, abduct or adduct it, and flex or extend it. (Inward and outward rotations of the upper arm are also called **pronation** and **supination**, respectively.) Your elbow has a hinge-like structure, which has only two degrees of freedom: You can flex or extend your lower arm around the elbow, and you can also pronate or supinate it. These movements have different names in various coordinate systems, but your shoulder always has three degrees of freedom and your elbow, two.

A degree of freedom is one dimension along which a limb segment can move. For each degree of freedom, an angle describes the state of each joint along that dimension. The muscles that attach to your forearm and upper arm change length as these joint rotations take place. Collectively, those muscle lengths uniquely identify the angles of each joint (figure 7.4). To simplify the presentation, ignore the fact that your arm's transducers mainly measure muscle length, and assume that they instead "sense" joint angles. This simplification is not strictly correct, but is close enough to reality for the present purposes.

Forward kinematics answers the following question: Given a set of joint angles, where is your hand? To describe your hand's location, you

9.3. Degrees of Freedom and Coordinate Frames

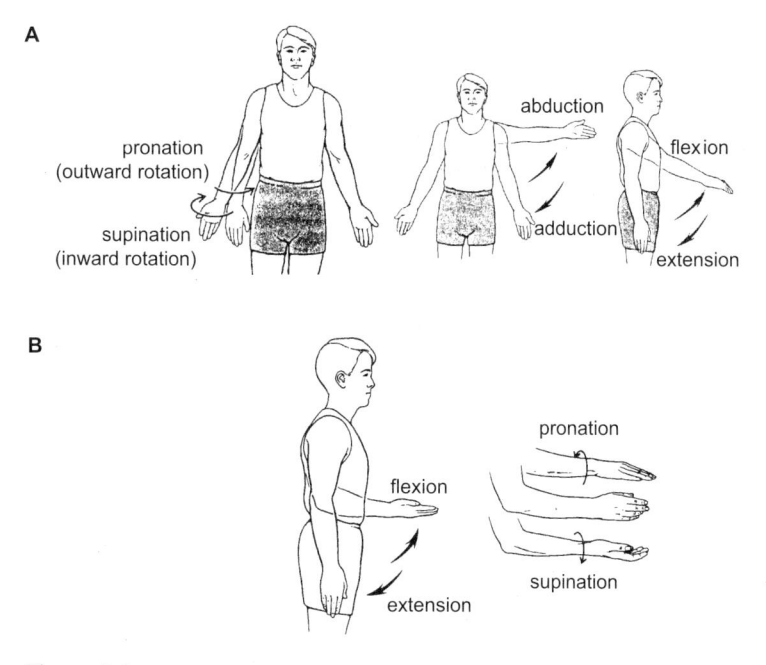

Figure 9.1
Degrees of freedom of the shoulder and elbow. (*A*) At the shoulder joint, three degrees of freedom are available to move the upper arm. (*B*) At the elbow, two degrees of freedom are available to move the forearm. Changes in limb configuration are often called rotations or joint rotations because they can be described as rotations of the limb around some axis. (From Abernethy et al.[1] with permission.)

might use a three-dimensional, Cartesian coordinate system centered on your shoulder. A vector specifies hand location in this coordinate system: It points from your shoulder to your hand, as illustrated by the dashed arrow in figure 9.2B. This coordinate system has three degrees of freedom: x, y, and z. Alternatively, you may describe hand location as a vector with respect to your head or with respect to a fixation point. How would you update these vectors as your hand moves? Because your two-joint arm has five degrees of freedom in all, you have five angles that collectively describe the configuration of your arm. When your arm has a particular posture, each of the five degrees of freedom has a specific angle.

These angles, therefore, describe a point in a five-dimensional space. You can imagine a vector that points from the origin of a five-dimensional coordinate system to this point. This vector describes arm configuration in **joint coordinates**, and every point in joint coordinates describes a potential posture for your arm. That is, if you specify the angle of each joint along each of the dimensions in which it can vary, you have the information needed to compute your hand's location in another coordinate system, for example, in Cartesian coordinates with respect to your shoulder. The computations that allow you to do this solve the problem of forward kinematics.

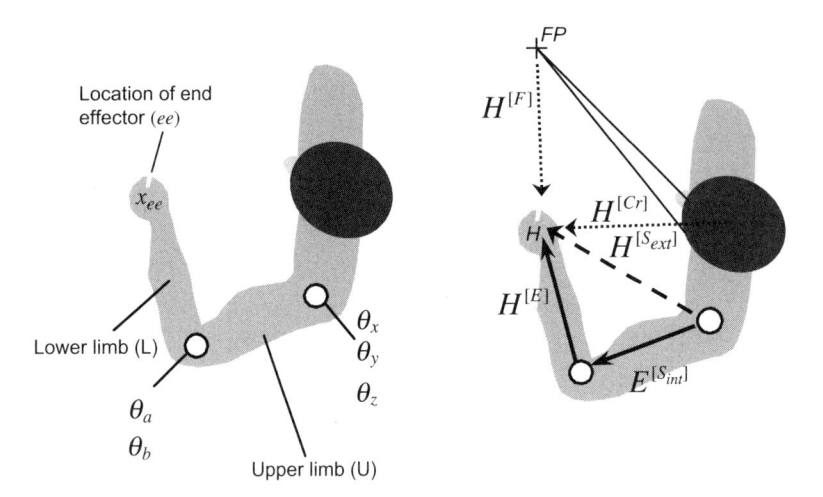

Figure 9.2
Schematic of a five-degrees-of-freedom, two-joint arm. (A) A view of a person's arm, from above. The angles θ_x, θ_y, and θ_z describe the orientation of the upper arm in the shoulder, and the angles θ_a and θ_b describe the orientation of the lower arm at the elbow. (B) The forward-kinematics computation results in an estimate of hand location (H, also more generally known as the location x of an end effector, x_{ee}). In this convention, vectors have their tails at a joint and the heads at the end of a limb segment. The + sign indicates the fixation point (FP). The lines to the fixation point depict the orientation of the fovea, also known as gaze or gaze angle. The hand's location (H) may be defined by several vectors, including one in fixation-centered coordinates [F] and another in craniocentric [Cr] coordinates. It may also be defined by a vector in an extrinsic (ext) frame with its origin at the shoulder, [S_{ext}], which has coordinates that do not move when the arm moves. The elbow's location (E) can be described by a vector in an intrinsic (int) frame with the same origin, [S_{int}], which has an axis that moves as the upper arm moves. The hand also has a coordinate in elbow-centered space [E].

9.4 End Effectors and Adaptive Mapping

The concept of forward kinematics applies much more broadly than simply computing the location of your hand. Consider reaching for a coffee cup (see figure 17.1). The CNS controls reaching for the cup with your hand, of course, but it can also control reaching with other things. For example, you might want to put a stick through the handle of a cup. The same control mechanism that moves your hand to the cup will guide the end of the stick. These mechanisms further allow you to aim a laser pointer attached to your elbow at the cup and to put a mouse-controlled cursor on a video image of a coffee cup. Collectively, you can call these controlled things—hands, sticks, laser pointers, and video cursors—end effectors. The general problem of forward kinematics involves the computation of end-effector location from sensors that measure the limb's joint angles and muscle lengths. The next sections discuss hands as a proxy for end effectors more generally.

As noted above, the process of computing motor commands depends crucially on information provided by sensory systems, particularly those of vision and proprioception. Engineers who design robots face a similar problem; they need to compute the location of the end effector of the robot's arm from transducers that measure joint angles or cameras that generate images of the end effector. However, unlike robots, your limbs grow as you develop, and growth changes the relationship between joint angles and hand location. In addition, you can control tools, skills that further change the relationship between joint angles and end-effector location. For these reasons, your computation of hand location must adapt: Your CNS must *learn* to map joint angles to an estimate of current end-effector location and to change that mapping as your limb develops and controls different objects.

9.5 Predicting the Location of an End Effector in Visual Coordinates

You can describe your hand's location in a variety of coordinate frames. For example, you might describe its location in five-dimensional joint-angle space, as mentioned above. You also might describe it in terms of a Cartesian coordinate system fixed on your shoulder. According to the model presented in this book, however, the primate CNS computes hand location in a particular coordinate frame, one based on vision. For some reason, vision predominates over other sensory modalities for the guidance of reaching and pointing in primates, and the coordinate frame used by the CNS for representing hand location reflects this dominance. (See section 4.3.5 for a glimpse into why visual coordinates might predominate in primates.) In order to reach for a target such as a coffee cup, your CNS first computes the current location of your hand in vision-based coordinates and compares its location to that of the target in the same coordinates. For now, think of vision-based coordinates as analogous to an image coming from a fixed video camera. The difference between *current* hand location and *desired* hand location is a vector, called a *difference vector*, which has a particular amplitude and a direction (see figure 17.1). The CNS uses this difference vector to compute the motor plan. Chapters 12–19 present evidence for these ideas, but the present chapter just deals with these concepts theoretically.

The idea that your CNS encodes hand location in vision-based coordinates seems intuitive enough, given the importance of vision in primates, but what if you cannot see your hand—or cannot see at all, as when you reach in the dark? If vision dominates the computation of target and hand location, how can you reach if you cannot see? The answer seems to be that you can predict both target location, based on memory, and the current location of your hand, based on forward kinematics. You can think of forward kinematics as a kind of *alignment* of information between two different sensory modalities: vision and proprioception.

Forward kinematics produces an estimate of where you would see your hand, even when you cannot see it.

Estimating your hand's location in vision-based coordinates has advantages besides reaching in the dark. One reflects the fact that your highest-resolution distance receptors—for any sensory modality—concentrate in the **fovea**. Resolution degrades dramatically for other parts of the visual field. When you prepare to reach, you usually look at the target, not your hand. Thus, the image of your hand usually falls outside central vision, on a part of your retina with inferior spatial resolution, and often outside the visual field entirely (see figure 17.1). Moving targets that need to be pursued by eye movements accentuate this problem. An important adage in baseball and other sports is "Keep your eye on the ball." No one ever suggested watching your hands. Forward kinematics allows high-quality proprioceptive information to substitute for low-quality visual input.

9.6 Predicting End-Effector Location with Proprioception: Virtual Robotics

How might the CNS solve the forward-kinematics problem? Begin by exploring how a robotics engineer might approach the problem. Many robotic arms have a gripper, which serves as the robot's end effector. Robotics engineers have devised many solutions to the problem of programming a robot arm to move its gripper to a target. Some begin with determining the location of the target with respect to the location of the end effector. If the robot has a video camera, the engineer can use its signal to determine the location of both the end effector and the target. Figure 9.3A denotes the end-effector location and target location as x_{ee} and x_t, respectively. These locations define vectors in "camera-centered" coordinates (i.e., a coordinate system that describes the pixels occupied by the end effector and the target relative to some origin). The difference between these two vectors is the difference vector x_{dv}. The present chapter deals with the estimation of x_{ee} in theory, and chapter 10 presents some relevant empirical evidence. Chapter 11 deals with the computation of x_t, and chapter 12 begins a consideration of x_{dv}.

Although the imaginary engineer in this example has to rely on the camera to detect the target's location, he or she might have sensors in addition to the camera that indicate the gripper's location. For example, the robot's joints might house sensors that measure their angle θ. If the engineer knew the lengths of each link of the robot's arm, he or she could use the readout from these joint-angle sensors to compute an estimate of gripper location. This computation is, of course, forward kinematics, but in what coordinate frame should the computation be made?

By comparing the estimate of end-effector location resulting from forward kinematics with the image coming from the camera, the engineer could write a computer program to ensure that the end-effector location, as estimated through forward kinematics, always matches the location recorded by the camera. By analogy with the CNS, the engineer's com-

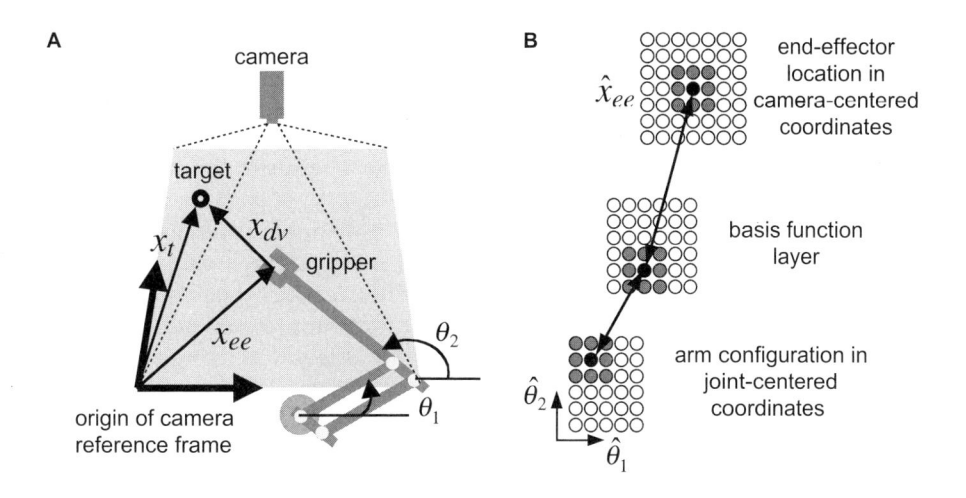

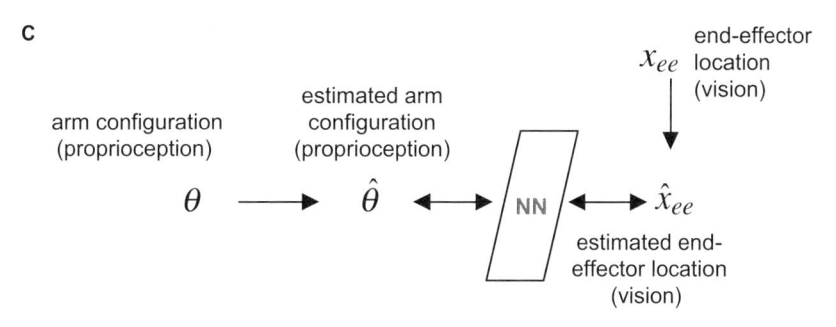

9.6. Predicting End-Effector Location with Proprioception: Virtual Robotics

Figure 9.3
Planning a reach in camera-centered coordinates. (*A*) A robotic arm that moves in two dimensions and a camera. The target *t* of the movement is the black ring. Gripper and target locations are specified in camera-centered coordinates with vectors x_{ee} and x_t. The difference vector x_{dv} is the distance and direction the gripper must move to reach the target. The joint-angle sensors on the robot measure joint angles θ_1 and θ_2. (*B*) Alignment of the joint-angle sensors and camera sensor via a neural network. The hat ^ over a variable indicates an estimation. (*C*) Depiction of a neural network (NN) that aligns joint angles (θ) to end-effector location (x_{ee}) and, importantly, vice versa.

puter program can be said to *align* the mappings of gripper location in *visual and proprioceptive coordinates*. Maintaining this alignment allows the engineer to use the joint-angle sensors on the robot to compute gripper location in camera-centered coordinates, rather than always relying on information from the camera. These adaptive alignments correspond to *mappings* of movement parameters and coordinate frames.

Rather than a symbolic computer program to align "proprioception" with "vision," the imaginary engineer might use one neural network based on feedforward connections to map proprioception to vision (forward kinematics) and another network to map vision to proprioception (inverse kinematics). Both of these networks would perform what is called a *function-approximation* computation, and this approach would work. However, a more parsimonious engineer might use a single neural network that finds the best alignment between the two sensory variables, rather than using two networks.[2]

Figure 9.3B illustrates such a "double-duty" network. It has connections both to and from each of its inputs, as indicated by double-headed arrows in figure 9.3C. The network has internal attractors (see boxes 12.1 and 15.1) that allow it to "complete its pattern of inputs." That is, given some initial estimate of hand location using *either* proprioception or vision, it computes whichever estimate is lacking. Such networks were first described by Sophie Deneve, Peter Latham, and Alex Pouget.[2] They can be termed *pattern-completion networks*, and this one aligns an estimate of joint angles $\hat{\theta}$ with an estimate of gripper location $\hat{x}_{ee}$. (The "hat" ^ over the x indicates an estimate.) The symbol $\hat{\theta} \leftrightarrow \hat{x}_{ee}$ denotes both the network and its mapping. If you look at this network from left to right, it appears to compute forward kinematics: $\hat{x}_{ee} = f(\hat{\theta})$. Alternatively, if you look at this network from right to left, it computes inverse kinematics: $f^{-1}(\hat{x}_{ee}) = \hat{\theta}$ (i.e., the network finds one of the many mappings between a given end-effector location and a possible arm configuration).

Another way to look at the neural network illustrated in figure 9.3B and 9.3C is that it computes a *location map* for the limb. Such a network does not explicitly compute either forward or inverse kinematics but—as a pattern-completion network—can do either and both. For heuristic purposes, you can say that forward and inverse kinematics emerge as the product of that network's computations. You can also say that the *mappings* reflect the product of an internal model of the relationship among the aligned variables. Later chapters take up the concept of internal models in considerable detail.

For the engineer's robot, x_{ee} represents an estimate of end-effector location in camera-centered coordinates. Importantly, in this example, the camera never moves. Your eyes do move, however, whenever they change their orientation within the orbit, your head rotates relative to your shoulders, or you move your head from one place to another. When your eyes and head move, the location of the end effector changes in *retinotopic* coordinates. That is, the part of the retina on which its image falls changes. Chapter 17 takes up this issue in detail (see figure 17.1). For now, think of the location map as aligning information about limb configuration—an **intrinsic** frame of reference—to the location of the end effector in camera-centered coordinates—an **extrinsic** coordinate frame. This aspect of forward kinematics computes where you would see an end effector if you could see it, even when you cannot.

How does the CNS solve this problem? The next section presents a partial mathematical solution to the forward-kinematics problem. For heuristic purposes, this development shows how to express end-effector location in terms of extrinsic coordinates centered on the *shoulder*. You can think of this computation as (1) taking joint angles from the limb (*intrinsic* coordinates) and (2) computing the target's location in *extrinsic*, Cartesian coordinates with your shoulder at the origin. Two additional transforms are not dealt with here: (3) moving the origin of the extrinsic coordinate frame to your head and (4) then to your fixation point. Figure 9.2B shows the results of these transforms as dotted lines. The additional transforms needed to transform joint angles (from muscle lengths) into fixation-centered coordinates do not differ conceptually from those presented here. If you understand how to compute the location of an end effector in terms of extrinsic coordinates based on the shoulder, the additional transforms need only to convert that coordinate into a head- and fixation-centered reference frame.

9.7 Predicting End-Effector Location with Proprioception: Computations

9.7.1 Rotation Matrices

Imagine that a splint at your left elbow keeps your arm fully extended and straight. In that condition, your arm can move through only the three degrees of freedom at your shoulder. The splint makes solving the problem of forward kinematics easier. The question is: Where is your hand, given the joint angles along each of the three degrees of freedom of the shoulder? Now imagine raising your splinted, left arm so that it is parallel to the ground and pointed directly in front of you. You can define a Cartesian coordinate system centered on your left shoulder. Imagine the x, y, and z axes shooting out of your shoulder, directly along your splinted arm (x), straight to the left (y), and straight up (z). This coordinate system is a frame of reference within which you can localize your left hand. Figure 9.4 depicts this frame as a coordinate system, symbolized by [A]. Later, the symbol [S_{ext}] substitutes for [A] to help you recognize that it is an

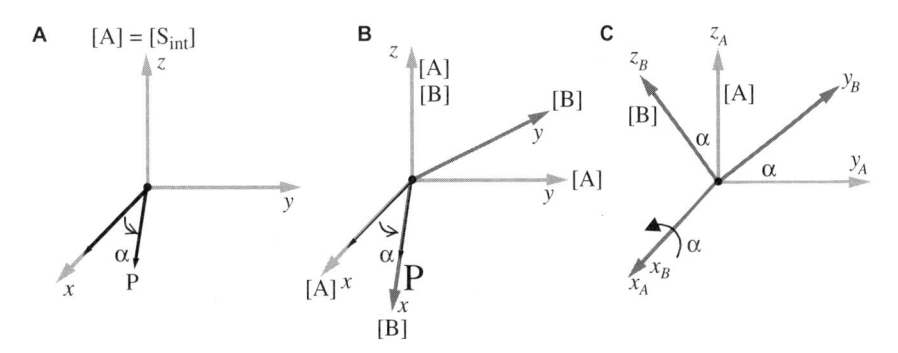

Figure 9.4
Rotations. (*A*) Rotation of a vector P in relation to the frame [A] by angle α. (*B*) Rotation of vector $P^{[B]}$ around the z-axis of frame [A]. (*C*) Rotation of vector $P^{[B]}$ around the x-axis of frame [A].

extrinsic reference frame (ext) with an origin at the shoulder (S), which remains unchanged when your upper arm moves.

Now imagine standing behind the page showing figure 9.4A, with your left arm straight along the x-axis. Your shoulder and hand define a vector that points from the center of rotation of your left shoulder (i.e., origin of frame [A]) to your left hand, which is at the end of your straight, splinted arm. Call this vector P.

$$P = \begin{bmatrix} P_x \\ P_y \\ P_z \end{bmatrix} = \begin{bmatrix} l \\ 0 \\ 0 \end{bmatrix}, \qquad (9.1)$$

where l is the length of your arm from the shoulder to your left hand. When you move your arm to your left, you are rotating your shoulder around the z-axis of coordinate frame [A]. As you can see in figure 9.4A, after this rotation, the vector P no longer aligns with the x-axis of [A]. If your limb rotated by amount α, P now becomes

$$P = \begin{bmatrix} P_x \\ P_y \\ P_z \end{bmatrix} = \begin{bmatrix} l \cos \alpha \\ l \sin \alpha \\ 0 \end{bmatrix}. \qquad (9.2)$$

That is, after rotation around the z-axis, your arm remains in the plane defined by the x and y axes; thus $P_z = 0$, and you can describe your hand location as a vector in that plane in terms of equation 9.2.

To go further, you might want to represent the change that occurred in P from equation 9.1 to equation 9.2 in some systematic way. To do this, note that in both equations P is a vector that points from your shoulder to your hand. Thus, in that sense, P has not changed, even though its description in equation 9.1 differs from that in equation 9.2. The equation has changed because your hand has moved, as described in an extrinsic coordinate system based on the shoulder. In a different coordinate system, an intrinsic one based on your arm, P never changes. You can think of that reference frame as one that is "attached" to your arm. As your arm moves, so do the axes of this intrinsic coordinate system, so that no matter where you move your arm, P in this coordinate system will remain as described in equation 9.1. The x-axis always points from your shoulder to your hand. You can call this coordinate frame [B] and refer to vector P in this coordinate system as $P^{[B]}$, as opposed to the frame described in equation 9.2, which you can call $P^{[A]}$. Later, the symbol [S_{int}] substitutes for [B] to help you recognize that it represents an intrinsic reference frame (int) with its origin at the shoulder (S). [B] and [S_{int}] move as your upper arm moves:

$$P^{[B]} = \begin{bmatrix} P_x^{[B]} \\ P_y^{[B]} \\ P_z^{[B]} \end{bmatrix} = \begin{bmatrix} l \\ 0 \\ 0 \end{bmatrix}.$$

As illustrated in figure 9.4B, equation 9.2 is in fact $P^{[A]}$. You get $P^{[A]}$ by rotating $P^{[B]}$ by an amount α around the z-axis of [A].

9.7. Predicting End-Effector Location with Proprioception: Computations

You can describe the change from $P^{[B]}$ to $P^{[A]}$ as a linear transformation. Call that matrix R because it performs a "rotation" and label it $R^{[B] \to [A]}$ because it rotates a vector from frame [B] into frame [A]:

$$R^{[B] \to [A]} P^{[B]} = P^{[A]}.$$

$R^{[B] \to [A]}$ is a 3×3 *rotation matrix*, and you can label each of its columns with a 3×1 vector as

$$R^{[B] \to [A]} = [x_B^{[A]}\ y_B^{[A]}\ z_B^{[A]}],$$

where $x_B^{[A]}$ is the x-axis of frame [B] written in terms of frame [A]. In the example, where [B] rotated around the z-axis of [A],

$$x_B^{[A]} = \begin{bmatrix} \cos \alpha \\ \sin \alpha \\ 0 \end{bmatrix} \quad y_B^{[A]} = \begin{bmatrix} -\sin \alpha \\ \cos \alpha \\ 0 \end{bmatrix} \quad z_B^{[A]} = \begin{bmatrix} 0 \\ 0 \\ 1 \end{bmatrix}$$

$$R^{[B] \to [A]} = \begin{bmatrix} \cos \alpha & -\sin \alpha & 0 \\ \sin \alpha & \cos \alpha & 0 \\ 0 & 0 & 1 \end{bmatrix} = ROT(z^{[A]}, \alpha). \quad (9.3)$$

The equation above relabels $R^{[B] \to [A]}$ as $ROT(z^{[A]}, \alpha)$ to indicate that it describes how to take $P^{[B]}$ to $P^{[A]}$ when [B] has rotated about the z-axis of [A] by amount α.

Now imagine that your left arm points straight out, as in figure 9.4A, but instead of moving your arm left around the z-axis, you twist it clockwise (figure 9.4C). In this case, you rotate your splinted limb around the x-axis of the shoulder joint, an inward rotation (also known as a supination). $R^{[B] \to [A]}$ takes a different form in this case. Inward rotation will not change the location of your hand with respect to reference frame [A]; your arm still points straight from your shoulder. So $P^{[A]}$ should not differ from $P^{[B]}$ in this case. You can check this by calculating the rotation matrix. Figure 9.4B shows the rotation of [B] around the x-axis of [A], from which you can calculate the rotation matrix $ROT(x^{[A]}, \alpha)$:

$$x_B^{[A]} = \begin{bmatrix} 1 \\ 0 \\ 0 \end{bmatrix} \quad y_B^{[A]} = \begin{bmatrix} 0 \\ \cos \alpha \\ \sin \alpha \end{bmatrix} \quad z_B^{[A]} = \begin{bmatrix} 0 \\ -\sin \alpha \\ \cos \alpha \end{bmatrix}$$

$$R^{[B] \to [A]} = \begin{bmatrix} 1 & 0 & 0 \\ 0 & \cos \alpha & -\sin \alpha \\ 0 & \sin \alpha & \cos \alpha \end{bmatrix} = ROT(x^{[A]}, \alpha). \quad (9.4)$$

This rotation does not affect $P^{[B]}$; when you multiply $P^{[B]}$ by this matrix, it stays the same.

But you know that your arm has rotated: Why does the rotation fail to have an effect on hand location? If you were holding a straight stick in your outstretched hand and had rotated your arm as depicted by the schematic in figure 9.4B, the location of the tip of the stick should not have changed. Rotating your arm produced a change in the angular

154 Chapter 9. Computing End-Effector Location I: Theory

configuration of the arm, and muscle lengths changed, but the net effect was no change in end-effector location as sensed by your vision. Similarly, if you had a robotic arm that could rotate with three degrees of freedom at its shoulder joint, a camera that imaged the robot would record no change in end-effector location. Although the location of your arm is sensed by proprioceptive and visual sensors, changes in one will not necessarily produce a change in the other.

9.7.2 A Shoulder and Elbow Joint System in a Plane

Of course, your arm usually bends at the elbow. In taking this fact into account, you need to consider a reference frame centered on the elbow. Figure 9.5 presents a view of your right arm from above. Imagine a coordinate frame [S_{int}] centered on the shoulder, a vector $E^{[S_{int}]}$ that points from your shoulder to your elbow, and a vector $H^{[E]}$ that points from your elbow to your right hand. Figure 9.2B illustrates these vectors in a different way. Previous sections discussed a reference frame that remains invariant when your splinted limb moves and, in that reference frame, a vector extended from your shoulder to your hand. The previous section called that vector $P^{[B]}$, but it could have used $H^{[S_{int}]}$ to recognize the fact that this reference frame has its origin at the shoulder and moves as the shoulder joint rotates. Instead of designating this frame [B], as above, this section uses [S_{int}]. If you consider the axis of your splinted limb to be the x-axis of a coordinate system [E], centered on the elbow, the description of hand location never changes: Your hand always has coordinate $l, 0, 0$, where b is the length of your lower arm. That is, regardless of hand location:

$$H^{[E]} = \begin{bmatrix} H_x \\ H_y \\ H_z \end{bmatrix} = \begin{bmatrix} b \\ 0 \\ 0 \end{bmatrix}.$$

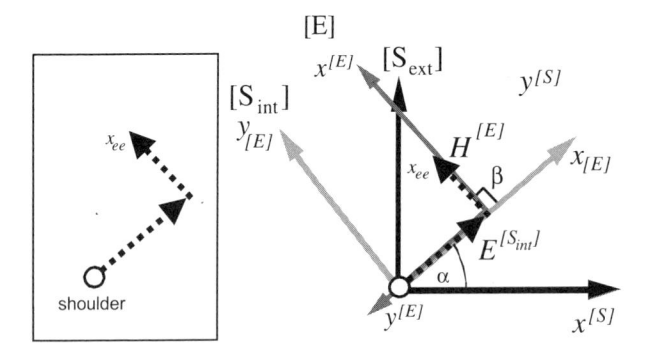

Figure 9.5
Left: A model of a two-link arm in the horizontal plane. Right: Reorientation of the coordinate frame with its origin at the elbow.

And if you assume that the upper arm has a length of a, you have

$$E^{[S_{int}]} = \begin{bmatrix} a \\ 0 \\ 0 \end{bmatrix}.$$

The elbow-centered frame [E] moves as the elbow moves. Neither [S_{int}] nor [E] gives you what you want: a description of hand location in a coordinate frame that does not move when *either* your hand or your elbow moves. There are many such coordinate frames, but the most proximate one has its origin at your shoulder. Because of its origin at the shoulder and the fact that its axes remain invariant as your hand moves, the symbol [S_{ext}] refers to this reference frame. Imagine its axes as Cartesian. The frame [S_{ext}] moves as the shoulder moves, but in the present model of a two-joint limb, the shoulder never moves. (This presentation ignores the additional degrees of freedom that arise from the fact that the shoulder girdle can move relative to the body axis, and uses other simplifications. If you are interested in a more complete presentation of forward kinematics for a five degree-of-freedom arm, consult the web document *kinematics*.)

In figure 9.5, the intrinsic shoulder reference frame [S_{int}] has rotated by amount α around the z-axis of [S_{ext}], and the elbow frame [E] has rotated by amount β with respect to the z-axis of frame [S_{int}]. The z-axes of these frames are pointing out of the page toward you and are not drawn. Your goal is to write the vector that expresses hand location in the frame [S_{ext}]. Start by expressing the vector that describes hand location in elbow-centered coordinates $H^{[E]}$. Notice that the origin of frame [E] has shifted by an amount specified by a vector described in $E^{[S_{int}]}$, your elbow's location in intrinsic shoulder-centered coordinates. To account for both of these, you have

$$H^{[S_{int}]} = R^{[E] \to [S_{int}]} H^{[E]} + E^{[S_{int}]}.$$

The rotation matrix in this case is

$$R^{[E] \to [S_{int}]} = [x_E^{[S_{int}]} \quad y_E^{[S_{int}]} \quad z_E^{[S_{int}]}] = \begin{bmatrix} \cos \beta & -\sin \beta & 0 \\ \sin \beta & \cos \beta & 0 \\ 0 & 0 & 1 \end{bmatrix},$$

which gives you an expression for the location of your hand with respect to your elbow:

$$H^{[S_{int}]} = \begin{bmatrix} a + b \cos \beta \\ b \sin \beta \\ 0 \end{bmatrix}.$$

The vector described by $H^{[S_{int}]}$ is not your goal, however. You want a description of hand location in extrinsic, shoulder-centered coordinates [S_{ext}], a frame that remains invariant as your arm moves. Because the origin of frame [E] has rotated around the z-axis of frame [S_{ext}], you can use

Table 9.1
The names and axes of various coordinate frames

Name	Symbol	Origin	Axes
Fixation-centered	[F]	fixation point	Cartesian, spherical
Retinotopic	[R]	fovea	Cartesian, spherical
Head-centered or craniocentric	[Cr]	midpoint between the eyes	Cartesian, spherical
Shoulder-centered, extrinsic	[S$_{ext}$]	shoulder	Cartesian, spherical
Shoulder-centered, intrinsic	[S$_{int}$]	shoulder	x, along upper limb (U)
Elbow-centered	[E]	elbow	x, along lower limb (L)
Hand-centered	[H]*	hand	Cartesian, spherical
Cartesian	[C]*	anywhere, but never moves	Cartesian

*[H] stands for hand, not head; [C] stands for Cartesian, not craniocentric, coordinates.

this information to transform $H^{[S_{int}]}$ into coordinates described in terms of frame [S$_{ext}$]:

$$H^{[S_{ext}]} = R^{[S_{int}] \to [S_{ext}]} H^{[S_{int}]}$$

$$R^{[S_{int}] \to [S_{ext}]} = [x^{[S_{ext}]}_{[S_{int}]} \quad y^{[S_{ext}]}_{[S_{int}]} \quad z^{[S_{ext}]}_{[S_{int}]}] = \begin{bmatrix} \cos\alpha & -\sin\alpha & 0 \\ \sin\alpha & \cos\alpha & 0 \\ 0 & 0 & 1 \end{bmatrix}.$$

Multiplying and simplifying gives you an expression for the location of your hand in extrinsic, shoulder-centered coordinates:

$$H^{[S_{ext}]} = \begin{bmatrix} H^{[S_{ext}]}_x \\ H^{[S_{ext}]}_y \\ H^{[S_{ext}]}_z \end{bmatrix} = \begin{bmatrix} b\cos\beta\cos\alpha + a\cos\alpha - b\sin\beta\sin\alpha \\ b\cos\beta\sin\alpha + a\sin\alpha + b\sin\beta\cos\alpha \\ 0 \end{bmatrix}$$

$$= \begin{bmatrix} b\cos(\alpha+\beta) + a\cos\alpha \\ b\sin(\alpha+\beta) + a\sin\alpha \\ 0 \end{bmatrix}.$$

What you have done is represent a vector that always points from your elbow to your hand ($H^{[E]}$) in terms of a coordinate system centered on the shoulder ([S$_{ext}$]), which remains stationary despite movements around the shoulder and elbow joints. However, the choice of a shoulder-centered representation was arbitrary. The camera-centered reference frame of figure 9.3A would have done, as well. In subsequent chapters, this camera-centered frame, rather than a shoulder- or body-centered frame, comes to the fore as the dominant frame for representing hand location for reaching. Table 9.1 presents some relevant coordinate frames.

References

1. Abernethy, B, Kippers, V, Mackinnon, L, Neal, RJ, and Hanrahan, S (1997) The Biophysical Foundations of Human Movement (Human Kinetics, Champaign, IL).
2. Deneve, S, Latham, PE, and Pouget, A (2001) Efficient computation and cue integration with noisy population codes. Nat Neurosci 4, 826–831.

Reading List

This presentation generalizes the ideas of Deneve et al.,[2] so you should consult the original for their specific ideas.

10 Computing End-Effector Location II: Experiment

Overview: The CNS computes an estimate of limb configuration through an alignment of information from various sensory modalities, including proprioception and vision. This computation appears to rely on neurons in which discharge varies monotonically and approximately linearly with location of the end effector in the workspace.

10.1 Role of Proprioceptive Signals in End-Effector Localization

Chapter 9 showed how hand (or some other end-effector) location could be computed from information about joint angles. This chapter presents some experimental data relevant to this computation. The equations presented in chapter 9 tell you how variables in one coordinate system relate to those in another, but not how the CNS solves the problem of forward kinematics. An important clue comes from experiments that manipulate the activity of muscle-spindle afferents.

10.1.1 Perception of Arm Configuration from Muscle Afferents

Figure 10.1 shows what happens when **muscle-spindle afferents** are artificially stimulated.[1] In this experiment, participants rested both elbows on a table, with their forearms nearly vertical. The experimenter asked participants to track the angle of their right elbow by flexing or extending their left one. As they maintained a steady, extensor force with their triceps, their right biceps tendon was vibrated. Figure 10.1 shows that the angle of the left, tracking elbow matched the angle of the right one before vibration. However, as vibration activated muscle-spindle afferents in the right biceps, the participants extended their tracking elbow. Vibration apparently caused the perception that the right biceps had lengthened, which corresponds to an extended elbow angle. This finding indicates that the CNS uses muscle spindles (section 7.6.2) to compute forward kinematics. Artificial activation of these sensory receptors by vibration increases the activity of muscle-spindle afferents and leads to a change in the perception of joint angles.

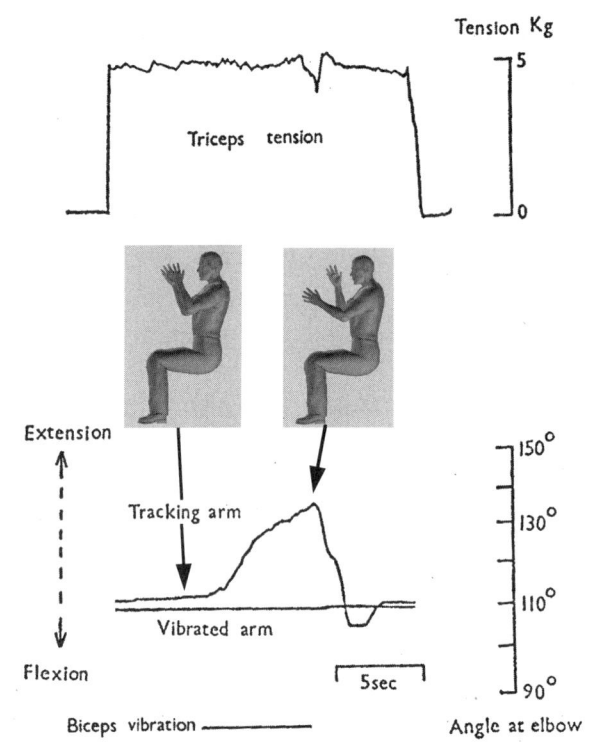

Figure 10.1
Vibration of biceps results in the perception of elbow extension. A participant's right elbow pulled down on a strain gauge to maintain a set level of force (equivalent to tension, top plot). Without vision of the right arm, the participant positioned the left arm (called the tracking arm) to indicate the angle of flexion in the right arm (bottom plot). Biceps vibration in the right arm (horizontal bar at the bottom) led to extension of the tracking (left) arm, indicating a sense of extension of the vibrated (right) arm. (From Goodwin et al.[1] with permission.)

10.1.2 Perception of End-Effector Location from Muscle Afferents

Stimulation of muscle-spindle afferents also affects the estimate of hand location. Importantly, this estimate of hand location extends to the location of objects held in the hand (i.e., to end effectors, generally). For example, if you are holding a small light source or audio speaker in your hand, biceps vibration produces the illusion that the light or sound moves. This percept occurs despite the fact that the visual system (in the case of light) or the auditory system (in case of the speaker) does not detect any motion of the object because, in fact, the object remains stationary.

Minna Levine and Jim Lackner[2] performed an experiment that revealed these properties and was followed by work of Lackner and Barbara Shenker.[3] They vibrated the right biceps of participants in a darkened room. In one condition, the experimenters taped a small light to the finger of the vibrated arm; in another, the participants held a small speaker in their hands.

Consider the condition in which participants held a small light source but it did not emit any light. The experimenters asked the participants to point with the index finger of their nonstimulated arm to the perceived location of the index finger of their stimulated arm. When the experimenters stimulated the right biceps, participants verbally reported that their forearm was extending. They pointed with their left hand to a location about 16° to the right of the right hand. At stimulation onset, their eyes pursued the illusory motion of the hand, then quickly returned to their initial location in a series of eye movements.

When the experimenters stimulated biceps as the source emitted light, the participants' eyes maintained fixation on the light, but they still perceived the elbow extending and the hand moving. Their left index finger pointed to a location about 12° to the right of the right hand. Participants reported that they saw the light in their hand moving to the right but that they were able to keep their eyes on it (i.e., they felt that they could visually track the moving light). Note, however, that the perceived displacement of the hand was smaller when visual information about hand location was available (12° versus 16°).

In the final condition, the participants held a small speaker. Biceps vibration again induced an illusion of hand motion, but it also induced an illusion of motion of the source of the sound. The participants reported that the sound source moved to the right.

These data suggest that your CNS relies on proprioception, at least in large part, to estimate the current location of your hand. As proprioceptive signals change through biceps stimulation, the current estimate of hand location changes. However, this estimate of hand location does not depend solely on proprioception. You can also see your hand, or, if you hold a sound source, you can localize it acoustically. The experiment suggests that you do not have three independent measures of hand location, as estimated through sound, vision, and proprioception. If you did, then biceps stimulation would affect only your proprioceptive sense of hand location and not induce visual or auditory illusions.

The results of this experiment further suggest that your CNS *aligns* visual, auditory, and proprioceptive cues to produce a single *estimate* of hand location. This alignment reflects contributions from each of those sensory systems. If proprioception indicates that your limb is extending, and your CNS receives no information from vision, then its estimate of hand location shifts accordingly. However, if vision indicates that your hand is stationary, then the effect of proprioception causes a smaller perceived shift of the hand.

Chapters 11 and 12 will review evidence which suggests that your CNS estimates hand location in visual coordinates—specifically, that it represents current hand location as a vector with its origin at your fixation point. Your CNS computes this vector through an alignment of sensory information available from proprioception, vision, and audition.

The estimate changes when any of these inputs indicates a change in hand location.

10.2 Introduction to Frontal and Parietal Neurophysiology

The next section, section 10.3, begins a fairly intensive presentation of neurophysiological data which continues, in one way or another, for the remainder of the book. To better understand those data, you need to know something about the relationship between the posterior parietal cortex (PPC) and the motor areas of the frontal cortex. You also need to gain some familiarity with concepts about cortical representations that have preoccupied many neurophysiologists for the past several decades. This section focuses on just one part of the frontal cortex, the primary motor cortex (M1), but the issues raised in that discussion pertain to many parts of the motor system.

10.2.1 Relation Between Frontal and Parietal Cortex

Most neurophysiologists agree that, together, PPC and the frontal motor areas of cortex play an important role in the transformation of spatial coordinate frames and the localization of visible targets and the limb in space. They may not play the same role in these computations, but the pattern of interconnections depicted in figure 6.3B suggests that they work together. You may need to refer to figure 6.3 and section 6.4 from time to time as you progress through this chapter and subsequent ones.

Ted Jones and Tom Powell first discerned the interrelationship of these two broad regions of cortex in the early 1970s.[4,5] More recent work by Roberto Caminiti and his colleagues[6,7] has confirmed and refined that view. As a general statement (see figure 6.3B), more rostral parts of the frontal motor cortex connect reciprocally with more caudal parts of the PPC. Thus, parietal and frontal areas nearest the central sulcus closely interconnect, as do parietal and frontal areas farthest from the central sulcus, including the prefrontal cortex.

Although great uncertainty remains about the division of labor between the various parietal and motor areas, as a first-level approximation you can think of the areas nearest the central sulcus as involved mainly in dynamics and those farthest from the central sulcus as involved mainly in kinematics.

10.2.2 Muscle versus Movements

You need to untangle several confusing and overlapping concepts in order to follow the neurophysiology presented in this and subsequent chapters. Some of these issues do not arise in full-fledged form until later, but some come up in section 10.3, and they are related. Figure 10.2 lays out some of the issues graphically.

For more than a century, experts on the primary motor cortex have argued (occasionally violently) about whether it "encodes," "controls," or

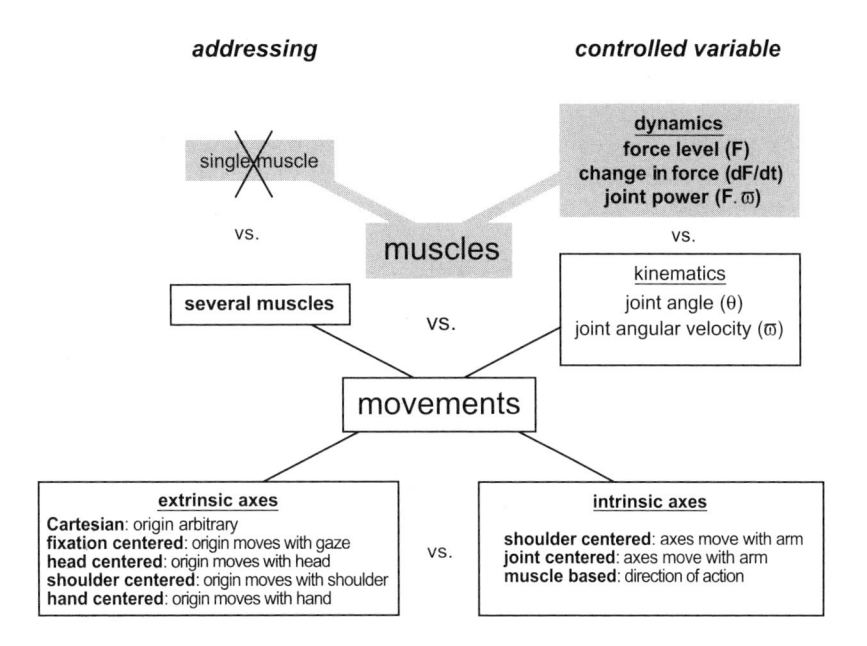

10.2. Introduction to Frontal and Parietal Neurophysiology

Figure 10.2
The "muscles-versus-movements" debate. Sketch of conceptual relations among relevant coordinate frames and control variables. The large "X" indicates a possibility that has been ruled out by empirical methods; boldface type for "dynamics" and "several muscles" indicates those having the most empirical support. The phrase **direction of action** refers to the joint-angle changes caused by the contraction of a muscle. See figure 9.2 for more on coordinate frames.

otherwise reflects the movements you make, as opposed to the muscle activations you need to make those movements. (You might reflect on how the terms "encode," "control," and "reflect" differ, and why you would choose one over the others.) This conflict is often called the "muscles-versus-movement" debate—and it has not yet been resolved. Progress has been made, however, by recognizing that early researchers reached different conclusions, in part, because they combined different aspects of the problem in a confusing way.

The reason that you need to learn about these issues now is that ideas about representations in M1—and other parts of PPC and frontal cortex—become entangled with the effort to understand whether they represent information in extrinsic coordinates versus intrinsic ones. Extrinsic coordinates correspond to an **absolute reference frame** based on the world; intrinsic coordinates correspond to a **relative reference frame** based on the body.

As depicted in figure 10.2, the "muscles-versus-movement" debate has two distinct components. The first involves the question of whether M1 controls the activity of single muscles or multiple muscles. You can call this aspect the *addressing* issue because it involves how many muscles an M1 neuron influences. Proponents of the "muscle" view claimed that small, focal groups of M1 neurons, sometimes called *columns* or *colonies*, control individual muscles. According to this view, most closely associated with Hiroshi Asanuma,[8] coordinating movements involves recruiting the columns in M1 needed to reach to a target. This idea has the merit of simplicity, but never seemed likely to most experts, nor was it based on sound evidence.

Those who held the alternative view, that individual columns—and even individual neurons in M1—controlled groups of muscles, considered their ideas to correspond to the "movement" view. This idea follows from the fact that groups of muscles act together in movements. In the end, empirical evidence of two kinds resolved the issue in favor of the "movement" view, at least for the addressing aspect of the debate. By filling single corticospinal axons with a marker, Yoshi Shinoda and his colleagues[9] showed that single cortical neurons send axonal branches to several motor pools. This anatomical finding was confirmed by Eb Fetz[10] and Roger Lemon[11,12] and their colleagues, working independently with a method called *spike-triggered averaging*. With this method, each discharge (or "spike") of a cortical neuron triggers a signal averager, which aligns EMG activity with neuronal activity. If an M1 neuron projects relatively directly to a motor neuron that innervates a particular muscle—and if there are few, if any, additional intervening synapses—the effect of cortical activity on the muscle can be measured. Although the inputs to hand and arm muscles vary to an extent, there is little doubt that the multiple-muscle view has prevailed. Individual M1 neurons each influence several muscles, which often work together (i.e., they tend to be **synergistic**). If you construe the recruitment of multiple, synergistic muscles as support for the "movement" side of the debate, then you might take that side. Figure 10.2 depicts this aspect of the debate in its upper left part.

Note, however, that figure 10.2—in its upper right part—depicts another aspect of the muscles-versus-movement debate, which involves kinematics and dynamics. You can call this aspect the *controlled-variable* issue, as opposed to the addressing issue. Whereas evidence on the addressing issue conclusively settled on the side of the "movement" view, the controlled-variable issue remains unsettled. In terms of kinematics versus dynamics, we think that the bulk of the evidence rests on the dynamics side. If you agree with this conclusion and equate the "muscle" side of the muscles-versus-movement debate with dynamics and the "movement" side with kinematics, then you might come down on the "muscle" side of the debate.

A related conceptual issue involves the coordinate frames in which the frontal motor areas and PPC encode locations. Chapter 9 discusses the concept of coordinate frames, laid out in figure 9.2B and table 9.1. Some coordinate frames describe intrinsic parameters such as joint angles, joint angular velocity (or other derivatives of joint angle), or the torques generated by muscles, as indicated in figure 10.2, bottom right. If you believe that M1 controls torques and only torques, then you might come down on the "muscle" side of the debate.

You can also describe kinematics in terms of extrinsic coordinates (figure 10.2, bottom left). For example, when your hand moves because of joint rotations, it also moves along a path in Cartesian coordinates, which you can base on any arbitrary origin in the external world. At the same time, your hand also moves in coordinates that have extrinsic axes, but have their origin at places such as your fixation point, your head, your shoulder, or your hand. Of course, your eye and head are "intrinsic" to you, but for these coordinates their axes are extrinsic, and that matters most. If you believe that M1 controls movements in extrinsic coordinates, then the "movement" side of the muscles-versus-movement debate is for you. Regardless, the issue of coordinate frames occupies a central place in the discussion of neurophysiological data in the remainder of the book.

10.3 Encoding of Limb Configuration in the CNS

Section 5.3 explains a little about the spinocerebellar system, and section 6.4 sketches the organization of the motor and parietal cortex, including the somatosensory cortex and PPC. The present section treats the cerebellar and neocortical systems separately, but bear in mind that they work together in many ways. For example, one of the principal corticofugal projections from the PPC goes to the cerebellum via the basilar pontine nuclei.[13,14]

10.3.1 Encoding of Limb Configuration in the Spinal Cord

When information from muscle-spindle afferents arrives in the spinal cord, information from several muscles appears to be integrated before it is sent to the brain. Dick Poppele and his colleagues[15] have investigated one particular group of spinal neurons that convey information about muscle length to the brain and project through the dorsal spinocerebellar tract. Poppele and his colleagues recorded from these spinal neurons as they changed the configuration of a cat's hindlimb. They initially considered the possibility that dorsal spinocerebellar tract neurons might simply encode information regarding single joint angles. Examples of two typical cells are shown in figure 10.3. In cell number 2080, they observed a strong, almost linear modulation of neural discharge with respect to hip angle, and in cell number 2146 they saw a linear modulation with respect to knee and perhaps ankle angle. Approximately 90% of cells had discharge modulation that was linear with respect to at least a single joint's angle, and very often this was the case for several joints.

Figure 10.3D plots the activity of the same two cells as a function of foot location, as measured in a polar coordinate system centered on the hip and specified by a length (L) and an orientation angle (o), as illustrated in figure 10.3B. The cat's foot is, of course, the end effector. Essentially, all of the cells that modulated linearly with respect to a joint angle also did so with respect to a polar representation of foot location.

A linear modulation of neural discharge for multiple joints might result from the cell's activity being dominated by inputs from one muscle

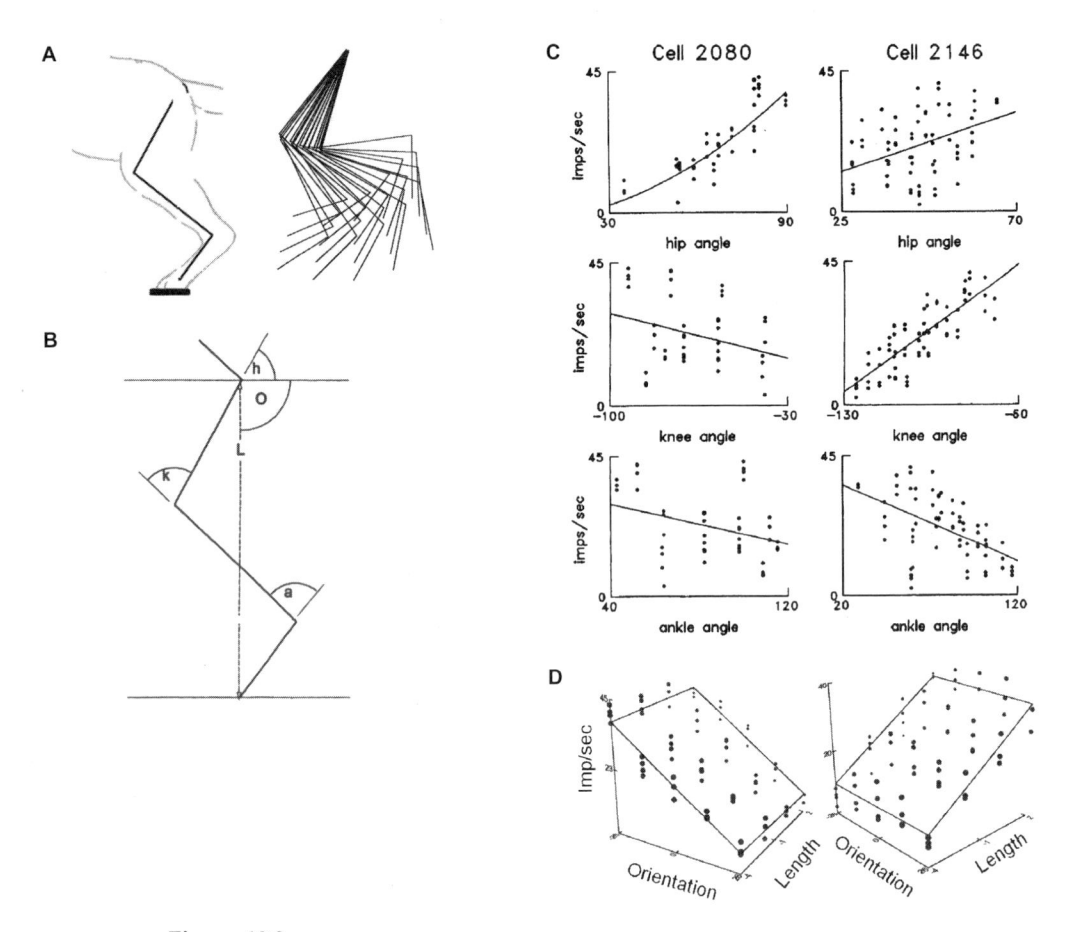

Figure 10.3

Neurons in the dorsal spinocerebellar tract show sensitivity to the configuration of the hindlimb. (*A*) Orientations studied for the hindlimb of an anesthetized cat. The experimenters positioned it at various points in the vertical plane. (*B*) Limb configuration was defined both in terms of a joint-based coordinate system (joint angles at the hip, knee, and ankle) and in terms of a polar coordinate system (an orientation angle o and a length L). (*C*) Firing rates of two spinocerebellar cells in terms of joint angles. Abbreviation: imp/sec, impulses per second, also known as discharge rate, activity rate, activity, and firing rate, sometimes rendered as Hz (cycles per second). (*D*) Firing rates of the same two cells plotted as a function of end-effector location in polar coordinates. (*A*, *B*, and *D* from Bosco and Poppele,[16] *C* from Bosco et al.[17] with permission.)

that spans more than one joint. Alternatively, length information from multiple muscles might converge on a single dorsal spinocerebellar tract neuron to produce the observed results. The anatomy of the spinal cord seems to favor the second interpretation. There is substantial convergence of sensory afferents upon dorsal spinocerebellar tract neurons. Poppele et al. hypothesized that these neurons computed foot location from convergent sensory inputs. However, they recognized that because foot location strongly correlates with joint angles, the apparent coding of foot location could also be due to joint angles.

Accordingly, they performed a second experiment to examine whether dorsal spinocerebellar tract cells encode the end-effector location or the limb's configuration in joint space.[15] The cat's hindlimb has three degrees of freedom for positioning the foot in the vertical plane (one degree of freedom each at the hip, knee, and ankle). The experimenters could place the foot at a particular location while having the joint angles in different configurations. In their experiment, Poppele et al. fixed a stiff plastic strip between the femur (thigh) and tibia (shin), so that the cat's knee angle would remain relatively fixed as they positioned its foot at different places. (Figure 3.3 shows a drawing of the hindlimb's skeleton in a different species.) Normally, positioning of the foot in this way would result in a 70° change in the knee angle (figure 10.3C). With the stiff plastic strip, however, the knee angle changed by only about 10°. The experimenters found that in as many as half of the cells, firing rates did not change from the unconstrained to the constrained condition (figure 10.4). From this result, they concluded that about half of the dorsal spinocerebellar tract neurons have activity that reflects end-effector location. The other half of the neuronal population appeared to encode the limb's configuration in joint coordinates.

Note especially what Poppele et al. did not observe. They did not see spinocerebellar tract neurons with a "preferred location" or "preferred configuration" of the limb within the tested space. That is, neurons did not discharge most for a particular limb configuration or end-effector location, and discharge less for all others. Rather, their discharge was *monotonically*—and often linearly—modulated as a function of limb configuration or end-effector location. Some neurons discharged as a linear function of one or more joint angles. Others did so as a linear function of foot location. For example, some neuronal activity was a planar function of end-effector location as expressed in a polar coordinate system centered at the hip (figure 10.4A and B).

The evidence obtained by Poppele et al. suggests that the spinal cord—and the spinocerebellar system in particular—represents both joint angles and end-effector location. It turns out that cells in S1, M1, and PPC do so as well, and in a similar way.

10.3.2 Sensitivity of S1 Neurons to Changes in Limb Configuration

Steven Helms-Tillery, Tim Ebner, and their colleagues recorded the activity of cells in S1 of monkeys, defined broadly to include areas 1 and 2, as

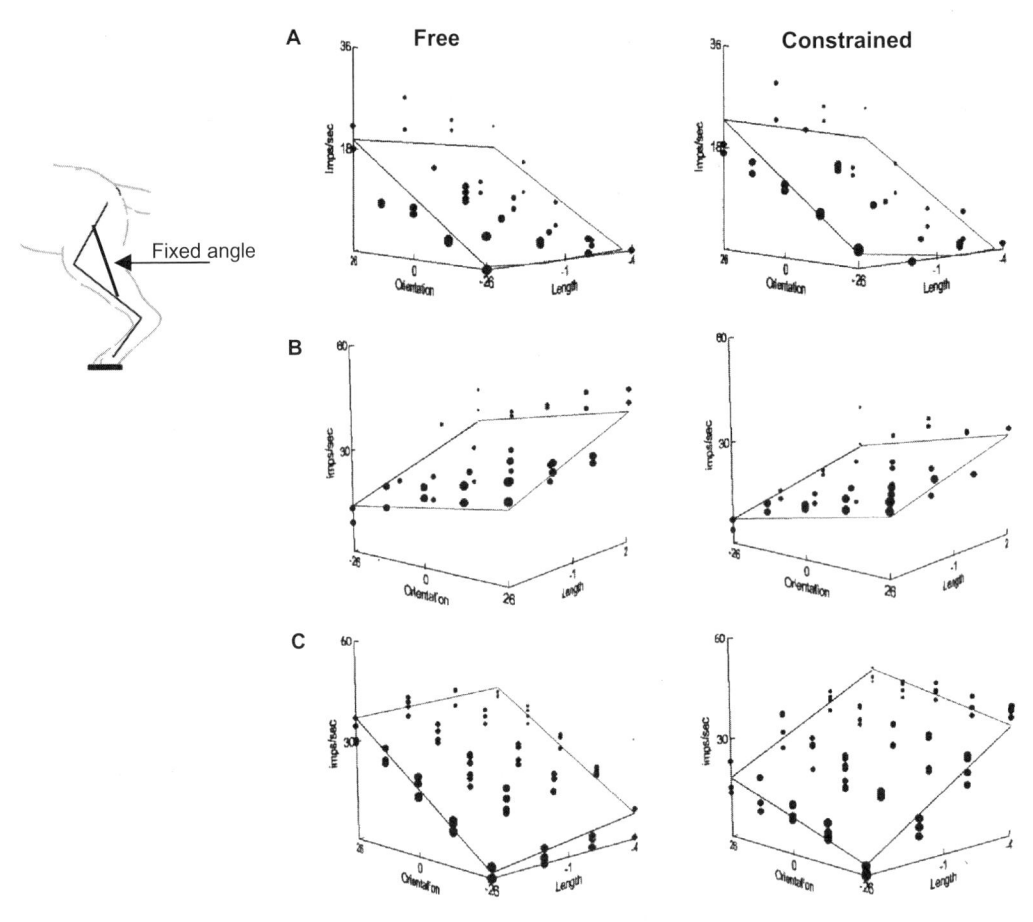

Figure 10.4
Discharge from three spinocerebellar tract neurons (A–C) as a function of limb configuration in unconstrained (free) and knee-constrained conditions. A plastic bar constrained the hindlimb of an anesthetized cat. Attached to both the femur and the tibia (see figure 3.3), it severely limited the changes in knee angle as the experimenters placed the cat's foot at various points on a plane. A polar coordinate system describes foot location, as in figure 10.3D. Each row shows the activity of a different cell. Left and right columns show the activity without and with the joint constraint, respectively. The two cells in A and B do not change their discharge significantly with the imposition of the joint constraint. The cell in C does. (From Bosco et al.[15] with permission.)

10.3. Encoding of Limb Configuration in the CNS

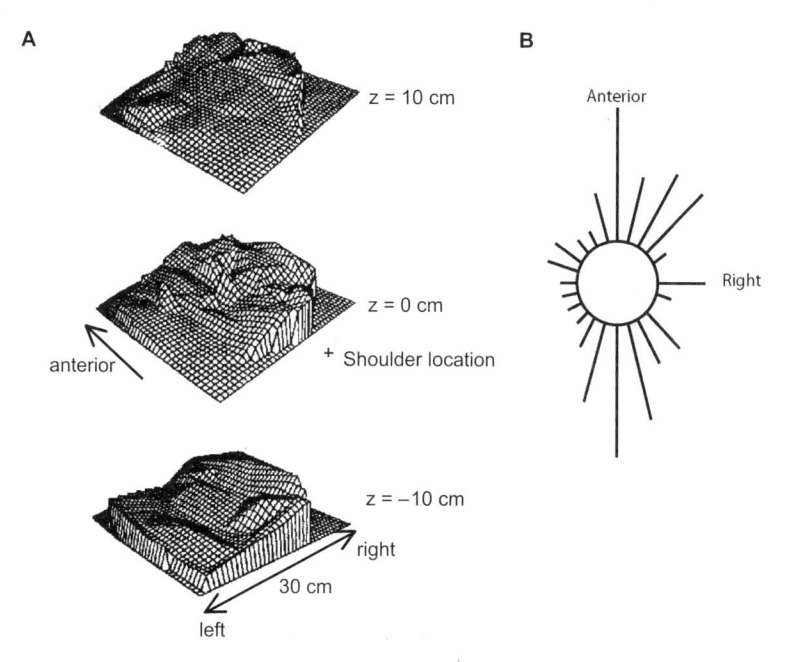

Figure 10.5
Discharge of neurons in the primary somatosensory cortex (S1) as a function of hand location. (A) The discharge of a single cell in the left somatosensory cortex of a monkey. The figure plots neural discharge as a function of hand location in three horizontal planes, top to bottom. The center plane was at the level of the monkey's shoulder, and the upper and lower planes were 10 cm above and below it, respectively. The location of the monkey's right shoulder is indicated. (B) The preferred displacements of a population of S1 neurons. The preferred displacement is the direction of hand movement associated with the largest change in the cell's activity level. The length of each line indicates the number of cells with their highest discharge in the direction of the line (from the center of the circle). More cells have preferred displacements along the toward–away dimension than for the orthogonal dimensions. (Adapted from Helms-Tillery et al.[18] with permission.)

well as area 3b. They wanted to determine how discharge varied as a function of hand location.[18] The monkey sat in a chair and held the handle of a robot with its hand. The experimenters slowly moved the robot to position the hand at various locations in the workspace. Figure 10.5A shows a cell's activity for hand locations along three horizontal planes. The center plane corresponded to the level of the monkey's shoulder, and the upper and lower planes were 10 cm above and below that plane, respectively. This neuron did not discharge for hand locations up or to the left. Neither did it respond to hand-location changes toward and away from the monkey. Rather, this cell displayed the most sensitivity to changes along the left–right axes. Its discharge gradually increased as the experimenters changed the monkey's hand location to the right, with the highest discharge rate associated with hand locations near the right boundary of reachable space.

Many S1 cells displayed this kind of gradual, monotonic change in discharge as hand location varied along a particular axis, but that axis varied from cell to cell. A linear function of hand position in Cartesian space, $f = c_0 + c_1 x + c_2 y + c_3 z$, often described this coding fairly accurately. In this formula, f represents the discharge rate of a cell (its "firing frequency"), and x, y, and z are the Cartesian coordinates of the hand. For the cell shown in figure 10.5A, the coefficient c_1 dominates because this cell's "preferred displacement" of the hand was along the left–right (x) axis. For a majority of the recorded cells in S1, c_2 dominated the cell's activity [i.e., location along the anterior–posterior axis (toward and away from the monkey) was most important (figure 10.5B)]. Thus, S1 appears to have more cells sensitive to the location of the hand along the anterior–posterior axis (immediately in front of the monkey) than along the left–right axis.

If, compared to other axes, more S1 neurons respond to changes in hand location toward or away from the body, you might predict that the CNS should be able to sense hand location more accurately along that axis. Robert van Beers and his colleagues[19] did an experiment testing this prediction. Blindfolded participants moved their right hand to a given location above an opaque table and attempted to move their other hand to the corresponding location below the table. The experimenters found that, at least near the midline, the CNS is better at computing hand location along the axis toward and away from the body than along the left–right axis.

Although the neurophysiological results of Helms-Tillery et al.[18] are intriguing, as is their correspondence with the psychophysics just mentioned,[19] it remains unknown whether S1 cells have a specific representation of hand location or if they instead respond to some combination of joint angles. Unfortunately, experiments like those of Poppele et al. for spinocerebellar tract neurons have not been done for S1. Testing whether S1 and other parts of parietal cortex encode joint angles versus end-effector location is especially important in monkey neurophysiology. In typical neurophysiological experiments, monkeys make such consistent and stereotyped reaching movements that to know hand location is virtually to know the joint angles[20] (see box 10.1). Despite the uncertainty about end-effector versus joint-angle coding, the data clearly demonstrate that cells both in the spinal cord and in S1 have activity that varies

Box 10.1
Stereotypy in reaching and pointing movements

> Section 19.4 examines why movements might be highly stereotyped in some circumstances. It suggests that the CNS attempts to minimize endpoint error in the face of noise. In this context, it is of interest that in patients with apraxia,[21,22] who may have damage to the parietal cortex,[23] these stereotyped relations break down.

10.3. Encoding of Limb Configuration in the CNS

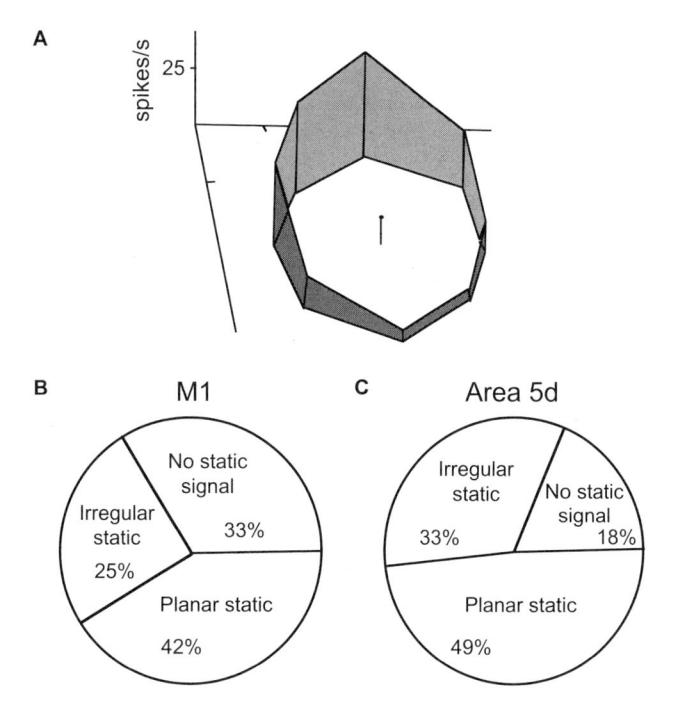

Figure 10.6
Neurons with hand-location signals in M1 and the PPC. (*A*) Each data point represents the static discharge rate for the neuron as a monkey holds its hand at the corresponding spatial location. The activity during holding at a central location is shown by the center data point. The three-dimensional surface shows the planar tuning of the cell for hand location. (*B*) Pie chart showing the proportions and kinds of hand-location signals in M1, for static locations of the hand. (*C*) Data for PPC (area 5d) in the format of *B*. Note that a large proportion of the sampled neurons show planar representations of hand location. (From Georgopoulos et al.[28] with permission.)

monotonically and approximately linearly with the location of an end effector in the workspace.

10.3.3 Sensitivity of M1 and PPC Neurons to Changes in Limb Configuration

Apostolos Georgopoulos, Roberto Caminiti, and John Kalaska[24] showed that, like spinocerebellar and S1 neurons, cells in M1 and PPC have activity that modulates roughly linearly as a function of hand location. For example, the part of PPC just caudal to S1, area 5d (see figure 6.3), has some of the properties expected for a system involved in the computation of forward kinematics. Figure 10.6A illustrates the discharge of an area 5d neuron that varied monotonically and approximately linearly with location of a monkey's hand in the workspace. As shown by the charts in figure 10.6B and 10.6C, both area 5d and M1 have many cells that change their discharge in this way. Later experiments by Steve Scott and John Kalaska[25-27] showed that joint angles also influence the activity of these neurons. That is, M1 and area 5d neurons do not simply encode hand location. Nevertheless, as shown in figure 10.6, nearly half of the neurons in M1 and PPC do show monotonic and approximately linear modulation with changes in hand location.

Work by Francisco Lacquaniti, Roberto Caminiti and their colleagues[29] has revealed some related properties of cells in area 5d. As explained in chapter 9, the shoulder has three degrees of freedom. In radial coordinates, they can be described in terms of azimuth, elevation, and distance. Azimuth is the angle along the left–right dimension, elevation is the angle along the up–down dimension, and distance measures the toward–away dimension. Lacquaniti et al. observed that different populations of neurons in area 5d encode azimuth, elevation, and distance. Thus, cells in PPC, like those in S1 and the spinal cord, have signals important to the computation of forward kinematics.

10.3.4 Encoding of the Location of Hand-Held Tools or Other End Effectors in PPC

If you hold a tool (a hammer, for example) and need to place part of it somewhere, the location of your hand, per se, matters only indirectly. The end effector's location—in particular, the part of the end effector most important to the job—matters a lot more. If you hold a tool consistently, over time your CNS might learn to associate proprioceptive information about your arm posture with visual information about end-effector location. This concept is exemplified by the plastic properties of certain PPC neurons as monkeys learn to use tools. Atsushi Iriki and his colleagues[30] trained monkeys to use a rakelike tool to extend their reach and pull food items close enough to obtain them. After training with the rake, the responses of neurons in area 5ip (see figure 6.3) showed remarkable properties. They summarized their observations thus:

In naïve animals, neurons in this area had essentially somatosensory responses, and thus they scarcely exhibited visual responses.... After training [with the rake] ..., there appeared a group of bimodal neurons that exhibited clear visual responses in addition to the somatosensory responses. Around the somatosensory receptive field was formed a visual receptive field defined as a territory in space where a neuron responded to the moving visual stimuli. Tool use induced an expansion of the visual receptive field only when monkeys intended to use tools to retrieve distant objects, but the modification was never induced when just holding it as an external object. (From Obayashi et al.,[30] p. 3499)

Thus, cells in area 5ip became more responsive to visual inputs when monkeys used an extended end effector, the rake. Indeed, Iriki and his colleagues have shown that the **activity field** of the cells expanded to include the head of the rake, and shrank back to the hand about 5 min after the monkey stopped using the tool. They interpreted their data as reflect-

10.3. Encoding of Limb Configuration in the CNS

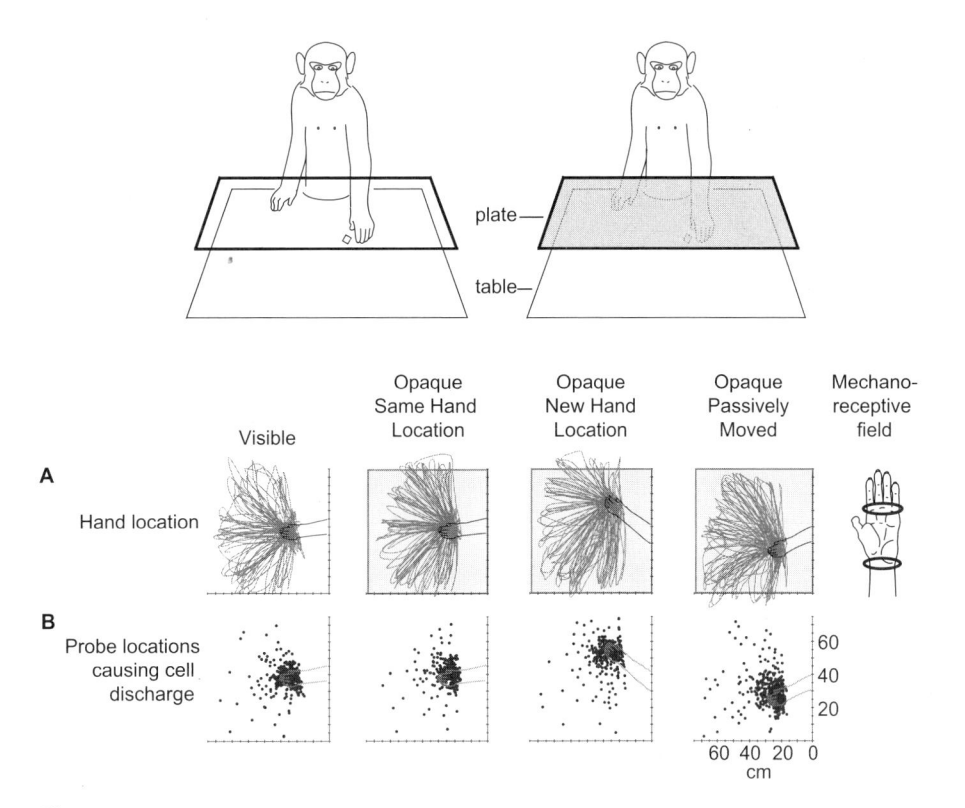

Figure 10.7
Response properties of a bimodal cell in area 5ip. (See figure 6.3 for the location of area 5ip.) Both visual and proprioceptive inputs activated this cell. As illustrated at the right, this neuron responded to flexion of the finger and wrist joints (ellipses). The monkey's hand was visible through a liquid-crystal plate (top left) or occluded (top right). (A) Hand locations and paths (lines) which led to activity in the neuron. (B) The neuron responded to visual probe stimuli as if it had a visual receptive field around the hand's location. The neuron also responded to a stimulus moving toward the proprioceptive field regardless of eye orientation (not shown). Note that when the liquid-crystal plate became opaque, the cell's activity field remained about the same, and that it did not matter if the monkey's arm was passively or actively moved. (From Obayashi et al.[30] with permission.)

ing something about the monkey's body image, as if the rake became an extension of the monkey's hand. Cells in area 5ip reflected both the location of the hand and visual stimuli near it. As illustrated in figure 10.7, the responses to visual stimuli persist even when the monkey can no longer see its hand. Further, the cell's activity field moves with the hand, suggesting that it encodes space in an extrinsic, hand-centered frame.[30] As mentioned later (section 14.1.1), cells in the motor areas of the frontal cortex have similar properties.[31]

These data suggest a gradual transition between a region, S1, that represents purely intrinsic, somatosensory information, and PPC, which

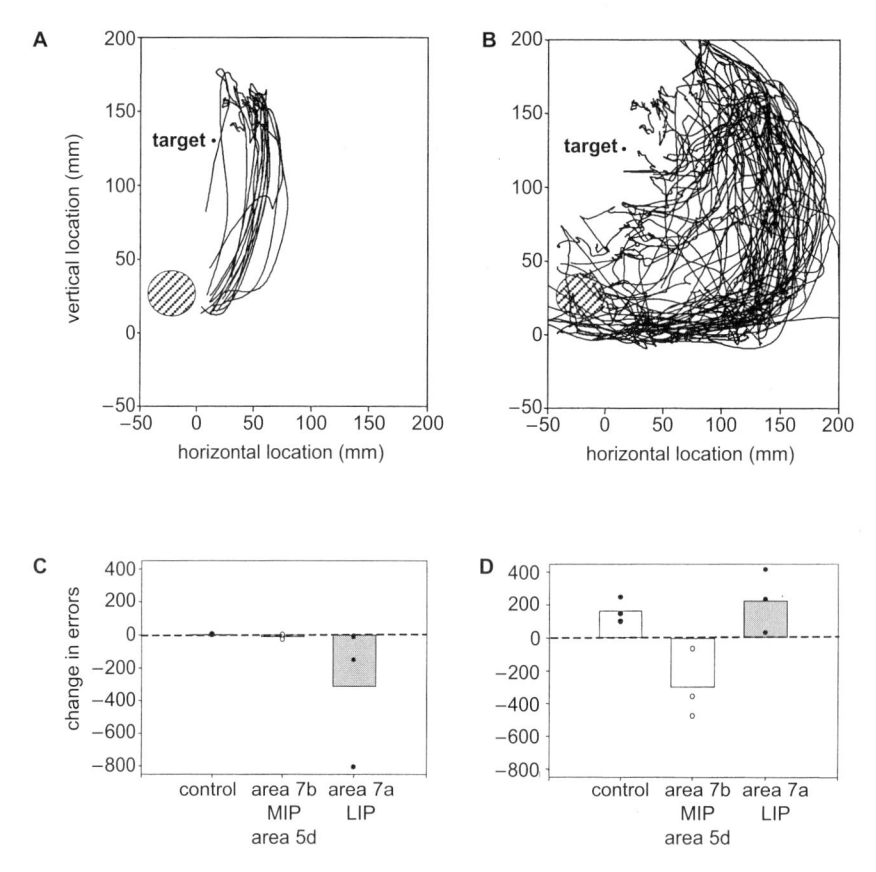

Figure 10.8
Reaching trajectories of a monkey with PPC damage. The starting location of the hand and the target location remain the same for all of these plots. (*A*) Trajectories for movements in which the initial hand location remained the same from trial to trial. The hatched circle marks the initial hand location, and the word "target" appears at the target location for each movement in *A* and *B*. (*B*) Trajectories for movements in which the monkey's initial hand location varied from trial to trial. (*C*) Area 7a and LIP lesions disrupted visually guided reaching in the light (more errors result in a negative score), but area 7b, MIP, and area 5d lesions did not affect this behavior. The bars show each group mean and the circles show the scores of individual monkeys. (*D*) Area 7b, MIP, and area 5d lesions disrupt reaching in the dark, but area 7a and LIP lesions did not. Format as in *C*. (A and B from Rushworth et al.[32]; C and D from Rushworth et al.[33] with permission.)

also receives information from vision about the location of the hand (and other end effectors). Thus, PPC has a special status: It receives *extrinsic*, spatial information from visual cortex and *intrinsic*, proprioceptive information from the somatosensory cortex.

10.4 Errors in Reaching due to Lesions of the PPC

Based on the information it receives, it seems likely that the PPC has visual information about target location, as well as both proprioceptive and visual information that it can align to estimate end-effector location. Without vision, the PPC would have to rely entirely on proprioception to estimate hand location and memory to estimate target location.

Matthew Rushworth, Phil Nixon, and Dick Passingham[32] have provided neuropsychological evidence that different parts of the PPC underlie reaching with versus without vision. They have shown that in order to reach in the dark, one part of PPC—specifically, areas 5d, 7b, and MIP (perhaps including area 5ip)—is necessary for reaching from variable hand locations to a fixed target. First, they trained monkeys to reach in the dark into a box containing a food item. As shown in figure 10.8, monkeys with lesions to those three areas reached very inaccurately if their initial hand location varied, even though the target was always in the same place. In contrast, if the experimenters allowed the monkeys to start their reaching movements from the same place on every attempt, their reaches were much more accurate. This result is what you would expect if the monkey had difficulty in computing hand location from proprioceptive inputs. Compare the results of this experiment against that of deafferentation described in section 8.4. Polit and Bizzi showed that by depriving the CNS of information about initial limb location, by cutting the connection between the sensory nerves and the spinal cord, their monkeys could not point accurately without vision of their arms. The effects of removing areas 7b, MIP, and 5d thus resemble those of deafferentation: In the dark, the monkey seems to need those parts of PPC to use proprioceptive information for computing hand location.

Lesions of other parts of PPC, specifically area 7a and LIP, did not have this effect. Instead, those lesions resulted in misreaching to visible targets (i.e., movement deficits in the light rather than in the dark). Thus, areas 7a and LIP might be necessary for computing the location of a visible target.

References

1. Goodwin, GM, McCloskey, DI, and Matthews, PBC (1972) The contribution of muscle afferents to kinaesthesia shown by vibration induced illusions of movement and by the effects of paralysing joint afferents. Brain 95, 705–748.
2. Levine, MS, and Lackner, JR (1979) Some sensory and motor factors influencing the control and appreciation of eye and limb position. Exp Brain Res 36, 275–283.

3. Lackner, JR, and Shenker, B (1985) Proprioceptive influences on auditory and visual spatial localization. J Neurosci 5, 579–583.

4. Jones, EG, and Powell, TPS (1970) An anatomical study of converging sensory pathways within the cerebral cortex of the monkey. Brain 93, 793–820.

5. Jones, EG, and Powell, TPS (1973). In Handbook of Sensory Physiology, eds Autrum, H, Jung, R, Loewenstein, WR, MacKay, DM, and Teuber, HL (Springer-Verlag, New York), pp 579–620.

6. Johnson, PB, Ferraina, S, Garasto, MR, et al. (1997). In Parietal Lobe Contributions to Orientation in 3D Space, eds Thier, P, and Karnath, HO (Springer-Verlag, Heidelberg), pp 221–236.

7. Johnson, PB, Ferraina, S, Bianchi, L, and Caminiti, R (1996) Cortical networks for visual reaching: Physiological and anatomical organization of frontal and parietal lobe arm regions. Cerebral Cortex 6, 102–119.

8. Asanuma, H (1975) Recent developments in the study of the columnar arrangement of neurons within the motor cortex. Physiol Rev 55, 143–156.

9. Shinoda, Y, Yokota, J-I, and Futami, T (1981) Divergent projection of individual corticospinal axons to motoneurons of multiple muscles in the monkey. Neurosci Lett 23, 7–12.

10. Fetz, EE, Cheney, PD, Mewes, K, and Palmer, S (1991). In Afferent Control of Posture and Locomotion—Progress in Brain Research, eds Allum, JHJ and Hulliger, M (Elsevier, Amsterdam), pp 437–449.

11. Maier, MA, Bennett, MB, Hepp-Reymond, M-C, and Lemon, RN (1993) Contribution of the monkey corticomotoneuronal system to the control of force in precision grip. J Neurophysiol 69, 772–785.

12. Bennett, KMB, and Lemon, RN (1994) The influence of single monkey cortico-motoneuronal cells. J Physiol (London) 477, 291–307.

13. Glickstein, M (2000) How are visual areas of the brain connected to motor areas for the sensory guidance of movement? Trends Neurosci 23, 613–617.

14. Glickstein, M, May, JG, and Mercier, BE (1985) Corticopontine projection in the macaque: The distribution of labelled cortical cells after large injections of horseradish peroxidase in the pontine nuclei. J Comp Neurol 235, 343–359.

15. Bosco, G, Poppele, RE, and Eian, J (2000) Reference frames for spinal proprioception: Limb endpoint based or jointlevel based? J Neurophysiol 83, 2931–2945.

16. Bosco, G, and Poppele, RE (2001) Proprioception from a spinocerebellar perspective. Physiol Rev 81, 539–568.

17. Bosco, G, Rankin, A, and Poppele, RE (1996) Representation of passive hindlimb postures in cat spinocerebellar activity. J Neurophysiol 76, 715–726.

18. Helms-Tillery, SI, Soechting, JF, and Ebner, TJ (1996) Somatosensory cortical activity in relation to arm posture: nonuniform spatial tuning. J Neurophysiol 76, 2423–2438.

19. van Beers, RJ, Sittig, AC, and Denier van der Gon, JJ (1998) The precision of proprioceptive position sense. Exp Brain Res 122, 367–377.

20. Helms-Tillery, SI, Ebner, TJ, and Soechting, JF (1995) Task dependence of primate arm postures. Exp Brain Res 194, 1–11.

21. Poizner, H, Clark, MA, Merians, AS, et al. (1995) Joint coordination deficits in limb apraxia. Brain 118, 227–242.

22. Clark, MA, Merians, AS, Kothari, A, et al. (1994) Spatial planning deficits in limb apraxia. Brain 117, 1093–1106.
23. Haaland, KY, Harrington, DL, and Knight, RT (2000) Neural representations of skilled movement. Brain 123, 2306–2313.
24. Kalaska, JF, Caminiti, R, and Georgopoulos, AP (1983) Cortical mechanisms related to the direction of two-dimensional arm movements: Relations in parietal area 5 and comparison with motor cortex. Exp Brain Res 51, 247–260.
25. Scott, SH, Sergio, LE, and Kalaska, JF (1997) Reaching movements with similar hand paths but different arm orientations. II. Activity of individual cells in dorsal premotor cortex and parietal area 5. J Neurophysiol 78, 2413–2426.
26. Scott, SH, and Kalaska, JF (1997) Reaching movements with similar hand paths but different arm orientations. I. Activity of individual cells in motor cortex. J Neurophysiol 77, 826–852.
27. Scott, SH, and Kalaska, JF (1995) Changes in motor cortex activity during reaching movements with similar hand paths but different arm postures. J Neurophysiol 73, 2563–2567.
28. Georgopoulos, AP, Caminiti, R, and Kalaska, JF (1984) Static spatial effects in motor cortex and area 5: Quantitative relations in a two-dimensional space. Exp Brain Res 54, 446–454.
29. Lacquaniti, F, Guigon, E, Bianchi, L, et al. (1995) Representing spatial information for limb movement: Role of area 5 in the monkey. Cerebral Cortex 5, 391–409.
30. Obayashi, S, Tanaka, M, and Iriki, A (2000) Subjective image of invisible hand coded by monkey intraparietal neurons. NeuroReport 11, 3499–3505.
31. Graziano, MS (1999) Where is my arm? The relative role of vision and proprioception in the neuronal representation of limb position. Proc Natl Acad Sci USA 96, 10418–10421.
32. Rushworth, MFS, Nixon, PD, and Passingham, RE (1997) Parietal cortex and movement. I. Movement selection and reaching. Exp Brain Res 117, 292–310.
33. Rushworth, MFS, Nixon, PD, and Passingham, RE (1997) Parietal cortex and movement. II. Spatial representation. Exp Brain Res 117, 311–323.
34. Soechting, JF, and Flanders, M (1989) Sensorimotor representations for pointing to targets in three-dimensional space. J Neurophysiol 62, 582–594.
35. Soechting, JF, and Flanders, M (1989) Errors in pointing are due to approximations in sensorimotor transformations. J Neurophysiol 62, 595–608.

Reading List

The work of John Soechting and Martha Flanders[34,35] is of interest with respect to the coordinate system used for forward kinematics. They showed that when people reach in the light, the distribution of errors suggests a coordinate framework in eye- and/or head-centered coordinates.

11 Computing Target Location

Overview: In order to control a reaching or pointing movement, your CNS computes the difference between the location of a target and the current location of the end effector. In computing target location, your CNS combines information about the location of the target on the retina with information about eye and head orientation. Neurons in the PPC encode this information in a multiplicative way.

Chapter 9 introduced an imaginary robot and an equally imaginary robotic engineer. If you ask him or her to explain how to program a robotic arm to move its end effector, a gripper, from one point to another, he or she might do so in steps. In one approach, the robot's **controller** first needs to figure out the current location of the gripper. Next, it needs to know the location of the target. By comparing these two vectors in some common coordinate system, the controller can compute a *high-level plan* about what the robot should do. After the controller performs this computation, it needs to decide the details of the plan. Will the robot make a fast movement or a slow movement? Will it move the gripper in a straight line to the target, or must it avoid obstacles along the way? All of these problems concern kinematics because they do not involve forces. As a final step, the engineer might consider the dynamics of the robotic arm (and whatever it may be holding in its gripper). Depending on the mechanics of the robot's actuators, the controller may have to compute the forces needed to make the movement so that when it sends the commands to the robot's torque motors, the arm moves its gripper to the target as planned.

How does your CNS move *its* end effectors from one location to another? For reaching and pointing movements, the CNS appears to compute the location of the target with respect to the end effector. In chapter 9 you saw how—at least in theory—the CNS could compute the location of an end effector from joint angles. Later, in chapter 12, you will begin to consider evidence that the high-level plan mentioned above depends on something called a *difference vector*, which points from the estimated end-effector location to the estimated target location. This chapter takes up the computation of target location. Later, in chapters 18 and 19, you will consider how the CNS transforms this desired difference vector into a trajectory that specifies the speed of the movement and its path. Then, chapters 20–22 take forces into account. They address how the CNS computes the forces needed to transform this trajectory from a plan into an action.

11.1 Computing Target and End-Effector Locations in a Common Frame

Evidence from psychophysical experiments indicates that before you reach for a target, your CNS computes the location of the target with respect to an estimate of the current location of your hand (section 12.1). The result of this computation corresponds to a difference vector (x_{dv} in figure 11.1A). How could target location be computed in such a way that a difference vector could emerge later?

The only transducer that can detect a visual target is the retina, and it can relay the location of the target's image only in its own coordinate system. However, the retina sits on a rather complicated set of structures that move. Eyes rotate within the orbit, the head rotates on the shoulders, and the body can move both by rotation (around the trunk) and by translation (by walking, etc.). Therefore, the location of the target's image on your retina does not provide your CNS with enough information to compute a target's location in any coordinate frame other than that of the retina. But the sensors that transduce the arm's configuration (proprioceptors) have intrinsic, joint-based coordinates, as do the actuators that move the limb (muscles). It makes sense to transform all of this information into a common coordinate frame to compute the difference vector, but which coordinate frame?

Early ideas about the common coordinate frame for target and end effector representations in the CNS focused on body-centered coordinates. This assumption makes sense because your body does the moving. To compute the location of a reach target in body-centered coordinates, your CNS needs to know the orientation of your eyes in the orbit and the angle of the head relative to the shoulder, as well as the location of the target on the retina. For example, with this sort of information, you could compute the location of the target as a vector with respect to a coordinate system based on the shoulder [S_{ext}] (see figure 11.1B and table 9.1). From arm proprioception, you could compute end-effector location in the same coordinate system [S_{ext}]. A simple vector subtraction could then yield the difference vector, also represented in [S_{ext}]. With this computational approach, the locations of the end effector and the target would be represented in a coordinate system that falls somewhere between those for vision and proprioception. (See box 11.1.)

However, you can think of the problem in a different way and still arrive at a difference vector that points from the hand to the target. Imagine that your right hand is under a table and that you are fixating an object on the table. If the tabletop is transparent glass, you see both your hand and the target. Therefore, you can represent the locations of the target and your hand as vectors with their tails at your fixation point. This

11.1. Computing Target and End-Effector Locations in a Common Frame

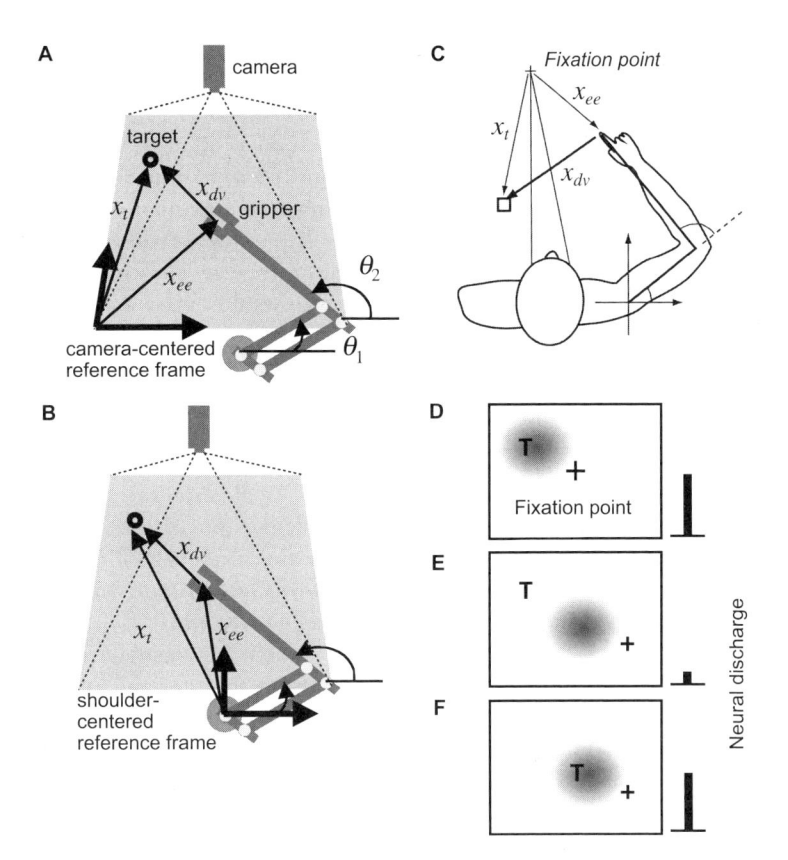

Figure 11.1

Encoding the location of an end effector and a target. (*A*) A virtual robot–camera system computes an end-effector location (*ee*, a gripper) and a target location (*t*) in camera-centered coordinates (pixel locations of the camera). The robot's controller then subtracts vector x_{ee} from vector x_t to produce a difference vector (*dv*). The thicker arrows at the lower left show the axes of the coordinate frame used to represent these vectors. θ_1 and θ_2 indicate the robot's joint angles. (*B*) The analogous vectors as computed in shoulder-centered coordinates. Format is as in *A*. (*C*) The analogous vectors represented with respect to the fixation point. (*D–F*) Retinotopic coordinates. A target (T) appears briefly on a screen while a monkey fixates the location indicated by the plus sign (+). The stimulus appears near the center of the cell's activity field (indicated by the fuzzy gray circle) and the neuron discharges robustly (vertical bar to the right). (*E*) The stimulus appears at the same location on the screen, but the monkey fixates a different location. As a result, the neuron discharges less than in *D* because the image falls outside the cell's activity field. (*F*) The images in *D* and *F* fall on the same retinal location, and therefore the cell discharges at the same rate as in D.

Box 11.1
The ancestral function of eye movements

> A common mistake is to view the world of other animals though your own eyes. For example, eye movements might seem to function primarily for scrutinizing objects with your **fovea**. Humans—and other primates—do indeed bring the fovea to bear on objects in the field of view. However, this fact can lead to serious misconceptions. As chapter 2 discusses, the earliest vertebrates likely evolved as mobile predators that used vision. One of the first problems their immediate ancestors had to solve was the blurring of their visual image whenever they moved their bodies or their heads. The problem is that retinal photoreceptors work relatively slowly, taking ~10–20 msec to reflect changes in illumination. Accordingly, movements of the head and body must be canceled for at least that long to prevent image blur. The mechanisms that stabilize the retina relative to the external world must have evolved in the earliest vertebrates. Saccadic eye movements and a saccade-and-fixate motor strategy thus both predate the development of a fovea. Indeed, many vertebrates do not have a fovea, at all. Therefore, foveation is not an ancestral function of the eye-movement system.
>
> **Primitive** brainstem mechanisms called the vestibulo-ocular reflex (VOR) and the optokinetic response perform the original function of the oculomotor system: blur reduction. Their purpose is not moving the eyes in intrinsic coordinates, but rather keeping the eyes still in extrinsic ones. Gordon Walls recognized in the 1940s that, at heart, eye movements function to "keep the visual field as nearly constant as possible during locomotion and during passive jogglings of the head and body." The VOR (see section 24.2.4) uses vestibular inputs to move the eyes at the same speed as head movements, but in the opposite direction. The optokinetic response uses retina motion detectors to move the eye smoothly in the same direction as that motion. These reflexes "clamp" the photoreceptor surface to the world to prevent "retinal slip," for at least long enough to allow signal transduction. Once those reflexes reach their limits, the saccadic mechanism minimizes the time the eye takes to reorient, so the image can again stabilize as quickly as possible.

would amount to a *fixation-centered* coordinate frame [F] (figure 11.1C), and much of this book is based on the idea that you indeed use this frame to guide reaching and pointing. (A fixation-centered frame resembles a *retinotopic* one, but there are some important differences that will be elaborated later.) A subtraction of one vector in this coordinate system from the other $(x_t - x_{ee})$ would yield a difference vector x_{dv} *similar* to the one computed in shoulder-centered coordinates. Note that the difference vector is *similar* in the two coordinate systems, but not necessarily exactly the same. Despite the fact that the difference vector always points from the estimated hand location to the target, in one case it has fixation-centered coordinates, and in another case it has shoulder-centered ones. If a given coordinate system distorts space differently than another, the difference vector will reflect that distortion.

Therefore, you do not have to represent the location of the target with respect to the *shoulder* in order to compute the difference vector. Indeed, any coordinate system will do. (Of course, if you think of a reaching movement as simply a process of minimizing a hand-to-target difference vector, then in the situation described here you would smash your hand against the underside of the glass table. Chapter 19 discusses the reasons that this does not happen in terms of planning and control *policies*.)

Now imagine a very different, and safer, situation: An opaque table top prevents you from seeing your hand. Your CNS might still estimate where your hand would fall on your retina if the table were transparent. That is, your CNS could still compute the location of your hand in fixation-centered coordinates, even though you cannot see your hand. It could do this by using proprioceptive information from the arm to compute hand location with respect to the shoulder [S_{ext}], and then use head orientation (relative to the shoulder girdle) and eye orientation (relative to the head) to compute hand location in fixation-centered coordinates (see figure 9.2). This computation results in a vector that estimates where the image of the hand would fall on the retina *if it could be seen*. In the end, the difference between hand and target locations would result in a similar difference vector for both transparent and opaque tabletops. The similarity of the two difference vectors—one computed with full vision, the other computed without it—depends on the accuracy of the fixation-centered estimate of hand location from proprioception.

Because the PPC combines information from various sensory modalities, most early work on PPC assumed that its neurons transform the reach target's location from retinotopic to shoulder-centered coordinates. More recent research, however, suggests that in parts of PPC, neural networks transform hand location in joint-centered coordinates into a fixation-centered frame. These data accord with psychophysical findings, presented here as well as in chapter 12, which indicate that the CNS plans reaching and pointing movements in an extrinsic, vision-based coordinate frame.

11.2 Computing Target Location in a Vision-Based Frame

11.2.1 Retinotopic Coordinates

When sensory neurophysiologists record from a neuron that appears to respond to a sensory stimulus, they typically characterize its discharge in terms of a *receptive field*. For example, the receptive field of a visually responsive neuron might describe the location on the retina where an image triggers a response. The cell would have a maximum discharge at the center of the receptive field, and its discharge would typically decline as the image appears progressively farther from that location. Such a receptive field, as traditionally defined, has a retinotopic frame [R]. A retinotopic receptive field moves with the eye (i.e., the location in external space that activates a neuron will change precisely as the orientation of the eye changes in the orbit).

Figure 11.1D illustrates this principle. Its basic features apply to many of the neurons that are located "early" in the visual pathway, such as those of retinal ganglion cells and cells in the thalamus that relay information to cortex. When you look at a stimulus, depicted in the figures as a "+" sign, you fixate it with your fovea. This location is the fixation point. Now, if someone flashes another stimulus somewhere in space (indicated by "T" in this figure) and that image falls within the receptive field of a cell, that neuron will discharge vigorously (figure 11.1D). When the fixation point moves but the location of the flashed stimulus remains constant, the cell's discharge decreases because the stimulus now falls outside the cell's receptive field (figure 11.1E). However, when the stimulus and fixation point change locations together (figure 11.1F), the discharge is the same as that displayed in figure 11.1D. Therefore, the cell's discharge rate signals the location of the stimulus on the retina and does so regardless of the orientation of the eye in the orbit.

For the control of eye movements, you might expect the encoding of target location in retinotopic coordinates. To understand this idea, consider the concepts of motor fields and activity fields. A *motor field* describes the targets of movement that activate a cell. An *activity field* simply describes the places associated with high activity levels, regardless of whether a cell seems mainly sensory or mainly motor in function. Imagine a cell with a motor field that moves the eyes 10° to the right of the fixation point. You might expect that such as cell would respond to stimuli 10° to the right of the fovea. That is, in a simple case, the cell's receptive field should correspond to its motor field. After the eyes have shifted 10°, so that the fixation point shifts 10° to the right of where it started, the cell's activity field should remain at 10° to the right. Thus, activating the neuron should contribute to a further shift in eye orientation of 10° to the right, and so forth. The traditional view of saccade-related activity in the superior colliculus follows this line of thinking.

11.2.2 Fixation-Centered Coordinates

Consider again the imaginary robot of figure 11.1A, and imagine that now you are the robotics engineer. You might choose to emulate the coding of target location in a retinotopic frame, which would roughly correspond to a situation in which you have a stationary camera above your imaginary robot and you represent target location in terms of pixel locations. One problem with this *camera-centered* coordinate system is that potential targets may appear outside the camera's field of view. So you buy a small motor and install the camera on a rotary joint that allows it to swivel around an axis. Your representation of space now depends not only on pixel locations in camera-centered coordinates but also on the angle of the camera "joint." Therefore, when a target appears, you will represent its location using both the pixel value on the camera's image plane and the angle of the camera.

Your retina resembles the camera's image plane in that both have a limited field of view and both can reorient. Your eyes can move in the orbit, and your head can move relative to your shoulder girdle. Imagine that you fixate an object as a spot of light briefly flashes in your peripheral field of view. If you wanted to move your eyes to the light spot immediately, then it would be sufficient to represent target location in retinotopic coordinates. However, what if something distracts you, and before you make a saccade to fixate the target, you fixate an intermediate location, say halfway to the target. If your CNS represented the target location only in terms of where the image of the light flash fell on the retina, then you could not compensate for this first saccade and your second saccade would overshoot the target by the amplitude of the first saccade. To make the second saccade accurately, you must represent the location of the target in terms of both the location of its image on your retina and the orientation of your eye in the orbit. When your head moves, the representation of target location also needs to include head orientation.

If you represent orientation of eye and head with vectors e and h, and an image location on the retina with vector r (with respect to the fovea), then any triplet $[e, h, r]$ establishes a unique location in Cartesian space $[x, y, z]$. You may be fixating some point $[x_0, y_0, z_0]$ with configuration $[e_0, h_0, 0]$. If a target light flashes in your peripheral field of view, its image may fall on some part of your retina r_t. The location of the target in fixation-centered coordinates would then be $[e_0, h_0, r_t]$. Now, if you were to fixate some other point by moving your eyes and head by amount $[\Delta e, \Delta h]$, the target's location, though it may be unseen, would become $[e_1, h_1, r'_t]$, where $e_1 = e_0 + \Delta e$, $h_1 = h_0 + \Delta h$, and r'_t equals the *remapped* location of the target with respect to the current fixation point. The value r'_t represents a fixation-centered representation of a potential target's location.

What might be the advantage of representing locations with respect to the fixation point rather than the shoulder or some other body part? Computationally, no compelling advantage presents itself. In all cases, the CNS needs to combine information about the orientation of the eye and head with the location of an image on the retina in order to represent—in some "internal" coordinate system—the location of a movement target. However, there may be biological advantages to fixation-centered coordinates. First, for fixation-centered coordinates, the representation of potential targets for reaching and pointing corresponds to the one used for planning eye movements to the same targets. Second, vision provides primates with the most accurate spatial information, especially for information at a distance. And third, maintaining a representation of space in terms closely aligned with vision might allow for easier calibration or error detection.

Accordingly, the model presented here relies on a fixation-centered coordinate scheme. Nevertheless, you should recognize that a lot of the experimental work described later in this chapter, especially that regarding the concept of a **gain field**, was based on the presumption that PPC computes target location in head-centered or shoulder-centered coordinates. The pioneers of the gain-field concept argued that retinotopic signals [R] and signals about eye and head orientation—called extra-retinal signals [xR]—combined in the PPC to produce a representation of target location in head-centered coordinates [Cr] or in extrinsic, shoulder-centered coordinates [S_{ext}]. (See table 9.1 for this notation and nomenclature of reference frames.) According to this concept, the CNS computes [R] + [xR] → [Cr] or [R] + [xR] → [S_{ext}].

Recent work, however—and the model used in this book—assumes that to control visually guided reaching and pointing movements, the CNS computes [R] + [xR] → [F], where [F] is based on [R] but extends beyond the visual field. You can call [R] a *retinotopic* coordinate frame and [F] a *fixation-centered* coordinate frame. The original view held, in essence, that your CNS needs to know [R] and [xR] in order to calculate target location in head-centered or shoulder-centered coordinates. This view assumed that the difference vector x_{dv} was computed from x_t and x_{ee} in frame [S_{ext}] or [Cr]. The present view holds that your CNS needs to know [R] and [xR] in order to recalculate target location with respect to the fixation point after the eyes move, including when they move so much that the target leaves the visual field entirely. On this view, the difference vector x_{dv} is computed from x_t and x_{ee} in frame [F].

11.2.3 Errors in Reaching due to Intervening Eye Movements

For reaching and pointing movements, people usually overestimate a visual target's retinal eccentricity (i.e., its distance from the fovea and, therefore, from central vision).[1,2] For example, if you look straight ahead and see a target to your left, when you reach without fixating the target, you will reach a little too far to the left (figure 11.2A).

Denise Henriques, Doug Crawford, and their colleagues used this observation to differentiate between a craniocentric representation of the target and one based on vision.[3] They performed an experiment that involved reaching to the remembered location of a target. The experimenters flashed a light, and after the participants fixated that point, the light was turned off and a second light appeared. After fixation on this second light, the participants reached to the remembered location of the first one. The experimenters argued that if people represented the target of movement in craniocentric, shoulder-centered, or some other body-centered coordinates, then it should not matter that the eyes had moved. On the other hand, if the target was represented in fixation-centered coordinates, then movement of the eyes required a remapping of that location. For example, imagine that after foveating the target of reach, your eyes moved 20° to the left (figure 11.2B). The remembered target location no longer corresponds to foveal, central vision, but now lies 20° to the right of the fixation point. Because of the tendency to overshoot targets in their direction relative to the fovea, this experiment can tell you something about how your CNS represents the target. If, after your eye move-

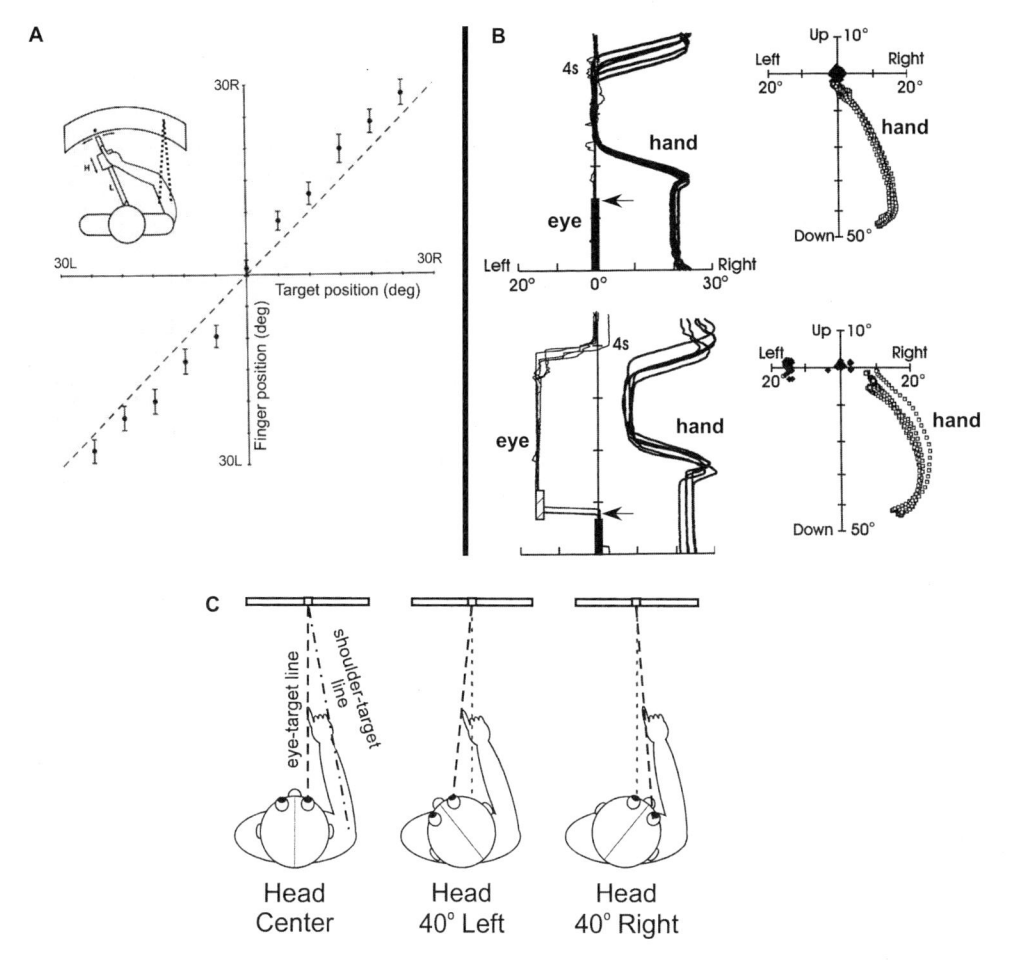

Figure 11.2
Fixation-centered remapping of remembered target location. The tasks involved reaching to a remembered target in the dark. (*A*) Participants fixated straight ahead (at target location 0, dotted lines) while a target flashed in the periphery. They then moved a lever to point at the perceived location of the target. The dashed line shows accurate pointing, and the circles show the lever's location at the end of movement, which overshot the target. (*B*) Two-dimensional view of hand trajectory (squares, from bottom to top) and eye orientation (black diamonds). Time runs up. Top: Trajectories of eye and hand in a control condition in which the eyes fixated the target location, although the target had disappeared (arrow). Note that the hand reached accurately to the target at 0° on the *x*-axis. Bottom: Fixation began at 0°, but then moved to a second fixation point, 15° to the left, after the reach target disappeared (arrow). Note that this put the representation of the target 15° to the right of the fovea. The participants reached ~10° too far to the right (squares). (*C*) With the head straight ahead (left), pointing toward a target involved positioning the fingertip on a line connecting the *viewing* eye (in this case, the right eye) to the target, not the shoulder to the target. However, with the head to the left (center) or right (right), the arm remained on the line from the viewing eye to the target (thick dashed line), although this required a slightly different finger location than for the left (thin dashed line). Pointing depended on aligning the fingertip so that its image coincided with the image of the target on the retina of the *viewing* eye. (A from Bock,[1] B from Henriques et al.,[3] C from Henriques and Crawford[6] with permission.)

ment, the memory of the target location remains represented in terms of its remapped location on the retina, then your reach will miss the target to the right.

The experimental results[3] confirmed this prediction (figure 11.2B). It appears that the representation of target location remains in fixation-centered coordinates even after intervening eye movements. Related results show that reaching remains inaccurate even when the binocular fixation point is displaced from the target in its depth dimension[4] and when the target is acoustic.[5]

Pointing also relies on a computation in fixation-centered coordinates. When you point to something in front of you, you orient your arm so that your fingertip falls on the line running from the target to the fovea of your dominant, or *viewing*, eye (typically your right eye).[6] Thus, for accurate pointing, the image of your fingertip falls on your fovea as you fixate the target (figure 11.2C). When your head moves, despite the fact that the target remains stationary, you will have to reorient your arm slightly so that the fingertip remains on the line from your viewing fovea to the target. Note that your fingertip does not fall on the line that connects your hand to your shoulder (figure 11.2C, right). Pointing, like reaching, depends on visual coordinates.

Figure 11.3 captures these important concepts. As the person waits for the fixated fly to come within reach (of the flyswatter), he hears the buzzing of another fly above and slightly behind him. Although this new target for his end effector remains out of sight, its acoustically detected location can be computed in a visual, fixation-centered coordinate frame. Accordingly, the difference vector x_{dv} results from computations in that frame.

11.3 Combining Retinal Location with Eye Orientation Through Gain Fields

Single neurons in the PPC often respond to both proprioceptive and visual information. The discharge of these neurons has a retinotopic activity field that modulates approximately linearly as a function of eye or head orientation.

11.3.1 Modulating the Gain of a Visual Activity Field as a Function of Eye Orientation

Recall that retinal ganglion neurons and neurons in the lateral geniculate nucleus of the thalamus code stimulus location in terms of a retinotopic coordinate system [R] (as in figure 11.1D). In contrast to these cells, Richard Andersen and his colleagues found that many cells in PPC had activity which reflected both the location of the stimulus on the retina and the orientation of the eyes in the orbit.[7]

Andersen and his colleagues trained monkeys to fixate a small light spot that appeared at different locations on a screen. The monkeys began

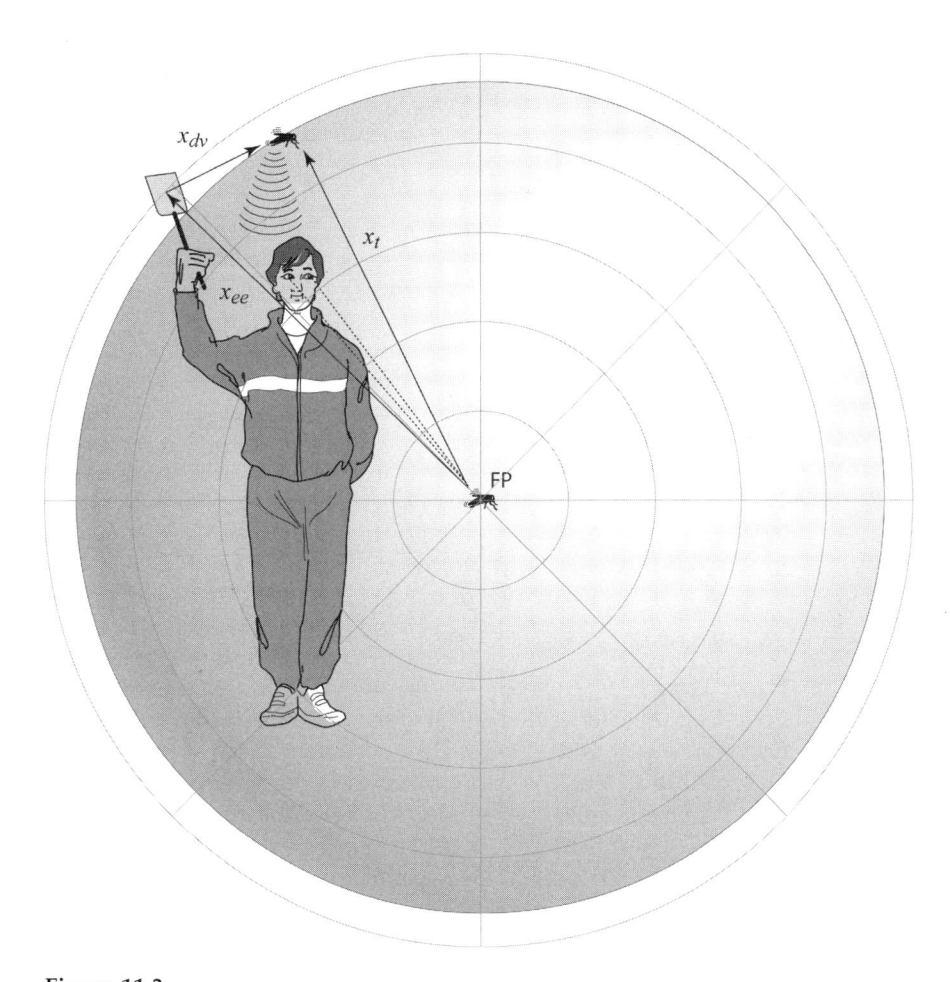

11.3. Combining Retinal Location with Eye Orientation Through Gain Fields

Figure 11.3
Fixation-centered representation of targets, end effectors, and difference vectors. The person fixates on a fly (at the center of the circle) while waiting for it to come near enough to swat. In the meantime, he hears another fly within swatting distance, localizes it acoustically, and plans a movement of the end effector (the flyswatter). Despite the fact that he cannot see the target, and despite its location a little behind and above his head, his CNS plans the movement in visually based, fixation-centered coordinates. Abbreviation: FP, fixation point. (Drawn by George Nichols.)

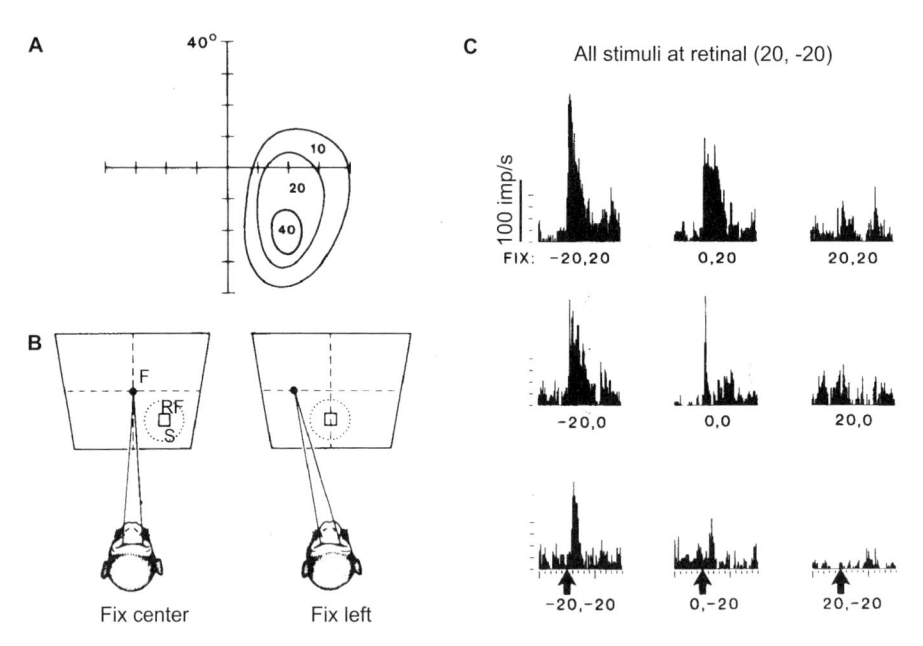

Figure 11.4

Gain fields. Recording from a cell in the PPC (area 7a) of a monkey. (*A*) Activity field of a neuron as the monkey fixates straight ahead at screen center (0°, 0°), as illustrated in *B*, left. The contours represent the mean increase in discharge rate from baseline in spikes per second. The center of the activity field for this cell is approximately at (+20°, −20°). (*B*) The monkey fixates at point F while a stimulus (S) appears for 500 msec in the region marked by the small box. The dotted circle shows the activity (or receptive) field (RF) of the cell as measured in retinotopic coordinates. The experiment involved moving the fixation point to several locations, such as left of center, as illustrated at the right. (*C*) Discharge of the cell as a function of fixation point. The arrow indicates the time of stimulus presentation. Despite the fact that the stimulus appears at the same retinotopic location in all nine plots, the cell's activity varies with fixation point (located at the coordinates given beneath each plot). (From Andersen et al.[7] with permission.)

each behavioral "trial" by fixating the center of the screen (coordinate 0, 0 in figure 11.4A). In this experiment, the experimenters recorded from area 7a of the PPC (see figure 6.3). As the monkeys maintained central fixation, a second visual stimulus appeared on the screen for 500 msec to map the cell's activity field. (Recall that you can use the term "activity field" when you do not want to distinguish among receptive fields, motor fields, and other kinds of fields.) Stimuli activated the cell described in figure 11.4A when they appeared to the right and below the fixation point, with maximal activity for screen coordinates +20°, −20°.

Next, a new fixation point appeared. Because the monkeys' heads were fixed, they could only move their eyes to fixate that location. At each new fixation point, a stimulus always appeared at the same retinotopic location (i.e., 20° to the right and below the fixation point in the center of the cell's activity field). Despite the fact that the stimulus's image fell on the same patch of retina, the cell's discharge differed depending on the location of the fixation point. Therefore, Andersen and his colleagues concluded that the cell's discharge reflected both where the stimulus fell on the retina in frame [R] (i.e., its retinotopic location) and the orientation of the monkeys' eyes in the orbit.

In the experiment just described, most of the cells behaved as if they combined information from retinotopic coordinates [R] with eye orientation (e). The remaining cells fell into two main classes: (1) cells that seemed to respond to changes in e but had no apparent activity field, and (2) cells that had an activity field in retinotopic coordinates [R] but were not influenced by e. The former population discharged as a Gaussian (i.e., bell-shaped) function of stimulus location in [R], whereas the latter population of cells changed their discharge rates linearly with e. For the majority of the cells, however, both e and stimulus location in [R] influenced discharge. Many parts of PPC have similar properties.[8–12]

You can see from figure 11.4A that when the monkeys fixated the center location, discharge approximated a Gaussian function of the location of the stimulus on the retina: It had a single peak at the *center* of its activity field, and activity fell off as distance increased from that peak. For cell i, if you label the location of the stimulus on the retina as r, a vector describing the location of the stimulus with respect to the fovea, the cell's discharge was proportional to

$$\exp \frac{-(r - r_i)^T (r - r_i)}{2\sigma^2},$$

where r_i is the center of the activity field in retinal coordinates (location $+20°, -20°$ for the cell in figure 11.4A) and σ describes the width of the Gaussian ($\sim 10°$ for that cell). Andersen and his colleagues discovered that as the fixation point changed, the discharge retained its Gaussian shape with respect to the stimulus location on the retina. However, for the same stimulus location in retinotopic coordinates, discharge changed approximately linearly as a function of the fixation point. That is, discharge scaled linearly as a function of e.

Figure 11.5 shows four neurons as examples. In each plot, each curve represents the stimulus-related activity of a neuron for a different fixation point. The figure plots the neuronal responses with respect to retinotopic location of the stimulus along either a horizontal or a vertical axis passing through the center of the cell's activity field. The x-axis of each plot represents stimulus location with respect to the center of the activity field. Note that the activity function *scaled* when e changed from one fixation point to another.

Note also that, at least for these cells, the response function did not simply shift up or down with changes in e, which would indicate an additive interaction between the eye orientation and retinotopic signals. The results indicate that the interaction of the stimulus location in frame [R] with e is a *multiplicative* one.

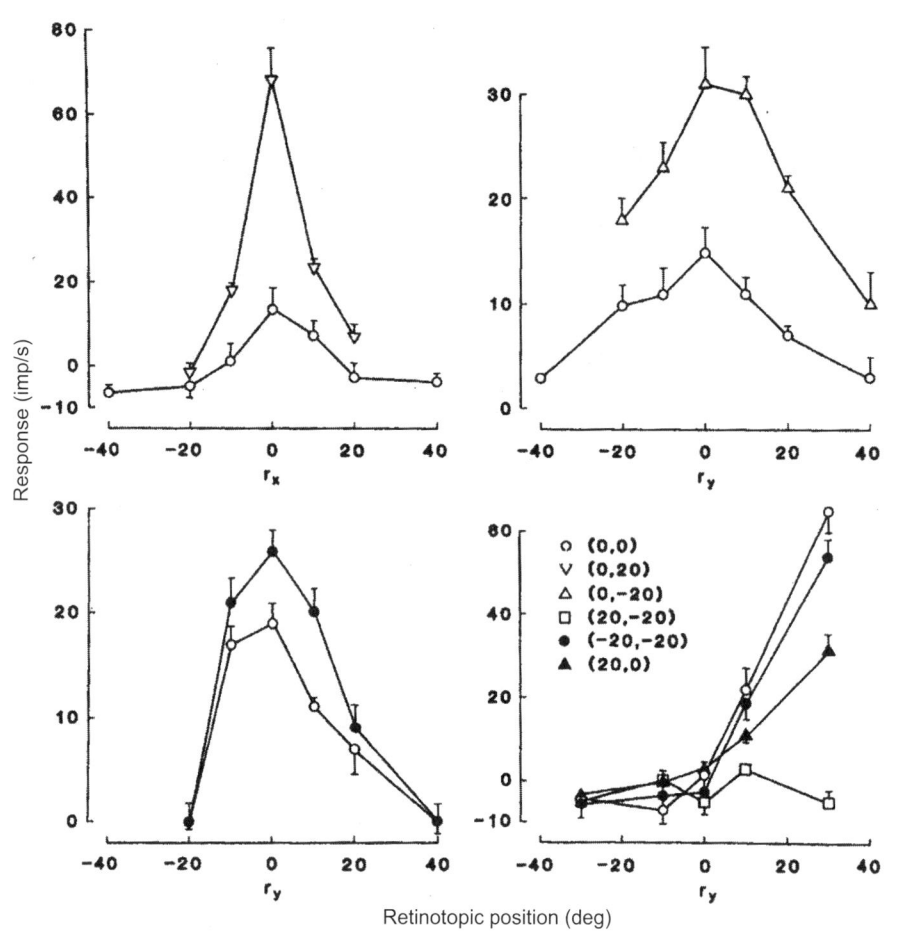

Figure 11.5
Gain fields. Mean discharge rates for different eye orientations plotted in retinotopic coordinates along horizontal (r_x) or vertical (r_y) axes passing through the centers of the activity fields of four neurons. Each graph shows the data for one neuron. The tuning curves have spontaneous, background activity subtracted. (From Andersen et al.[7] with permission.)

These properties characterize a gain field, and its components often have collinearity in their preferences. For example, if the activity field falls to the right of the fixation point, then the cell's response scales up for eye orientations to the right. Collinearity, in this sense, has been observed for smooth-pursuit neurons in LIP[13] and for visual responses in area V3a.[14] Collinearity promotes coordinate transformations,[15,16] and computational models[17,18] indicate how this property could be used in neural networks that perform multidirectional computations such as comparisons between sensory modalities, sensorimotor transformations, and transformations from motor to sensory coordinates.[5]

11.3.2 Modulating Gain by Eye Orientation: Theory

Andersen and his colleagues showed that if you represent eye orientation e as a vector that has two angular components (for example, an angle with respect to a horizontal plane and an angle with respect to a vertical plane), you can describe a cell's discharge rate p_i as follows:

$$p_i = (k_i^T e + b_i) \exp \frac{-(r - r_i)^T (r - r_i)}{2\sigma^2}, \qquad (11.1)$$

where k_i and b_i represent the cell's gain and bias parameters. For example, k_i would scale both the horizontal and the vertical eye-orientation signals. This formula represents the activity "evoked" by the stimulus. That is, p_i represents the change in discharge with respect to baseline activity (measured over a short period before the stimulus appears).

Ignore for the time being the fact that baseline discharge may change as the monkey fixates different points. Indeed, Andersen and his colleagues found that in their largest group of cells, baseline activity remained constant, but for many other neurons the baseline varied as a function of e. *Gain modulation* of this type provides a mechanism by which a neuron could simultaneously encode two variables (see section 11.3.4 for elaboration of this point). But how could your CNS use that information to "internally" represent the location of a visual target?

To investigate this question, David Zipser and Richard Andersen[19] took a pioneering step: They developed a computer model with which to compare activity in PPC. (See box 11.2.) They modeled a group of 64 cells in which each cell had an activity field that was a Gaussian function of stimulus location. That is, each cell had a preferred location on this two-dimensional screen where its discharge peaked and dropped off with $\sigma = 15°$. They distributed the preferred locations at $10°$ intervals in their model.

Zipser and Andersen reasoned that these cells composed the output of a neural network and suggested that this network should have two kinds of inputs: orientation of the eye in the orbit and stimulus location on the retina. By analyzing how this network transformed these two inputs into a representation of the stimulus that accounted for both factors, Zipser and Andersen explored whether the particular coding of equation 11.1 plays a role in the representation of target location.

Recall that Andersen and his colleagues found that a minority of PPC cells lacked a retinotopic activity field, and instead discharged solely as a linear function of eye orientation. They also found another minority of cells that had a retinotopic activity field but no sensitivity to eye orientation. Zipser and Andersen reasoned that these were examples of how neurons in the first "stage" of the computation coded eye orientation and retinotopic stimulus location. They assumed that in the last stage of computation, neurons should code stimulus location in a way that combined these two sources of information. Therefore, they created a model with neurons between these two stages that combine and transform the signals.

Box 11.2
Models in Biology 1

> Because this chapter introduces the first detailed discussion of a neural network model, a comment on models in general may be helpful. The models presented in this book have many gaps and errors, as do all models. Any departure from veridicality in modeling inevitably draws criticism, but this book takes a different view, one based on the knowledge that models have a long and productive history in biology. Take, for example, the model developed in the 1930s by Hugh Davson and James Danielli.[20] They knew that hydrophobic substances pass through membranes more easily than water-soluble (hydrophilic) ones, and that a membrane composed entirely of fatty molecules called phospholipids did not account for these data. Accordingly, they modeled the membrane of living cells as a two-layered sheet, with each sheet composed of two layers, one consisting of proteins and the other of phospholipids. They reasoned that if two of these sheets composed a cell's membrane, with the fatty sides touching, this architecture would form a stable barrier between the inside and the outside of a cell. The proteins, being more hydrophilic, would interact well with the inside of the cell (the cytoplasm) and the outside of the cell, both of which are mainly aqueous. Their model seemed to account for most of the observations about cell permeability. Of course, the model was highly flawed—it did not take into account ion channels, transport mechanisms, and other properties now known to be part of biological membranes. Nevertheless, their model represented a stunning advance in understanding the structure of cell membranes, and the model's estimate of membrane thickness proved remarkably accurate. The Davson–Danielli model held up for over 30 years and remains a landmark accomplishment in the history of biology, one showing that models, however flawed and incomplete, can nevertheless contribute greatly to the understanding of complex biological systems.

These "in-between" neurons form a second (or "hidden") layer in a three-layer neural network.

In their original way of thinking about this model, Zipser and Andersen theorized that PPC computes $[R] + [xR] \rightarrow [Cr]$, that is, it combines retinotopic and extraretinal (eye-orientation) signals in order to compute target location in head-centered coordinates. The model presented here (and developed on the way to figure 14.1) assumes that the CNS computes $[R] + [xR] \rightarrow [F]$ (i.e., it combines retinotopic and extraretinal signals to compute target location in fixation-centered coordinates). This change in orientation does not alter the fundamental principles of the Zipser–Andersen model; it merely changes the reference frame of its output. For the present purpose, you can think about their model either way.

When Zipser and Andersen simulated the three-layer neural network, 64 cells had a Gaussian activity field defined retinotopically and a second group of 4 cells in the first layer reflected eye orientation, each

11.3. Combining Retinal Location with Eye Orientation Through Gain Fields

Box 11.3
Models in Biology 2

> Unlike neurophysiological data, neural network simulations come with the certainty that the individual neurons in the network causally link the network's inputs to its outputs. An attempt to understand the functional operations of a model network corresponds with ideal neurophysiological experiments involving simultaneously recording from the entire population of neurons that contribute to a given behavior. No a priori assumptions need be made about the means by which a network will transform an input signal into a specified output. Of course, this stance does not imply that artificial neural networks veridically replicate the biological system. Nor must you accept the idea that the back-propagation learning algorithm resembles the mechanisms of learning in your CNS. However, the properties of a model network's hidden units during input–output mapping has proven useful in contemplating similar properties in cortical neurons and their potential function in biological networks.[19,21-26]

coding the eye's angular position in the horizontal or vertical axis linearly (one cell for each dimension with increasing discharge as an angle increased and the other with a decreasing discharge). Therefore, the first layer of the network had 68 model neurons. You can label the discharge of any cell i in this first layer $p_i^{(1)}$. Twenty-five cells composed the network's second, hidden layer. Each of these cells received a weighted input from all the cells in the first layer; summed this input, and discharged as a sigmoidal function of its input:

$$p_i^{(2)} \equiv \frac{1}{1 + \exp\left(-\sum_j w_{ij}^{(1)} p_i^{(1)}\right)}. \tag{11.2}$$

In the above expression, $w_{ij}^{(1)}$ refers to the "synaptic" weight that connected the output of cell i in layer 1 with cell j in layer 2. These second-layer cells each projected to all of the cells in the third layer such that

$$p_i^{(3)} \equiv \frac{1}{1 + \exp\left(-\sum_j w_{ij}^{(2)} p_i^{(2)}\right)}.$$

The equations above characterize the architecture of the network; only the synaptic weights remain unspecified. The goal of the theorist involves finding the weights that transform the input to the desired output, much like the network model of kinematics discussed in section 9.6. (See box 11.3.) To find these weights, Zipser and Andersen trained the network with examples. They presented a stimulus to the first layer, resulting in an activation pattern in the output layer. Their computer

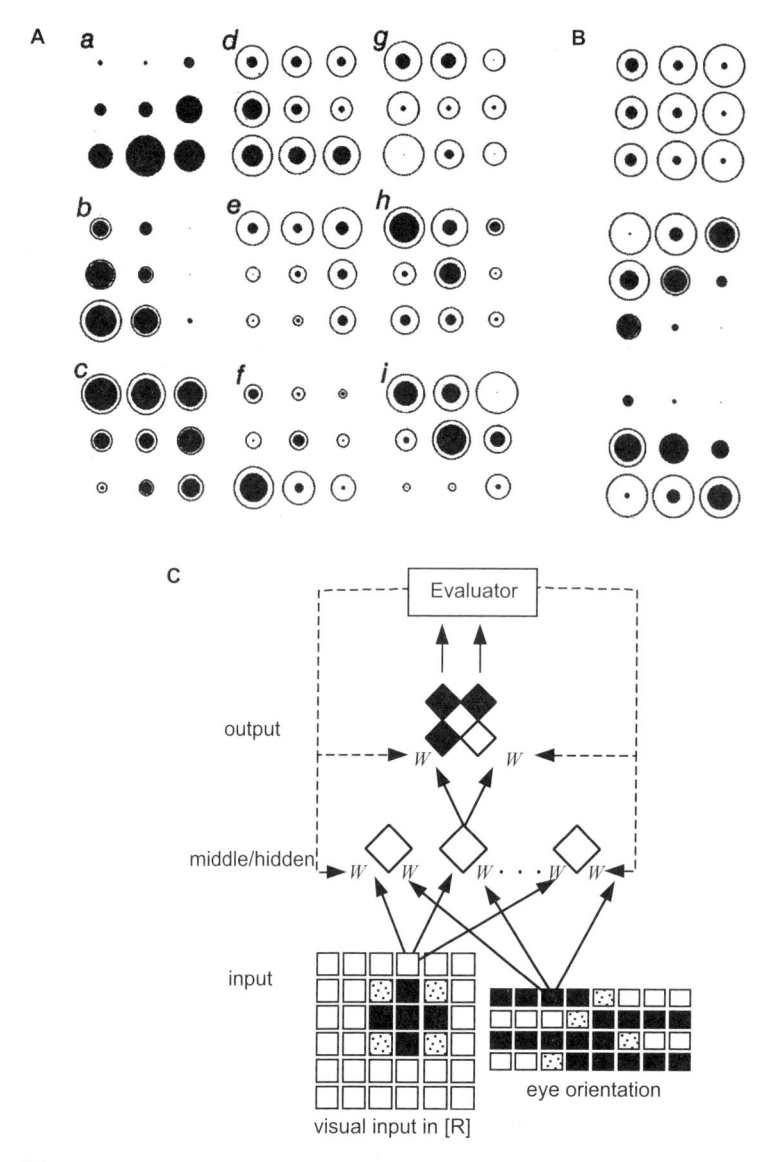

Figure 11.6

Gain fields for model units in an artificial neural network versus PPC neurons. (*A*) A graphical method to illustrate data like that in figure 11.4C, which plots a neuron's discharge for a stimulus presented at the center of its activity field for nine different eye orientations. The outer circle represents the total discharge of the neuron after stimulus presentation. The filled circle shows the cell's modulation (evoked activity minus background). Thus, the difference between the diameters of the filled circle and the ring around it reflects the amount of background activity. In about half of the cells, background activity remained constant although stimulus-evoked activity changed. (*B*) Activity of hidden units in the simulated network. Each unit's activity field was mapped for a straight-ahead eye orientation. The network was then provided with an eye-orientation input for each of program then compared that activation against the expected activation that should result, which reflected the location of the target in head-centered coordinates. The resulting error for each output cell was calculated and then back-propagated to the network to change its synaptic weights.

After a large number of examples, the network converged on a set of synaptic weights that generated the desired output. Zipser and Andersen investigated the discharge pattern of the model cells in the hidden layer and found that they behaved like equation 11.1 and like cells in PPC. Cells in the hidden layer had activity that was generally a linear function of eye orientation, with each cell having a Gaussian activity field. Therefore, the hidden layer combined these two factors. Figure 11.6B shows the discharge of three of these hidden layer cells, and you can compare them with the neuronal activity recorded from PPC cells in figure 11.6A. This figure illustrates data similar to that in figure 11.4C, but in a different format. Here, the plots show both the baseline activity before presentation of the stimulus (white annulus) and the total activity during stimulus presentation (black circle) for each eye orientation. The model cells and the actual neurons both discharge as a linear function of eye orientation.

These results of the simulation demonstrate that to compute the location of a target in head-centered coordinates, eye orientation and retinotopic information are sufficient and may be combined in an intermediate representation similar to that found in the PPC. This representation involves a linear sensitivity to eye orientation that modulates a retinotopic activity field. However, as Alex Pouget and Terry Sejnowski[27] have pointed out, the multiplicative mechanism with which the PPC neurons combine proprioceptive and visual information in equation 11.1 allows the brain to internally represent spatial locations in essentially any coordinate system.

11.3.3 Modulating Gain by Head Orientation

Changes in head orientation affect the neural discharge as a gain field, much like the orientation of the eye in the orbit.[28] Figure 11.7 shows an example. When the experimenters oriented a monkey's head to the right (with the eyes roughly centered in the orbit), the maximum discharge occurred for a target that appeared at ~225° in radial coordinates (i.e., down and to the left of the current fixation point). When the experimenters presented the stimuli at the same retinotopic locations but with the head oriented to the left, the gain of the tuning function increased.

Figure 11.6 (continued)
nine fixation points, without simulated visual input, which estimated background activity. Finally, the input layer of the network received a visual stimulus at the peak position of the unit's activity field for all nine eye orientations. (C) Architecture of the network. Abbreviations: W, synaptic weight; [R], retinocentric coordinates. Diamonds and squares indicate model units, with darker fill representing higher activity. (From Zipser and Andersen[19] with permission.)

Together with the data in figure 11.5, this finding suggests that PPC neurons receive both eye and head orientation signals, both of which affect cell discharge when a target stimulus appears. Assuming that both of these signals contribute to "background" activity, then the following equation presents one model of how these two different signals affect the cell's discharge:

$$p_i = (k_i^T e + c_i^T h + b_i) \exp \frac{-(r - r_i)^T (r - r_i)}{2\sigma^2}.$$

That equation describes the discharge rate p of cell i as a function of the location of the target's image on the retina r; the orientation of the eyes in the orbit e; and the orientation of the head with respect to the shoulder girdle h, its preferred retinotopic location r_i, and the slope of the gain field for eye and head orientation, k and c, respectively. Different cells have different gains and biases for eye and head orientation, and different retinotopic activity fields.

The formulation above assumes that the discharge-rate modulation due to the orientation of the eyes adds to that due to head orientation. Alternatively, these signals might combine multiplicatively:

$$p_i = (k_i^T e + a_i)(c_i^T h + b_i) \exp \frac{-(r - r_i)^T (r - r_i)}{2\sigma^2}.$$

As discussed below, multiplication allows more powerful combinations than addition. For now, keep both possibilities in mind.

11.3.4 The Neural Basis of Gain Modulation

How does gain modulation take place? In the Zipser–Andersen model, the hidden-layer neurons simply added their two inputs (eye orientation and retinal location) and submitted that sum to a nonlinear activation function. This network function will not really give you a multiplication of the two signals, however. In the Zipser–Andersen model, the inputs to the middle layer are summed, not multiplied. How could you account for the apparent multiplication seen, for example, in figure 11.7?

Emilio Salinas and Larry Abbott[29] discovered a possible mechanism for multiplicative combination. They suggested that gain modulation arises not from the way that a PPC cell responds to its eye orientation and retinotopic signals, but rather from the way that the neuron interacts with its neighboring neurons through robust interconnections. In the Salinas–Abbott model, each PPC neuron receives a signal that encodes eye orientation linearly, as well as a signal that encodes stimulus location in a retinotopic frame with a Gaussian function. For simplicity, assume that eye orientation e is coded using a scalar variable (that is, assume that the fixation point varies along only one dimension). Assume further that stimulus location on the retina also varies along only one dimension, specified by the variable r. Salinas and Abbott proposed that if you could isolate a single PPC neuron and deprive it of connections to other parietal neurons, you would see that its discharge is actually

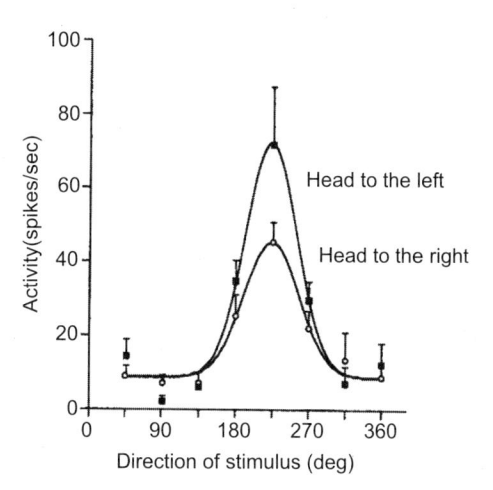

11.3. Combining Retinal Location with Eye Orientation Through Gain Fields

Figure 11.7
Effect of head orientation on a PPC neuron. The experimenters recorded activity from the PPC as a monkey responded to the appearance of a light spot by making a saccade to that location. The plot shows the mean discharge rate immediately before the saccade as a function of angle of the stimulus from the initial fixation point, in polar coordinates. Eye orientation relative to the head did not vary, but the cell's tuning curve differed when the monkey's head was turned to the right (unfilled circles) versus the left (filled circles). (From Brotchie et al.[28] with permission.)

$$p_i = s\left[(k_i e + b_i) + \exp\frac{-(r-r_i)^2}{2\sigma^2} - \beta\right]_+ .$$

Here, β corresponds to a bias, indicating that the input must exceed this threshold before the neuron will discharge; s represents a gain that specifies how quickly the discharge increases as a function of the neuron's input; and the symbol $[\ ]_+$ means that the neuron will have a discharge only when the quantity in the bracket exceeds 0 (due to the threshold requirement). (In another model, s takes the form of a sigmoid function that saturates the firing rate above a certain value.) You can imagine a collection of such cells and choose the parameter of each from some reasonable distribution. For example, Salinas and Abbott chose r_i to range from -5 to 5 (in arbitrary units), e to vary from -1 to 1, $s = 0.2$, and $\beta = 1$. Parameters k_i and b_i were distributed around the mean values of 1.0 and 0.5, respectively.

Now assume that each neuron in your collection connects to all other neurons in the collection so that it receives their output. If cell i receives an input from another cell j, the discharge of cell i at any given time step t depends on the discharge of the other cell at time step $t - \Delta$, such that

$$p_i(t) = s\left[(k_i e + b_i) + \exp\frac{-(r-r_i)^2}{2\sigma^2} + \sum_j w_{ij} p_j(t-\Delta) - \beta\right]_+ , \quad (11.3)$$

where a weight w_{ij} specifies the strength of connection from cell j to cell i.

Chapter 11. Computing Target Location

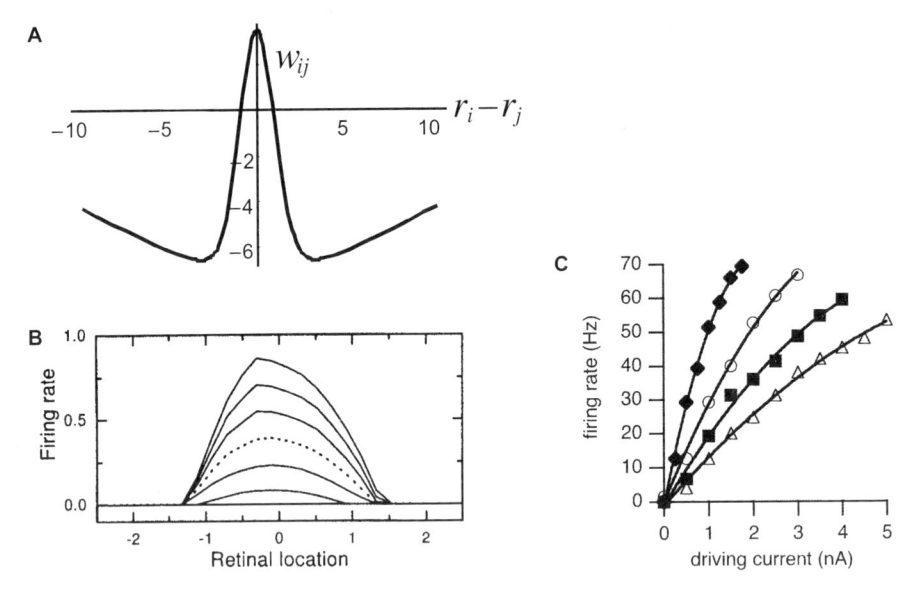

Figure 11.8
Computational mechanisms for producing a multiplicative interaction between two variables. (A) A plot of the function described in equation 11.4. It represents the connection weight between two neurons as a function of the distance between their activity-field maxima. Simulation parameters: $a_E = 10.5$, $a_I = 7.0$, $\sigma_E = 1$, $\sigma_I = 10$. (B) A gain-field mechanism. A model cell's response as a function of stimulus location on the retina. The dotted line shows the cell's tuning curve for eye orientation at 0. As eye orientation varied from −0.4 to 0.6 (bottom to top curves), gain modulation resulted. (C) Firing rates of a rat somatosensory cortex neuron, in a slice preparation, plotted as a function of a driving current. The curve with the filled diamonds shows the cell's response without background synaptic input. As the background input increased in three steps (unfilled circles, filled squares, and unfilled triangles), the response function scaled down. (B from Salinas and Abbott,[29] C from Chance et al.[30] with permission.)

Salinas and Abbott observed that if they set the network's weights so that neurons with overlapping activity fields excited each other, and neurons with nonoverlapping activity fields inhibited each other, then a multiplicative response developed in every neuron. They set the weights as follows:

$$w_{ij} \equiv a_E \exp \frac{-(r_i - r_j)^2}{2\sigma_E^2} - a_I \exp \frac{-(r_i - r_j)^2}{2\sigma_I^2}, \qquad (11.4)$$

where $a_E > a_I$ and $\sigma_I > \sigma_E$. The subscripts E and I refer to excitatory and inhibitory interactions, respectively. This weight exceeds 0 only for neurons that have similar activity fields (i.e., for neurons where $r_i \approx r_j$). Figure 11.8A illustrates an example of such a weight function.

Salinas and Abbott simulated a network of 500 units with discharge rates specified by equation 11.3 and connective weights determined by equation 11.4. Figure 11.8B plots the discharge of one representative cell as a function of stimulus location in retinotopic coordinates. This cell had a preferred stimulus location at retinotopic location 0 when eye orientation also equaled 0. You can think of this as a light spot at the fovea as the network "looked" straight ahead. The cell's tuning curve for that condition corresponds to the dotted line in figure 11.8B. As eye orientation changed from −0.4 to 0.6, the tuning curve changed multiplicatively.

The Salinas–Abbott model demonstrates that while individual neurons may sum their inputs, in a network with recurrent connections their activity can become the product of those inputs. Multiplicative combination requires that cells with similar activity fields excite each other, an often observed feature of cortical circuits.[31] Gain modulation appears as an emergent property of such a network.

An alternate theory for gain modulation involves intrinsic mechanisms of neuronal synaptic integration rather than a network mechanism. Frances Chance, Larry Abbott, and Alex Reyes[30] noted that neurons typically receive converging excitatory and inhibitory inputs. Often, the sum of these inputs approximately equals 0. The absolute level of the input (i.e., the sum of the unsigned value of inputs) may go up or down, but the background discharge should not change if the sum remains near 0. You can think of a barrage of balanced (excitatory and inhibitory) inputs as a noisy background input. Now imagine that a driving signal increases neuronal discharge. Simulations show that when the background "noise" increases, neurons become less responsive to a driving signal (figure 11.8C). According to this idea, the gain-field properties of PPC neurons might arise because the eye-orientation signal creates a noisy background input. Changes in eye orientation vary the sum of unsigned background input to PPC neurons, but not the signed value. Therefore, baseline discharge would not change much as a function of eye orientation, but the gain of the response to a visual stimulus would. Slight changes in the parameters of the model determine whether the discharge activity of the neuron reflects the background input.

References

1. Bock, O (1986) Contribution of retinal versus extraretinal signals towards visual localization in goal-directed movements. Exp Brain Res 64, 476–482.
2. Enright, JT (1995) The non-visual impact of eye orientation on eye-hand coordination. Vision Res 35, 1611–1618.
3. Henriques, DYP, Klier, EM, Smith, MA, et al. (1998) Gaze-centered remapping of remembered visual space in an open-loop pointing task. J Neurosci 18, 1583–1594.
4. Henriques, DYP, Medendorp, WP, Gielen, CC, and Crawford, JD (2003) Geometric computations underlying eye-hand coordination: Orientations of the two eyes and the head. Exp Brain Res 152:70–78.
5. Pouget, A, Ducom, JC, Torri, J, and Bavelier, D (2002) Multisensory spatial representations in eye-centered coordinates for reaching. Cognition 83, B1–11.

6. Henriques, DYP, and Crawford, JD (2002) Role of eye, head, and shoulder geometry in the planning of accurate arm movements. J Neurophysiol 87, 1677–1685.

7. Andersen, RA, Essick, GK, and Siegel, RM (1985) Encoding of spatial location by posterior parietal neurons. Science 230, 456–458.

8. Colby, CL, Gattass, R, Olson, CR, and Gross, CG (1988) Topographical organization of cortical afferents to extrastriate visual area PO in the macaque: A dual tracer study. J Comp Neurol 269, 392–413.

9. Battaglia-Mayer, A, Ferraina, S, et al. (2001) Eye-hand coordination during reaching. II. An analysis of the relationships between visuomanual signals in parietal cortex and parieto-frontal association projections. Cerebral Cortex 11, 528–544.

10. Ferraina, S, Battaglia-Mayer, A, Genovesio, A, et al. (2001) Early coding of visuomanual coordination during reaching in parietal area PEc. J Neurophysiol 85, 462–467.

11. Duhamel, JR, Bremmer, F, Ben-Hamed, S, and Graf, W (1997) Spatial invariance of visual receptive fields in parietal cortex neurons. Nature 389, 845–848.

12. Cohen, YE, Batista, AP, and Andersen, RA (2002) Comparison of neural activity preceding reaches to auditory and visual stimuli in the parietal reach region. NeuroReport 13, 891–894.

13. Bremmer, F, Distler, C, and Hoffmann, KP (1997) Eye position effects in monkey cortex. II. Pursuit- and fixation-related activity in posterior parietal areas LIP and 7A. J Neurophysiol 77, 962–977.

14. Galletti, C, and Battaglini, PP (1989) Gaze-dependent visual neurons in area V3A of monkey prestriate cortex. J Neurosci 9, 1112–1125.

15. Xing, J, and Andersen, RA (2000) Memory activity of LIP neurons for sequential eye movements simulated with neural networks. J Neurophysiol 84, 651–665.

16. Xing, J, and Andersen, RA (2000) Models of the posterior parietal cortex which perform multimodal integration and represent space in several coordinate frames. J Cog Neurosci 12, 601–614.

17. Mitchell, J, and Zipser, D (2001) A model of visual-spatial memory across saccades. Vision Res 41, 1575–1592.

18. Pouget, A, and Sejnowski, TJ (1994) A neural model of the cortical representation of egocentric distance. Cerebral Cortex 4, 314–329.

19. Zipser, D, and Andersen, RA (1988) A back-propagation programmed network that simulates response properties of a subset of posterior parietal neurons. Nature 331, 679–684.

20. Davson, H, and Danielli, JF (1952) The Permeability of Natural Membranes (Hafner, Darien, CT).

21. Fetz, EE, and Shupe, LE (1990). In Advanced Neural Computers, ed Eckmiller, R (Elsevier, Amsterdam), pp 43–50.

22. Zipser, D, Kehoe, B, Littlewort, G, and Fuster, J (1993) A spiking network model of short-term active memory. J Neurosci 13, 3406–3420.

23. Moody, SL, Wise, SP, di Pellegrino, G, and Zipser, D (1998) A model that accounts for activity in the primate frontal cortex during a delayed matching-to-sample task. J Neurosci 18, 399–410.

24. Moody, SL, and Zipser, D (1998) A model of reaching dynamics in primary motor cortex. J Cog Neurosci 10, 35–45.

25. Moody, SL, and Wise, SP (2001) Connectionist contributions to population coding in the motor cortex. Prog Brain Res 130, 245–266.

26. Torres, EB, and Zipser, D (2002) Reaching to grasp with a multi-jointed arm. I. Computational model. J Neurophysiol 88, 2355–2367.

27. Pouget, A, and Sejnowski, TJ (1997) Spatial transformations in the parietal cortex using basis functions. J Cog Neurosci 9, 222–237.

28. Brotchie, PR, Andersen, RA, Snyder, LH, and Goodman, SJ (1995) Head position signals used by parietal neurons to encode locations of visual stimuli. Nature 375, 232–235.

29. Salinas, E, and Abbott, LF (1996) A model of multiplicative neural responses in parietal cortex. Proc Natl Acad Sci USA 93, 11956–11961.

30. Chance, FS, Abbott, LF, and Reyes, AD (2002) Gain modulation from background synaptic input. Neuron 35, 773–782.

31. Gilbert, CD, and Wiesel, TN (1989) Columnar specificity of intrinsic horizontal and corticocortical connections in cat visual cortex. J Neurosci 9, 2432–2442.

12 Computing Difference Vectors I: Fixation-Centered Coordinates

*Overview: According to the model presented here, in order to control a reaching or pointing movement, your CNS compares an estimate of end-effector location against an estimate of target location. Neural networks subtract the former from the latter estimate to represent a desired difference vector for the end effector. The difference vector represents a movement plan for reaching the target. For reaching and pointing movements in primates, the CNS represents both targets and end effectors in a visual coordinate frame—with the **fovea** as its origin—termed* fixation-centered.

12.1 Planning Reaching and Pointing with Difference Vectors

Figure 12.1A provides a kind of road map for the presentation that follows. It outlines a model in which neural networks receive information from vision and proprioception. Visual information includes the retinal location of a target. Proprioceptive information includes information about the configuration and orientation of the arm, head, and eyes. From this information, neural networks estimate target and end-effector locations in fixation-centered coordinates, and then another neural network computes a difference vector that points from the end effector to the target. The computations taken up in chapters 9 and 10 correspond to those performed by network 1 in figure 12.1A, called a *location map*. That hypothetical network transforms proprioceptive information from the arm, head, and eyes into an estimate of end-effector location $\hat{x}_{ee}$. Chapter 11 dealt with some aspects of network 2 in figure 12.1A. Network 2 takes information from visual, auditory, or other sensory receptors and combines it with information about eye and head orientation to estimate target location $\hat{x}_t$ in fixation-centered coordinates. A third neural network, network 3, computes a difference vector, x_{dv}, by subtracting those two vectors $x_t - x_{ee}$. Figure 12.1B shows the analogous situation for controlling a robot with a movable camera.

12.1.1 Estimating Current End-Effector Location

Consider an experiment in which participants move one of their hands to a visually presented target, but cannot see either their arm or their hand.

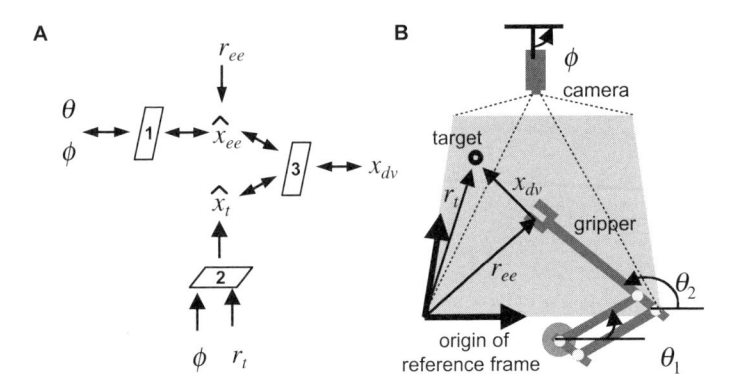

Chapter 12. Computing Difference Vectors I: Fixation-Centered Coordinates

Figure 12.1
A model system that computes a displacement vector from **proprioceptive** and visual information. (A) Neural network sketch. ϕ represents camera angle, as shown in B (top), r indicates retinal inputs for the target (t) and end effector (ee), and θ represents proprioceptive inputs. (B) Imaginary robot. Format as in figure 11.1A.

Philippe Vindras and his colleagues[1] hypothesized that in order to make such a movement, the CNS estimates the location of the target from visual information and then compares that location against an estimate of hand location computed from proprioceptive information. To make this comparison, the locations of the target and the end effector need to be represented as vectors in a common coordinate system. Section 11.2 laid the theoretical foundation for considering some alternative coordinate frames that the CNS might use for this computation, and section 12.2 addresses empirical evidence about which frame it probably uses. At this point, you do not need to commit to any given coordinate system, but assume that such a coordinate frame exists.

Vindras and his colleagues reasoned that if the CNS's estimate of the end-effector location, sensed through proprioception, was in error, then both the desired difference vector and the movement would reflect this error. To test this hypothesis, they covered the right hand of each participant and brought it to a starting location. The experimenters then presented a target for a reaching movement. After completion of the movement, they measured the error in hand location with respect to the target. Figure 12.2A shows the error pattern for a typical participant. Movements always began at the center of the workspace, and the targets appeared along the circumference of two concentric circles. Nearly all the error vectors pointed to the left, meaning that the hand consistently ended up to the left of the target. A vector corresponding to the average of the errors, labeled E, appears at the center.

The experimenters then asked the participants to estimate the location of their right hand. They did this with their left hand, using a joystick to position a light on top of their unseen right hand. Figure 12.2B shows the average of these estimated locations as a vector drawn from the actual

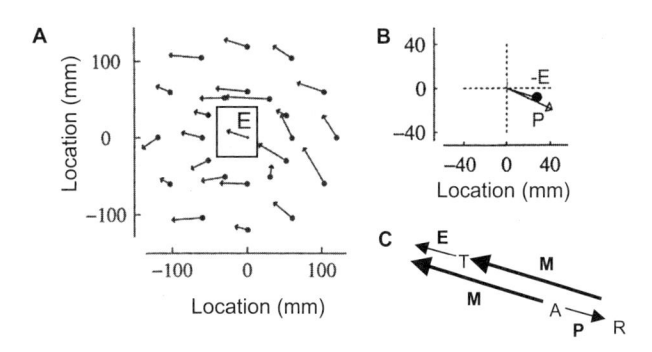

12.1. Planning Reaching and Pointing with Difference Vectors

Figure 12.2
Reaching to targets without vision of the hand produces errors. (*A*) A participant made reaching movements from a central location to two concentric rings of peripheral targets (filled circles) without visual feedback. The end location of each movement relative to each target is shown as an error vector. The average-error vector (**E**) appears in the box at the center of the plot. (*B*) The participant used a laser pointer to provide an estimate of hand location at the central location. The vector **P** indicates the average difference between this estimate and the actual hand location. Note that P ≅ −E. (*C*) The vector from the actual hand location (A) to the reported hand location (R) corresponds to the perceptual-error vector **P**. The movement vector (**M**) to target (T) has a motor-error vector **E** due to the misestimation of A. (*A* and *B* adapted from Vindras et al.[1] with permission.)

location of the right hand to the reported one. The participants reported that their hands were farther to the right than was actually the case. Thus, for a target to the left of center (figure 12.2C), the participants thought that their hand started farther from the target than it was, and therefore they made an excessively large movement.

What does the model depicted in figure 12.1A predict about the quantitative relationship between the perceptual-error vector **P** and the movement-error vector **E**? It says that the CNS computes the target location and initial hand location with respect to a common coordinate system. Call these estimates $\hat{\mathbf{T}}$ and $\hat{\mathbf{H}}_i$, respectively. The model predicts that the CNS computes the desired difference vector for the hand:

$$\mathbf{D} = \hat{\mathbf{T}} - \hat{\mathbf{H}}_i. \tag{12.1}$$

This vector represents a high-order motor plan. At the end of movement, the hand will settle at a final location $\mathbf{H}_f$ specified by the difference vector added to the initial hand location:

$$\mathbf{H}_f = \mathbf{D} + \mathbf{H}_i.$$

Substituting for **D** from equation 12.1, you have

$$\mathbf{H}_f = \hat{\mathbf{T}} - \hat{\mathbf{H}}_i + \mathbf{H}_i.$$

Vindras and his colleagues measured hand location at the end of the movement and measured the error **E** between that location and the target:

$$\mathbf{E} = \mathbf{H}_f - \mathbf{T} = \hat{\mathbf{T}} - \hat{\mathbf{H}}_i + \mathbf{H}_i - \mathbf{T}. \tag{12.2}$$

If you assume that in estimating target location, the CNS relies on fairly accurate visual information, whereas the estimate of hand location from proprioceptive information involves a fair amount of noise, then you can also assume that the error results largely from misestimations of $\hat{\mathbf{H}}_i$. Stated somewhat differently, assume that $\mathbf{T} \approx \hat{\mathbf{T}}$. Therefore, the error in equation 12.2 results from errors in estimating the initial location of the hand:

$$\mathbf{E} \approx \mathbf{H}_i - \hat{\mathbf{H}}_i.$$

Vindras et al. had their participants report an independent estimate of this error with the vector $\mathbf{P}$. Recall that $\mathbf{P}$ measured the difference in reported and actual hand locations:

$$\mathbf{P} = \hat{\mathbf{H}}_i - \mathbf{H}_i.$$

Therefore, the model illustrated in figure 12.1A predicts that the error in the perceived location of the hand should be approximately equal but opposite in direction to the error in reaching movements:

$$\mathbf{P} \approx -\mathbf{E}.$$

And this is what Vindras and his colleagues found (figure 12.2B). Participants made errors in their reaching movements that reflected their misestimation of initial hand location. This finding is consistent with the idea that to generate a reaching movement, the CNS computes hand and target locations in a common reference frame and subtracts the former from the latter vector in order to arrive at a difference vector. Without visual information about the location of the hand, the CNS will estimate that parameter from proprioception. However, if disease or some experimental manipulation eliminates proprioception and vision, the CNS has no reliable information regarding current location of the hand and cannot generate an accurate reaching movement (see sections 8.4 and 8.9).

12.1.2 Errors Accumulate in a Sequence of Movements

If the CNS plans each reaching movement in terms of a difference vector, then systematic errors in computing that difference vector should accumulate when people perform a sequence of movements. This idea led Otmar Bock and Rolf Eckmiller[2] to perform an experiment in which participants were seated in a cylindrical enclosure and held a device that rotated around a single axis (figure 12.3A). The experimenters projected target lights onto the plane of the cylinder, and the participant rotated his or her arm to point to the target, without vision of the arm or hand.

Bock and Eckmiller observed that when participants made movements from a given initial location to a visual target, they tended to undershoot the target slightly (figure 12.3B). The larger the movement, the larger the amount of undershoot. Therefore, they predicted that if the CNS specifies each movement in terms of a difference vector, the error

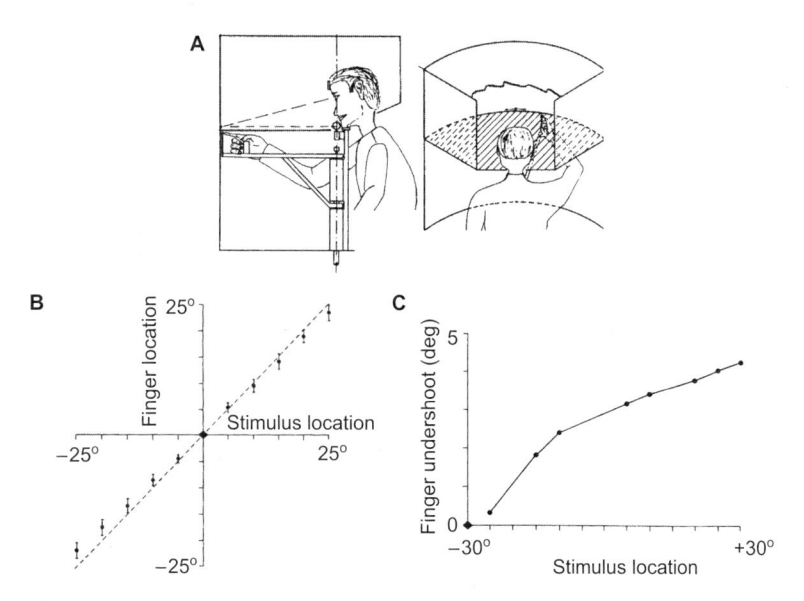

Figure 12.3
Reaching movements without visual information concerning hand location. (*A*) Experimental setup from side and rear. A participant points to a visual target by moving a single-degree-of-freedom armature and pointing to the target. (*B*) Movements tend to undershoot the target location. The filled circles and error bars show the means and standard deviations, respectively, for final finger location at the end of movement. Broken line indicates perfect accuracy; 0° is straight ahead. Diamond indicates starting location of all movements. (*C*) Movements began with the hand at −30°. Targets were displayed one at a time, with the last target at +30°. Error in final finger location increased monotonically as a function of target number. (Adapted from Bock and Eckmiller[2] with permission.)

for each movement should add to the error in the previous movement. If all the movements had the same direction, then the result should be an increase in movement error as a function of the number of individual movements.

To test this hypothesis, Bock and Eckmiller had the participants start their movements from a location in the far left of the cylinder (−30°). They then placed the target at −25°, and when the participant completed that movement, the first target disappeared as a second target appeared at −15°. The sequence repeated for a series of targets until the last target appeared at +30°. Bock and Eckmiller observed that the error at each target grew with movement number (figure 12.3C), as the model predicts.

12.2 Shoulder-Centered Versus Fixation-Centered Coordinates

Consider two hypotheses regarding the computation of a movement plan. One suggests that the CNS transforms the target's location from retinotopic coordinates into a body-centered coordinate frame and compares

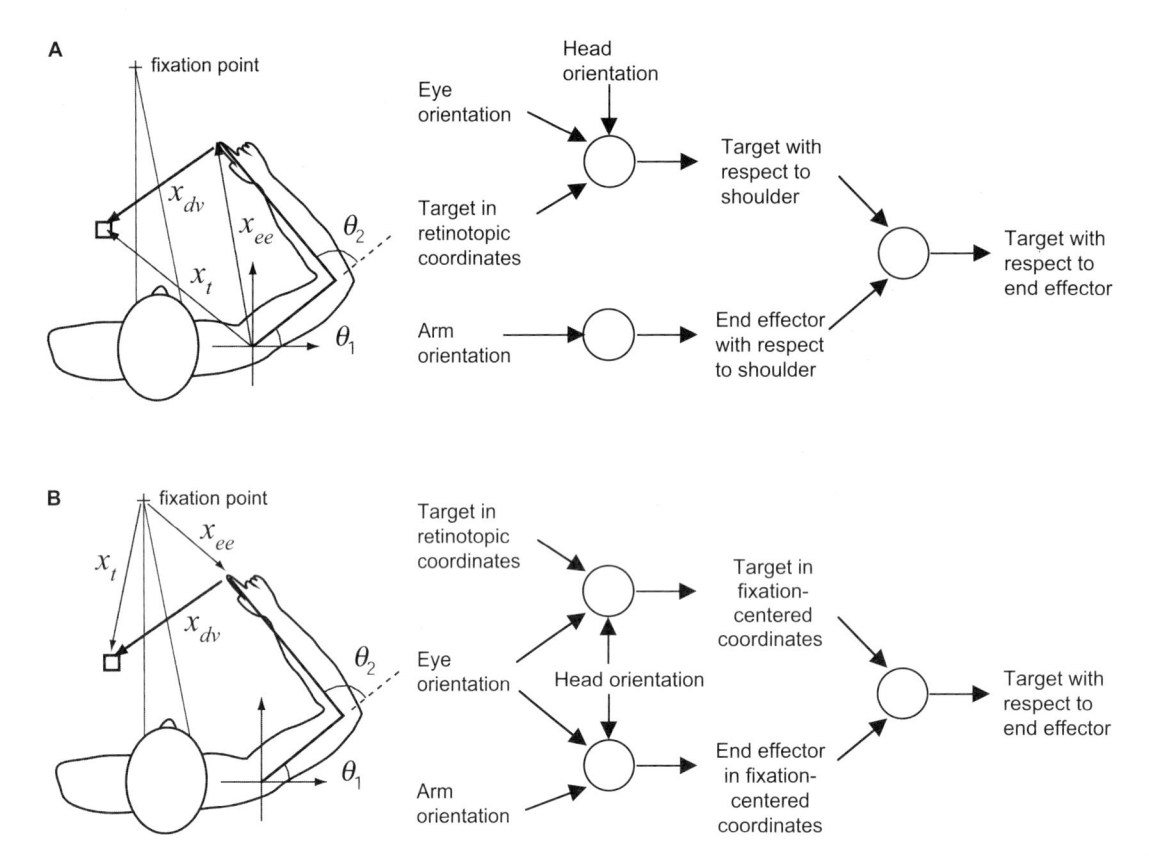

Figure 12.4
Computing a plan for a reaching movement in terms of a difference vector. (A) Joint angles of the arm can be used to compute end-effector location with respect to the shoulder. The CNS can combine target location on the retina, eye-orientation, and head-orientation signals to compute target location with respect to the shoulder. These two vectors specify the initial and final locations of the end effector, and the subtraction $x_t - x_{ee}$ yields a desired difference vector x_{dv}. (B) Alternatively, the same signals can be combined to compute end-effector location in fixation-centered coordinates. Subtraction of this vector from the target location in fixation-centered coordinates produces the same difference vector.

that location against the end effector's location in order to compute a difference vector in body-centered coordinates (figure 12.4A). The second hypothesis suggests that the CNS maintains target location in fixation-centered coordinates but transforms end-effector location into fixation-centered coordinates using information from both proprioception and vision. According to this second hypothesis, the movement plan corresponds to a difference vector for the end effector in fixation-centered coordinates (figure 12.4B). Imagine that the figure depicts you as you read this book. As illustrated in figure 17.1B, your fixation point falls on one of the book's many, many pages. Now you decide to reach for a pencil to make a note in the margin. Without taking your eyes off the book, you reach for the pencil. What computations does your CNS have to perform in order to accomplish this simple task? Figure 12.4 presents two scenarios for performing this computation. In both, you fixate a point that does not correspond to the end effector, your hand, or the reach target, the pencil.

According to the first hypothesis (figure 12.4A), the CNS computes the location of the target in body-centered coordinates. It does this by combining the information regarding the target's location on the retina with information regarding the orientation of the eye in orbit and the orientation of the head on the shoulder girdle. The output of this computation is the location of the target with respect to the shoulder. The **gain fields** discussed in section 11.3 and the neural networks that produce them appear to be well suited for performing such a computation. Simultaneously, the CNS uses proprioceptive information from the arm to compute the end-effector location with respect to the shoulder, then subtracts these two vectors; a vector that points from the end effector to the target is the successful result.

According to the second hypothesis (figure 12.4B), the CNS maintains the target's location in fixation-centered coordinates and does not compute its location with respect to an intermediate coordinate system such as one based on the shoulder. Rather, it uses visual and proprioceptive information about the arm, hand, and eyes to compute the location of the end effector in fixation-centered coordinates. If the image of the hand falls on the retina, the CNS does not need to use proprioceptive information to compute its location in this coordinate system. However, if this end effector falls outside the field of view, the CNS could use information about head and eye orientation to compute where the image would be in the same coordinate frame. It would then subtract end-effector location from target location to arrive at a difference vector in fixation-centered coordinates. This vector, as in the first scenario, points from the end effector to the target.

The key difference between these two hypotheses involves the coordinate system in which the CNS represents end-effector location. The first hypothesis (figure 12.4A) predicts that the CNS encodes end-effector location with respect to the shoulder, a shoulder-centered coordinate system. In the second (figure 12.4B), the CNS encodes end-effector location with respect to the fovea, a fixation-centered coordinate system.

You can use a simple network, analogous to the one that was proposed by Zipser and Andersen (see section 11.3.2), to examine how the CNS can compute a difference vector that represents the location of the target with respect to the end effector. In figure 12.5, two groups of input neurons each code for target and end-effector location. In this case, assume that each group of artificial neurons computes location in fixation-centered coordinates. These neurons map onto a second layer of neurons. As you will see below, experimental evidence suggests that this second group of neurons might function like neurons in area 5d of the PPC. By combining the target and end-effector locations, this layer then maps onto

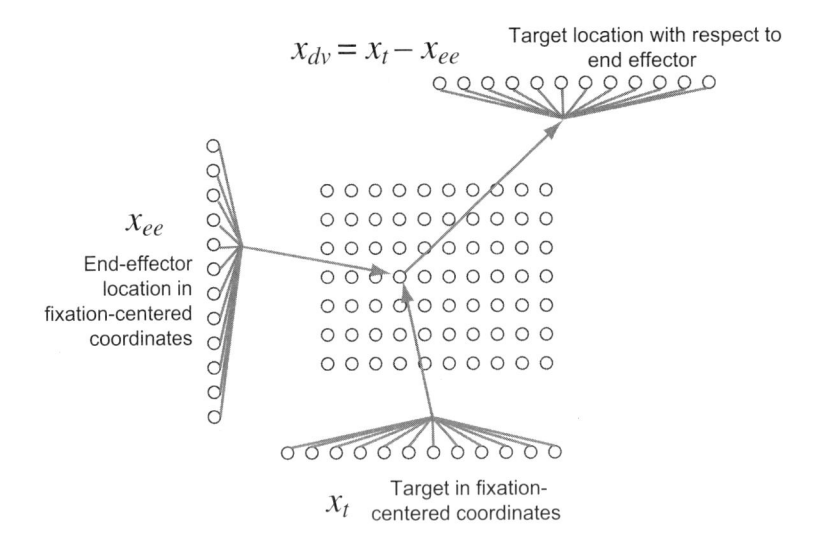

Figure 12.5
A neural network that transforms target and hand locations in fixation-centered coordinates to target location with respect to the hand (or some other end effector). If you assume that the activity in a region of the CNS represents the cells in the middle, hidden layer (the square array of cells), then activity should remain invariant as long as hand and target locations do not change in fixation-centered coordinates.

an output layer that represents the target locations as a vector with respect to the hand. This parameter is the difference vector.

In the example shown, the network simply subtracts target location from end-effector location. However, note that if end-effector and target locations do not change in fixation-centered coordinates, then activity of the hidden-layer neurons should not change. That is, the activity in the hidden-layer units should not change as long as the target and end-effector locations remain constant with respect to the *fixation point*. This prediction contrasts sharply with what you would expect if the input layer encodes target and end-effector locations in a body-centered coordinate system, such as one centered on the shoulder. In this situation, the activity in the hidden layer should not change when the target and end effector remain fixed with respect to the *shoulder*.

12.3 Planning in Fixation-Centered Coordinates: Experiment

Section 11.2C has already presented some evidence pointing to fixation-centered coordinates, but in order to test these two hypotheses neurophysiologically, Chris Buneo, Murray Jarvis, Aaron Batista, and Richard Andersen[3] recorded from area 5d of a monkey. They examined how the discharge rates of cells in this area changed as a function of the location of the target and the hand. The idea was to keep hand and target locations

12.3. Planning in Fixation-Centered Coordinates: Experiment

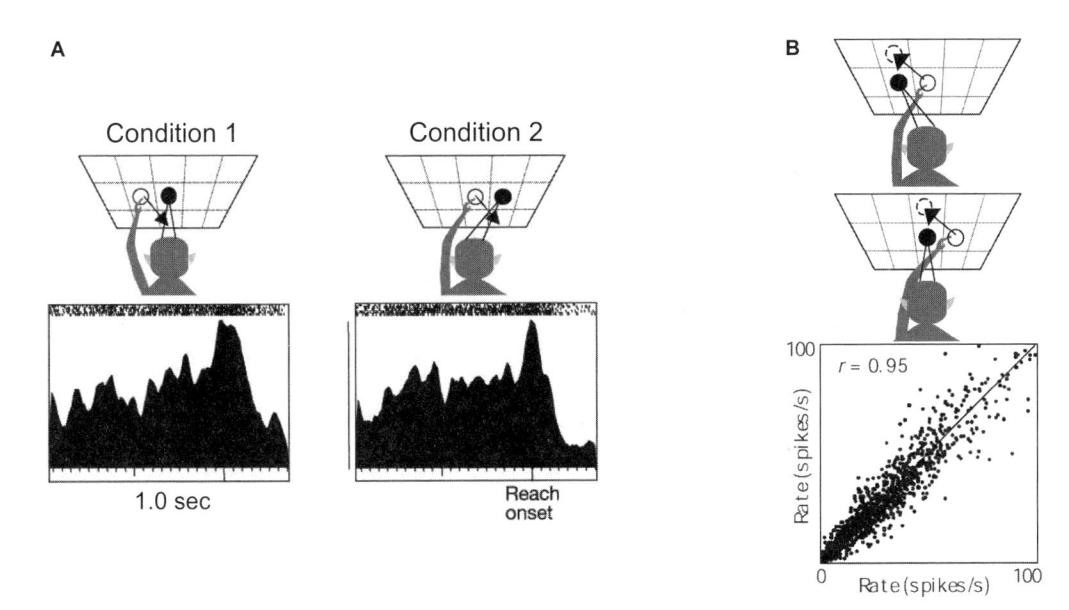

Figure 12.6
Activity of a neuron in area 5d. (*A*) Histograms of the activity of a single neuron in two experimental conditions. In those conditions, the fixation point (filled circle) changed location (one unit to the right from condition 1 to condition 2), but hand location (unfilled circle) and target location (arrowhead) remained constant with respect to the fixation point. Discharge of the cell varied little between the two conditions. The vertical bar between the two histograms shows a scale at 140 spikes/s. (*B*) Top and middle: Two additional conditions in which the fixation point (filled circle), hand location (unfilled circle), and target location (dashed circle) all shift one unit to the right (from top to middle). Bottom: Cell discharge for a 400-msec window centered on the beginning of reach. Fixation point, target location, and hand location varied systematically, and neuronal discharge was recorded from 80 cells. In pairs of trials in which hand and target locations remained constant with respect to the fixation point, the cell's discharge varied little. (From Buneo et al.[3] with permission.)

constant with respect to the fixation point and determine whether the discharge of the cells remained the same.

In their task, a touch-sensitive panel was placed in front of a monkey (figure 12.6). The panel housed an array of locations that had both a green and a red light. The red light instructed the monkey where it should fixate, and the green light instructed where the monkey should place its hand. The trials began with the illumination of both a red and a green light, which established the initial locations of both fixation and the monkey's hand. A green (reach target) light then appeared briefly at another location. After a delay period, the lights instructing the initial hand location disappeared as a signal to begin movement, and the monkey reached to the remembered location of the reach target while maintaining stable fixation.

Buneo et al. observed that when both hand and target locations remained constant with respect to the fixation point, the discharge of a typical area 5d cell changed little. For example, as shown in figure 12.6A, conditions 1 and 2 do not vary in terms of hand location in fixation-centered coordinates. In both conditions, the hand's location is one square to the left of the fixation point, although the fixation point has shifted between conditions (straight ahead in condition 1, to the right in condition 2). These two conditions also have the same target location in fixation-centered coordinates. The monkey always moves its hand to the square beneath the fixation point.

However, if you choose to represent hand and target locations in shoulder-centered coordinates, then the two conditions represent different locations for both the target and the monkey's hand. That is, the monkey's shoulder does not move between conditions, but both the target and the initial end-effector location shift one square to the right in condition 2. The cell's response changes very little across these two conditions. This finding suggests that the cell does not reflect hand and target locations in a shoulder-centered coordinate system. Rather, in accord with the model in figure 12.5, this cell could well represent one of the elements in the "hidden layer" of a neural network that transforms inputs from cells that encode hand and target locations in fixation-centered coordinates into the difference vector.

To quantify the similarity in discharge between different conditions for hand and target location, Buneo et al. analyzed the neuronal discharge during a 400-msec window centered on the start of the reaching movement. They varied fixation, target, and hand locations, and picked pairs of trials which had the same hand and target locations in fixation-centered coordinates. They found that the discharge of neurons in area 5d varied little for these pairings (figure 12.6B). The cell's discharge varied more if pairs of trials were picked in which the hand or target locations varied with respect to the fixation point. Their result agreed with the second of the two hypotheses mentioned above, and with the model depicted in figure 12.5: It appears that the CNS represents end-effector and target locations in fixation-centered coordinates, and the discharge of the area 5d neurons resembles the activity of units in the hidden layer of a network that performs the postulated computation.

The model in figure 12.5 resembles that described in section 11.3 for using **gain fields** to update target location in fixation-centered coordinates (or some other coordinate frame). It also seems consistent with the data from PPC presented so far. However, it does not capture an important part of those data. Closely examine the model in figure 12.5 and ask how activity should change in the hidden layer as you systematically change target and hand locations while keeping the fixation point constant. Consider the circumstance shown in figure 12.6 in which possible hand and target locations are confined to a plane. In the input layer of figure 12.5, cells that encode target location have an **activity field** in fixation-centered coordinates. You can assign each unit in the target input layer a Gaussian distribution with its center at a random location on the retina, much like the imaginary receptive field illustrated in figure 11.1D. This Gaussian distribution defines the activity of that unit, when a target appears at that location, with respect to the fixation point. Similarly, you can assign each unit in the hand input layer a Gaussian that defines the activity of that cell when the hand appears somewhere relative to the fixation point.

The output layer of the model in figure 12.5 also encodes space in two dimensions, but unlike the input layers, the output layer encodes target location with respect to the current location of the hand (rather than with respect to the fixation point). Each unit in the output layer will have a discharge described by a Gaussian distribution that has its center at a preferred target location with respect to the hand. To transform the inputs to this output, a hidden layer cell would have to have weights that connect input layer cells preferring hand locations at x_{ee}, and input layer cells that prefer target locations at x_t with output layer cells that prefer $x_t - x_{ee}$. Assume that such weights have been assigned to this network or that the network has learned those weights.

Now imagine that the fixation point and initial hand location remain constant, but target location changes. An example of this situation is shown in figure 12.7A. In the network of figure 12.5, a hidden-layer unit will have a maximum discharge at a particular target location and dis-

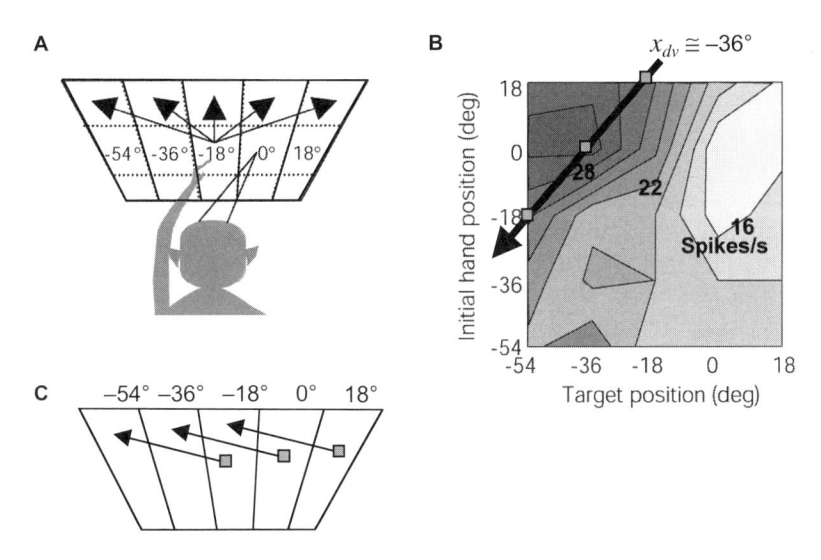

Figure 12.7
Discharge of a reach-related cell in area 5d, as a function of target location and initial hand location. (*A*) The monkey fixates a point (at 0°) and reaches to different targets from a given initial hand location (at −18°). (*B*) A cell's discharge, plotted as a function of target and initial hand locations. With the hand beginning at 0°, the cell's tuning function peaks for a target at about −45°. However, as the hand begins at −54° or −36°, the cell's tuning function peaks at more positive values. (*C*) Approximately equivalent difference vectors (−36° to the left), corresponding to the small squares in *B*. (*A*, *B* from Buneo et al.[3] with permission.)

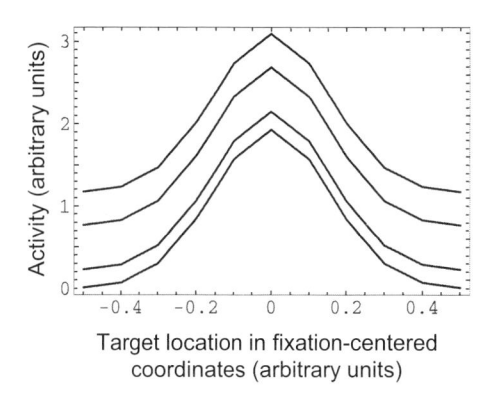

Figure 12.8
Simulation result describing the activity of a hidden-layer unit in the network of figure 12.5. Each line represents the activity in the cell as a function of target location for a constant hand location. As hand location changes, the function shifts up or down, but the center of the activity field does not shift.

charge will decline at other locations. How should this tuning function change as hand location changes? The simulation illustrated in figure 12.8 shows that as you change hand location, the tuning function simply shifts up and down. Importantly, the peak of the tuning function always appears at the same target location with respect to the fixation point. Therefore, this simple model predicts that as hand location changes, units in the hidden layer should not exhibit a shift in the peak location of their tuning function for target location.

Buneo et al. examined how the tuning function of single cells changed as a function of hand location. They kept hand location constant as the target of the movement was systematically varied. This manipulation produced a tuning curve as a function of target location. They then moved the hand to a new starting location and again mapped the cell's tuning function. They observed that as the hand moved, the tuning function of the cell shifted its peak location (figure 12.7B). In the simple network of figure 12.5, changing hand location resulted not in a shift of the tuning function but only in a change in its bias (figure 12.8). Therefore, this simple model cannot account for the observed data. How could you explain shift in tuning as initial hand location varies?

12.4 Planning in Fixation-Centered Coordinates: Theory

The neural network in figure 12.5 was of the feedforward variety that Zipser and Andersen[4] had used in their gain-field analysis. Its network combined hand and target locations in fixation-centered coordinates and computed the difference between these two values. However, because the input to the hidden layer simply summed, the model failed to predict that as hand location changed, the tuning curves of the hidden layer cells should also shift. In addition, the tuning curves moved up and down instead of changing amplitude (gain).

Theoretical work by Sophie Deneve, Peter Latham, and Alex Pouget explains why the model in figure 12.5 failed and how to remedy the situation.[5] In their work, Deneve et al. noted that networks such as that shown in figure 12.5 perform a computation called *function approximation*. This particular network has inputs that provide it with an estimate of hand location x_{ee} and target location x_t. It needs to compute an output that combines the information provided by each of the inputs. A function $f(x_{ee}, x_t)$ represents this combination. In this case, the function represents target location with respect to the hand, which results from a linear combination of the two input variables (subtraction). If you assume that noisy neurons represent each of the input variables, then the trained network needs to estimate the location of the target with respect to the hand despite the noise in the representation of target and hand locations. This noise will result in errors in estimating target location with respect to the hand.

In order to perform the estimation optimally, the network needs to compensate for the noise. A feedforward network like the one in figure 12.5 makes no allowance for noise. It passes on any noise in the inputs to the output. To solve this problem, Deneve et al. observed that a network with feedback connections from the hidden layer to both the input and the output layers has a remarkable property: It can "clean up" the noise and perform function approximation optimally.

Figure 12.9 shows an example of such a network. The neural network in figure 12.9D resembles the one in figure 12.5. As in the latter, the input layers representing the target and hand locations in fixation-centered coordinates connect to the hidden layer, and the hidden layer connects to the output layer (representing target location with respect to the hand). In the figure 12.9D network, however, the hidden layer also connects with weights to the input layers, and the output layer connects with weights to the hidden layer. Therefore, the hidden layer has recurrent connections to both the input and the output layers.

Now imagine that a target appears. Each cell in the target-location layer x_t has a preferred target location and will respond according to the location of the target with respect to the cell's receptive field. Because of the neurons' noisiness, the pattern of activity in the cells might look something like that shown in figure 12.9B. This figure plots the activity of each cell in the target input layer as a function of the location of the center of its receptive field. In figure 12.9C, you see what the activity among these cells would have looked like without noise in the system. You can imagine a similar "noisy hill" for the representation of the hand location in the input layer.

In order for the network to perform function approximation optimally and compute target location with respect to the hand, it needs to figure out the target and hand locations from the noisy hill exemplified by figure 12.9B. It "cleans up" the activity of the noisy neurons, as illustrated

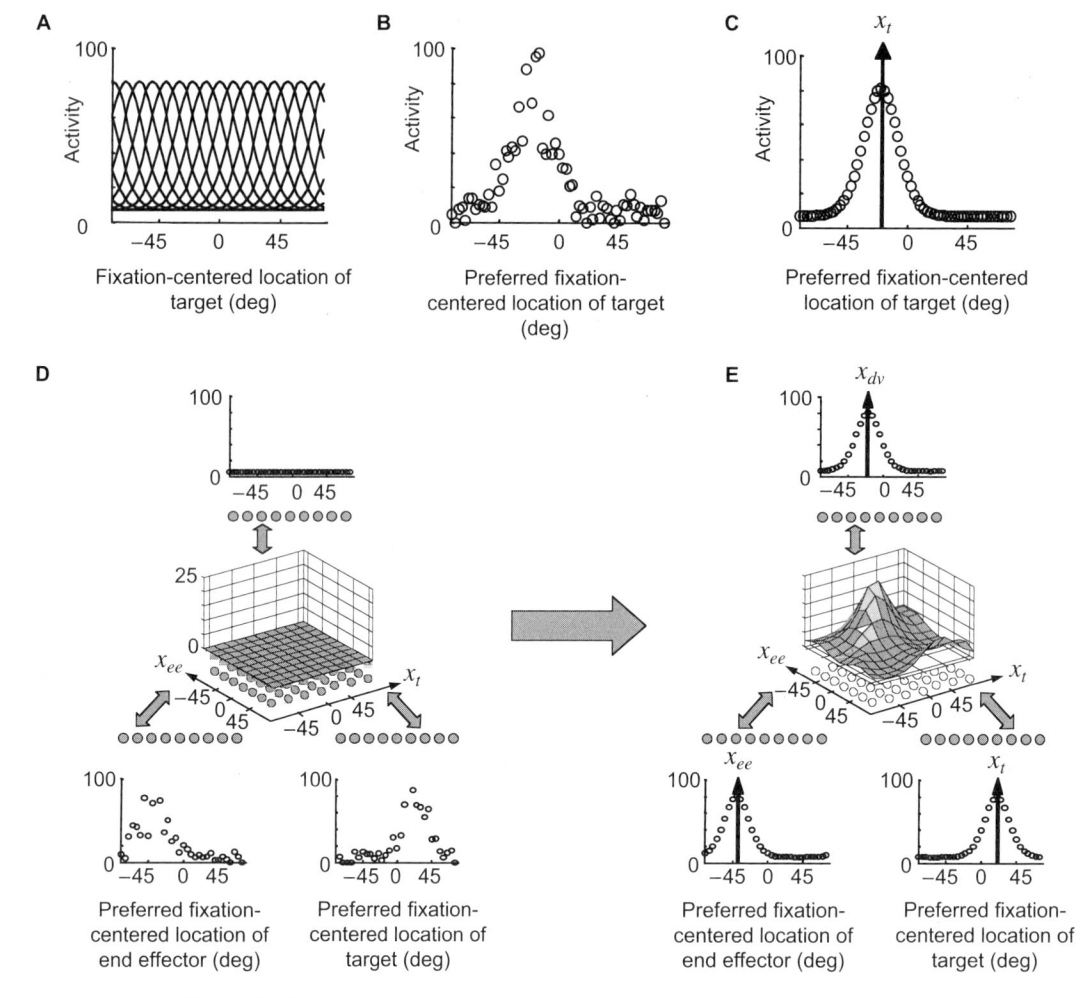

Figure 12.9
A neural network that performs function approximation in the presence of noise. (A) Population of model neurons with bell-shaped tuning curves for the fixation-centered location of a visual target stimulus. (B) Noisy response across the population for an object at $-20°$. (C) Response without the noise. (D) Network architecture and activity in the input and output layers at t = 0. The two input layers at the bottom encode the target and hand locations in fixation-centered coordinates. The output layer encodes target location in hand-centered coordinates. The hidden-layer units connect reciprocally to all three layers. The activity of each unit is obtained at each iteration by computing the weighted sum of its input, followed by a divisive normalization within each layer. Note that at t = 0 the network has not yet learned the mapping and the inputs have a substantial noise component. (E) Network activity after three iterations of learning. The noisy inputs relax into smooth estimates due to the network's dynamics, and the output layer forms an equally smooth estimate of target location in hand-centered coordinates. (After Deneve et al.[5] with permission.)

in the change from figure 12.9B to figure 12.9C. The peak of the noiseless hill in figure 12.9C corresponds to $\hat{x}_t$, the optimal estimate of target location.

If you were given the data illustrated in figure 12.9B and had to estimate the peak of activity, the best you could do would be to fit the noisy hill to one of the functions plotted in figure 12.9A. That is, you would take a template function and slide it along the x-axis of figure 12.9B until you got a good fit. In using this template-matching approach, you would, in effect, be using a statistical technique called *maximum-likelihood estimation*. You can use a maximum-likelihood estimate of the two noisy input variables to clean up the noise, estimate the target and hand locations in fixation-centered coordinates, and then subtract the latter from the former to locate the target with respect to the hand.

Deneve et al. showed that by using recurrent connections to wire up a hidden layer to two noisy input layers and an output layer, it, too, performed like a maximum-likelihood estimator. Figure 12.9D shows an example of this computation. In this figure, the two input layers receive noisy estimates of target and hand locations in fixation-centered coordinates. At first, the output layer has no activity. Because the network has recurrent connections, activity in all the layers changes with time. In a few iterations, the network's activity evolves to that shown in figure 12.9E. Here, each input layer has become noiseless, and the output layer has become $x_{dv} = x_t - x_{ee}$.

The key to the model devised Deneve et al. is the inclusion of reciprocal connections. (To learn more about how to construct such networks, see the web document *recurrentnetworks*.) The weights in these connections cause the network to perform three kinds of transformations simultaneously. It maps x_t and x_{ee} onto x_{dv}; it maps x_t and x_{dv} onto x_{ee}; and it maps x_{ee} and x_{dv} onto x_t. That is, if the network is given any two of the three variables, it can estimate the third, missing one. Thus, like the network illustrated in figure 9.3B, this system functions as a *pattern-completion network*. Each element in the hidden layer receives a set of connections from x_{ee}, x_t, and x_{dv}. In turn, each hidden layer element is connected with a weight to each of the other layers. Therefore, the "input" layers also are "output" layers. When the input is assigned to any two layers, the connections are set so that the third layer's activity will be consistent with the function that needs to be computed. If the variables are noisy, the dynamics of the network settles into an attractor (see box 12.1 and box 15.1) defined by the connecting weights. This attractor allows the network to clean up the noise in each variable, so that at the end of a few iterations, the estimates of each variable become optimal.

Now recall the problem with the model depicted in figure 12.5. It did not provide an account for the finding that area 5d cells shifted in their activity field as a function of hand location.[3] The feedforward network in figure 12.5 cannot explain this finding, but the reciprocally connected network described in figure 12.9 can. Deneve et al. showed that if they

Box 12.1
Attractor states in dynamical systems

> One way to think about an attractor is to use the analogy of a bowl. The bottom of the bowl serves as an attractor for a marble due to the force of gravity. Small forces applied to the marble do not lead to long-term changes in its resting state, as long as the force does not kick it out of the bowl. The bottom of the bowl represents a "low-energy" state, and you must pay some energy cost to get the marble out of its attractor state. Given enough time, the marble returns to the bottom of the bowl, with some variation due to "noise." A system, including a neural network, may have one or several attractors, and the network may have a different probability of "falling into" each one. Take, for example, a roulette wheel. Each of the numbered wells serves as a basin of attraction, and if the game is honest, each has the same probability of capturing the ball. If, however, you place a magnet under some of the numbers, a steel ball will fall into those wells preferentially. The bowl and the roulette wheel serve as examples of *point attractors* because you can characterize the state or states finally adopted by the system as a location in space. Another kind of attractor is often known as a line attractor or, more generally, a continuous attractor. Consider marbles rolling down a roof toward a gutter. Obstacles may divert the marbles to the left or right, but once they reach the gutter, they stop. (Assume that the house has a lousy gutter, one perfectly parallel to the ground.)
>
> As an example of a different kind of attractor, consider a clock's pendulum. Based on the pendulum's mass and length, when you push on it, it adopts a characteristic period of oscillation and, within a broad range, the precise force that you apply to it does not matter much. The pendulum oscillates with a given period, its periodic (or harmonic) attractor. Other kinds of attractors have yet more complications (e.g., chaotic attractors). In chaotic attractors, initial conditions matter very much and small changes in those conditions send the system off to very different intermediate and end states. Nevertheless, the system cannot adopt all possible states with the same probability, so you can consider the states it can adopt as chaotic or "strange" attractors.

kept one of the input-layer variables constant while changing one of the other variables, the hidden-layer elements exhibited a shift in their tuning curves. Figure 12.10 shows an example in which the location of the target remains constant in fixation-centered coordinates. By analogy to the experiment illustrated in figure 12.7, the point of fixation and the target's location remain the same, and starting hand location varies.

Because the target does not change with respect to the point of fixation, x_t remains the same. Therefore, the different hand locations x_{ee} do not affect input-layer elements that code for x_t (figure 12.10A). However, the different x_{ee} values correspond to different target locations with respect to the hand, x_{dv}, and figure 12.10B portrays this coding as different peaks of

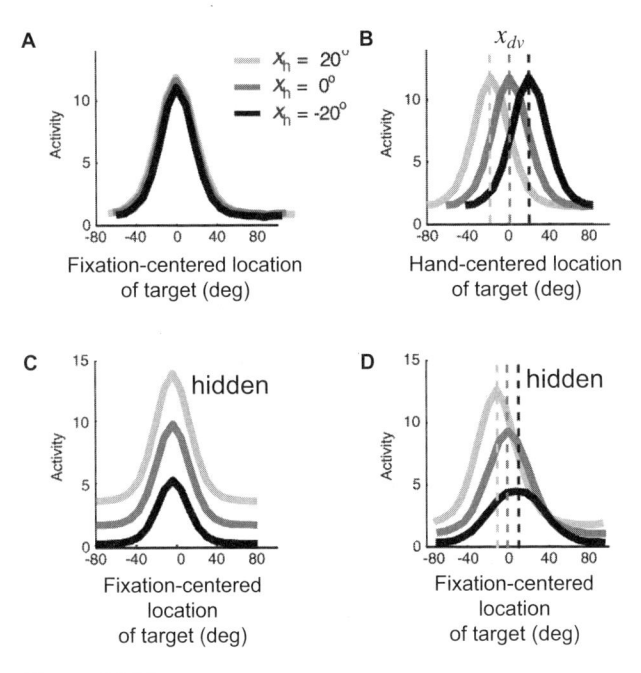

Figure 12.10
Activity field of network units for three different hand locations. (*A*) For a typical unit in the layer that codes for target location x_t, a change in hand location x_h does not affect its activity. (*B*) The change in hand location is reflected in a change in the activity of units in the "output" layer that codes for target location with respect to the hand. (*C*) If the network had only feedforward connections, as in figure 12.5, the activity for a typical unit in the hidden layer would show a modulation but no shift in peak location. (*D*) When the network has reciprocal connections between the hidden layer and other layers, both the location and the height of the fixation-centered activity field of the hidden-layer unit change. (Redrawn from Deneve et al.[5] with permission.)

activity. If the hidden layer connected to the other layers with only feedforward connections, the activity in an element in the hidden layer would look like figure 12.10C. However, the reciprocal connections between the hidden layer and other layers cause something closer to a multiplication than to addition of the signals. This multiplicative property produces a pattern of activity in the hidden layer that shifts its peak of activation as the hand location changes, along with changing the gain (figure 12.10D).

12.5 Localizing an End Effector in Fixation-Centered Coordinates

The data of Buneo et al.[3] suggest that in a region of PPC (area 5d), neuronal activity reflects target and hand locations in fixation-centered coordinates. The theoretical work of Deneve et al.[5] suggests that the neuronal properties observed in this region are consistent with a network that computes target location with respect to the hand. However, if you cannot see your hand, how can you compute hand location in fixation-centered coordinates in order to make a reaching movement?

Without visual information about hand location, the CNS uses proprioceptive inputs to compute hand location. Chapter 9 presented a framework for computing end-effector location based on information about joint angles and muscle lengths. This information could be used to estimate end-effector location in any of several coordinate frames, but evidence suggests the use of fixation-centered coordinates. One can construct a network similar to figure 12.9D to perform this operation, as shown in figure 12.11. Concentrate on the bottom part of the network, where the input layers receive proprioceptive cues regarding the configuration of the arm and the orientation of the eye and the head. The hidden layer uses a recurrent set of connections to map the joint angles of the arm—along with eye and head orientation in proprioceptive coordinates—into hand location in fixation-centered coordinates. Thus, even if you cannot see your hand, this neural network can nevertheless compute its location in fixation-centered coordinates. It does so from proprioceptive inputs, through forward kinematics; it tells you where you would see your hand if your could see it. Because of its bidirectional connections, the hidden layer also maps hand location in those fixation-centered coordinates "back" into proprioceptive coordinates. If you can see your hand, you can use visual information to calibrate the mapping from proprioceptive to fixation-centered coordinates.

According to this model, making a reaching movement relies on kinematic maps that transform various sensory inputs into a fixation-centered representation. It predicts that the retinal image of the hand (and other end effectors) plays an important role in the neural mechanisms underlying reaching and pointing movements. The location of that image should correspond to the estimation derived from proprioceptive cues. (Chapters 15 and 16 take up this topic in considerable detail.)

12.6 Encoding End-Effector Location in Fixation-Centered Coordinates

Section 10.3D described data showing that area 5ip encodes end-effector location for both hands and tools. Other parts of PPC also reflect the appearance of an end effector in the visual field. This evidence comes from studying the discharge of cells in areas PO and 5d.

Roberto Caminiti and his colleagues[6] discovered that some neurons in area PO lack a sensory-like activity field. That is, these cells did not respond when a target light appeared or a small bar of light moved in a monkey's field of view. However, these cells often discharged vigorously if the monkey positioned its hand somewhere in its field of view. For many cells, their activity strongly reflected the location of the hand with respect to the fixation point. These cells appear to be sensitive to a visually observed location of the hand (and not the target) in a fixation-centered coordinate frame.

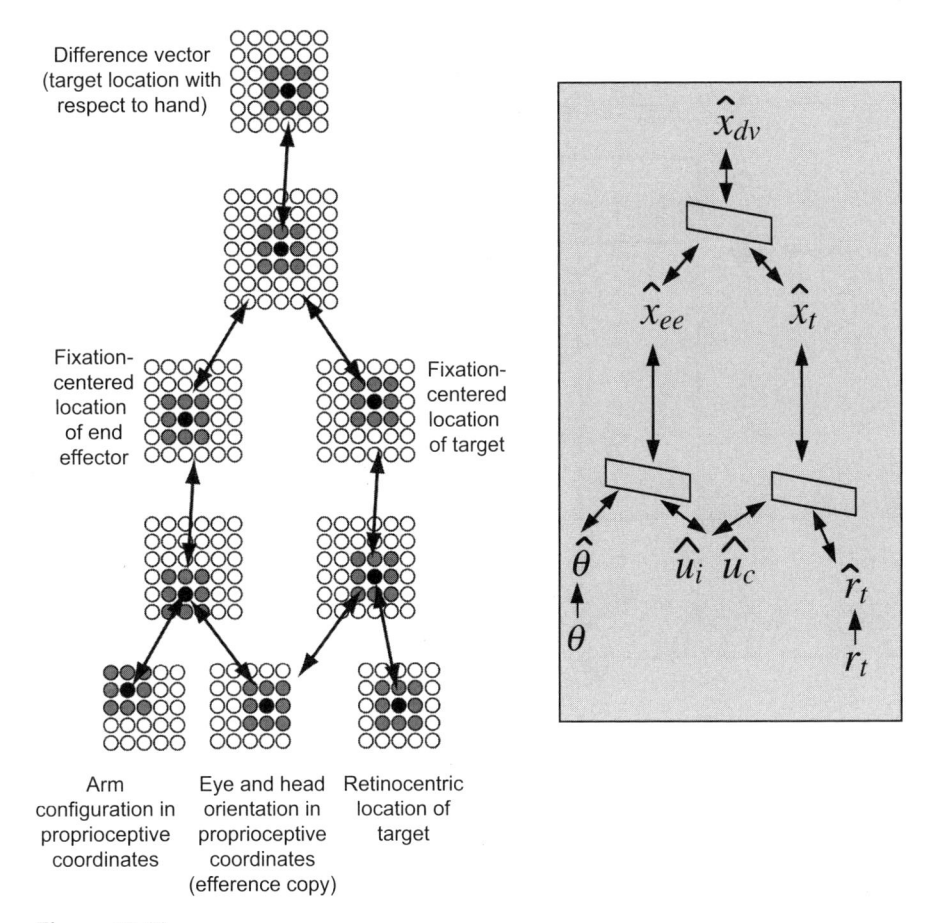

12.6. Encoding End-Effector Location in Fixation-Centered Coordinates

Figure 12.11
A network that performs the computation illustrated in figure 12.4B. The network computes the location of the target with respect to the hand from proprioception (θ) and **efference copy** (u) for the eyes (i), head (c), and arm (not shown), and from retinocentric (r) information about target (t) location.

In another experiment, Michael Graziano and his colleagues[7] investigated the relationship between the visually observed location of the hand and its proprioceptively computed location. They covered the monkey's arm but displayed a realistic arm prepared by a taxidermist. In this experiment, the experimenters recorded from area 5d, as well as areas 1 and 2, while the animal fixated a point in space. The experimenters covered the animal's arm and moved it from one side to another. In doing so, they noted that cells in areas 5d, 1, and 2 responded to the movement. Next, they placed the fake (but realistic) arm near the monkey's actual arm. The monkey could see the former but not the latter. The question was: If the monkey saw an arm at a particular location, but felt its own arm in another location, did the cortex register the visual cue or the proprioceptive one? Graziano et al. found that the sight of the fake arm

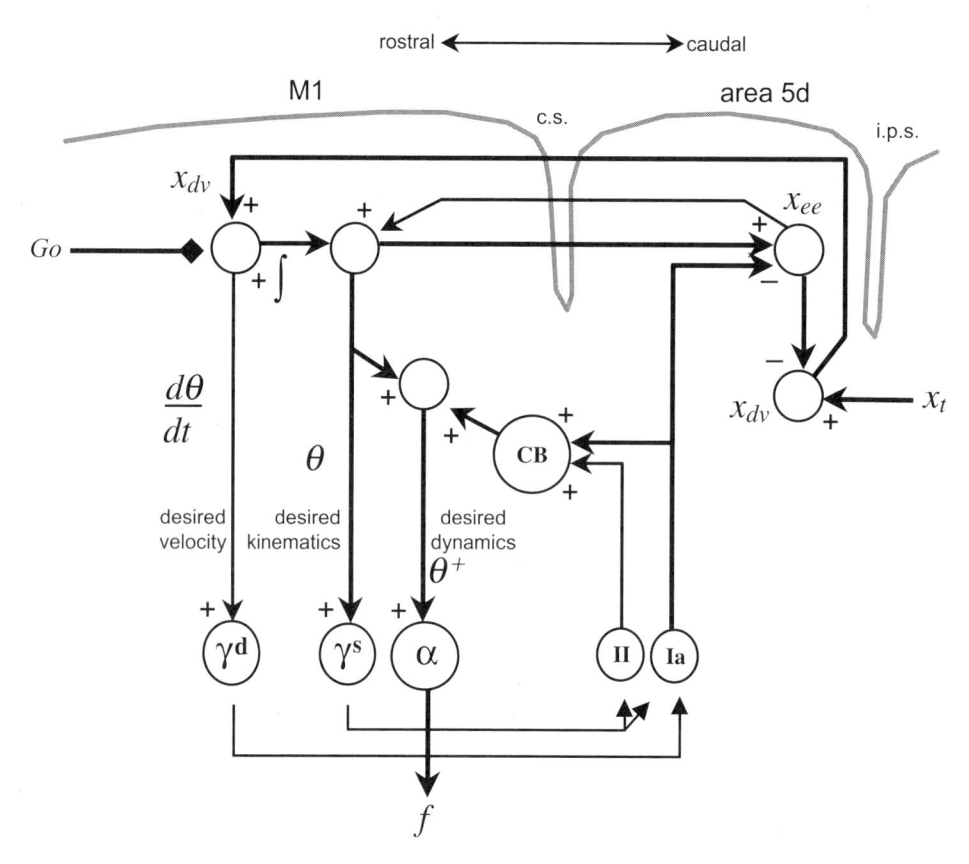

Figure 12.12
The vector integration to endpoint (VITE) model, incorporating feedback assistance. Note that the authors have assigned the various computations to particular cortical areas. Their paper[12] explains the basis for these assignments. The diamond-ended arrow from the "go" signal indicates a scalar (multiplicative) interaction, as taken up in more detail in section 19.2. Abbreviations: CB, cerebellum; x_{dv}, difference vector; x_{ee}, end-effector-location vector; x_t, target-location vector; γ^d, dynamic commands from γ-motor neurons; γ^s, static commands from γ-motor neurons; α, motor commands from α-motor neurons; Ia, primary muscle-spindle afferents; II, secondary muscle-spindle afferents; f, force output; i.p.s., intraparietal sulcus; c.s., central sulcus. (From Cisek et al.[12] with permission.)

affected the discharge of cells in area 5d, but not in areas 1 and 2. This activity did not occur when the monkey saw other interesting objects or an unrealistic fake arm, even for area 5d. Neurons in area 5d apparently respond to both proprioceptively sensed location of the limb and its visual representation.

12.7 Issues Concerning Fixation-Centered Coordinates

An important question involves the distinction between fixation-centered coordinates and retinotopic ones. Fixation-centered coordinates take into account eye and head orientations in addition to retinotopic information. Aside from that, consider how space around the fixation point differs from space around the fovea. For example, certain locations cannot be represented in retinotopic coordinates because the images of those locations never fall on the retina. For example, if you wish to scratch your back, it is difficult to imagine how the location on your back might be represented in retinotopic space. The unseen fly depicted in figure 11.3 falls into that category as well. If all reaching movements are represented in fixation-centered coordinates, then that coordinate system must have boundaries that exceed the field covered by the retina at any given time, or ever. For example, you can never—no matter how fast you turn around—see the back of your head. But it could be represented as a point in fixation-centered coordinate space. Note also the three-dimensional nature of the fixation-centered coordinate frame (figure 11.3). Unlike the eye per se, the fovea, and the retina, the fixation point lies "out there" in space: It has a distance as well as an orientation.

It is also interesting to ask whether a fixation-centered coordinate system for reaching can be generalized beyond reaching and pointing, to all movements. For example, consider movements of your tongue as you speak. Could your CNS plan movements of your tongue in fixation-centered coordinates? We doubt it. It seems unlikely that the framework described here for the planning of reaching movements generalizes to all movements.

References

1. Vindras, P, Desmurget, M, Prablanc, C, and Viviani, P (1998) Pointing errors reflect biases in the perception of the initial hand position. J Neurophysiol 79, 3290–3294.
2. Bock, O, and Eckmiller, R (1986) Goal-directed arm movements in the absence of visual guidance: Evidence for amplitude rather than position control. Exp Brain Res 62, 451–458.
3. Buneo, CA, Jarvis, MR, Batista, AP, and Andersen, RA (2002) Direct visuomotor transformations for reaching. Nature 416, 632–636.
4. Zipser, D, and Andersen, RA (1988) A back-propagation programmed network that simulates response properties of a subset of posterior parietal neurons. Nature 331, 679–684.

5. Deneve, S, Latham, PE, and Pouget, A (2001) Efficient computation and cue integration with noisy population codes. Nat Neurosci 4, 826–831.
6. Battaglia-Mayer, A, Ferraina, S, Mitsuda, T, et al. (2000) Early coding of reaching in the parietooccipital cortex. J Neurophysiol 83, 2374–2391.
7. Graziano, MS, Cooke, DF, and Taylor, CS (2000) Coding the location of the arm by sight. Science 290, 1782–1786.
8. Bullock, D, Grossberg, S, and Guenther, FH (1993) A self-organizing neural model of motor equivalent reaching and tool use by a multijoint arm. J Cog Neurosci 5, 408–435.
9. Bullock, D, and Grossberg, S (1988). In Dynamic Patterns in Complex Systems, eds Kelso, JAS, Mandell, AJ, and Shlesinger, MF (World Scientific Publishers, Singapore).
10. Grossberg, S, Guenther, FH, Bullock, D, and Greve, D (1993) Neural representation for sensorimotor control. 2. Learning a head-centered visuomotor representation of 3-D target positions. Neural Networks 6, 43–67.
11. Guenther, FH, Bullock, D, Greve, D, and Grossberg, S (1994) Neural representations for sensorimotor control. 3. Learning a body-centered representation of a three-dimensional target position. J Cog Neurosci 6, 341–358.
12. Cisek, P, Grossberg, S, and Bullock, D (1998) A cortico-spinal model of reaching and proprioception under multiple task constraints. J Cog Neurosci 10, 425–444.
13. Andersen, RA, and Buneo, CA (2002) Intentional maps in posterior parietal cortex. Annu Rev Neurosci 25, 189–220.
14. Mitchell, JF, and Zipser, D (2003) Sequential memory-guided saccades and target selection: A neural model of the frontal eye fields. Vis Res 43, 2669–2695.
15. Torres, EB, and Zipser, D (2002) Reaching to grasp with a multi-jointed arm. I. Computational model. J Neurophysiol 88, 2355–2367.
16. Moody, SL, and Zipser, D (1998) A model of reaching dynamics in primary motor cortex. J Cog Neurosci 10, 35–45.
17. Burnod, Y, Baraduc, P, and Battaglia-Mayer, A, et al. (1999) Parieto-frontal coding of reaching: An integrated framework. Exp Brain Res 129, 325–346.
18. Burnod, Y, Grandguillaume, P, Otto, I, et al (1992) Visuomotor transformations underlying arm movements toward visual targets: A neural network model of cerebral cortical operations. J Neurosci 12, 1435–1453.
19. Kuperstein, M (1988) Neural model of adaptive hand-eye coordination for single postures. Science 239, 1308–1311.
20. Mel, BW (1991) A conncectionist model may shed light on neural mechanisms for visually guided reaching. J Cog Neurosci 3, 273–292.
21. Rosenbaum, DA, Engelbrecht, SE, Bushe, MM, and Loukopoulos, LD (1993) Knowledge model for selecting and producing reaching movements. J Mot Behav 25, 217–227.
22. Jordan, MI, and Rumelhart, DE (1992) Forward models: Supervised learning with a distal teacher. Cog Sci 16, 307–354.

Reading List

The DIRECT model by Dan Bullock and Steve Grossberg provides a thorough discussion of the issues raised in considering a difference vector.[8] The account of reaching and pointing presented here owes much to DIRECT and other previous models of visually guided reaching. Many pioneering models of reaching and pointing came from Steve Grossberg and his colleagues.[9-12] We especially direct your attention to the model devised by Paul Cisek, Steve Grossberg, and Dan Bullock[12] called VITE, for *v*ector *i*ntegration *t*o *e*ndpoint (figure 12.12). This model, a development of the DIRECT model, incorporates feedback assistance and, very importantly, seeks to account for the established neurophysiological data in an anatomically realistic framework. Like the present model and that of Andersen and Buneo,[13] VITE models the difference vector as the product of parietal computations, which are passed on to the frontal cortex to compute estimations about joint-angle changes, forces, and feedback. The models of David Zipser and his colleagues[4,14-16] also have a lot to offer. Zipser's approach uses neural networks to gain insight into the information processing functions underlying visuomotor transforms, with a minimum of prior assumptions. Other vector-based models[17,18] are also worth exploring for a balanced view of this topic. Like the present model, previous ones have dealt with computing inverse kinematics and motor learning.[8,19-22] You should look into these and other models to appreciate the comparisons and contrasts with the ideas presented here.

13 Computing Difference Vectors II: Parietal and Frontal Cortex

Overview: The difference vector represents a high-level plan for movement, which specifies a displacement of an end effector from its current location to the target's location. However, several questions remain about the nature of this plan. Does it correspond to a movement that your CNS will make with the hand, with the eye, or with some other part of your body? Does it reflect a movement the CNS might make or definitely will make? And what parts of the CNS play the most direct role in formulating this plan? This chapter presents some evidence that areas in the PPC, acting in close concert with the frontal motor areas, participate in computing the motor plan.

13.1 Computing a Movement Plan

Evidence that the PPC plays a role in planning movements comes, in part, from examining deficits in patients with damaged cortex. Such patients often suffer from *optic ataxia*, a condition that prevents the accurate localization of a reach target.[1] Section 10.4 explained that lesions of the PPC cause inaccurate reaching in monkeys as well. Although lesions of the frontal motor cortex (sparing M1 and maybe parts of premotor cortex) leave reaching movements relatively intact,[2] assume for the sake of this presentation that the frontal motor areas and PPC interact in the computation of motor plans. The neurophysiological evidence shows that cells in the PPC play an important role in computing the difference vector and that frontal cortex reads out this signal to plan, prepare, and control the movement.

13.1.1 Directional Tuning and Delay-Period Activity in PPC and Frontal Cortex

Ed Evarts and Jun Tanji[3,4] took the first steps toward understanding motor planning when they presented a monkey with either a red light, which instructed the monkey to pull a handle, or a green light, which meant it should push the handle. To receive a reward, the monkey needed to withhold movement until a subsequent time, seconds later, when a force pulse delivered to the handle triggered the instructed movement. The red or green light can be called an *instruction stimulus*, the time between the instruction and the force pulse can be termed a *delay period* or *instructed-delay period*, and the force pulse can be called the *trigger stimulus* because it served as a "go" signal.

Evarts and Tanji found that neurons in M1 began discharging shortly after the colored light appeared and continued to discharge until the movement occurred. This observation suggests that the neuronal discharge reflects the plan to make a movement. In order not to prejudice the ultimate interpretation of this kind of activity in functional terms, you might say that the activity reflects the *preparation* for a subsequent movement. "Preparation" means that neural signals do not lead directly to movements in a causal sense, but instead participate less directly in computations relevant to some future, potential action. Many M1 neurons showed activity modulation for only one of the two future movements during the delay period, and some showed changes in modulation of opposite sign. For example, a given M1 neuron might show an increased discharge rate for pulling movements and a decreased rate for pushing movements.

Since those pioneering experiments, neurophysiologists have studied the activity of frontal and parietal neurons in a variety of tasks, all of which involved monkeys preparing to make a reaching movement. Mike Weinrich and Wise[5] conducted an early experiment on dorsal premotor cortex (PMd) during motor preparation. They designed a task in which the monkey either saw a visual instruction stimulus or heard an auditory one indicating the target of reach, but had to wait until a subsequent visual trigger stimulus to make the movement. The instruction stimuli usually consisted of lights inside buttons, and those buttons served as targets for the reaching movement. Alternatively, tones emitted from a small speaker directly behind the button could serve as the instruction stimulus. The monkey could not predict the time between the instruction and trigger stimuli—the instructed-delay period. The experimenters observed that cells in PMd reflected the instructed (and presumably planned) direction of reach. For many PMd cells, this delay-period activity did not differ for visual and acoustic targets. Although they did not think of it this way at the time, the signal Weinrich and Wise observed probably reflected the difference vector. Some neurons preferred preparation of movements to the right and others preferred movements to the left, often regardless of the sensory modality in which the target appeared.

Similar experiments in M1, PMd, and PPC extended these results. In M1, Apostolos Georgopoulos and his colleagues[6] devised an experiment to measure a cell's spatial "tuning" properties. Donald Crammond and John Kalaska[7] did a similar experiment and recorded from area 5d and PMd (see figure 6.3).[8] Their monkeys began each trial by holding a handle that could move in the horizontal plane. The handle began each trial over a central light located just below the plane of the handle. One of the eight peripheral lights, which were arranged in a circle around the central one, turned on to serve as the instruction stimulus. They served as the targets for reaching movements. The target remained on for a variable period (1–3 sec), after which its color changed, triggering the reaching movement.

The experimenters observed that during the delay period, cells in M1, PMd, and area 5d were *tuned* to the direction of the upcoming movement. This means that the cell discharged maximally for one of the targets. They called that target direction the cell's **preferred direction**. In the next chapter, figure 14.9 illustrates the concept of preferred direction, and it applies to both movement-related activity (i.e., neuronal discharge immediately before and during movements) and activity during instructed-delay periods. As is typical for directional tuning of this kind, a neuron's discharge declines gradually as a function of angle from that direction, so that the cells show minimal discharge when the target appears in the direction opposite to the preferred direction. Therefore, the discharge during the delay period appeared to be related in some way to the target's location.

These investigators made an important additional observation. In area 5d, after the trigger stimulus, the discharge remained much the same as it was during the delay period, at least for a while. These experimenters called the time between the trigger stimulus and the onset of movement the "reaction time," often abbreviated RT. They called the time between the onset of reaching and target acquisition the movement time (MT), and any time spent at the target they called the target hold time (THT). The observation that activity continues from the delay period into the RT period suggests that whatever the cell encoded during the delay period did not change immediately after the trigger ("go") signal. Because the kinematic plan for the movement also did not change when the trigger signal was given, it seems likely that the cells encoded something related to planning the ensuing movement.

Interestingly, although the experimenters found—over a series of studies—that delay-period activity occurred prominently in the discharge of cells in both area 5d and PMd, M1 showed these signals less frequently and area 2 never did. Area 2 composes a part of the somatosensory cortex rather than part of the PPC. Crammond and Kalaska[7] found that activity in area 2 mostly reflected sensory feedback from the moving limb rather than a planning-related signal. These cells probably receive indirect inputs from muscle-spindle afferents, perhaps from area 3a. Therefore, taken together with the results from PMd, the results of Crammond and Kalaska suggest that area 5d and PMd play a role in planning the kinematics of an intended movement, but area 2 does not. Studies differ on the degree of delay-period activity in M1, but they generally agree that M1 shows less of it than does either PMd or area 5d. Subsequent studies have shown delay-period activity in the ventral premotor cortex (PMv) as well. Thus, area 5d, PMd, and PMv have activity consistent with a role in computing or reading out the difference vector: the kinematic plan for a reaching movement.

13.1.2 Maintaining Movement Plan After a Target Disappears

In many experiments that had instructed-delay periods, the instruction stimulus remained on during the entire delay period. However, in some the target disappeared before the trigger stimulus, and the monkey made a movement to a remembered location. If neurons participate in a neural network that plans the kinematic goal of the upcoming movement, then their activity should not change if the instruction disappears before the trigger stimulus. Wise and Karl Mauritz[9] tested this hypothesis. A monkey sat in front of a board with depressible buttons that could be illuminated from the inside. The experimenters compared neuronal activity in PMd in three conditions: (1) as in the experiments described above, the instruction stimulus remained on throughout the delay period; (2) the light inside the target button was turned off after 1 sec, although the delay period lasted 0.5–2.0 sec longer; and (3) during the delay period the target jumped from a location on one side of the monkey's hand to the other side. The activity of cells in PMd was unaffected by the disappearance of the instruction light, as judged either on a cell-by-cell basis or over the entire sampled population of neurons.

Figure 13.1 shows this result for a population of neurons collected by Crammond and Kalaska.[8] Activity during the delay period does not depend on cue visibility, which is consistent with the idea that it reflects a difference vector. What about when the target changed from one side of the monkey's hand to the other? Wise and Mauritz observed that this target switch caused dramatic change in cell activity in PMd. In this case, the cell reacted in about 140 msec to the changed direction of the difference vector. As a general statement, the cells became highly active 140 msec after a target appeared in their preferred direction.

PMd and M1 activity has additional similarities to that expected of a difference vector. If the end effector moves, even passively because of some external force, delay-period activity should change in the same manner as described by Wise and Mauritz for target displacements. Take, for example, a cell with a preferred direction to the right. Assume that this preferred direction represents something akin to a motor field. Now imagine moving the monkey's hand relative to a fixed target so that this end effector begins to the left of the target but ends up to the right. Difference-vector neurons should discharge as the target approaches the cell's preferred direction. Erhard Bauswein and Christof Fromm[10] performed this experiment. They reported that cells in PMd (and maybe a few in M1) changed their activity to reflect the reoriented difference vector, as expected. (Because they passively moved the hand with a motor, the experimenters needed to rule out the possibility that this passive proprioceptive stimulus accounted for the cell's activity, and they did so.) These findings show that neural discharge in area 5d and PMd represents aspects of the movement plan, and not merely a passive, sensory response to visual or proprioceptive information.

13.1.3 Planning in Terms of Kinematics, Not Dynamics

Consider what the term "planning a movement" might mean. Is the plan limited to the kinematics of the task (e.g., a difference vector), or does it also involve dynamics? For example, do the cells take part in computing

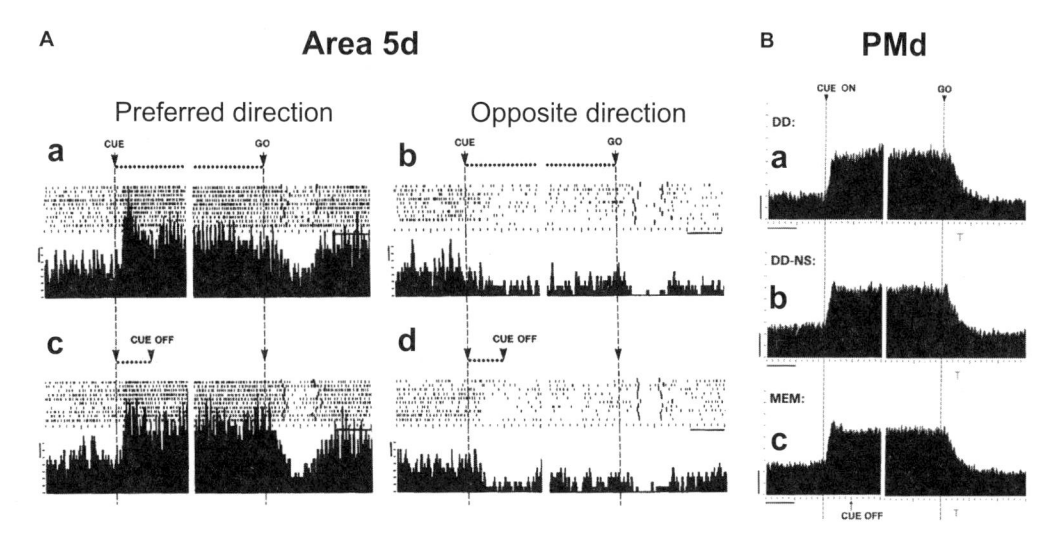

13.1. Computing a Movement Plan

Figure 13.1
Difference vectors in the absence of a visible target in area 5d and PMd. (*A*) Activity of a reach-related area 5d cell during an instructed-delay task. (*a*) The instruction stimulus (cue) indicates that the subsequent movement will be in the cell's **preferred direction**, and the cue remains visible during the entire delay period (dotted horizontal line), which ends with the trigger ("go") stimulus. The heavy marks on each raster line show the onset of movement (left mark) and target acquisition (right mark). (*b*) The cue instructs a movement opposite to the cell's preferred direction. As in *a*, the cue remains visible throughout the delay period. The time scale shows a 500-msec duration. (*c* and *d*) Memory trials. The instruction cue appears for only 500 msec. Note that the cell continues to show directionally tuned activity throughout the delay period. (*B*) The same result obtained for PMd, illustrated in average histograms for a population of neurons. (*a*) The population activity for the preferred direction of each cell, during an instructed-delay task with a directional trigger stimulus (DD stands for direct delay, which means that the monkey moved *direct*ly to the instruction cue after a *delay* period). The "T" beneath the histogram shows the time of target acquisition, on average. (*b*) Same population, with a *nondirectional* trigger *stimulus* (NS). (*c*) Same population, when the stimulus disappeared after 500 msec, leading to a memory (MEM) requirement. (From Kalaska and colleagues[11,12] with permission.)

the location of the target with respect to the hand? Do they also take part in computing the forces needed to move the hand to the target?

To answer these questions for area 5d, John Kalaska and his colleagues[13] performed an experiment in which a monkey held on to a handle and reached to a target while a pulley imposed a load on the handle (figure 13.2). Therefore, the monkey always reached to the same target along the same trajectory, but by varying the load placed on the handle, the monkey needed to produce different **torques** to make the movement. Thus, the movement kinematics remained more or less the same, but the dynamics differed. The experimenters found that the activity of area

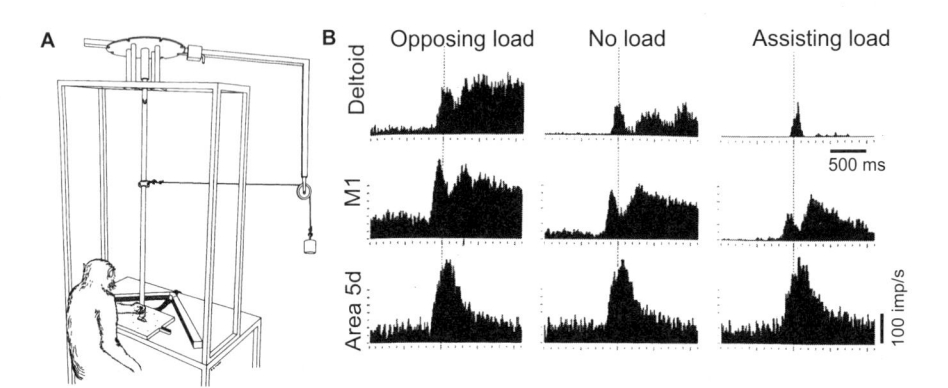

Figure 13.2
Load insensitivity of area 5d neurons. (A) Experimental setup. The monkey made a reaching movement to a constant target under three different conditions: with an opposing load on the hand, with no load, and with an assisting load pulling the hand toward the target. (B) The discharges of an M1 cell, an area 5d neuron, and the deltoid muscle are aligned on the start of the movement. Loads affected muscle and M1 activity during a reaching movement, but not area 5d activity. (A from Kalaska et al.,[14] B from Kalaska[15] with permission.)

5d cells did not depend on the magnitude of the load or its direction. This observation contrasted sharply with what they saw in area 2 and M1, where discharge strongly depended on the load. This finding shows that the dynamics of a task influence the activity of at least some cells in M1 and area 2, but appear to be less important in area 5d. Similar experiments with loads have not yet been done for PMd, PMv, or other frontal motor areas, although some relevant evidence has been obtained with other experimental methods (see section 14.1.2).

13.1.4 Planning Saccades Versus Reaching Movements

At least to an extent, separate regions in the frontal motor cortex control reaching movements versus eye movements. For example, the frontal eye field (FEF) plays an important role in saccadic eye movements, and a nearby area called the frontal pursuit area plays an analogous role in smooth-pursuit eye movements. These areas lie near the parts of PMd and PMv that play an important role in reaching movements. Other areas play a role in both eye and arm movements, but different regions within those areas seem to specialize for one or the other.

What about the PPC? To answer that question, Larry Snyder, Aaron Batista, and Richard Andersen[16] designed an experiment involving the control of both reaching and eye movements. They trained a monkey to move its hand to a central button on a screen while fixating the same location. At the beginning of each trial, the end-effector location and the fixation point were co-localized at that spot. Approximately 500 msec after the monkey touched the button, either a red or a green light appeared for 300 msec at one of the eight locations surrounding the center button. That stimulus instructed both the target of movement and the end effector that the monkey needed to use. After it disappeared, another 800 msec or so passed, and then the center light turned off. This event served as the trigger stimulus, after which the monkey either made a saccade, if the cue had been red, or a reaching movement, if the cue had been green. If the monkey made a saccade, it did not move its hand, and if it reached, it maintained fixation on the central location.

The experimenters found that some cells showed delay-period activity only before saccades (figure 13.3A), whereas others did so only before reaching movements (figure 13.3B). Neurons that signaled an intended saccade occurred in a different part of the PPC than neurons that signaled an intended reach (figure 13.3C). During the instructed-delay period, neurons in LIP discharged preferentially before saccades, whereas neurons in MIP and the dorsal parts of PO did so before reaches (see figure 6.3). Andersen and his colleagues named the regions encompassed by this latter group of neurons the *parietal reach region* (PRR).

These results indicate that PPC cells have activity that reflects more than the information on the retina. These cells not only continue to discharge when the target has been removed, as in the experiments described above for PMd and area 5d,[9,11] but they also carry information about the plan for a specific kind of movement to the target. A sensory response—indicating a "receptive field"—should not differ when the stimulus signals an eye movement or a reaching movement. In the main, these experiments show that frontal motor areas and PPC have activity that reflects much more than a simple sensory response, although there remains some controversy about whether the neural signals in area LIP reflect the plan to saccade to a given target, as opposed to directing attention there.

To further explore the idea that activity in PRR specifically signals a plan to reach, Snyder et al.[18] considered a condition in which the movement plan changed because the instruction did, as in the experiment by Wise and Mauritz[9] described above. In this task, a red target initially appeared as an instruction to make a saccade. However, during the ensuing delay period, a green target appeared at the same location to indicate that the monkey must instead reach to that location. Of course, the animal maintained fixation at the center location as the cues appeared and disappeared. After a further delay period, the central fixation target disappeared, acting as the trigger stimulus. Figure 13.3D shows the activity of a reach-specific neuron in PRR. Immediately after the first instruction appeared, the activity of the cell increased—a kind of nonspecific sensory-like response. Importantly, however, its activity then declined in a specific way, and its delay-period activity for a reach instruction (R1) exceeded that for a saccade instruction (S1). After the second instruction appeared at the same location, the cell's activity increased for a reach instruction (R2) but fell for a saccade cue (S2). These results indicate that PRR neurons carry information about both the target of movement and what kind movement the monkey plans.

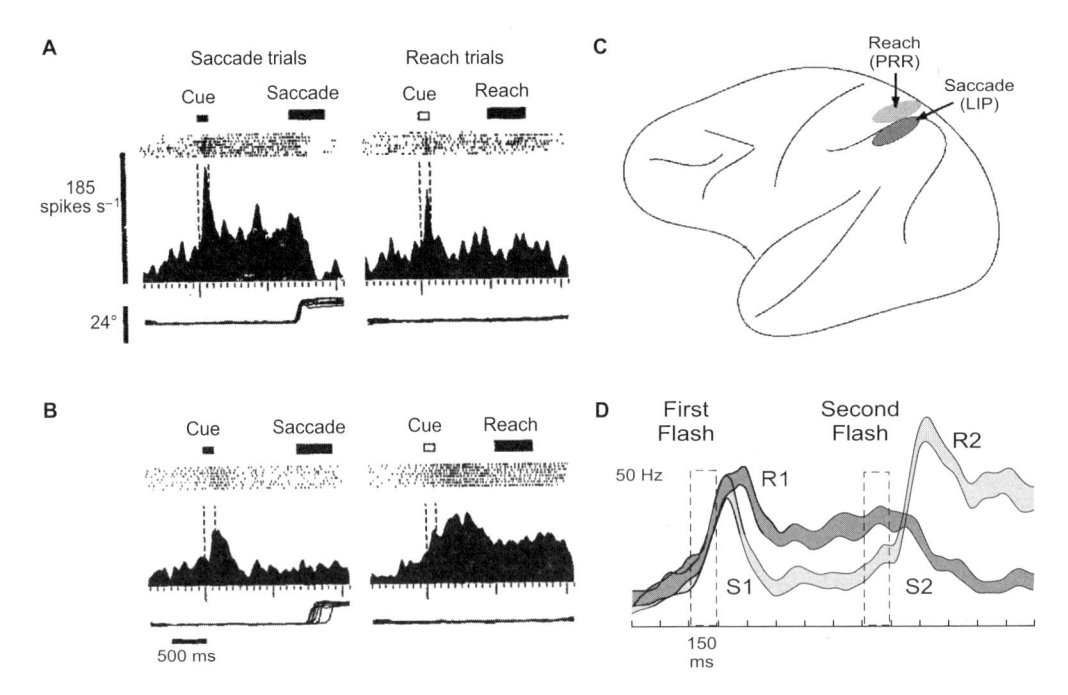

Figure 13.3

Activity of two PPC neurons in the delayed-saccade and delayed-reach tasks. An instruction stimulus (cue) appeared to guide, by its color, a saccade (filled bar) or a reach (unfilled bar) to that location. Rasters and histograms show neuronal activity; gaze angle (head fixed) appears below. Neuronal activity during the instruction-on period (bar beneath "Cue") was not specific to the color of the cue, but activity during the instructed-delay period depended on whether the monkey planned a saccade or a reaching movement. (*A*) A cell preferring saccades. (*B*) A cell preferring reaching movements. (*C*) Schematic drawing of the regions in PPC where cells with predominantly reach-related activity or saccade-related activity were found. PRR stands for the parietal reach regions; LIP, for the lateral intraparietal area. (*D*) Activity of a neuron in the PRR when instruction changed from reach to saccade or vice versa, without changing the target location. Activity resulting from an instruction to plan a reach (R1, meaning that the first stimulus instructed a reach) was reduced when a second stimulus changed the instruction to a saccade (S2). The reverse occurred when the first stimulus (S1) instructed a saccade and the second instruction, a reach (R2). Each thick line represents the mean response ± 1 standard error. (*A* and *B* from Snyder et al.,[16] *C* from Andersen and Buneo,[17] *D* from Snyder et al.[18] with permission.)

13.1.5 Planning a Reach with an End Effector

In **psychophysical** and neurophysiological experiments, participants often move a *cursor* on a video monitor indirectly through a **manipulandum**, such as a computer mouse. In this condition, the location of the hand corresponds to the location of the cursor, and cursor velocity has a relationship to hand velocity. To move the cursor, the participant needs to learn these relationships (section 16.2C returns to this point). For now, assume that these relationships have been learned.

In one such experiment, Paul Cisek, Donald Crammond, and John Kalaska[19] trained a monkey to move a cursor with either its left or its right hand. With the cursor as the end effector and the target as a spot on the same video monitor, the difference vector was the same for right- or left-hand movements. Consistent with this idea, the experimenters found that the preferred direction of PMd cells did not differ very much for reaching movements of the left versus the right hand. Jun Tanji and Eiji Hoshi[20,21] have recently confirmed this result for PMv. The majority of neurons in that area showed selectivity for target location only, regardless of which hand the monkey used to reach. These results agree with the idea that the difference vector contributes importantly to activity in these premotor areas.

13.1.6 Planning Movements Away from a Stimulus

Movements often target a sensory cue or object, but the same stimuli can direct movements elsewhere. Crammond and Kalaska[22] recorded from cells in area 5d in both kinds of tasks. In one condition, the monkey reached to a visual cue, as usual. In the other, the monkey had to reach toward a target away from the cue (figure 13.4). They found that except for a brief period immediately after the appearance of the cue, the cells signaled the intended movement direction. That is, these cells indicated the target location, even when it differed from the location of the visual cue. Similar results have been obtained from PMd.[23] Thus, PPC and PMd use available information to compute the goal for the upcoming movement—perhaps the difference vector—even if the current visual information does not directly provide a target. These findings agree with the idea that the CNS transforms the visual cues into a fixation-centered representation of the target location. (Section 25.5 takes up this topic in more detail.)

13.2 Planning Potential Movements but Not Executing Them

Because many cells in PPC appear to encode the location of reach targets, you might wonder whether, by looking at neuronal activity, you could discern whether the monkey would make a movement in the near future and where it might direct that movement. For example, in the delay-period activity of cells presented in figure 13.3, you can tell whether

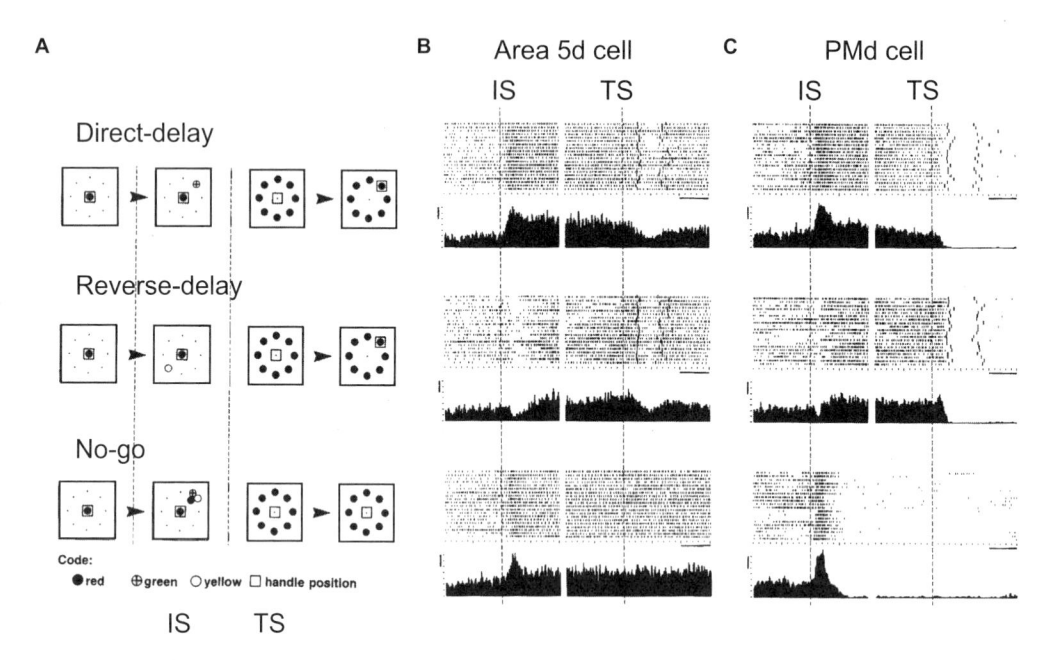

Figure 13.4
Planning movements toward or away from a visual stimulus. (*A*) Experimental design. In the direct-delay condition, a green instruction stimulus (IS) appeared, instructing a reaching movement to that location, to be initiated after all eight potential target locations illuminated as a trigger (go) stimulus (TS). The square in the right panel shows the correct handle (and hand) location at the end of movement. In the reverse-delay condition, a yellow cue appeared, instructing a movement in the opposite direction (of equal magnitude). In the no-go condition, red, green, and yellow cues appeared at the same location, instructing the animal to do nothing when the TS later appeared. (*B*) The activity of an area 5d cell in the three conditions. The cell was active in the direct-delay condition for an IS and movement in its **preferred direction**. In the reverse-delay condition, despite the fact that the IS was opposite to its preferred direction, the cell showed high activity levels because the intended direction of movement corresponded to the cell's preferred direction. In the last row, the IS instructed the withholding of movement, but the cell remained active. (*C*) Response of a neuron in PMd during the same conditions. The main difference between this cell and the cell in area 5d was in the no-go condition. (From Kalaska[12] with permission.)

the monkey planned to make an eye or an arm movement. In fact, at any given time the monkey may have several motor plans, involving eye movements, reaching movements, and others. The monkey may plan several movements simultaneously.[24] But can you tell from monitoring single neurons whether the monkey will make a movement of a particular kind?

Kalaska and Crammond[25] found that area 5d activity reflects potential targets and movements, and Cisek and Kalaska[24] extended this result by showing that PMd activity does so as well, but only as long as a given location remains a possible next-movement target. As soon as some subsequent cue eliminates a potential target, PMd neurons cease to represent its location. Taken together, these two findings suggest that PMd more closely reflects whether a potential plan has been selected for execution.

In the experiment of Kalaska and Crammond,[25] green cues instructed the monkey to reach to their location; yellow cues instructed a movement in the opposite direction; and yellow, green, and red cues appearing at a single location instructed the monkey to withhold movement when the "trigger" signal later appeared (figure 13.4A). The experimenters found that many area 5d cells showed strong, sustained discharge during the delay period in all three conditions (figure 13.4B). The discharge of cells in PMd resembled that of area 5d cells when the monkey planned to move toward the cue or away from it (figure 13.4C, top and middle, respectively). However, when the spatial cues delivered the instruction to withhold movement on that trial, PMd cells quickly ceased their discharge (figure 13.4C, bottom). This finding suggests that PMd, but not area 5d neurons, reflects the monkey's selected motor plan—not to reach to a given location. Area 5d and PMd neurons both reflect the plan for a possible movement in their preferred direction.[24] Once some cue eliminates this possibility, PMd neurons cease their delay-period activity, but PPC neurons do not. PPC (area 5d, at least) continues to represent potential targets.

Results from a different experiment by Snyder et al.[16] reinforce this conclusion. Recall that the delay-period activity of some cells in PRR and LIP showed selectivity for reaching versus saccadic eye movements. However, Snyder et al. found other cells that had high levels of delay-period activity regardless of whether stimuli instructed a reach or a saccade. For example, figure 13.5A and 13.5B show a cell that discharged strongly for both reaches and saccades. Because a monkey may have covert plans to make a movement, even though that movement actually does not take place, interpretation of this kind of activity requires further investigation. Accordingly, Snyder et al.[16] trained a monkey to reach to one target while making a saccade to a different one. At the trigger signal, the monkey had to move its eyes in one direction while moving its hand in the opposite direction. In that condition, the cell shown in figure 13.5A and 13.5B had delay-period activity only when the monkey planned to *reach* to a target in its **activity field**. It did not discharge before upcoming *saccades* into its activity field (figure 13.5C). Thus, when visual cues instructed both a reach and a saccade, delay-period activity in this cell reflected the planned reach, not the planned saccade.

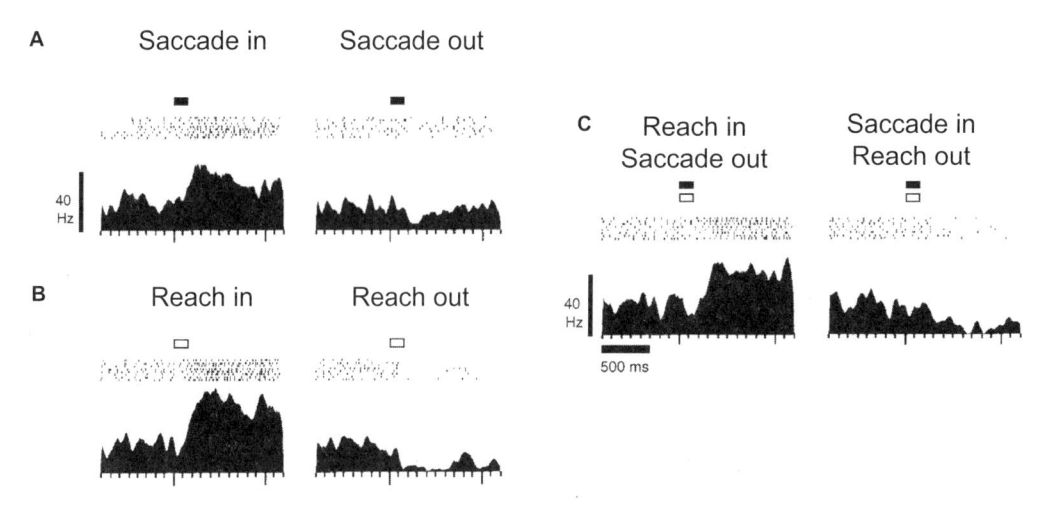

Figure 13.5
Responses of a PPC neuron in three different conditions. (*A*) In one condition, a cue instructed a saccade. The cue fell either in the cell's activity field (saccade in) or outside it, in an opposite direction (saccade out). (*B*) In a second condition, a different cue instructed a reaching movement. As in *A*, the cue fell either inside (reach in) or outside (reach out) the cell's preferred field. (*C*) In the third condition, the monkey reached to one target while making a saccade to another, in the opposite direction. Left: The reach cue appeared in the cell's activity field with the saccade cue outside (reach in, saccade out). Right: The reach cue fell outside the cell's activity field, with the saccade cue inside (saccade in, reach out). (From Snyder et al.[16] with permission.)

The fact that the cell in figure 13.5A discharged before a saccade suggests that the monkey covertly planned to reach to that target, even though at the trigger signal it actually performed a saccade. As with the results from Kalaska and his colleagues, PPC seems to encode potential targets and actions regardless of whether those plans progress to execution. According to this idea, when the trigger appeared in the experiment by Snyder et al., some other part of the CNS decided not to perform the covertly planned reach. It seems likely that the decision not to move the arm depended on neural networks involving the frontal cortex, as reflected in the discharge of cells in PMd.

Further support for this idea comes from studies by Jeff Schall and his colleagues, using an experimental design called the countermanding task.[26,27] (Chapter 25 takes up the results of these experiments in more detail; see figure 25.3.) For now, note that a cue appears after the target, which indicates that the monkey should not make the previously instructed saccade. Cells in the FEF show a signal that accounts for making or withholding the saccade. These cells may participate in the decision to make (or not to make) a planned, potential movement. Interestingly, cells in the SEF and in the anterior cingulate cortex show a similar signal,[28–30] but too late to causally affect the response (or veto) decision.

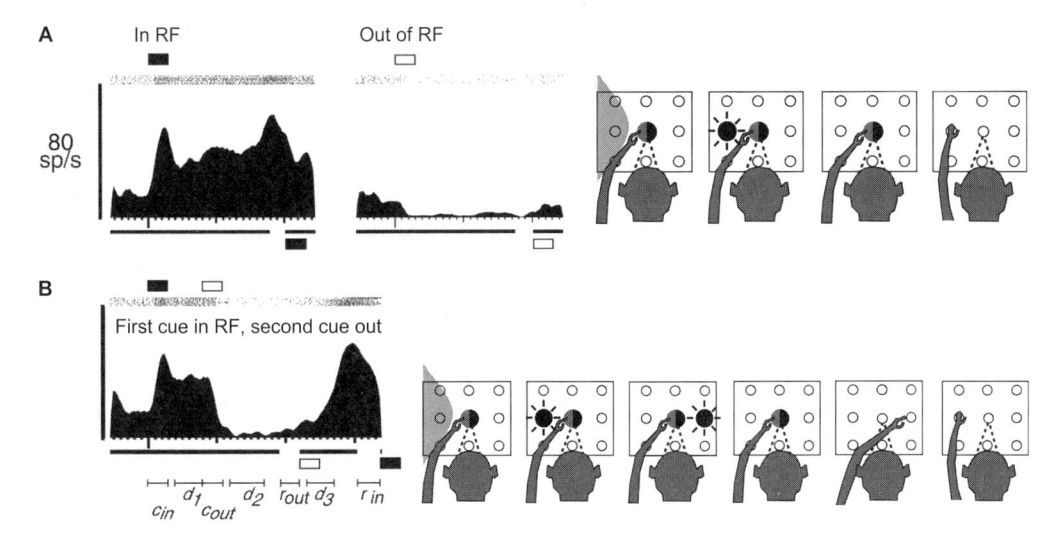

13.3. Planning the Next Movement in a Sequence

Figure 13.6
Behavior of a PPC neuron in two conditions. (*A*) In the first condition, a single target appeared either in the cell's activity field (gray region in the drawing to the right) or outside it. After a delay period, the monkey reached to that target. Left: Above each raster and activity average, a filled bar indicates that a stimulus appeared in the cell's activity field for the period under the bar, and an unfilled bar indicates a stimulus outside it. Tick marks indicate 100-msec intervals. The lines beneath each activity average show the period when the hand remained stationary. The bars below those lines indicate times when the monkey touched a button in the cell's activity field (filled bar) or a button outside it (unfilled bar). (*B*) In a second condition, two targets appeared in sequence, and the monkey reached to each target. For the data shown, the first stimulus fell in the cell's activity field, but the second one fell outside it. After viewing the stimulus sequence, the monkey first reached to the remembered location of the second stimulus, and then to the remembered location of the first stimulus. Format as in *A*. (From Batista and Andersen[34] with permission.)

Schall and his colleagues suggest that such activity reflects the monitoring of behavior by those parts of the frontal cortex.

13.3 Planning the Next Movement in a Sequence

If a task involves multiple reaching movements, do the cells that reflect the motor plan code for all aspects of the reaching movements simultaneously, or do they code for each part in sequence? Work by Jim Ashe, Apostolos Georgopoulos, and their colleagues[31] indicated that cells in M1 encode the next movement in a sequence.[32,33]

To explore this question for PPC, Aaron Batista and Richard Andersen[34] trained monkeys to make two reaching movements in sequence. Recall that cells in PPC increase activity for targets that appear in their spatial activity field. The experimenters first measured the activity field of neurons in PRR. They then chose cells that had increased activity in the delay period before a reach but not before a saccade. As in the experiments described above, each trial began with the monkey pressing a button at a central location and fixating the same place. Two reach targets then appeared and disappeared prior to any reach. By design, the first reach target fell in the activity field of the cell, after which a delay period of 600 msec ensued (labeled d_1 in figure 13.6B). Then a second target appeared outside the activity field of the cell, followed by another delay period (d_2), after which the monkey reached to the remembered location of the second target. Another delay period (d_3) followed, after which the monkey reached to the remembered location of the first target.

Figure 13.6 shows data from one neuron recorded during this experiment. In the simple instructed-delay task, in which only one target appeared, the response for this cell increased only when the stimulus image fell in the cell's activity field (gray region in figure 13.6A), which you can also consider its preferred direction. In the two-stimulus paradigm (figure 13.6B), the neuron's activity increased when the first cue appeared (period marked c_{in}), then decreased when the second cue came on (period marked c_{out}). This result makes sense because the first cue fell in the cell's preferred direction but the second cue did not. Prior results show the same properties for PMd.[9]

In the PPC experiment, the monkey first reached to the remembered location of the second stimulus, the one in its nonpreferred direction, then reached back to the first. Accordingly, the discharge remained at a low level until the monkey completed the first movement (period marked r_{out}). Then, the cell increased its discharge and maintained that increase until the end of the second reach (period marked r_{in}). Thus, the cell reflects the target location only for the upcoming reaching movement, the next planned state of the limb.

The PPC cells studied by Batista and Andersen[34] coded only the location of the target for the next reaching movement. You might wonder about the other, future targets in a sequence. Section 25.3 takes up the topic of how the CNS codes for sequences of actions. For the purposes of this chapter, note that to make a reaching movement to each target in a sequence, the CNS has to compute the end effector's location with respect to the target and to compute a difference vector, which changes whenever the end effector moves or the target changes location. Chapter 19 discusses this operation in terms of a next-state planner, which establishes "policies" for action.

References

1. Rondot, P, de Recondo, J, and Dumas, JL (1977) Visuomotor ataxia. Brain 100, 355–376.
2. Rea, GL, Ebner, TJ, and Bloedel, JR (1987) Evaluations of combined premotor and supplementary motor cortex lesions on a visually guided arm movement. Brain Res 418, 58–67.

3. Evarts, EV, and Tanji, J (1974) Gating of motor cortex reflexes by prior instruction. Brain Res 71, 479–494.

4. Tanji, J, and Evarts, EV (1976) Anticipatory activity of motor cortex neurons in relation to direction of an intended movement. J Neurophysiol 39, 1062–1068.

5. Weinrich, M, and Wise, SP (1982) The premotor cortex of the monkey. J Neurosci 2, 1329–1345.

6. Georgopoulos, AP, Kalaska, JF, Caminiti, R, and Massey, JT (1982) On the relations between the direction of two-dimensional arm movements and cell discharge in primate motor cortex. J Neurosci 2, 1527–1537.

7. Crammond, DJ, and Kalaska, JF (1989) Neuronal activity in primate parietal cortex area 5 varies with intended movement direction during an instructed-delay period. Exp Brain Res 76, 458–462.

8. Crammond, DJ, and Kalaska, JF (2000) Prior information in motor and premotor cortex: Activity during the delay period and effect on pre-movement activity. J Neurophysiol 84, 986–1005.

9. Wise, SP, and Mauritz, K-H (1985) Set-related neuronal activity in the premotor cortex of rhesus monkeys: Effects of changes in motor set. Proc Roy Soc London B 223, 331–354.

10. Bauswein, E, and Fromm, C (1992) Activity in the precentral motor areas after presentation of targets for delayed reaching movements varies with the initial arm position. Eur J Neurosci 4, 1407–1410.

11. Crammond, DJ, and Kalaska, JF (1996) Differential relation of discharge in primary motor cortex and premotor cortex to movements versus actively maintained postures during a reaching task. Exp Brain Res 108, 45–61.

12. Kalaska, JF (1996) Parietal cortex area 5 and visuomotor behavior. Can J Physiol Pharmacol 74, 483–498.

13. Kalaska, JF, Cohen, DA, Prud'homme, M, and Hyde, ML (1990) Parietal area 5 neuronal activity encodes movement kinematics, not movement dynamics. Exp Brain Res 80, 351–364.

14. Kalaska, JF, Cohen, DAD, Hyde, ML, and Prud'homme, M (1989) A comparison of movement direction-related versus load direction-related activity in primate motor cortex, using a two-dimensional reaching task. J Neurosci 9, 2080–2102.

15. Kalaska, JF (1988) The representation of arm movements in postcentral and parietal cortex. Can J Physiol Pharmacol 66, 455–463.

16. Snyder, LH, Batista, AP, and Andersen, RA (1997) Coding of intention in the posterior parietal cortex. Nature 386, 167–170.

17. Andersen, RA, and Buneo, CA (2002) Intentional maps in posterior parietal cortex. Annu Rev Neurosci 25, 189–220.

18. Snyder, LH, Batista, AP, and Andersen, RA (1998) Change in motor plan, without a change in the spatial locus of attention, modulates activity in posterior parietal cortex. J Neurophysiol 79, 2814–2819.

19. Cisek, P, Crammond, DJ, and Kalaska, JF (2003) A comparison of neural activity in primary motor and dorsal premotor cortex during reaching tasks with the contralateral versus the ipsilateral arm. J Neurophysiol 89, 922–942.

20. Hoshi, E, and Tanji, J (2002) Contrasting neuronal activity in the dorsal and ventral premotor areas during preparation to reach. J Neurophysiol 87, 1123–1128.

21. Hoshi, E, and Tanji, J (2000) Integration of target and body-part information in the premotor cortex when planning action. Nature 408, 466–470.
22. Crammond, DJ, and Kalaska, JF (1994) Modulation of preparatory neuronal activity in dorsal premotor cortex due to stimulus-response compatibility. J Neurophysiol 71, 1281–1284.
23. di Pellegrino, G, and Wise, SP (1993) Visuospatial vs. visuomotor activity in the premotor and prefrontal cortex of a primate. J Neurosci 13, 1227–1243.
24. Cisek, P, and Kalaska, JF (2002) Simultaneous encoding of multiple potential reach directions in dorsal premotor cortex. J Neurophysiol 87, 1149–1154.
25. Kalaska, JF, and Crammond, DJ (1995) Deciding not to GO: Neuronal correlates of response selection in a GO/NOGO task in primate premotor and parietal cortex. Cerebral Cortex 5, 410–428.
26. Schall, JD, Hanes, DP, and Taylor, TL (2000) Neural control of behavior: Countermanding eye movements. Psychol Res 63, 299–307.
27. Hanes, DP, Patterson, WF, and Schall, JD (1998) Role of frontal eye fields in countermanding saccades: Visual, movement, and fixation activity. J Neurophysiol 79, 817–834.
28. Schall, JD, Stuphorn, V, and Brown, JW (2002) Monitoring and control of action by the frontal lobes. Neuron 36, 309–322.
29. Stuphorn, V, and Schall, JD (2002) Neuronal control and monitoring of initiation of movements. Muscle Nerve 26, 326–339.
30. Stuphorn, V, Taylor, TL, and Schall, JD (2000) Performance monitoring by the supplementary eye field. Nature 408, 857–860.
31. Ashe, J, Taira, M, Smyrnis, N, et al. (1993) Motor cortical activity preceding a memorized movement trajectory with an orthogonal bend. Exp Brain Res 95, 118–130.
32. Kettner, RE, Marcario, JK, and Port, NL (1996) Control of remembered reaching sequences in monkey. II. Storage and preparation before movement in motor and premotor cortex. Exp Brain Res 112, 347–358.
33. Kettner, RE, Marcario, JK, and Clark-Phelps, MC (1996) Control of remembered reaching sequences in monkey. I. Activity during movement in motor and premotor cortex. Exp Brain Res 112, 335–346.
34. Batista, AP, and Andersen, RA (2001) The parietal reach region codes the next planned movement in a sequential reach task. J Neurophysiol 85, 539–544.
35. Wise, SP, Boussaoud, D, Johnson, PB, and Caminiti, R (1997) The premotor cortex: Combinatorial computations and corticocortical connectivity. Annu Rev Neurosci 20, 25–42.

Reading List

If you want to look into the literature concerning the neurophysiology of the premotor cortex, the review article by Wise et al.[35] provides an overview up to the mid-1990s.

14 Planning Displacements and Forces

Overview: It seems likely that motor areas of the frontal cortex—functioning as parts of cortical–cerebellar and cortical–basal ganglionic modules—transform the high-level motor plan for reaching and pointing, corresponding to a difference vector, into joint-angle changes and force commands.

Return to the imaginary robot that appeared in figures 9.3, 11.1, and 12.1 to review the process of computing a difference vector and producing something from it. The gray region of figure 14.1 schematically reiterates the computational process of planning a reach by estimating a difference vector from sensors that read and transduce the joint configuration and visual sensors. The neural networks that perform these computations align information from various sensory modalities to estimate the location of the end effector and the target in fixation-centered coordinates. The bidirectional neural networks perform pattern completion. For example, if your camera cannot see the end effector, the networks estimate its location in camera-centered coordinates from joint sensors. If the camera provides an image of the end effector but that image does not correspond to the information from the joint sensors, network 1 finds the most likely estimate of end-effector location, given the two pieces of information. Each network's **synaptic weight** matrix describes its dynamics, and each network with its weights corresponds to an **internal model** of the relationship among the aligned variables.

Once this system has computed a difference vector in camera-centered coordinates, it can use the results to compute commands that actually move the robot's arm. The robot's actuators move the links of its arm by putting torques on its joints. However, rather than directly computing the torques needed to move the end effector along the difference vector, it might be a good idea to compute how the camera-centered difference vector corresponds to a joint-coordinate representation of the movement. This intermediate transformation describes what changes in joint angle would accompany movements along a given difference vector.

There are a number of reasons for performing this intermediate transformation. First, depending on the joint configurations of the robot, a

Chapter 14. Planning Displacements and Forces

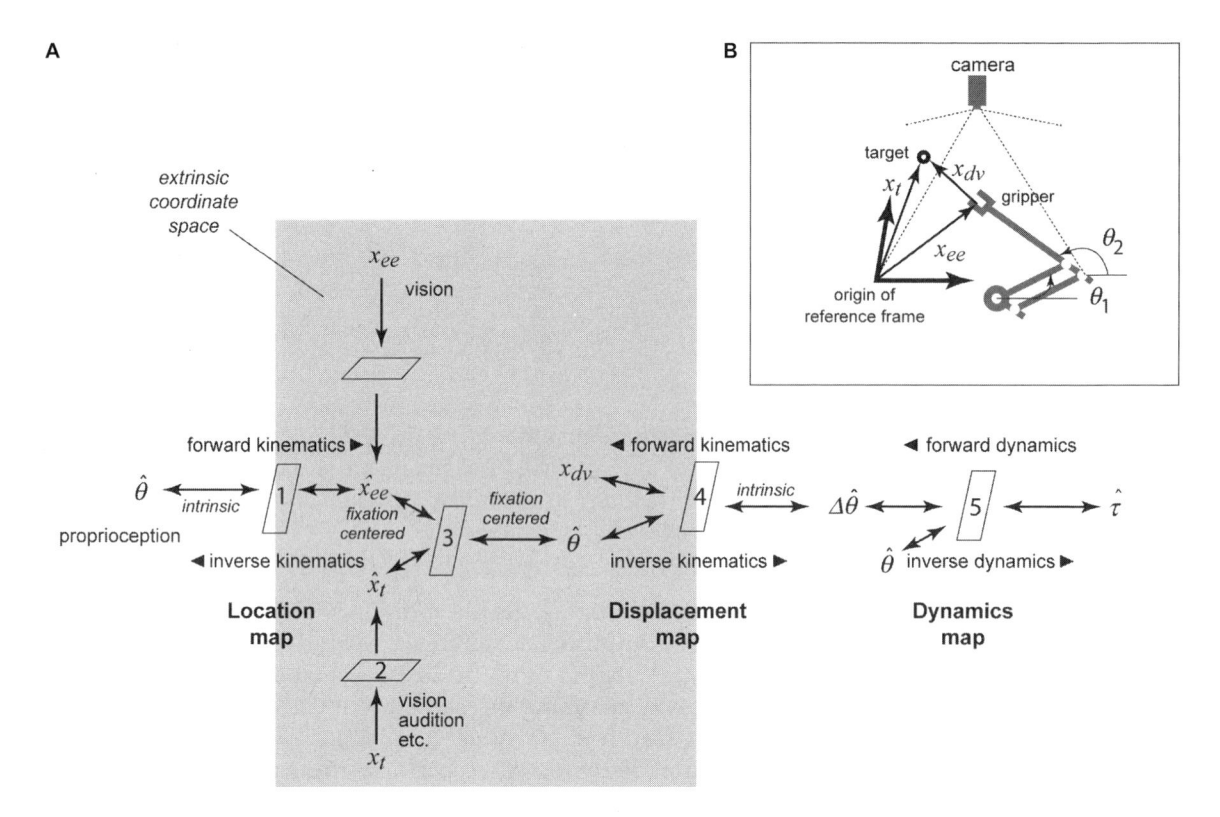

Figure 14.1
Virtual robotics. (*A*) A schematic of a series of computations for planning a reaching movement, in the format of figures 9.3C, 12.1A, and 12.11, with additions. Joint sensors provide a measure of arm configuration θ and a camera provides an estimate of hand and target location, x_{ee} and x_t, in camera-centered coordinates. A network with bidirectional connections aligns joint- and camera-based estimates of end-effector location $\hat{\theta} \leftrightarrow \hat{x}_{ee}$. From the estimates of target and end-effector location, another neural network computes a difference vector x_{dv}: target location with respect to the end effector. An additional network aligns this difference vector with a joint-rotation vector $\Delta\hat{\theta}$. This final transformation depends on the arm's configuration θ. In some cases, you might want to move to a target that you cannot see, but that relates in some way to a visible target. The network that computes $x_t \to \hat{x}_t$ serves this purpose. (*B*) Depiction of the virtual robot, in the format of figure 12.1B. (*A* drawn by George Nichols.)

given difference vector maps to several different joint rotations and forces, often known as the problem of *redundancy*. And second, as the robot moves, joint sensors provide information about the motion. Your robot might have low-level position and velocity controllers near the motors—something that corresponds roughly to stretch reflex mechanisms of the spinal cord—which can respond quickly to unforeseen events. This transformation from camera-centered to joint-centered coordinates would help you program these adjustments. It might also help when your camera views a video monitor that represents end-effector location as a cursor. (This situation corresponds to controlling a cursor with a computer mouse.) For these and other reasons, you might want to align the difference vector x_{dv} with joint rotations $\Delta\hat{\theta}$. Unlike network 1 in figure 14.1, which aligns information regarding end-effector *location*, network 4 aligns information regarding end-effector *displacements*. Accordingly, in figure 14.1 network 1 is labeled as the *location map* and network 4 as the *displacement map*. This displacement map depends on $\hat{\theta}$, and $x_{dv} \stackrel{\hat{\theta}}{\leftrightarrow} \Delta\hat{\theta}$ expresses its computation.

Reaching and pointing involves an additional mapping, labeled as network 5, which involves the alignment of joint rotations and force commands. Figure 14.1 labels this transform a *dynamics map*. The expression $\Delta\hat{\theta} \stackrel{\hat{\theta}}{\leftrightarrow} \hat{\tau}$ refers to this computation and the networks that perform it. As motor commands reach the motors and movement begins, the sensor readings change. For example, joint sensors and visual sensors indicate new arm configurations during movement, which in turn leads to a new x_{dv}. To reach the target, the process continues until $x_{dv} \to 0$.

The presentation in this chapter assumes that the frontal motor areas collectively play a role in the displacement and dynamics maps. Some experts disagree about the role of frontal motor areas in these transformations. For example, John Stein and Mitch Glickstein[1,2] believe that PPC, acting in concert with the cerebellum and its motor outputs, mediates visually guided movement without involving the frontal motor cortex. For now, assume that both these mappings involve the frontal motor areas as well as the cerebellum.

What would you expect to observe if you recorded from the neurons that compose network 4, the displacement map? From the model, you might expect to observe signals that reflect joint angles, changes in joint angles (i.e., movements), and the difference vector, which network 4 receives in a fixation-centered coordinate frame. Because of the latter property, you might predict that some of the signals in network 4 should reflect extrinsic coordinates. Since the difference vector has an amplitude as well as a direction, you might also expect to find neurons that have activity correlated with the amplitude of an upcoming or ongoing movement. Moreover, you might predict that many of the signals should be independent of gaze angle (i.e., it should not matter where you are looking). This chapter presents evidence in support of these predictions and, at the end, takes up network 5 in the context of the concepts of **preferred direction** and population coding in M1 cortex.

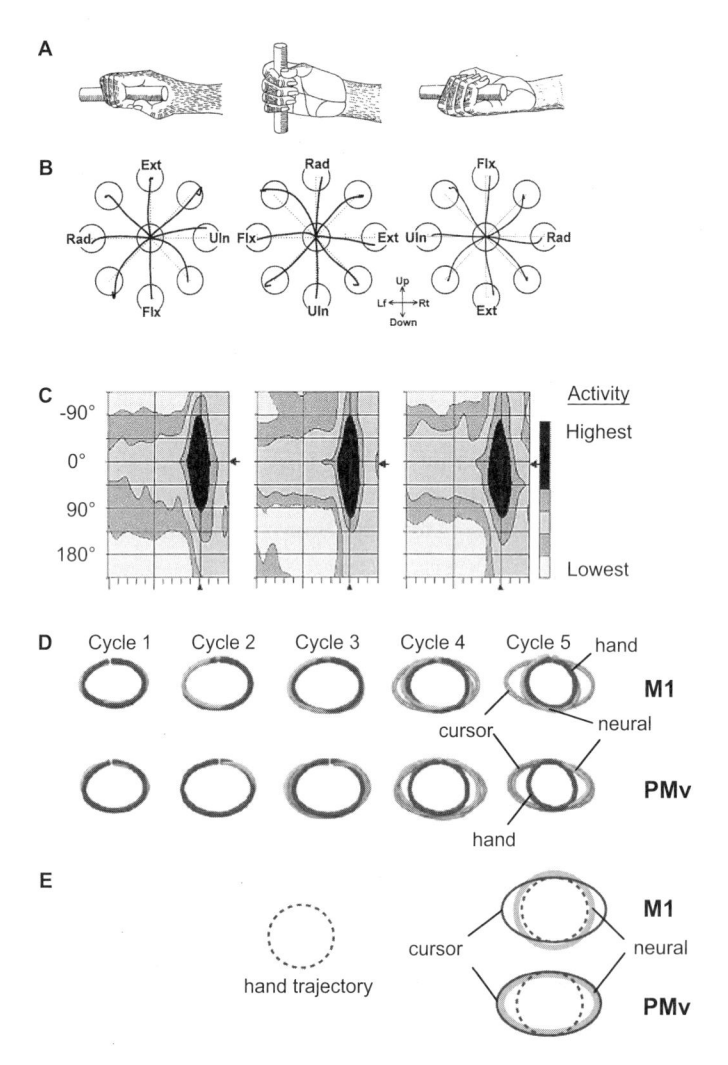

Figure 14.2

Coding in extrinsic coordinates in PMv. (*A*) A monkey's hand as it grasps a handle in three different postures. (*B*) Movement trajectories for the hand, displayed in an extrinsic coordinate frame, with the joint rotations needed to place the cursor on a target labeled for the cardinal directions from center. Abbreviations: Ext, extension of the wrist; Flx, flexion; Rad, radial deviation (wrist movement in the direction of the radius); Uln, ulnar deviation (see figure 3.1 for reference to radius and ulna). (*C*) Activity of a neuron encoding movement direction in extrinsic coordinates. The darkest areas on the plot represent the movement direction (*y*-axis) associated with the most activity, as a function of time (*x*-axis). Note that the cell shows most activity shortly before the onset of movement (upward arrowhead) for movements of the handle at approximately the 0° angle, regardless of posture. (*D*) Population vectors and both cursor and hand trajectory. In the first cycle of movement, a monkey tracks an elliptically moving cursor by making an elliptical hand movement. Then the experimenters change the relationship between hand movement and cursor movement. By the fifth cycle, the monkey must make a circular hand

14.1 Representing the Difference Vector in the Motor Areas of the Frontal Lobe

14.1.1 Varying Hand and Gaze Locations

Recall the idea that for retinotopic coding of location, the **activity field** of neurons should move as the retina moves (see figure 11.1D–F). Clearly, if the CNS represents target location with respect to the hand, the neurons that represent this information should not have spatial activity fields that move with the eyes. Rather, given a stationary target, their activity fields should move with the hand. Based on some earlier findings to the effect that activity fields in PMv did not change qualitatively when a monkey's eye orientation changed, PMv seemed a likely place to find such neurons. To test this hypothesis, Michael Graziano and his colleagues[3] examined the neuronal activity in PMv when a monkey fixated a point and a robot brought objects close to its arm. The experimenters mapped a neuron's tactile activity field by touching various parts of the arm. As the fixation point changed, the discharge did not vary much. When the experimenters moved the monkey's arm to the left, the activity field also shifted to the left. That is, the cell's activity field moved with the arm and not with the orientation of the eyes, as expected for a network encoding target location with respect to the location of the end effector.

14.1.2 Varying Hand Orientation

If the CNS computes the difference vector by comparing the current location of the end effector with the target's location in fixation-centered coordinates, and if the location of the target remains invariant with respect to the location of the end effector, the representation of the difference vector should also remain unchanged.

Shinji Kakei, Donna Hoffman, and Peter Strick[4,5] tested this hypothesis for PMv and M1. They recorded neuronal discharge in a monkey trained to move a cursor on a video monitor by moving its hand at the wrist[5] (figure 14.2A). The monkey held a device that translated hand movements into cursor motion. When a target appeared on the screen, the monkey moved its hand to bring the cursor to the target. The monkey performed this task for three different initial orientations of its hand. For example, when the monkey **pronated** the forearm so that the palm faced down, the muscle-activation patterns needed to move the cursor to the

Figure 14.2 (continued)
movement to make the required elliptical cursor movement. In M1 cortex, the population vector (neural) closely resembles the hand trajectory, consistent with a role in control of joint rotations and forces. In PMv, the population vector closely resembles the cursor trajectory, consistent with the computation of a difference vector. (E) Artist's depiction of the data for Cycle 5 in D. (A, B from Kakei et al.[5]; C from Kakei et al.,[4] D from Schwartz et al.[6] with permission.)

target at 45° differed from those required when the monkey moved the cursor to the same target with its arm **supinated** (palm up).

Kakei et al. found that PMv cells had activity that reflected the direction of the target with respect to the cursor, and not the initial orientation of the monkey's hand. Interestingly, during the instructed delay period—the planning phase of the trial—PMv cells showed this property most robustly. During the actual movement, the properties became somewhat more like the muscle activity patterns (figure 14.3). Thus, especially in the delay period, PMv cells reflected an extrinsic coordinate frame, encoding the location of the target with respect to the end effector, rather than either joint rotations or patterns of muscle activity.

This apparent representation of the difference vector in PMv contrasts with the properties observed in M1. In M1, activity during the instructed-delay period resembled that in PMv during movements. A few cells seemed to encode the difference vector, but most showed properties somewhere between extrinsic coordinates and those of the muscles that move the wrist. During movements, M1 activity shifted further to become more clearly intermediate in its reference frame, with one population of neurons appearing to reflect the same reference frame as the muscles and another resembling that seen in PMv during the movement.

Note that in this experiment, no cell or muscle reflected joint angles per se. Recall that a description of limb configuration in terms of joint angles does not match the transducers, which measure muscle length instead. The muscles, when described in terms of joint angles, showed properties intermediate between extrinsic, hand-centered coordinates and joint-centered coordinate frames. Nevertheless, the conclusion that M1 neurons reflect mainly **intrinsic** coordinates seems to capture the data best, notwithstanding some "intermediate" properties.

Recently, Andy Schwartz, Dan Moran, and Tony Reina[6] reported a related result from PMv. They measured the **population vector** in both PMv and M1 as monkeys performed a "virtual reality" reaching task. (The population vector is explained in section 14.2 and, in more detail, in section 21.2. For now, just think of the population vector as a "statement" made by a group of neurons about the direction in which something moves.) In the experiment by Schwartz and his colleagues, the monkeys saw an elliptical track, along which a target moved. The monkeys' hand path controlled the movement of a cursor, and the monkeys' goal was to match the cursor and target locations. At first, as shown in figure 14.2D, the hand, cursor, and target trajectories all had the same elliptical shape. Over the course of a few movement cycles, the experimenters altered the relationship between hand and cursor movement so that the monkeys needed to move their hands in a circular hand trajectory to keep the cursor moving along the required elliptical path. The question they posed was: Does the neuronal activity, as measured by the population vector, match what the monkeys see or what they do? As shown in figure 14.2D for their data and in figure 14.2E in idealized form, the answer differed for the two cortical areas. In M1, the trajectory of the population vector matched the *performed* hand trajectory; in PMv it matched the *viewed*

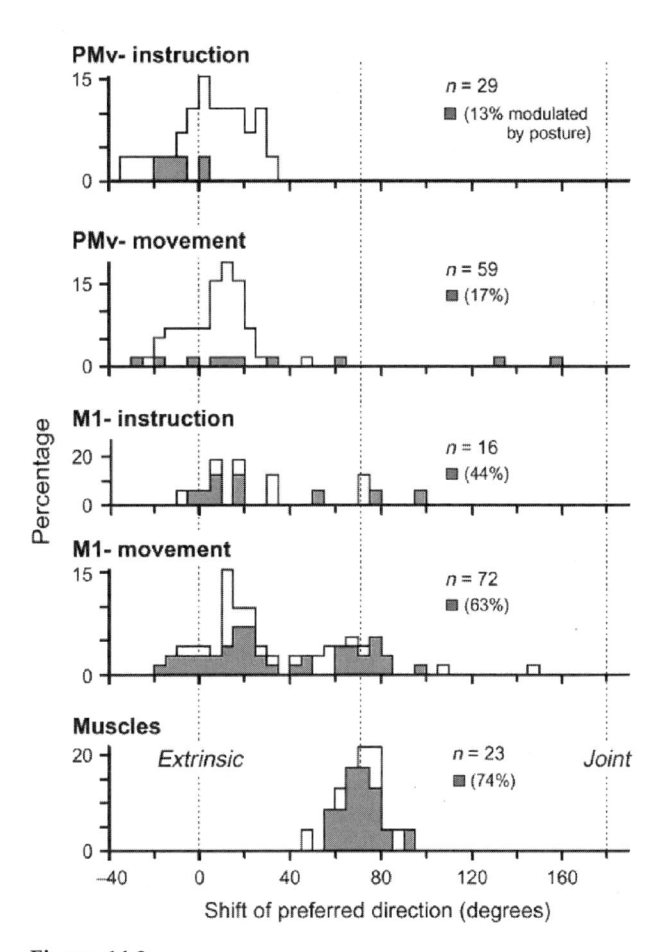

14.1. Representing the Difference Vector in the Motor Areas of the Frontal Lobe

Figure 14.3
Coding in extrinsic coordinates in PMv, but a mixture in M1. Cells in PMv encoded the target's location relative to the end effector (a cursor), and did so independent of arm configuration. This conclusion follows from the finding, shown in the top two plots, that their preferred directions did not change substantially as a function of the three postures depicted in figure 14.2A. In PMv, there was little change in preferred direction either in the period after the instruction stimulus or in the movement period. Cells in M1, however, showed properties intermediate between an extrinsic reference frame (left vertical dashed line) and the coordinate frame of muscles (middle vertical dashed line), especially during the movement period (fourth plot from the top). Note that in the bottom plot, the activation patterns of muscles did not reflect a joint-angle reference frame (right vertical dashed line), in accord with the idea that the transducers for computing hand location involve muscle receptors, not joint receptors. (From Kakei et al.[4] with permission.)

cursor trajectory. This finding supports the idea that activity in PMv reflected, in part, the computation of a difference vector (the direction the cursor had to move to reach the target), whereas activity in M1 reflected changes in joint angles and forces. (Schwartz et al. interpret their result somewhat differently, as explained in their paper.[6])

14.1.3 Varying Visual Feedback

Another study, this one by Tetsuji Ochiai, Hajime Mushiake, and Jun Tanji,[7] examined whether neuronal activity in a different premotor area, PMd, reflects visually presented motion or the actual movement of the monkey's hand. They trained a monkey to capture a target on a video screen with part of the image of a hand, as displayed on a video screen. Sometimes the right side of the image had to capture the target, sometimes the left side did. The monkey could not see its own hand and, therefore, had to guide movements by the inputs from vision. The experimenters established two conditions: one involving right–left inversion of the image on the screen, the other with the hand displayed normally. They found that neuronal activity in PMd predominantly reflected the video image of hand movement rather than the actual hand movement. Thus, given the choice between reflecting the difference vector in extrinsic, visual coordinates versus intrinsic, joint-based coordinates, the cells in PMd reflected the difference vector in extrinsic coordinates.

14.1.4 Coding the Direction and Amplitude of the Difference Vector

After some initial studies of amplitude coding in the PMd,[8,9] Tim Ebner[10,11] and John Kalaska,[12] and their colleagues, investigated this issue in more detail. Both groups of investigators found an amplitude signal, but it had less prominence and appeared later during a trial than activity reflecting the direction of movement. Figure 14.4 shows the results obtained by Julie Messier and Kalaska.[12]

14.1.5 Changing the Directional Mapping Between Hand Movement and Difference Vector

Liming Shen and Gary Alexander[13,14] performed an experiment that examined whether PMd and M1 neurons code for a difference vector with respect to an end effector presented on a video monitor or with respect to the actual direction of hand movement. They trained a monkey to move a handle that controlled a cursor on a video screen. In one condition, termed the 0° mapping condition, the relation between handle movement and cursor movement was straightforward. Movement of the handle to the monkey's left caused movements of the cursor to the left, and so forth. In the other condition, termed the 90° mapping condition, handle movements to the left moved the cursor down on the screen. In both conditions, the cursor was the end effector. A given difference vector in visual coor-

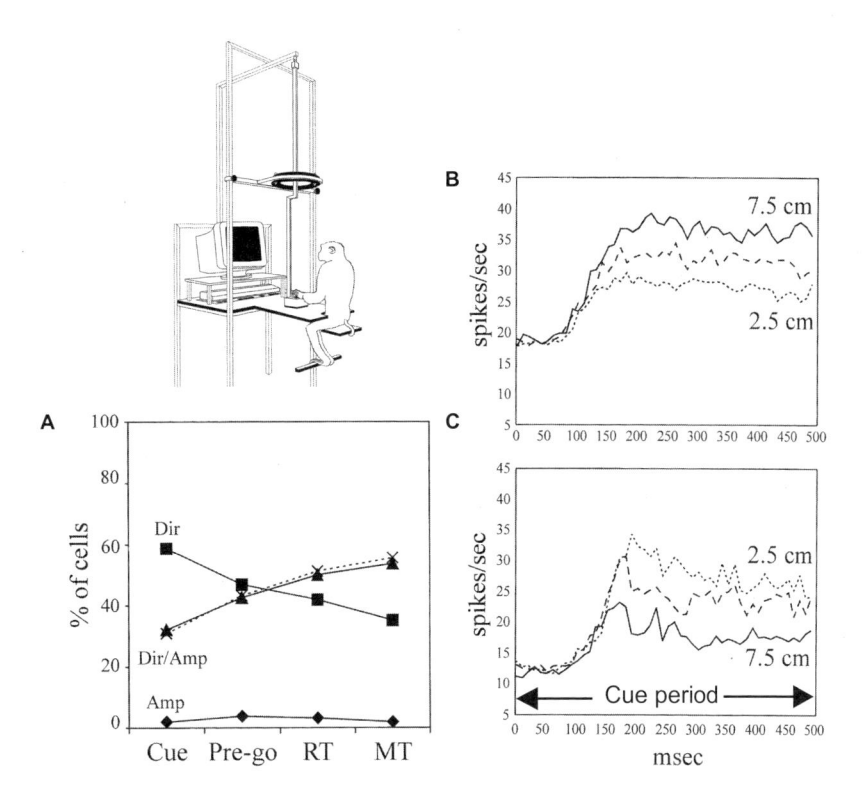

14.1. Representing the Difference Vector in the Motor Areas of the Frontal Lobe

Figure 14.4
Coding of the direction and amplitude of the difference vector in PMd. Left: The end effector and target both appear on a video display in front of the monkey. (A) Percentage of PMd cells in which discharge correlated with movement direction (dir), amplitude (amp), or both, in each of four task periods: the period that the instruction cue appeared (cue period), the instructed-delay period (pre-go), the period after the trigger (go) stimulus but before movement (called the reaction-time period, RT), and the period during the movement (called the movement-time period, MT). (B) Population histograms normalized to the preferred direction of each cell, for cells with a positive correlation of activity with movement amplitude. The amplitude of the largest and smallest movements appear above the top histogram, and below the bottom one. (C) Cells with negative correlations in the format of B. (From Messier and Kalaska[12] with permission.)

dinates would require two different joint rotations (and hand movements) in these two conditions.

Shen and Alexander found that neuronal discharge in PMd primarily reflected the visual difference vector for the end effector, and only to a lesser extent reflected the hand movements or joint rotations of the limb. For example, the cell illustrated in figure 14.5A and 14.5B showed high delay-period activity when the target for the end effector was down from the origin, regardless of whether the joint-angle changes yielded a hand movement down (figure 14.5A) or mainly to the left (figure 14.5B). From this result, Shen and Alexander concluded that this cell encoded the difference vector for the end effector (in their terminology, "target cells") rather than changes in joint angles (which they referred to as "limb cells").

The experimenters then compared the properties of neurons in PMd and M1 (figure 14.5C). As a general rule, cells in M1 reflected joint-angle changes more than cells in PMd at every stage during a trial. In both areas, there was a trend toward a greater reflection of the target location in extrinsic, visual coordinates early in a trial (e.g., during the presentation of the instruction stimulus) and, especially for PMd, during the delay period. In terms of figure 14.1, you might expect this property for x_{dv}, the output of network 3 and the input of network 4. Later in the trial, the dominant coordinate frame appeared to shift from an extrinsic frame, representing the difference vector in fixation-centered coordinates (as you might expect for x_{dv}), to an intrinsic frame representing the changes in joint angle needed to acquire the target (as you might expect for $\Delta\hat{\theta}$). This shift occurred most prominently in the delay period for M1, but only later—during the movement—for cells in PMd. Perhaps these shifts reflected the conversion of a difference vector in extrinsic coordinates to a joint-rotation vector in intrinsic coordinates $x_{dv} \overset{\theta}{\leftrightarrow} \Delta\hat{\theta}$. Note that the timing of this transition in PMd resembles that described in section 14.1.2 and illustrated in figure 14.3 for the relationship between M1 and PMv.

14.1.6 Varying Arm Configuration but Keeping a Constant Difference Vector

In the Shen and Alexander experiment, the monkey's hand moved in two different directions while the cursor moved more or less in the same direction. It appeared that during the delay period, PMd activity mainly reflected the difference vector for the cursor. Is this sensitivity to the difference vector maintained when the planned hand and cursor movements are similar but the planned joint rotations differ? Steve Scott, Lauren Sergio, and John Kalaska,[15,16] performed an experiment to answer this question. They established two conditions, based on the posture of the monkey's arm. In both conditions, the monkey made reaching movements from a given initial handle location to a set of targets. But in one condition, the arm posture had a much greater degree of upper arm abduction than in the other (figure 14.6A). Therefore, in this experiment the handle made the same movement, but the joint rotations varied.

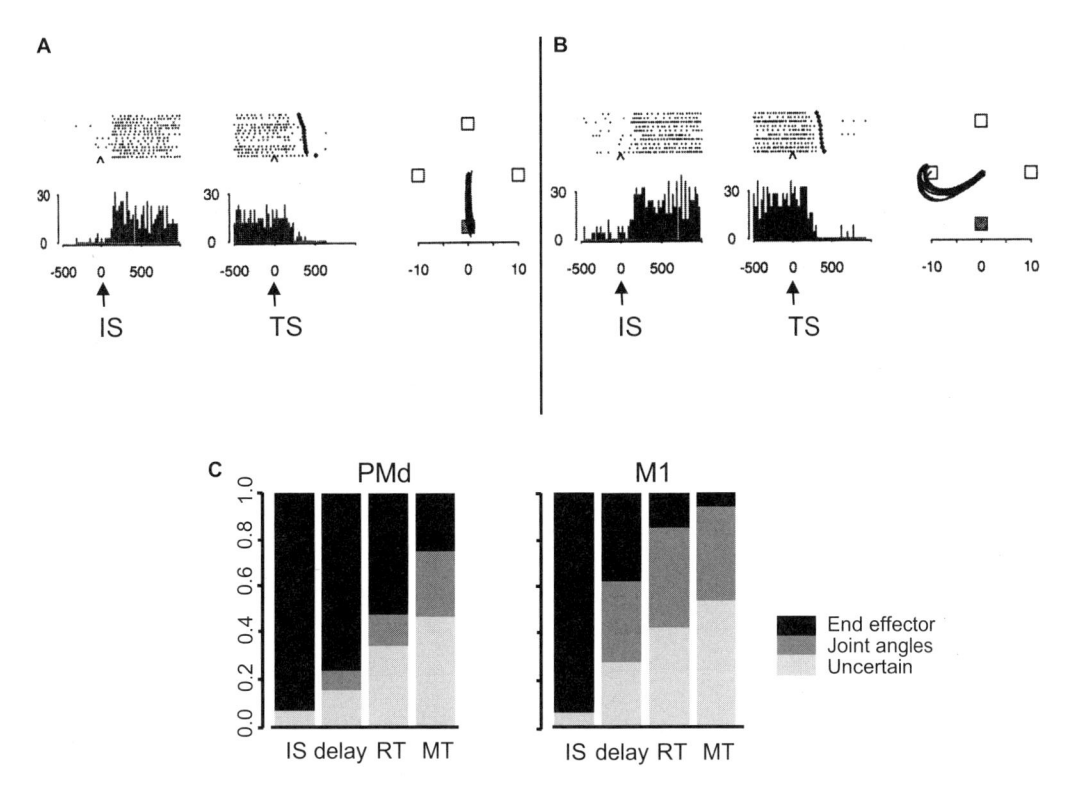

14.1. Representing the Difference Vector in the Motor Areas of the Frontal Lobe

Figure 14.5
A PMd cell that coded the orientation of the visual difference vector, not the direction of hand movement. (A) The 0° rotation condition, in which downward handle movements caused downward cursor movements (right). The cell showed dramatic delay-period activity following the instruction stimulus (IS), which continued after the trigger (go) stimulus, until near the time of target acquisition (square on each raster line). Time scale in msec. (B) The 90° rotation condition, in which leftward handle movement (right) caused downward cursor movements. The cell showed a preferred direction for downward targets in visual coordinates, regardless of whether the monkey's hand moved downward or mainly to the left. (C) Comparison between PMd and M1 in this experiment. The bars show the proportion of cells with activity that reflected limb trajectory (joint-angle changes, gray bars) versus a difference vector for the visual end effector, the cursor (black bars). Task periods include the time that the monkey could see the visual instruction stimulus (IS), the instructed delay period (delay), and the RT and MT periods, as defined in the legend to figure 14.5A. (From Shen and Alexander[14] with permission.)

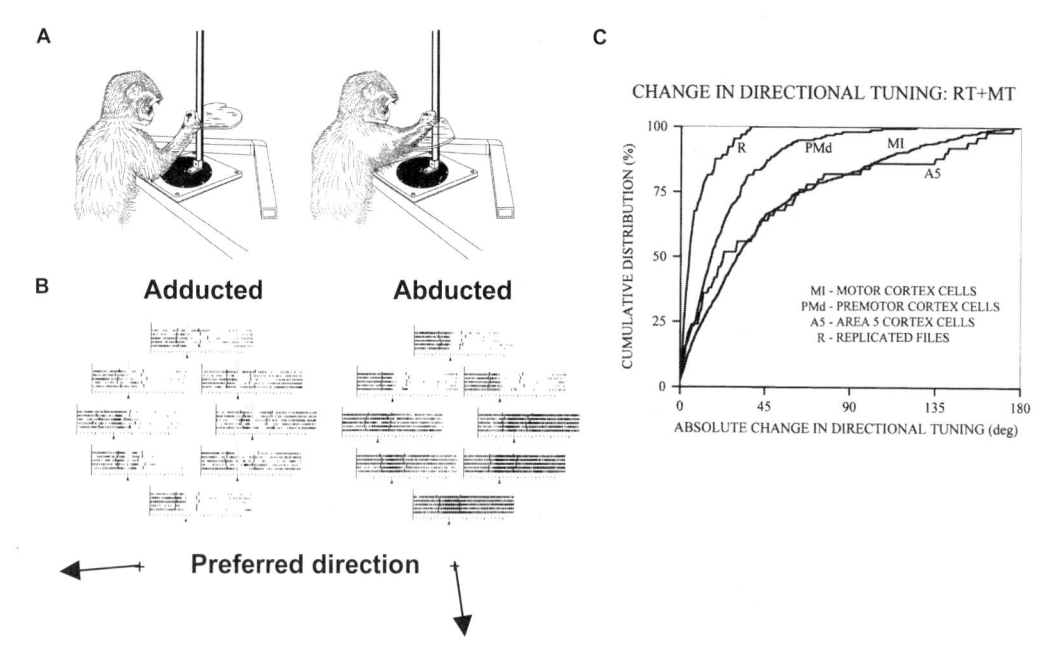

Figure 14.6
Effect of varying arm configuration on M1 activity during reaching. (*A*) The monkey made a reaching movement while adopting either an adducted or an abducted arm posture. (*B*) Neural discharge for both arm configurations, aligned (arrowhead) on the onset of movement for each of eight movement directions. Marks before and after the arrowhead on each raster line indicate the time of appearance of the target light and the end of movement, respectively. The vectors beneath each set of rasters show the cell's preferred direction just before and during movement (i.e., during the reaction time, RT, and movement time, MT), for the two postures. (*C*) Cumulative frequency distribution of the changes in preferred direction due to posture for PMd, MI (equivalent to M1), and area 5d. In addition, to see whether these changes occur by chance, the experimenters examined the variation that occurs for replicated (R) blocks in which arm configuration remained constant. R-curves, of course, show the least change, as indicated by the faster rise of cumulative distribution at low *x*-axis values. (*A* and *B* from Scott and Kalaska[15]; *C* from Scott et al.[16] with permission.)

The experimenters observed that the activity of cells in M1 reflects the difference in joint rotations (figure 14.6B). While the difference vector was constant, the preferred direction of this cell rotated as the arm posture changed. The result has been somewhat controversial, but the disagreements relate to whether these postural effects change the output of the population, not whether they occur. Apostolos Georgopoulos and Andy Schwartz[17–19] maintain that these variations "wash out" of the population. Therefore, in their view, M1 encodes reaching in an extrinsic, hand-centered coordinate space. Regardless of your opinion on that subject, note that the model illustrated in figure 14.1 predicts the effect reported by Scott and Kalaska for both network 4 and network 5; they both receive inputs conveying arm configuration (θ). Figure 14.6C shows that just before and during movement, M1 and area 5d cells show comparable degrees of shift in preferred directions, in both cases more than observed in PMd cells.

14.1.7 Reaching Without Vision of the Hand

Wise, Mike Weinrich, and Karl Mauritz[20] did an experiment testing another prediction for network 4 in figure 14.1. As predicted from the input for joint rotations $\Delta\theta$ and the idea of pattern-completion networks, some cells in PMd discharged before and during movements around the elbow, even when the monkey made spontaneous movements in total darkness. Thus, changes in joint angle per se affect neuronal activity in PMd.

Along these lines, Michael Graziano and his colleagues[21] noted that when they placed items near a monkey's arm, the activity evoked in PMv neurons persisted even when the monkey could not see them. This kind of activity presumably supports reaching in the dark.

14.1.8 Effects of Sensory Cues That Instruct a Movement

The experiment by Shen and Alexander showed that, for a given visual difference vector, most cells in PMd had similar activity regardless of the joint-angle changes. The experiments by Scott et al. showed that PMd activity reflects more than just the visual difference vector or the trajectory of the hand. Joint-angle changes also matter. Other experiments also show that PMd activity reflects more than just the visual difference vector.

An experiment by Giuseppe di Pellegrino and Wise[22] addressed some important aspects of the signal found in PMd. Early work on the PMd suggested that some of its activity reflected inputs from visual receptors. Erroneously, these early neurophysiologists referred to this activity as a "visual response." Di Pellegrino and Wise tested the hypothesis inherent in that term. In their experiment, the monkey began each trial with its hand directly beneath a light at the center of a ring of lights, all of which turned off at the onset of each trial (figure 14.7A). In

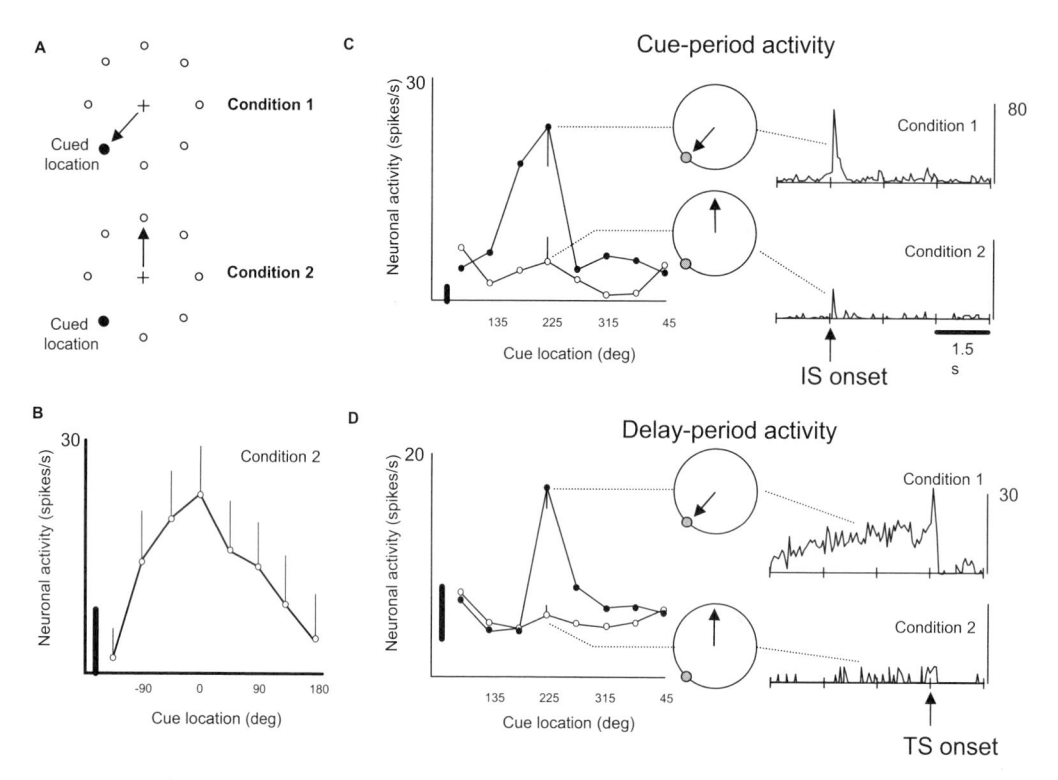

258   CHAPTER 14. Planning Displacements and Forces

Figure 14.7
PMd cells affected both by the stimulus location, for a given movement (*B*), and by the movement plan, for a given stimulus location (*C* and *D*). (*A*) The experimental design. A ring of lights surrounded a central location. The monkey began each trial by placing its hand beneath the central location, which was also the fixation point (+). The filled circle shows an example visual cue for a given trial; the unfilled circles show the other seven potential targets. The cue location varied from trial to trial. The arrows show the reaching movement required for that cue, in each of two conditions. In condition 1, the monkey reached to the cue wherever it appeared. In condition 2, the monkey always reached to the top location, regardless of the cue's location. The condition varied in blocks of trials. For a given movement direction, the kinematics and the dynamics of movement did not vary at all. (*B*) Delay-period activity of a PMd cell in condition 2, in which the monkey always made the same reaching movement. The thick vertical bar shows baseline activity, and the circles show the mean activity (bars, the standard deviation) for each of the eight cue locations. Despite the fact that the movement was always the same, the cue's location strongly affected the cell's activity. (*C*) Left: Tuning curves during the cue period for a different PMd cell. The filled circles show the cell's activity for condition 1; the unfilled circles, for condition 2. Right: Average activity histogram for the 225° cue location, for both conditions. (*D*) Activity during the instructed-delay period for another PMd cell, in the format of *C*. Note that in *C* and *D*, the motor plan instructed by the cue had a profound effect on the cell's activity, despite the fact that the cue appeared in the same location in all relevant coordinate systems. (From di Pellegrino and Wise[22] with permission.)

condition 1, a light turned on to signal the location of the reach target for that trial. After a delay, the same stimulus reappeared as the trigger stimulus, and the monkey reached directly toward the cue's location. In condition 2, the same sequence of sensory inputs occurred, but the monkey reached to a fixed, constant location, regardless of the cue's location. The two conditions were performed in blocks of trials, but in both the monkey continued fixating the center of the display from the beginning of each trial until movement onset. Accordingly, the location of all visual stimuli remained constant in fixation-centered coordinates (as well as in all other spatial coordinate frames).

For most PMd neurons, the results contradicted the idea that the visual stimulus—alone—or the hand trajectory—alone—accounted for their activity. For a given movement trajectory, the location of the guiding cues mattered significantly. Figure 14.7B shows a PMd cell's tuning curve for reaching movements along one trajectory to one target, for cues at the eight different locations. Most cells in PMd, like the one illustrated, reflected the location of the cues, not just the direction of hand movement. Similarly, for a given cue location, PMd cells showed high sensitivity to the reaching movement guided by the cue. Figure 14.7C illustrates activity during the cue's presentation, and figure 14.7D does so for the delay period. Note that when a given cue guided the movement, the cell discharged only when that cue served as the reach target and only in the cell's preferred direction. Thus, PMd activity does not simply reflect the location of a visual cue, even when the monkey must attend to that cue for the guidance of its reaching movements (figure 14.7C and 14.7D). Nor does PMd activity reflect simply the limb trajectory (figure 17B).

In contrast to the results in PMd, di Pellegrino and Wise found that in the prefrontal cortex, visual inputs and attention alone could account for most neuronal activity. Therefore, whereas in the premotor cortex di Pellegrino and Wise found signals reflecting the motor significance of the visual stimuli, in large part probably the difference vector, in the prefrontal cortex they found more signals that corresponded to the location of the visual stimuli and not their motor significance.

Taken together with the results of Shen and Alexander and those of Scott, Sergio, and Kalaska, the results of di Pellegrino and Wise show that a diverse set of influences affects PMd activity. Figure 14.1 helps explain this diversity. One influence appears to be the visual difference vector x_{dv}, as emphasized by the results of Shen and Alexander. Another influence involves the joint configuration $\hat{\theta}$, as emphasized in the results of Scott, Sergio, and Kalaska. Other influences, emphasized by the results of di Pellegrino and Wise, involve the alignments or "mappings" between the variables noted for networks 2–4 in figure 14.1. One such influence involves location of the cues x_t that guide a given movement $\Delta\hat{\theta}$, as indicated by figure 14.7B. And yet another influence on the PMd signal, which figure 14.7C and 14.7D show to be a very dramatic one, involves whether the visual difference vector x_{dv} determines the movement $\Delta\hat{\theta}$ or, alternatively, the monkey selects the movement in some other way. Section

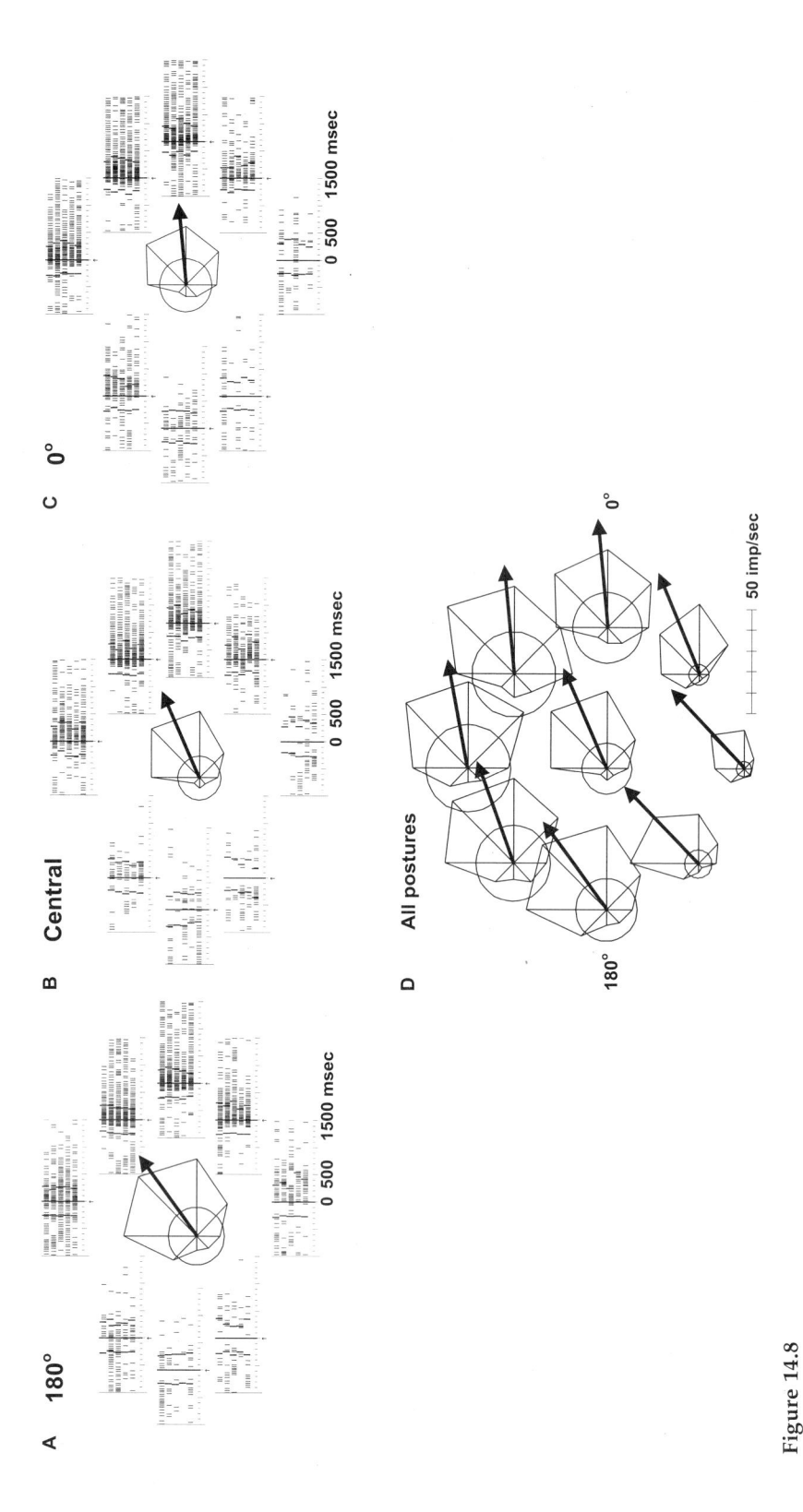

260   Chapter 14. Planning Displacements and Forces

Figure 14.8
Directional tuning for force in M1. The activity of an M1 cell during isometric force production. The monkey pushed against a fixed handle at leftward (*A*), central (*B*), and rightward (*C*) locations, among six others. Each of the eight rasters in each panel (*A–C*) illustrates activity for one force direction. The rasters are arrayed to correspond to the direction of force. The solid vertical line at time 0 indicates force onset. Each panel also shows a polar plot of activity as a function of force direction. The circle in the polar plot shows the mean discharge rate before the beginning of the force pulse. The arrow corresponds to the preferred direction of the cell. (*D*) Polar plots and preferred directions of the cell for nine hand locations. (From Sergio and Kalaska[35] with permission.)

25.5 takes up situations in which targets at one place instruct movements elsewhere.

Despite the diversity reflected in PMd's signal, note the importance of the visual difference vector for all of the data. Earlier work by Mike Weinrich and Wise[23] had shown that for many cells in PMd, difference vectors established by auditory cues resulted in the same activity as those established by visual cues. Thus, PMd probably participates importantly in the computations identified with network 4 in figure 14.1.

14.1.9 Effects of Gaze

The difference vector describes the target's location with respect to the hand, but according to the model depicted in figure 14.1, it does so in fixation-centered coordinates. If the fixation-centered coordinate system is isotropic (i.e., it contains no regions where resolution is better than in other regions), then the representation of x_{dv} should be independent of gaze angle. To a first approximation, the difference vector should not depend on the orientation of the eye. In support of this prediction, neurophysiological studies based on qualitative examination—in the style of classic visual receptive-field mapping—have reported that activity fields in PMv appear to be gaze independent.[24,25]

Quantitative examination of those activity fields, however, has consistently revealed significant, if modest, effects of gaze angle in both PMd and PMv neurons.[26-30] The present model can help resolve the apparent conflict between these two kinds of data. Psychophysical studies indicate that movement accuracy depends on target location in fixation-centered coordinates.[31-33] You saw earlier that movement accuracy depends on the fixation-centered location of targets and end effectors. Figure 11.2 shows that you misreach in proportion to the distance from the fixation point to the target. It is reasonable, then, to expect a quantitative influence of gaze angle on the neuronal activity encoding the difference vector.

14.2 Population Vectors, Force Coding, and Coordinate Frames in M1

The presentation to this point has focused to a large extent on kinematics and the computations to the left of network 5 in figure 14.1. The evidence indicates that both difference vectors in extrinsic, fixation-centered coordinates and joint rotations are important to cells in PMd and PMv. M1, however, seems different (figures 14.3 and 14.5). Note that network 5 has torque as one of its outputs. Torque production is important to the activity of neurons in M1: They discharge during **isometric** contractions in much the same way as they do during movement.[34] The remainder of this section addresses some topics pertinent to the function of M1: force coding, population coding, and various issues concerning the coordinate frame reflected in M1 activity.

14.2.1 Force and Torque Coding in M1

Figure 14.8 shows the results of a recent study by Lauren Sergio and John Kalaska.[35] In their experiment, a monkey held a stationary handle and generated a force vector of a given magnitude along one of eight directions on each trial. The task was then repeated for the other seven directions. Next, the handle was moved to another location, producing a change in arm configuration, and the task was repeated until the experimenters had collected data for a central location and eight peripheral ones (figure 14.8D). Figure 14.8A–C each show eight raster displays of M1 activity. At the center, a polar plot shows the cell's force-related activity (the polygon) and its resting level of discharge (the circle). Figure 14.8C shows the cell's activity for a hand location to the right (0° in polar coordinates); figure 14.8B, for a central hand location (at the origin); and figure 14.8A, for a hand location to the left (180°). Figure 14.8D shows the polar plots for those three hand positions (left, middle, and right plots), along with similar plots for six other hand locations. Notice three features of the results. First, although no movement occurred, the cell showed clear modulation in activity. Second, that activity showed directional selectivity (for torques up and to the right). Third, the preferred direction for the torque depended on initial hand location, as did the amplitude of the cell's tuning for torque direction. How could this be?

Emo Todorov[36] has resurrected the idea that activity in M1 mainly reflects the forces and torques that limb muscles generate during a movement. He argued that many of the appearances to the contrary result from insufficient attention to the length–tension relationship that you learned about in section 7.2.3 and section 8.2. For example, you can imagine, looking at figure 14.8, that this cell drives muscles that become stretched for hand locations at the lowest position on the figure (the posture with the hand nearest the monkey). You can also imagine that this could be the triceps muscle, given that the cortical cell prefers extension torques. From the length–tension relation, you would predict that the cortical drive needed to produce a given force would decrease as the triceps muscle lengthens. You *really* need to use your imagination, because there is no evidence that this cell mainly drives the triceps, but this account is plausible.

In an earlier study, Jim Ashe, Apostolos Georgopoulos, and their colleagues[37] showed that neuronal activity in M1 varied according to the degree of tonic isometric force needed to maintain a cursor at the central point on a video screen and, on top of this tonic discharge, the phasic discharge of the cells seemed to correlate best with the change in force. Many studies of M1, most notably by Marie-Claude Hepp-Reymond and her colleagues,[38,39] also suggest that activity of M1 cells correlates best with the level of force and the time derivative of force (df/dt). Why, then, do many experts claim that M1 encodes kinematic variables such as the speed and direction of hand movement, as described in an extrinsic coordinate frame?

A lot depends on whether you view M1 as corresponding to network 5 in figure 14.1 or as a combination of networks 4 and 5. Because the activity of a population of M1 neurons accurately predicts the trajectory of movement, some experts have concluded that M1 encodes movement trajectories in extrinsic, hand-centered coordinates, much as postulated for the inputs to network 4. This conclusion depends, for the most part, on a measure of population activity in M1 known as the **population vector**.

14.2.2 Population Coding in M1

Despite some objections, the population vector has proven to be a useful and robust measure of activity in a wide variety of cortical areas, including M1. Figure 14.9 shows how this computation works. Cells in M1, as elsewhere in the CNS, show broad tuning for direction, in this case the direction of a reaching movement. The cell depicted in figure 14.9B has its most intense discharge for movements toward 315° (its preferred direction), with progressively less modulation as the movement diverges from that direction. The following equation describes its activity:

$$a(\omega) = b_0 + b_1 \cos(\omega - \omega_{pd}),$$

where $a(\omega)$ is the cell's activity when the monkey reaches in direction ω; b_0 is the cell's resting discharge rate; b_1 is a scalar; and ω_{pd} is the cell's preferred direction (pd). Apostolos Georgopoulos and his colleagues[40,41] devised one method for computing the population vector, a weighted vector sum, although other methods yield comparable results. That method is simply

$$\vec{p} = \sum a(\omega) \cdot \vec{\omega}_{pd},$$

where $\vec{\omega}_{pd}$ is a unit vector that points in each cell's preferred direction.

There have been many studies showing that the population vector describes hand velocity accurately in a wide variety of circumstances. Andy Schwartz and his colleagues, along with Apostolos Georgopoulos and his colleagues, have shown that the population vector predicts the direction of hand movement for straight movements,[43] ellipses, ovals, and other complex trajectories, such as figure eights.[18,44] Many experts have taken this finding as evidence that M1 codes movements in hand-centered coordinates. Alternative computational methods, some of which make fewer assumptions about particular tuning functions, also predict hand-movement trajectory and lead to similar conclusions.[45]

The population vector has generated some controversy since its introduction by Georgopoulos et al. in the early 1980s. (See box 14.1.) Some experts have referred to this idea as a "population-vector hypothesis," which has generated several objections: (1) as a hypothesis it is not falsifiable because, given certain assumptions—all of which pertain to M1 activity—the population vector will always point in the direction of movement; (2) there might be other, better ways to describe a population and, if so, perhaps the CNS uses them instead of the population-vector computation; (3) no experiment has produced evidence that M1 cells excite muscles that produce forces in their preferred direction or their **di-**

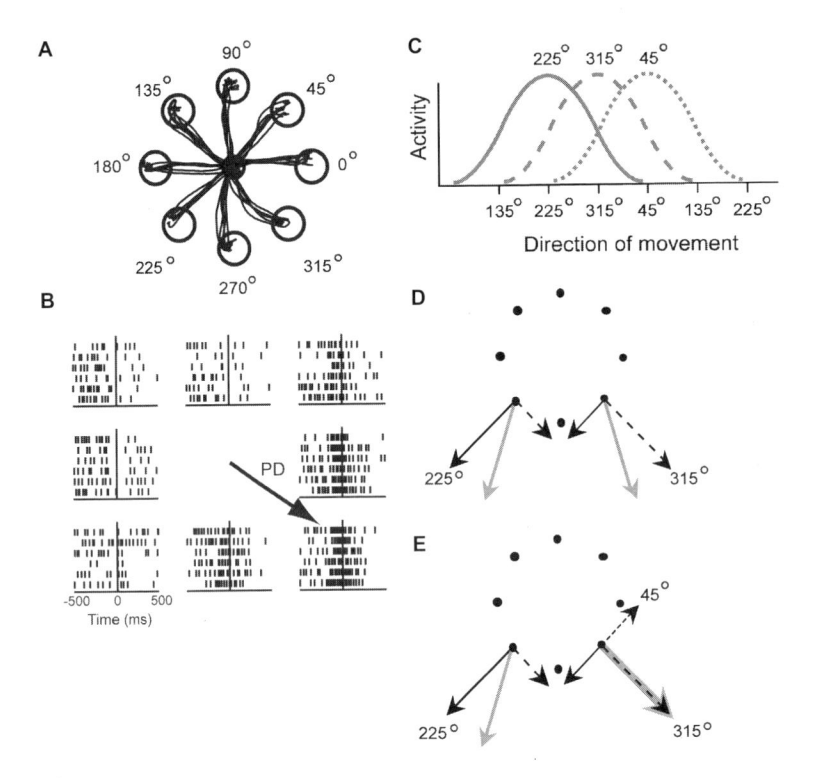

Figure 14.9

The population vector. The figure shows the principle underlying the population-vector computation. (*A*) Trajectories of movement in eight directions from the center. (*B*) Rasters of neuronal activity from an M1 neuron for each of the eight directions of movement, aligned on the onset of movement (solid vertical bar). This neuron had a preferred direction (PD) of approximately 315° (arrow). (*C*) Ideal tuning curves for three neurons: the one like *B* (dashed line), and two fictitious neurons with different preferred directions (solid and dotted lines). (*D*) A two-neuron population vector (solid arrow), computed as a vector sum from one cell with a preferred direction at 315° (down and to the right) and another with a preferred direction at 225° (down and to the left). Each cell contributes to the population vector in its preferred direction, and does so in proportion to its activity for a given direction of force or movement. In this example, note the movement made at 315°. Note that the two-neuron population vector points reasonably close to the actual movement direction (315°), but not perfectly. (*E*) The addition to the population of a third neuron, which has a preferred direction of 45° (dotted line in *C*), makes the population vector accurate for the 315° movement. Note, however, that because the third cell has virtually no activity for movements made at 225° (see part *C*), its inclusion does not improve the accuracy of the population vector for that movement direction. (Adapted from Scott[42] with permission.)

> **Box 14.1**
> History of population coding
>
> > Studies of the motor system contributed to the earliest understanding of population coding in the nervous system. Based on work begun in the 1960s, Don Humphrey and his colleagues[46] showed that the capacity to predict force from the activity of neurons in M1 was improved dramatically by combining the activity of several neurons.

rection of action; and (4) given that changes in joint angle at the shoulder reorient the reference frame centered on the elbow, many muscle activations have no fixed relationship with the direction of end-effector movement in extrinsic, hand-centered coordinate space. An example demonstrates the latter point. Imagine holding a stick in your right hand, with your arm straight out. When you excite flexor muscles, the end effector moves left when your palm faces down, and up when you palm faces up. If M1 neurons control muscle activation, how could this encode movements in extrinsic coordinates?

As Tony Reina, Andy Schwartz, and his colleagues[47] have emphasized, at the neuronal level you cannot distinguish between a joint-centered (intrinsic) coordinate frame and a hand-centered (extrinsic) coordinate frame. Through **forward kinematics** (chapter 9), the two correlate very closely: To know the changes in joint angle (joint velocity) is essentially to know the hand movement (hand velocity). Thus you can accept the view that the population vector reflects hand movements accurately without rejecting the notion that M1 primarily controls movements in intrinsic (joint- or muscle-based) coordinates, as indicated by the data from isometric contractions.

14.2.3 Coordinate Frames in M1

Now return to a consideration of some of the conceptual issues introduced in section 10.2. Figure 10.2 emphasizes a distinction between the "coordinate-frame" issue and the "dynamics versus kinematics" issue, but these two issues are related. A control system based on intrinsic coordinates could operate either in purely kinematic terms (controlling changes in joint angle) or in a mode that also takes into account the forces generated in an intrinsic coordinate frame.

Such a concept has been championed by Steve Scott and his colleagues,[42] who have argued that neurons in M1 are affected by a variable called joint-angular power. Recall (from appendix D, if necessary) that $P_a = \tau \cdot \omega$, where P_a is angular power, τ is torque, and ω is angular velocity. You can think of power this way: If you move a mass from point A to point B, you have done a certain amount of work (force times distance, whether translational or rotational). If you move it twice as fast, you have done the same amount of work, but you have clearly done something different. That difference corresponds to *power*, the amount of work per unit time or, in angular terms, angular power.

Steve Scott and his colleagues developed a device that allows independent control of forces on the shoulder and elbow joints. In their experiment, monkeys made 16 hand movements from a central location at a fixed amplitude and a relatively constant velocity. More cells showed tuning for directions with higher peak joint-angular power, which reflects the importance of *both* joint-angle changes and torques in M1 function. The model depicted in figure 14.1 predicts just that.

So, given all of these findings, what frame of reference does M1 use? The question remains open, but a mathematical model devised by Rob Ajemian, Steve Grossberg, and Dan Bullock[48] indicates that the data fit best to an intrinsic coordinate frame based on the limb. In the terminology presented in table 9.1, you might think of this coordinate system as frame $[E + S_{int}]$, without implying linear summation. The alternatives rejected by Ajemian et al. include extrinsic coordinates based on the shoulder alone, called $[S_{ext}]$ in table 9.1, as well as hand-centered or Cartesian frames.

Ajemian et al. set out to explain some findings obtained by Shraga Hocherman and Wise[49] for M1. In that experiment, monkeys made movements with curved or straight trajectories (figure 14.10E). Surprisingly, cells in M1 had more activity for curved movements to a given target than for straight ones. Ajemian et al. wanted to understand whether this phenomenon was related to the fact that M1 neurons change their preferred directions as a function of hand location.

Figure 14.10 shows how their models explored these ideas. Take the Cartesian frame, which has extrinsic axes and an arbitrary origin. If M1 cells encoded movement direction in Cartesian coordinates, then the cell's preferred direction should not change for hand locations in different parts of the workspace (figure 14.10B). According to their models, an extrinsic frame centered on the shoulder $[S_{ext}]$ should lead to predictable changes in preferred direction, rotating more or less with shoulder angle. Note, however, that this coordinate frame would not lead to changes in the amplitude of directional coding. Neither of these assumptions seemed to fit the data well. Ajemian showed that the best match to the neurophysiological data occurred for intrinsic, joint-centered coordinates. This coordinate frame results in a progressively larger amplitude of tuning with increasing distance from the body, as well as rotations in preferred direction with changes in shoulder angle (figure 14.10C).

How can this reference frame help explain the data of Hocherman and Wise? Figure 14.10F shows how. The changes in each cell's preferred direction as the various movements progress through the workspace match best for curved movements. Notice how, for the joint-centered coordinates $[E + S_{int}]$, the cell's preferred direction rotated to maintain the best alignment with the movement trajectory (figure 10F, right). This alignment exceeds that for straight movements. When a movement occurs in a cell's preferred direction, it has more activity. Thus, M1 neurons show greater activity for curved trajectories because the rotation of

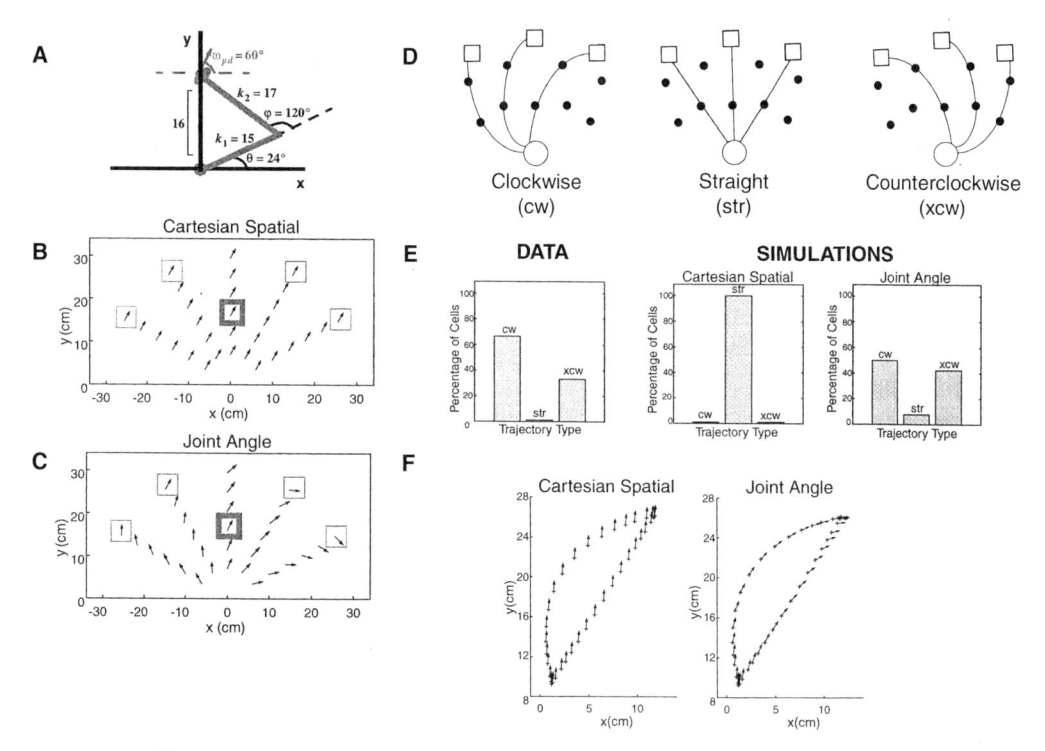

14.2. Population Vectors, Force Coding, and Coordinate Frames in M1

Figure 14.10
A model suggesting that M1 activity best fits an intrinsic, joint-angle reference frame. (*A*) A two-joint arm. (*B* and *C*) Predictions about how the preferred direction of an M1 neuron should vary with hand location in the workspace, for two different coordinate frames: Cartesian and joint-angle coordinates. (*D*) Movement trajectories made by a monkey in an experiment by Hocherman and Wise.[49] They found that in M1, many cells discharged most for the curved trajectories. (*E*) Left: Data from Hocherman and Wise showing that M1 neurons prefer clockwise (cw) and counterclockwise (xcw) trajectories to straight (str) trajectories. Right: Simulations, using the coordinate systems in B and C, for the three kinds of movement trajectories illustrated in *D*. (*F*) The plots show how an M1 cell should change its preferred direction over the course of a movement, as the monkey's hand moves through the workspace, for both coordinate systems. Under the assumption of Cartesian coordinates (left), most simulated cells discharged preferentially for straight trajectories. For joint-angle coordinates (right), the vast majority of model cells showed the empirically observed selectivity for curved trajectories. Note that for the joint-angle coordinate simulation, the cell's preferred direction has better alignment with the trajectory for curved versus straight trajectories because the cell's preferred direction changes as a function of hand location. (From Ajemian et al.[48] with permission.)

their preferred directions in the workspace matches the hand's trajectory through that workspace. Neither of the other coordinate frames makes this prediction.

References

1. Stein, JF, and Glickstein, M (1992) Role of the cerebellum in visual guidance of movement. Physiol Rev 72, 967–1017.
2. Glickstein, M (2000) How are visual areas of the brain connected to motor areas for the sensory guidance of movement? Trends Neurosci 23, 613–617.
3. Graziano, MS, Hu, XT, and Gross, CG (1997) Visuospatial properties of ventral premotor cortex. J Neurophysiol 77, 2268–2292.
4. Kakei, S, Hoffman, DS, and Strick, PL (2001) Direction of action is represented in the ventral premotor cortex. Nat Neurosci 4, 1020–1025.
5. Kakei, S, Hoffman, DS, and Strick, PL (1999) Muscle and movement representations in the primary motor cortex. Science 285, 2136–2139.
6. Schwartz, AB, Moran, DW, and Reina, GA (2004) Differential representation of perception and action in the frontal cortex. Science 303, 380–383.
7. Ochiai, T, Mushiake, H, and Tanji, J (2002) Effects of image motion in the dorsal premotor cortex during planning of an arm movement. J Neurophysiol 88, 2167–2171.
8. Riehle, A, and Requin, J (1995) Neuronal correlates of the specification of movement direction and force in four cortical areas of the monkey. Behav Brain Res 70, 1–13.
9. Kurata, K (1993) Premotor cortex of monkeys: Set- and movement-related activity reflecting amplitude and direction of wrist movements. J Neurophysiol 69, 187–200.
10. Fu, Q-G, Suarez, JI, and Ebner, TJ (1993) Neuronal specification of direction and distance during reaching movements in the superior precentral premotor area and primary motor cortex of monkeys. J Neurophysiol 70, 2097–2116.
11. Fu, Q-G, Flament, D, Coltz, JD, and Ebner, TJ (1995) Temporal encoding of movement kinematics in the discharge of primate primary motor and premotor neurons. J Neurophysiol 73, 836–854.
12. Messier, J, and Kalaska, JF (2000) Covariation of primate dorsal premotor cell activity with direction and amplitude during a memorized-delay reaching task. J Neurophysiol 84, 152–165.
13. Shen, L, and Alexander, GE (1997) Neural correlates of a spatial sensory-to-motor transformation in primary motor cortex. J Neurophysiol 77, 1171–1194.
14. Shen, L, and Alexander, GE (1997) Preferential representation of instructed target location versus limb trajectory in dorsal premotor area. J Neurophysiol 77, 1195–1212.
15. Scott, SH, and Kalaska, JF (1997) Reaching movements with similar hand paths but different arm orientations. I. Activity of individual cells in motor cortex. J Neurophysiol 77, 826–852.
16. Scott, SH, Sergio, LE, and Kalaska, JF (1997) Reaching movements with similar hand paths but different arm orientations. II. Activity of individual cells in dorsal premotor cortex and parietal area 5. J Neurophysiol 78, 2413–2426.
17. Georgopoulos, AP (1991) Higher order motor control. Annu Rev Neurosci 14, 361–377.

18. Moran, DW, and Schwartz, AB (1999) Motor cortical activity during drawing movements: Population representation during spiral tracing. J Neurophysiol 82, 2693–2704.
19. Schwartz, AB, and Moran, DW (1999) Motor cortical activity during drawing movements: Population representation during lemniscate tracing. J Neurophysiol 82, 2705–2718.
20. Wise, SP, Weinrich, M, and Mauritz, K-H (1986) Movement-related activity in the premotor cortex of rhesus macaques. Prog Brain Res 64, 117–131.
21. Graziano, MS, Yap, GS, and Gross, CG (1994) Coding of visual space by premotor neurons. Science 266, 1054–1057.
22. di Pellegrino, G, and Wise, SP (1993) Visuospatial vs. visuomotor activity in the premotor and prefrontal cortex of a primate. J Neurosci 13, 1227–1243.
23. Weinrich, M, and Wise, SP (1982) The premotor cortex of the monkey. J Neurosci 2, 1329–1345.
24. Graziano, MS, and Gross, CG (1998) Visual responses with and without fixation: Neurons in premotor cortex encode spatial locations independently of eye position. Exp Brain Res 118, 373–380.
25. Fogassi, L, Gallese, V, di Pellegrino, G, et al. (1992) Space coding by premotor cortex. Exp Brain Res 89, 686–690.
26. Boussaoud, D, Barth, T, and Wise, SP (1993) Effects of gaze on apparent visual responses of frontal cortex neurons. Exp Brain Res 93, 423–434.
27. Boussaoud, D, and Bremmer, F (1999) Gaze effects in the cerebral cortex: Reference frames for space coding and action. Exp Brain Res 128, 170–180.
28. Jouffrais, C, and Boussaoud, D (1999) Neuronal activity related to eye-hand coordination in the primate premotor cortex. Exp Brain Res 128, 205–209.
29. Cisek, P, and Kalaska, JF (2002) Modest gaze-related discharge modulation in monkey dorsal premotor cortex during a reaching task performed with free fixation. J Neurophysiol 88, 1064–1072.
30. Mushiake, H, Tanatsugu, Y, and Tanji, J (1997) Neuronal activity in the ventral part of premotor cortex during target-reach movement is modulated by direction of gaze. J Neurophysiol 78, 567–571.
31. Bock, O (1986) Contribution of retinal versus extraretinal signals towards visual localization in goal-directed movements. Exp Brain Res 64, 476–482.
32. Henriques, DYP, Klier, EM, Smith, MA, et al. (1998) Gaze-centered remapping of remembered visual space in an open-loop pointing task. J Neurosci 18, 1583–1594.
33. Pouget, A, Ducom, JC, Torri, J, and Bavelier, D (2002) Multisensory spatial representations in eye-centered coordinates for reaching. Cognition 83, B1–11.
34. Sergio, LE, and Kalaska, JF (1998) Changes in the temporal pattern of primary motor cortex activity in a directional isometric force versus limb movement task. J Neurophysiol 80, 1577–1583.
35. Sergio, LE, and Kalaska, JF (2003) Systematic changes in motor cortex cell activity with arm posture during directional isometric force generation. J Neurophysiol 89, 212–228.
36. Todorov, E (2000) Direct cortical control of muscle activation in voluntary arm movements: A model. Nat Neurosci 3, 391–398.
37. Taira, M, Boline, J, Smyrnis, N, et al. (1996) On the relations between single cell activity in the motor cortex and the direction and magnitude of three-dimensional static isometric force. Exp Brain Res 109, 367–376.

38. Hepp-Reymond, MC, Kirkpatrick-Tanner, M, Gabernet, L, et al. (1999) Context-dependent force coding in motor and premotor cortical areas. Exp Brain Res 128, 123–133.
39. Wannier, TMJ, Maier, MA, and Hepp-Reymond, M-C (1991) Contrasting properties of monkey somatosensory and motor cortex neurons activated during the control of force in precision grip. J Neurophysiol 65, 572–589.
40. Georgopoulos, AP, Caminiti, R, Kalaska, JF, and Massey, JT (1983). In Neural Coding of Motor Performance, eds Massion, J, Paillard, J, Schultz, W, and Wiesendanger, M (Springer-Verlag, Heidelberg), pp 327–336.
41. Georgopoulos, AP, Lurito, JT, Petrides, M, et al. (1989) Mental rotation of the neuronal population vector. Science 243, 234–236.
42. Scott, SH, Gribble, PL, Graham, KM, and Cabel, DW (2001) Dissociation between hand motion and population vectors from neural activity in motor cortex. Nature 413, 161–165.
43. Georgopoulos, AP, Schwartz, AB, and Kettner, RE (1986) Neural population coding of movement direction. Science 233, 1416–1419.
44. Moran, DW, and Schwartz, AB (1999) Motor cortical representation of speed and direction during reaching. J Neurophysiol 82, 2676–2692.
45. Sanger, TD (1996) Probability density estimation for the interpretation of neural population codes. J Neurophysiol 76, 2790–2793.
46. Humphrey, DR, Schmidt, EM, and Thompson, WD (1970) Predicting measures of motor performance from multiple cortical spike trains. Science 170, 758–762.
47. Reina, GA, Moran, DW, and Schwartz, AB (2001) On the relationship between joint angular velocity and motor cortical discharge during reaching. J Neurophysiol 85, 2576–2589.
48. Ajemian, R, Bullock, D, and Grossberg, S (2000) Kinematic coordinates in which motor cortical cells encode movement direction. J Neurophysiol 84, 2191–2203.
49. Hocherman, S, and Wise, SP (1991) Effects of hand movement path on motor cortical activity in awake, behaving rhesus monkeys. Exp Brain Res 83, 285–302.
50. Taylor, DM, Helms-Tillery, SI, and Schwartz, AB (2002) Direct cortical control of 3D neuroprosthetic devices. Science 296, 1829–1832.

Reading List

For alternative views on these topics, see Taylor et al.,[50] Schwartz et al.,[6] and Stein and Glickstein.[1]

III Skills, Adaptation, and Trajectories

15 Aligning Vision and Proprioception I: Adaptation and Context

Overview: To represent end-effector location in fixation-centered coordinates, your CNS must align proprioceptive inputs about joint angles and muscle lengths with visual inputs about where the end effector appears in fixation-centered space for that pattern of proprioceptive inputs. The CNS needs to recalibrate these computational maps whenever something alters the visual feedback of end-effector location.

Put yourself once again in the shoes of the imaginary robotics engineer you met in section 9.6, who suggested that in order to program a robot to move its gripper from one point to another, the robot's controller needed to compute target location with respect to the current location of the gripper. You might proceed by using the image from the camera to determine the locations of the gripper and the target (figure 15.1A). You could describe the gripper and target locations as vectors in "camera-centered" coordinates (i.e., a coordinate system that provides the pixel locations for both objects). The difference between these two vectors describes the desired difference vector.

As mentioned in section 9.6, although you must use the camera to detect the target location, you may have other sensors besides the camera that (indirectly) tell you the gripper's location. If the joints of the robot housed angle sensors and you knew the length of each link in the robotic arm, you could use the readout from the angle sensors to compute an estimated gripper location, the **forward-kinematics** computation presented in chapter 9. By comparing this estimate with the gripper's location as sensed by the camera, you could write a computer program that ensured the alignment of gripper location in these two coordinate frames. In terms of the robot, you can describe those frames as joint-angle and camera coordinates for describing gripper location. In terms of reaching movements in biological systems, you can say that this program aligns *visual* and *proprioceptive* signals about end-effector location.

You need to maintain this alignment because you still want to reach even when you cannot see your hand. For the robot, this means that you have to use the angle sensors on the robot to compute gripper location in camera-centered coordinates. For your hand, it means that you have to

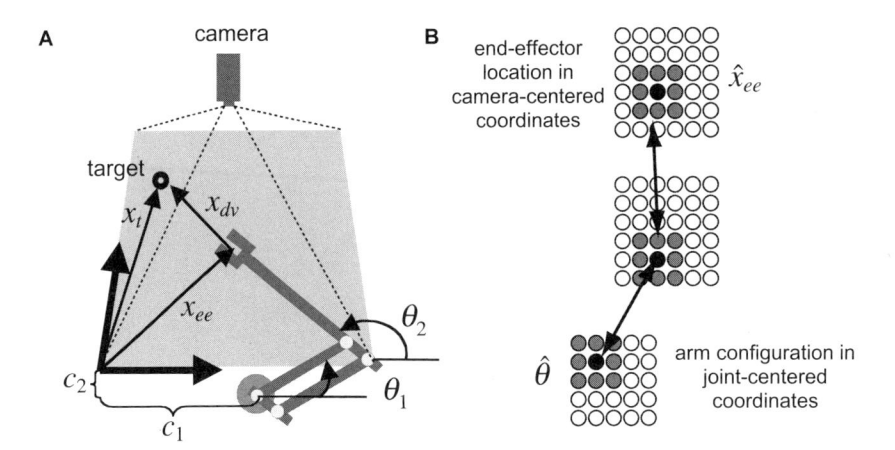

Figure 15.1
Virtual robotics. (*A*) A robotic arm that moves in two dimensions and a camera. The black ring represents the target of movement. The gripper and target locations have camera-centered coordinates with vectors x_{ee} and x_t from the origin of the camera's frame of reference; c_1 and c_2 give the offset of that origin from the "shoulder" of the robot's arm. The desired difference vector for the gripper is given by the subtraction of x_{ee} from x_t, which yields x_{dv}. The joint-angle sensors on the robot measure joint angles θ_1 and θ_2. (*B*) The alignment of information from the joint sensors and the camera via a recurrent network.

use proprioceptive signals to compute hand location in fixation-centered coordinates. To perform this computation, you might use a recurrent network such as that in figure 15.1B, which aligns joint angle with visual information.

The robot's controller would have to adjust the connections of this network (i.e., its synaptic weights) through trial and error. Each time the gripper moves, you could measure where the image of the gripper falls on the camera and what the joint-angle sensors read. Then you could adjust the weights in the network. Based on the theory of Deneve et al.,[1] you should eventually have a network that can perform pattern completion. That is, given a noisy reading from the joint sensors, the network would reject the noise and estimate gripper location in camera-centered coordinates. Similarly, given a reading from the camera, the network would estimate the corresponding readings from the joint sensors.

Once your computer program had calibrated the system, you could stop the learning process and keep the network's weights fixed. But if someone changed the camera lens or an accident bent a link in the robot's arm, the robot would reach to the wrong place. It makes sense for the calibration process to continue, in case the system changes.

In planning a reaching movement, the evidence presented in chapter 12 suggests that your CNS uses fixation-centered coordinates to compute a difference vector from the end effector to the target. Does your CNS also align proprioceptive information with visual information and maintain that calibration as postulated for the imaginary robot? This chapter considers experiments that have altered the relationship between proprioception and vision. In these studies, experimenters alter the path of light to the retina, a perturbation of the input that resembles changing the lens on the robot's camera.

15.1 Newts Cannot Adapt to Rotation of Their Eyes

In the 1940s, Roger Sperry[2] performed an interesting experiment on newts. He knew that the human CNS demonstrated remarkable **adaptation**, but he also knew that it did not always happen. To test the degree of adaptation in amphibians, he rotated the eyes of red-spotted newts. Rotation of the eye by 180° left the optic nerve intact, but retinal cells that received light from above the newt prior to rotation, received light from below after the rotation. Similarly, cells that were normally active when light from the left hit the retina responded to light from the right after rotation. Within an hour after eye rotation, as soon as the newts recovered from anesthesia, Sperry tested their response to moving objects. He found that when something in the visual field moved to the right, they moved their head to the left. If a lure with meat attached wiggled above them, they rotated their head downward. As Sperry put it:

When the piece of meat was moved back and forth in the water several centimeters above and a little to one side of the animals, they tilted their heads downward on that side and began to move toward the bottom of the aquarium. Even though the newts happened to be resting on the bottom when the lure was thus waved above them, they cocked their heads down under them and began pushing about among the pebbles of the bottom with the nose and forefeet. If the lure was placed below the animals, the head and forebody were tilted upward and the newts started toward the surface. (From Sperry,[2] pp. 271–272)

This misdirected behavior never changed. In Sperry's words: "No correction of the maladaptive visuomotor responses was observed in the course of time." After 4–5 months of testing, the newts remained as misguided as on the first day. In fact, blind newts behaved better.

By analogy with the imaginary robot (figure 15.1A), consider what would happen if you sent the camera out for repair but when it was returned, you installed it upside down. The information from the camera tells your computer that the object lies to the left when in fact it lies to the right, up when down, and so on. Unless you have an adaptation mechanism that notes the errors in alignment of the robot's "proprioception" and "vision"—and somehow realigns them—your robot will have a lot of trouble reaching to targets.

Sperry's experiment in newts suggested that some vertebrates have serious limitations in this recalibration mechanism, at least for rotations of as much as 180°. That is, the weights in their neural networks appear to be relatively hardwired once the animal reaches a certain stage of development. Sperry[2] speculated that the greater adaptive flexibility of adult mammals, including humans and other primates, could occur as a result of "greater differential control through adjustable cortical processes" (p. 277).

15.2 Primates Adapt to Rotation of the Visual Field

In contrast to Sperry's newts, adult monkeys and humans can adapt to radical changes in visual inputs. Yoichi Sugita explored this ability in monkeys. He fitted the monkeys with goggles and replaced the glass over each eye with a dove prism.[3] Dove prisms invert an image along a plane that runs parallel to the largest surface (figure 15.2A). Sugita installed the prisms on the goggles so that they reversed the image along the horizontal (left–right) plane. In this way, things to the left of a monkey's fixation point appeared to be on the right.

The monkeys wore these goggles continuously for about 4 months. For the first 2 weeks, they could barely move by themselves. However, by the fourth week, they began to move and play. Each day, the experimenters brought the monkeys to the laboratory and set them in front of a touch screen. The monkeys had previously learned to touch a sequence of targets that appeared on the screen. For the first few weeks, a target to the right resulted in a reach to the left. After about a month and a half, the monkeys began to reach straight to the target. The time it took to hit the target steadily decreased and approached baseline levels in 30–50 days (figure 15.2A). Therefore, the monkeys had the capacity to adapt to this radical change in visual **feedback**.

According to the model put forward in figure 14.1,[8] the monkey's CNS maintains a map that aligns proprioceptive cues regarding the arm's configuration with visual cues regarding the hand's location (figure 15.1B). This map must have changed during the period of adaptation. What happened to the map that aligned proprioception and vision prior to the experiment? Was that information completely lost, or did the new alignment (caused by the goggles) somehow add more "attractors" to the network without significantly degrading the previous ones? Many neural networks have dynamics that converge when their internal states fall into an attractor, which is something like a "preferred state" of a network (see boxes 12.1 and 15.1). According to the model in figure 14.1, as new attractors form in the networks, the system as a whole gains a new capability, and the monkey acquires a new motor skill (see section 4.3.1). If the new information "overwrote" the previous alignment, it should take the monkeys as long to adjust to removal of the prisms as to their introduction.

After 2 months, the experimenters removed the monkeys' goggles and, for a day or so, the animals again had trouble moving. When presented with a target to the left, they reached to the right of it. These errors are *aftereffects* of adaptation. However, unlike the many weeks

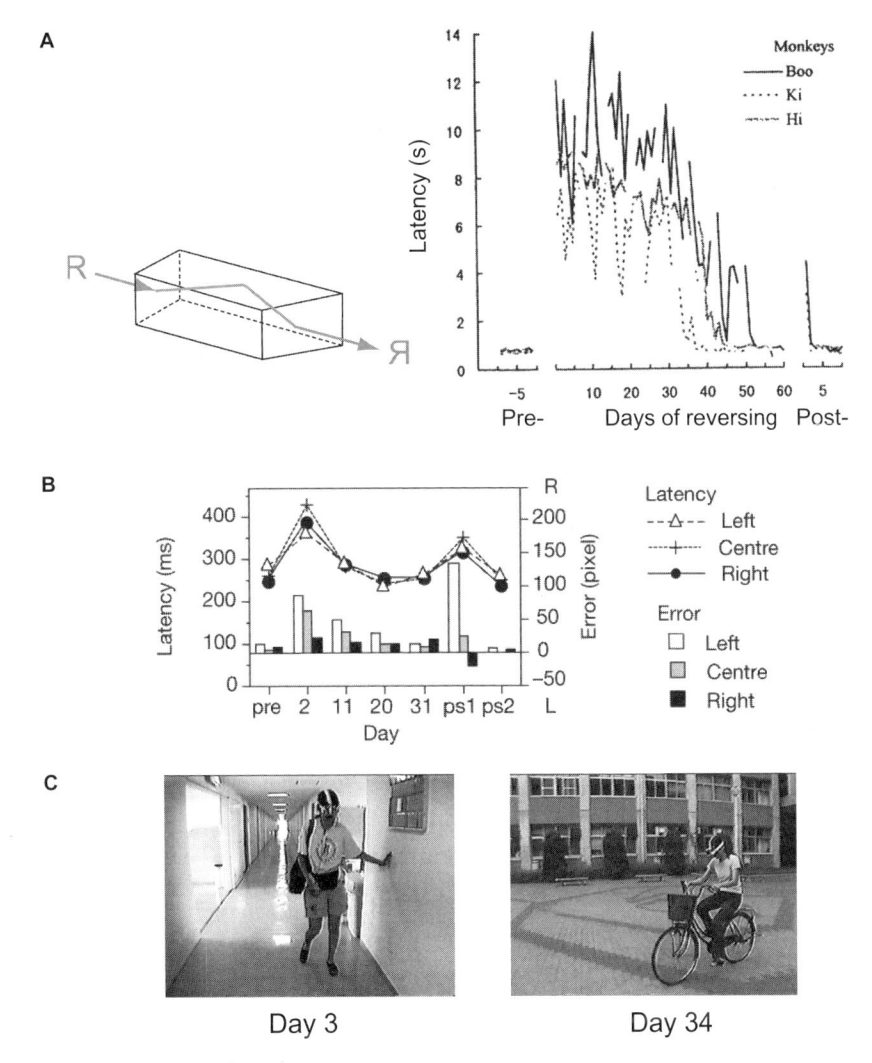

Figure 15.2

Reaching movements while wearing dove prisms. (*A*) Left: A dove prism, which inverts the visual image along the left–right dimension. Right: The response latency (the time from target presentation to target acquisition) in monkeys named Boo, Ki, and Hi. Abbreviations: pre, before experience with prisms; post, after wearing prisms. (*B*) Performance of people in a localization task, with and without dove prisms. Participants reported the side of a screen on which a target appeared by pressing a left or right key. The graph shows mean reaction time (latency) for the left (unfilled triangles), center (+signs), and right targets (filled circles), as well as mean errors for the left (unfilled bars), center (gray bars), and right (black bars) targets, in terms of the horizontal distance between the target and reach location in screen pixels. Positive errors indicate reaching to the right of the target. Abbreviations: ps1: post1, within an hour of removing the prisms; ps2: post2, the day after removal. (*C*) Examples of performance in a typical human participant. On day 3, the participant could not walk down a hall without maintaining tactile contact with the wall. However, on day 34 the same participant not only could walk normally but also could ride a bicycle. (Right part of *A* from Sugita,[3] *B–C* from Sekiyama et al.[9] with permission.)

Box 15.1
Attractor states in the context of skill acquisition

> Box 12.1 introduced the concept of attractor states in dynamical systems. Attractor networks have been used to model the ability to maintain gaze at a particular orientation[4] and maintain items in a short-term memory buffer,[5–7] but this presentation treats the concept a little differently. For the neural networks that learn skills, a skill corresponds to a given set of transformations (or *alignments*) of motor parameters and coordinate frames. These transforms depend on the weights of the connections within the network that computes the transform, collectively termed its weight matrix. If you think of the activity of each element in a neural network as one dimension of a vector having as many dimensions as the network has elements, then the attractor represents a particular vector that the network is relatively likely to adopt. This attractor is a function of the synaptic weight matrix of the network. According to the idea presented here, adaptation involves a change in the weights so that the attractor shifts from one location to another. Skill acquisition involves the formation of new attractors, often along with a change in the input to the network so that a contextual cue pushes the system toward one attractor or another.

that it took the monkeys to learn to reach while wearing the dove prisms, adjusting to their removal took much less time. The aftereffects washed out by the third day (figure 15.2A). This slow speed of adaptation and fast washout suggest that when the monkeys wore the goggles, the network changed in a way that slowly accommodated the new alignment between proprioception and vision, without losing the information regarding the previous alignment.

In a similar experiment on people,[9] participants wore left–right reversing prisms continuously for about 40 days. During the first days of wearing the prisms, they could barely move down a hallway without bouncing off the walls (figure 15.2C). They also had trouble eating with utensils because they reached to the wrong location. However, by the end of the experiment, they could perform ordinary functions such as riding a bicycle or using a knife to finely slice a cucumber. Occasionally, they came to the laboratory and reached to targets. Both the amount of reaching error and the latency to hit the target returned to pre-exposure levels after about 3 or 4 weeks (figure 15.2B). Their CNS had adapted to the altered visual field. As with the monkeys, it took only a day or so for the aftereffects of wearing the prisms to wash out. When the participants returned to the laboratory on subsequent days and again wore the prisms, they quickly readapted. These findings suggest that the participants acquired information during adaptation that did not wash out when they took off the prisms. Rather, their CNS retained separate, altered maps for the prisms and for unperturbed vision.

15.3 Prism Adaptation Requires Modification of Both Location and Displacement Maps

Now return to the model presented in schematic form in figure 14.1. This exercise in virtual robotics helps explain prism adaptation and its aftereffects. It turns out that the association between displacements $x_{dv} \stackrel{\hat{\theta}}{\leftrightarrow} \Delta\hat{\theta}$ computed by network 4 in the model has a relationship with the association between locations $\hat{\theta} \leftrightarrow \hat{x}_{ee}$ computed by network 1. They both involve a transformation between intrinsic, proprioceptive coordinates and extrinsic, vision-based ones. To see this relationship mathematically, consider figure 15.1A and the forward-kinematics computation $\hat{\theta} \leftrightarrow \hat{x}_{ee}$. The end effector's location in camera-centered coordinates is a function f of the joint angles θ_1 and θ_2, link lengths l_1 and l_2, and constant offsets c_1 and c_2:

$$x_{ee} = f(\hat{\theta}) = \begin{bmatrix} l_1 \cos(\hat{\theta}_1) + l_2 \cos(\hat{\theta}_2) + c_1 \\ l_1 \sin(\hat{\theta}_1) + l_2 \sin(\hat{\theta}_2) - c_2 \end{bmatrix}.$$

This equation describes how the end effector's location—as computed from the robot's joint-angle sensors—corresponds to camera-centered coordinates. Network 1 in figure 14.1, the *location map*, computes the relationship between locations in these two coordinate systems. Network 4 computes a *displacement map*. The function that relates joint-angle changes to end-effector displacements is the derivative of f with respect to the joint angles, the Jacobian J:

$$\Delta x_{ee} = \frac{df}{d\hat{\theta}} \Delta \hat{\theta}$$

$$= [J(\hat{\theta})][\Delta\hat{\theta}]$$

$$= \begin{bmatrix} -l_1 \sin(\hat{\theta}_1) & -l_2 \sin(\hat{\theta}_2) \\ l_1 \cos(\hat{\theta}_1) & l_2 \cos(\hat{\theta}_2) \end{bmatrix} \begin{bmatrix} \Delta\hat{\theta}_1 \\ \Delta\hat{\theta}_2 \end{bmatrix}.$$

Because network 4 in figure 14.1 computes a rotation in joint coordinates from a desired displacement of the end effector, you have

$$\Delta\hat{\theta} = (J(\hat{\theta}))^{-1} \Delta x_{ee}.$$

Therefore, for a desired end-effector displacement x_{dv} at a particular arm configuration $\hat{\theta}$, you can compute the desired joint rotations with function g:

$$\Delta\hat{\theta} = g(x_{dv}, \hat{\theta}) = (J(\hat{\theta}))^{-1} x_{dv}.$$

The map $x_{dv} \stackrel{\hat{\theta}}{\leftrightarrow} \Delta\hat{\theta}$ summarizes these equations. Recall that function f relates end-effector *locations* in the robot's joint- and camera-based coordinates: $\hat{\theta} \leftrightarrow \hat{x}_{ee}$. The equation immediately above tells you that function g, which relates *displacements* in joint- and camera-based coordinates, is a function of f. Thus, the map $\hat{\theta} \leftrightarrow \hat{x}_{ee}$ computes function f, and the map $x_{dv} \stackrel{\hat{\theta}}{\leftrightarrow} \Delta\hat{\theta}$ computes function g. So if you change the lens of the robot's camera, or rotate the camera, you change function f. However, that change also affects function g. In summary, a change in the characteristics of the camera causes a change both in a map that aligns location $\hat{\theta} \leftrightarrow \hat{x}_{ee}$ and in a map that aligns displacement in joint- and camera-based coordinates $x_{dv} \stackrel{\hat{\theta}}{\leftrightarrow} \Delta\hat{\theta}$.

Now imagine that someone has rotated the camera in figure 15.1A by 180°, by analogy with Sperry's experiment in newts (section 15.1). The rotation changes the relationship between θ and x_{ee}. You now have

$$x_{ee} = f_1(\hat{\theta}) = \begin{bmatrix} -(l_1 \cos(\hat{\theta}_1) + l_2 \cos(\hat{\theta}_2) + c_1 - k_1) \\ -(l_1 \sin(\hat{\theta}_1) + l_2 \sin(\hat{\theta}_2) - c_2 - k_2) \end{bmatrix}.$$

To relate displacements in the camera- and joint-based coordinates, you can compute the derivative of f_1 with respect to $\hat{\theta}$ and invert the resulting matrix to arrive at function g_1.

Recall that the location map $\hat{\theta} \leftrightarrow \hat{x}_{ee}$ computes function f and the displacement map $x_{dv} \stackrel{\hat{\theta}}{\leftrightarrow} \Delta\hat{\theta}$ computes function g. When someone comes along and puts prisms on your robot's camera or rotates it, the robot's arm will not move correctly unless you adapt both maps. However, rotating the camera or changing its lens does not require you to change the mapping of network 3 in figure 14.1, which computes the difference vector $x_{dv} = \hat{x}_t - \hat{x}_{ee}$.

Now consider what sorts of information you need to adapt $\hat{\theta} \leftrightarrow \hat{x}_{ee}$ and $x_{dv} \stackrel{\hat{\theta}}{\leftrightarrow} \Delta\hat{\theta}$. Every time the camera images the robot's gripper location $\hat{x}_{ee}$, and the robot's controller reads joint angles $\hat{\theta}$, it has a data point to use for aligning $\hat{\theta} \leftrightarrow \hat{x}_{ee}$. A simple, static view of the limb suffices to adapt this map. However, before the robot can make accurate reaching movements, the controller also needs to align gripper displacements x_{dv} with joint rotations $\Delta\hat{\theta}$. To adapt $x_{dv} \stackrel{\hat{\theta}}{\leftrightarrow} \Delta\hat{\theta}$, the robot has to *move* the gripper and observe how displacements in camera-centered coordinates correspond to rotations in joint-based coordinates.

15.4 Long-Term Memories and Learning to Switch on Context

Wearing other kinds of prisms can lead to much faster adaptation than with dove prisms. Figure 15.3 shows adaptation to wedge prisms, which differ from the dove prisms in that they shift light, typically to the left or right, by 5°–25°, rather than inverting the visual field. Tom Thach and his colleagues[10] used a task that Dick Held, Nigel Daw, and Torsten Wiesel had developed for teaching medical students about motor adaptation. In that experiment, participants wore prisms that shifted the visual field by approximately 17°. After putting on the prism glasses, they tried to throw clay balls at an 8 cm^2 visual target, ~2 m away (figure 15.3). The investigators asked the participants not to look at their hands and measured the accuracy of their throws. (This experimental design prevents participants from adjusting the movement after they release the ball. In con-

15.4. Long-Term Memories and Learning to Switch on Context

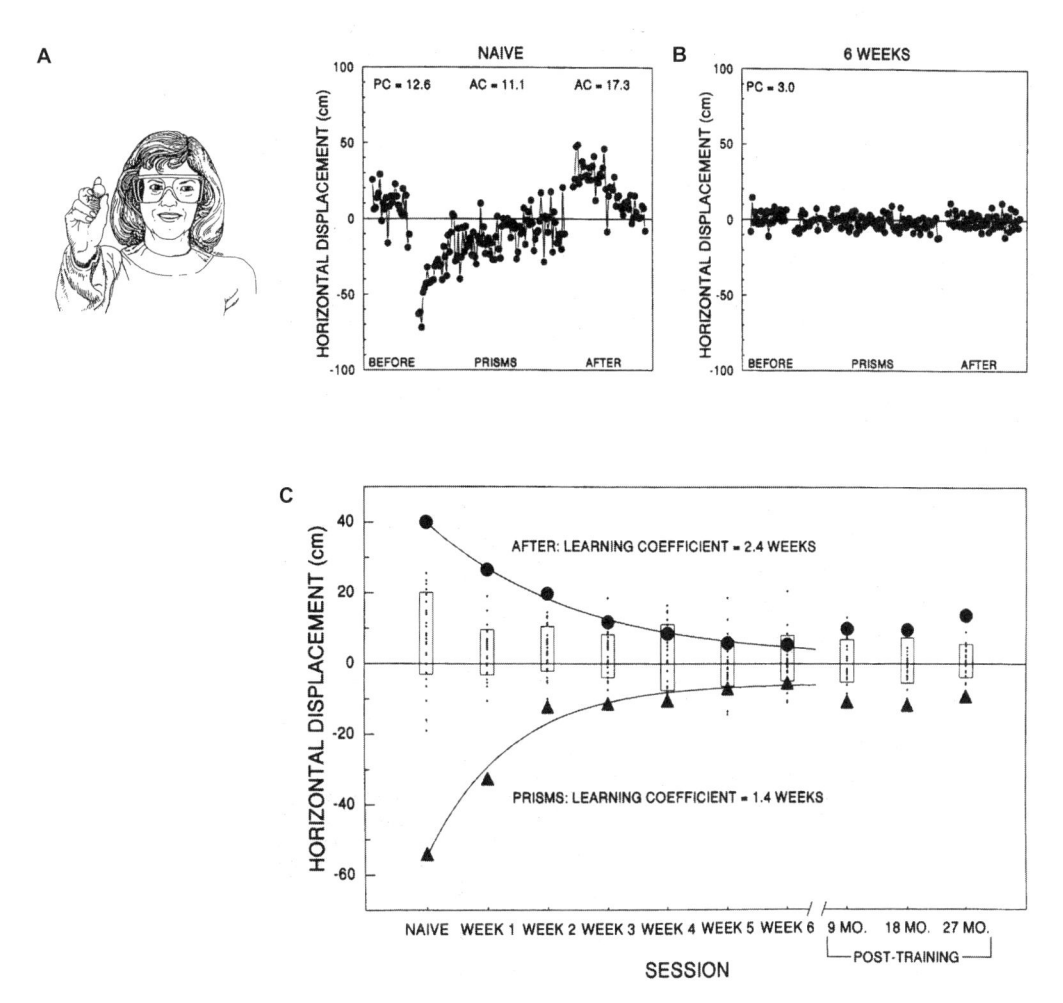

Figure 15.3

Adaptation to a 17° wedge prism that bent the path of light to the right. (*A*) Throwing errors in a naïve participant before, during, and after wearing prisms, plotted sequentially. The time constants for the curves during the prism and after-prism conditions are 11 and 17 trials, respectively. (*B*) Throwing errors before, during, and after wearing prisms for a skilled participant, with six weeks of training. (*C*) The initial throwing errors after putting on the prisms (triangles) and aftereffects (circles) resulting from their removal, for the same skilled participant, during the progression from A to B over the course of 2–4 weeks of practice, and for more than two years thereafter. (From Martin et al.[10] with permission.)

Box 15.2
Neural networks and skills

> Artificial neural networks most resemble the implementation level of a computational problem. For the present purpose, consider the skill required to switch to a new displacement map after putting on prisms. At the computational level, the problem to be solved involves the production of joint rotations needed to bring the end effector to the target. At the level of an algorithm, this amounts to computing a $\Delta\theta$ in joint-centered coordinates from a given difference vector in fixation-centered coordinates. The algorithm level involves the Jacobian matrix as described in section 7.5.4. At the level of implementation, this algorithm could result from the network's falling into an attractor state for computing the appropriate transform. The network falls into the attractor state through the effects of its inputs (including contextual inputs) on the activity of its computational elements.

trast, during a reaching movement, participants can adjust their reach if they see that it begins inaccurately.)

The adaptation period began with inaccurate throws. After a few throws, however, accuracy improved to near baseline levels (figure 15.3A). In this experiment, unlike the one with dove prisms, the adaptation period was relatively short, and the washout of aftereffects required about the same number of trials as the initial adaptation. This finding suggests that the wedge prisms caused a shift in existing attractors in the neural networks in figure 14.1, not the formation of new ones (see boxes 12.1 and 15.2). According to the model, this is a fundamental distinction between motor adaptation and skill acquisition. In adaptation, existing attractors change; in learning skills, new attractors emerge. As discussed in section 4.3, motor adaptation results in the motor system's regaining a former level of performance; skill acquisition leads to new capabilities.[11]

The discussion to this point might lead you to think that dove prisms induce new skills but wedge prisms do not. However, with extensive experience, participants can learn the *wedge-prism skill* as well. Thach and his colleagues[10] trained a few of the participants about a month and a half with the prisms. During each experimental session, participants wore the prism glasses, made a few throws, and then removed the glasses and made a few more throws. Figure 15.3C shows that it took about 2.5 weeks for the participants to be able to throw accurately on the first try with the prisms on and, importantly, to throw accurately on the first try with the prisms off. In a month or so they reached a plateau. Recall that it took participants many weeks to learn the dove-prism skill, and it required extensive practice to learn the wedge-prism skill, as well. The CNS needs to learn distinct kinematic mappings and gain the ability to switch between them in order to attain a new skill. Thus, after several weeks of practice, involving hundreds of movements, participants eventually develop the ability to put on and remove wedge prisms on demand, with minimal adaptation times and aftereffects. Effectively, they can use context (wearing prisms or not) to switch between the alignments $\hat{\theta} \leftrightarrow \hat{x}_{ee}$ and $x_{dv} \overset{\hat{\theta}}{\leftrightarrow} \Delta\hat{\theta}$ appropriate for the prism and those appropriate for unperturbed vision. In terms of the model in figure 14.1, the location and displacement maps learned two alignments between proprioception and vision based on a context: wearing of—or freedom from—prisms.

Experiments on monkeys led to similar conclusions. Kiyoshi Kurata and Eiji Hoshi[12] put wedge prisms in front of monkeys' eyes and trained them to reach with the right arm. Reach targets appeared on a touch screen, and each day the monkeys began reaching to those targets without prisms. After these baseline measures, the experimenters perturbed the visual input with two kinds of prisms. One kind shifted the light 10° to the right; the other, an equal amount to the left. The location of the targets on the touch screen changed so that the monkeys saw the targets at the same location with the prisms and without them. With the 10°-left prisms in place, the monkeys reached to the left of the targets, and when looking through the 10°-right prisms, they reached too far to the right. With the prisms, it took about ten trials for the monkeys to reduce reaching errors to baseline levels. After about 50 trials, the experimenters removed the prisms, resulting in aftereffects. After washout of these aftereffects, the experimenters switched to the other set of prisms and the monkeys continued to reach. Therefore, unlike the experiments with the dove prisms, in which monkeys wore the same prisms for many weeks, here the monkeys wore two "opposing" prisms on the same day.

Despite repeated exposure to the same sets of prisms, the rate of adaptation did not change significantly over the period of the study. Recall that reaching movements while looking through dove prisms took many weeks of experience (figure 15.2), although the aftereffects washed out relatively quickly. Here, with prisms that only shifted the image, the monkey adapted within a few trials, and both adaptation and washout took about the same amount of time.

According to the model, adaptation to the dove prisms required the monkeys to learn new kinematic mappings specific to the prisms. With wedge prisms, the mappings changed to accommodate the prisms and changed back upon their removal: The existing maps adapted. Thus, for dove prisms, the monkeys learned to *switch* from one mapping to another, with each mapping corresponding to a different skill. On the other hand, switching does not occur when the same network repeatedly adapts. Remember that in the experiment of Kurata and Hoshi, minutes after they removed one prism, they inserted the opposite kind of prism. As you will see in section 22.1.2, on dynamics, when the learning of one kind of map quickly follows the learning of an "opposite" map, skills do not **consolidate** in memory. However, when the learning of one kind of map and its "opposite" occur at widely spaced time intervals (~4–6 hours), the CNS can eventually consolidate each memory. Thus, the fact that the monkeys of Kurata and Hoshi never learned a "prism skill" probably reflects the continual application of conflicting visual-field distortions.

15.5 Prism Adaptation in Virtual Robotics

The imaginary robot in figure 15.1A helps explain prism adaptation. It serves as a model for how the CNS copes with this kind of altered visual feedback. Imagine that someone, for some nefarious reason, inserts wedge prisms in front of your robot's camera (figure 15.4A). To simplify matters, you have programmed the camera to record the location of the gripper only when it completes a reaching movement. Assume that the prism shifts the path of light to the left, so that, from the point of view of the camera, the visual world appears displaced to the left.

As illustrated in figure 15.4A, after insertion of the wedge prism, a target activates pixels on the camera to the left of the pixels that "should be" activated. The black ring indicates the viewed target location, as distorted by the prisms; the dotted circle shows the target's actual location, often called "real" space. Before your tormentor introduced the prism, the robot had adaptively aligned the location map $\hat{\theta} \leftrightarrow \hat{x}_{ee}$ and the displacement map $x_{dv} \stackrel{\hat{\theta}}{\leftrightarrow} \Delta\hat{\theta}$ in "visual" and "proprioceptive" coordinates. In this exercise, the camera cannot "see" the robot's gripper, so the controller must estimate its location $\hat{x}_{ee}$, in camera-centered coordinates, from "proprioception." (Why? Assume that, for some reason, you must conserve your source of room light to the greatest possible extent. Therefore, you cannot illuminate the field of view until the very end of movement.) In the darkness, the *target* has a small light-emitting diode that registers on certain camera pixels. The camera tells the central controller that the target has location x_t, and the system computes a difference vector x_{dv} based on an estimate of gripper location from the joint-angle sensors in the robot's arm.

Figure 15.4A illustrates this difference vector—in camera-centered coordinates—beneath the drawing of the robot. Using the displacement map $x_{dv} \stackrel{\hat{\theta}}{\leftrightarrow} \Delta\hat{\theta}$, the controller computes a joint-rotation vector. This vector flexes the robot's "shoulder" (increasing θ_1) and extends its "elbow" (decreasing θ_2), bringing the arm and gripper to the location indicated by the gray arm in figure 15.4B. After the movement ends, you turn on the room lights for a brief period and the camera records the gripper at the location shown by the black arm. The gripper missed the target to the left.

Note that the prism induced two kinds of errors, as indicated by the arrows and numbers in figure 15.4B. First, for the displacement map (network 4 in figure 14.1), the desired difference vector produced joint rotations that moved the gripper too far to the left. Therefore, in planning the next movement, the controller should reduce the amount of flexion for the "shoulder." Second, for the location map (network 1 in figure 14.1), the camera recorded the gripper's location farther to the left than it "should have been," according to the robot's joint angles. Therefore, the controller should associate a given pair of joint angles with a gripper location slightly to the left of its location in "real" space. The system requires visual feedback for such recalibrations.

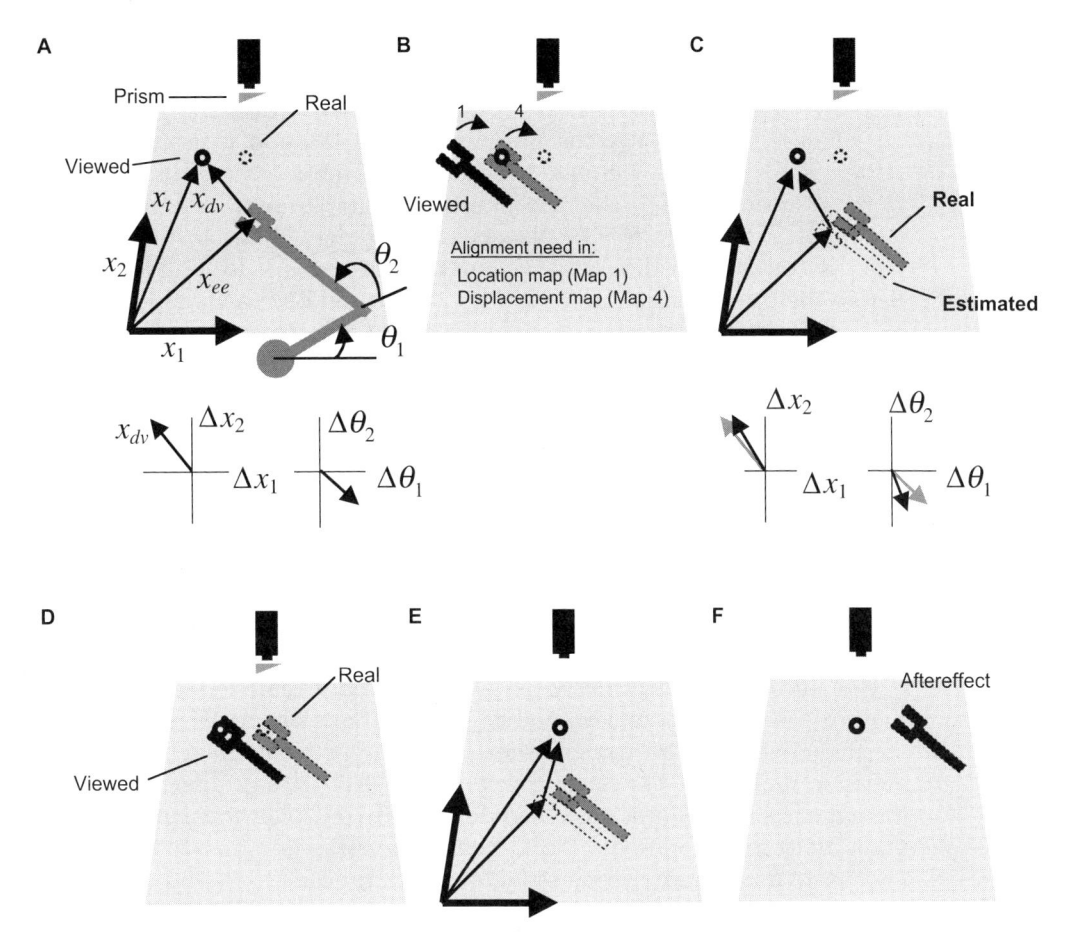

Figure 15.4
Virtual robotics for prism adaptation and aftereffects. Someone puts a wedge prism, which shifts the path of light to the left, in front of the camera. (*A*) Before adaptation condition. Assume that the camera cannot see the arm (the room is dark), and therefore the robot's controller must estimate the gripper's location from "proprioception." Assume further that the camera records gripper location only at the end of each movement (the room lights flash on). The dotted circle shows the target's location in real space, and the black ring shows the location viewed by the camera. In drawings of the robot's arm, black arms indicate viewed locations, gray arms indicate real locations, and dotted outlines of the arm indicate estimated locations. The plot below the robot arm shows the desired end-effector displacement in camera-centered coordinates (left) and joint-centered coordinates (right). (*B*) End result of the joint rotations after the first movement with the distorted visual feedback. At the end of the movement, the camera sees the arm (black) miss the target to the left. The numbered arrows correspond to the mappings that need realignment due to the mismatch (see figure 14.1). (*C*) After adaptation to the visual distortion, both the estimated gripper location and the difference vector in joint coordinates have changed. The plot below the robot has the format of A. Right: The original rotation in joint-centered coordinates and the new one, after prism adaptation. Left: The difference vector in camera-centered coordinates. (*D*) Arm and gripper location at the end of the reaching movement after adaptation. (*E*) Presentation of the target on the first trial after removal of the prism. (*F*) The resulting reaching movement displays the aftereffect of adaptation.

Figure 15.4C shows the gripper's actual location in gray, and the dotted line shows its estimated location in camera-centered coordinates from "proprioception." The system estimates the gripper's location to the left of its location in "real" space because of the change in the location map. It computes a difference vector x_{dv} and maps it to joint rotations. Before the prisms were inserted, this x_{dv} required both a shoulder flexion and an elbow extension. Now, because of the error in the last movement, the controller produces a smaller flexion in the shoulder. As the trials continue, the shoulder flexion becomes small enough so that along with the elbow extension, the gripper reaches the target. When the robot performs these joint rotations, the gripper moves to the gray location in figure 15.4D, straight ahead, and the camera records it at the black location. In both "real" and camera-centered coordinates, the gripper acquires the target. Moving straight ahead produces success, because the target in fact lies straight ahead. After adaptation, the somewhat leftward target maps to joint rotations generating movements straight ahead. Importantly, the trials change both $\hat{\theta} \leftrightarrow \hat{x}_{ee}$ and $x_{dv} \overset{\hat{\theta}}{\leftrightarrow} \Delta\hat{\theta}$.

Figure 15.4E illustrates what happens if you remove the prism and require the robot to make another reaching movement from the same starting location. The camera senses the target location at x_t, straight ahead in real space. However, the controller estimates gripper location to the left of its actual location. When the robot's controller computes x_{dv} and maps it to joint rotations, the system now estimates that to move the gripper to a target appearing straight ahead, it needs to extend the elbow according to the mappings it learned when the prism distorted the camera's input. This computation results in a movement away and to the right. This mapping causes the gripper to miss the target to the right: the *negative aftereffect* of adaptation (figure 15.4F).

The present model of prism adaptation has many similarities to a previous one,[13] but differs in that the former emphasized representations in body-centered coordinates, in contrast to the present emphasis on fixation-centered coordinates for representation of the target, end effector, and difference vector.

15.6 Consequences of Planning in Vision-Based Coordinates

The computations elaborated in the virtual robot have a number of interesting consequences, many of which have been tested in psychophysical experiments.[14] For example, after participants in those experiments adapted to prisms, when they closed their eyes and tried to point straight ahead, they showed aftereffects. These aftereffects arose because the participants imagined target $\hat{x}_t$ at the same location (in fixation-centered space) as before adaptation, but displacement map $x_{dv} \overset{\hat{\theta}}{\leftrightarrow} \Delta\hat{\theta}$ had changed. Further, when participants prism-adapted for reaching to a spatial auditory stimulus, and later removed the prisms, their movements to visual targets showed aftereffects for the same reason. Finally, if participants prism-adapted with their right hand, they showed misreaching due to aftereffects when they tried to reach to their right hand with their left hand. That is, their estimate of the location of the right hand $\hat{\theta} \leftrightarrow \hat{x}_{ee}$ had changed.

Because planning of reaching movements depends on estimating end-effector location in fixation-centered coordinates, it would make sense that the actual sight of the hand has more influence over adaptation than the sight of something that represents hand location. That is, reaching movements should adapt more rapidly if you can see your hand, as opposed to seeing a video image of your hand. Indeed, experiments have confirmed this prediction,[15,16] and section 25.5.1 resumes a discussion of this topic.

Another idea captured by the virtual robot involves the timing of visual input. Shegiru Kitazawa and Ping-Bo Yin varied the time of end-effector feedback during adaptation to wedge prisms in monkeys.[17] They found adaptation and large aftereffects as long as the feedback signaling a mismatch between the end effector and the target arrived within 10 msec or so. Longer delays resulted in significantly less adaptation. A delay of about 100 msec reduced the amount of recalibration by half, and a delay of 500 msec almost eliminated it.

Given the importance of visual inputs to reaching, you might wonder what would happen in a youngster who had never seen his or her hands. In the 1970s Dick Held and Joseph Bauer[18] examined this issue in monkeys. They reared monkeys in the dark so that they could not see their hands or body during the first month of life. Monkeys lacking experience with visually guided movements did not reach accurately even when they could later see perfectly normally, but Held and Bauer found this phenomenon difficult to quantify. To examine it more carefully, they seated each monkey inside a box, with its head outside. For a control group, a clear box allowed vision of the hand, but for the visually deprived group, an opaque box prevented such feedback. Each monkey could, however, see a stick of candy above the box and, by pulling a ball inside the box, it could move the candy to its mouth and lick it. Later, all the monkeys were tested in clear boxes, but only the control monkeys had had substantial previous experience with moving the ball-and-stick assembly when they could see their hands. Control monkeys could reach to the ball very easily. Monkeys that had reached one month of age without seeing their hands, however, did poorly. Nevertheless, with practice, the visually deprived monkeys learned to grasp and pull the ball.

Held and Bauer then tested spatial generalization. Control monkeys generalized well: They could see the stick of candy to the left or right of the initial training location and adjusted their reaches accordingly. Visually deprived monkeys did not guide their reaches according to the location of the candy, but instead groped for the ball much as blind animals do. Held and Bauer[18] concluded that "the deficiencies of the [visually deprived] animals result from the absence of a calibration system for visual guidance of the direction of reach." These data reinforce the idea that control of reaching movements involves vision of the hand and alignment of that information with proprioceptive input.

15.7 Moving an End Effector Attached to the Hand

The framework described in figure 14.1 deals with the problem of moving your hand to a target; it also applies to moving something attached to or grasped by your hand, such as a computer mouse or a stick. You need to consider these conditions for at least two reasons. First, you often hold tools in your hand with the goal of placing some part of that tool at a target location. For example, imagine holding a hammer and using it to pound a nail. In this situation, the nail head serves as the target and the hammer's head as the end effector. Second, with the development of low-cost video displays, much of the research on motor control involves cursors on video monitors. In these tasks, the participant holds a **manipulandum** and a computer translates the motion of that device into cursor movement. The person or monkey moving the manipulandum often cannot see the hand or any part of the arm. Consider how reaching movements under these kinds of conditions might be related to the "classical" situation in which the participant has direct visual feedback from the hand during reaching.

To do so, return to the imaginary robot, but now imagine that the gripper holds a tool and, as the engineer in charge, you want to program the robot to move the tool to a target location. The vector x_{dv} that you need to compute describes a displacement for the end of the tool, not the gripper. To compute this vector, you need to know the location of the tool x_{ee} and the location of the target x_t in camera-centered coordinates. The computations that the robot's controller needs to perform differ little from what it used to move the robot's gripper, but the controller can simply replace gripper location with tool location.

According to the model in figure 14.1 (and to others, as well[19]), to use a tool, the controller needs to learn two kinds of kinematic maps: a location map $\hat{\theta} \leftrightarrow \hat{x}_{ee}$ that relates joint-angle signals to the location of the tool in fixation-centered coordinates, and a displacement map $x_{dv} \stackrel{\hat{\theta}}{\leftrightarrow} \Delta\hat{\theta}$ that relates motion of the tool to joint rotations. When the controller learns these maps, it can generate a plan for reaching with the tool. Notice that the controller need not compute the location of the gripper, nor need it align vision of the gripper with joint-angle signals. Indeed, for using the tool, the camera must record its location, not that of the gripper.

Now imagine that you hold a computer mouse in your hand and wish to move a cursor on a video screen. This skill takes practice. The motion of the mouse clearly relates to the motion of the cursor: The cursor has become the end effector. Joint rotations clearly affect the movement of the cursor. Thus, you need to learn a displacement map $x_{dv} \stackrel{\hat{\theta}}{\leftrightarrow} \Delta\hat{\theta}$ for this new end effector. For example, the displacement map specifies that in order to move the cursor to the right while holding the mouse with your right hand, you could extend your elbow. (You probably use your wrist more, but you could use your forearm to move the cursor while keeping your wrist stiff.) To move the cursor up, you flex your shoulder and extend your elbow. Therefore, to move the cursor to a target, you need to use vision to estimate cursor location on the screen $\hat{x}_{ee}$, subtract that from the target location $\hat{x}_t$, and arrive at a desired cursor difference vector $\hat{x}_{dv}$. This estimated displacement then maps to joint rotations $x_{dv} \overset{\hat{\theta}}{\leftrightarrow} \Delta\hat{\theta}$. Thus, to control the cursor, you must learn the displacement map $x_{dv} \overset{\hat{\theta}}{\leftrightarrow} \Delta\hat{\theta}$, computed by network 4 in figure 14.1.

However, learning the location map $\hat{\theta} \leftrightarrow \hat{x}_{ee}$, network 1 in figure 14.1, plays a much less important role in using a mouse. The location of the cursor has no fixed relationship to the location of your hand. Consider that you can pick up the mouse and place it somewhere else, causing a large change in proprioceptive signals, without seeing much change in cursor location on the screen. Therefore, you should not expect a computation based on joint angles (or muscle lengths) to yield the location of the cursor. Network 1 and its forward kinematics are not so important for "mousing." Instead, you need to rely almost entirely on vision to determine the cursor's location. *Changes* in joint angles do cause *movements* of the cursor, however, so network 4 plays a large role in "mousing."

15.8 Internal Models of Kinematics

Up to this point, the model in figure 14.1 presents a framework for understanding the relations among kinematic variables and reference frames, and does so in terms of *mappings*, computed by neural networks, which bring kinematic variables into *alignment* though changes in the networks. You can also think of each neural network in the model, such as network 1, as several partially overlapping networks rather than a single network. The mappings depend on the weight matrix of the network, and the network has an attractor (i.e., its activity tends toward a stable state that depends on its weight matrix). If you think of several partially overlapping networks, you can also think of several attractors which correspond to different skills and which can be instantiated by different contextual inputs.

When you switch between various tools, for example, their different shapes can radically change the relationships among the kinematic variables in the location and displacement maps. (The dynamics also differ, often much more than the kinematics, but neglect that for now.) Someone who can wield a hammer skillfully has learned a location map and a displacement map for that hammer (i.e., adaptive weight matrices for networks 1 and 4). The CNS of a skilled carpenter can accurately compute the joint rotations needed to put the end effector (the head of the hammer) on the target (the head of the nail). A novice, on the other hand, has a significant chance of smashing his or her hand rather than hitting the nail on the head.

You can think of the location and displacement maps for a particular end effector (e.g., a hammer) as a kind of internal model of kinematics that your CNS learns for that end effector. When you switch end effectors, your CNS must learn to *de-instantiate* one internal model and *instantiate*

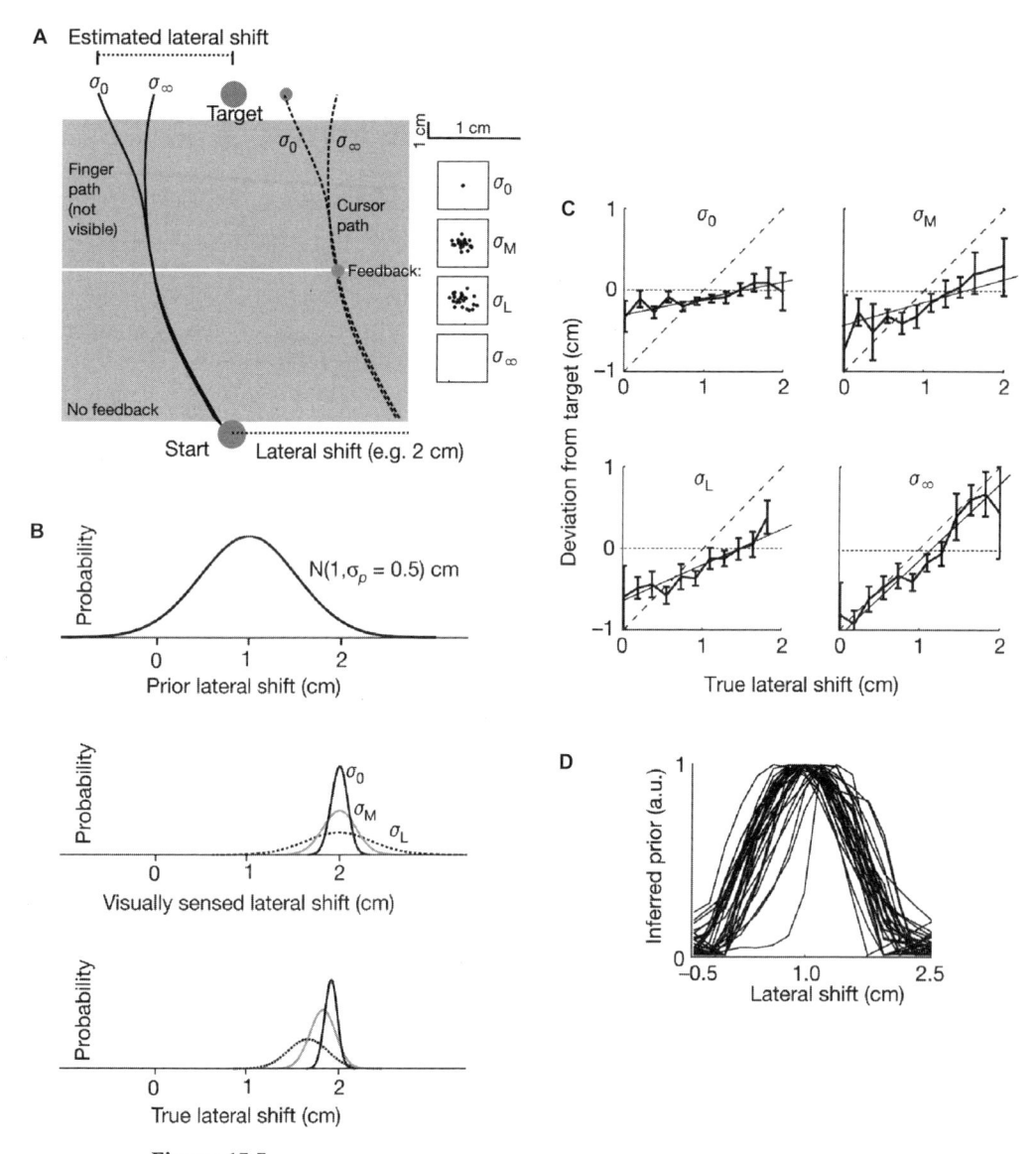

290   CHAPTER 15. Aligning Vision and Proprioception I: Adaptation and Context

Figure 15.5
Response to visual feedback depends on both the noise in the feedback and its prior likelihood. (*A*) As the participant's finger moves from the starting circle, a cursor is extinguished and shifted laterally. Halfway to the target, feedback is briefly provided clearly (σ_0), with different degrees of blur (σ_M and σ_L), or withheld entirely (σ_∞). The paths illustrate typical trajectories for a displacement of 2 cm. (*B*) Top subplot: For 1000 trials, participants trained with the illustrated distribution of lateral shifts, i.e., a Gaussian distribution with a mean of 1 cm. This distribution constitutes the participants' prior knowledge, $P(x_{ee})$. Middle subplot: A diagram of various probability distributions associated with the current evidence, i.e., $P(\hat{x}_{ee}|x_{ee})$. This distribution is shown for the clear feedback condition and for the two blurred feedback conditions. On the trial illustrated, the true shift

another.[20] The term "internal model" thus refers to networks in your CNS that relate the variables of kinematics and dynamics. These networks compute estimations and mappings to overcome noise and perform coordinate transforms.

Your CNS can learn and store multiple internal models. For example, in the dove-prism experiments described in this chapter, participants took weeks to adapt their reaching, but only days to "de-adapt" to their removal. When they started coping with the prisms, the participants presumably had an internal model of the kinematics appropriate for unperturbed vision. Over the long period of practice with the prisms, however, they learned a new internal model specific to the prisms. In a sense, they learned a prism skill. When the participants removed the prisms, they retained the internal model for unperturbed vision, and therefore could switch rapidly between the prism skill and skills for everyday life. What about misalignments caused by noise rather than perturbed vision?

15.9 Estimate of Limb Location Is Influenced by the Likelihood of the Sensed Variables

Return to the example of the robot in figure 15.1. What happens when the camera reports one location for the end effector but the joint sensors reports a different location? If the two inputs are close, the pattern-completion network will interpolate between the two estimates and form a new estimate somewhere in between. If the two inputs differ a lot, it will effectively reject one of the estimates and converge toward the estimate provided by the other (i.e., the winner takes all). When should the network interpolate, and when should it ignore one of the sensed variables? The answer should depend on how confident the network is regarding the estimate provided by each sensory system, that is, their likelihood.

For example, assume that the camera imaged the end-effector at location $\hat{x}_{ee}$, and the encoders sensed joint angles at $\hat{\theta}$. Both sensors are noisy, so their estimates are likely to be different from the true values, x_{ee} and θ. The noise in each sensor can be represented as a probability distribution. For the camera, you have the **conditional probability** $P(\hat{x}_{ee}|x_{ee})$, and for the joint sensors you have $P(\hat{\theta}|\theta)$. Clearly, if noise in one sensor

Figure 15.5 (continued)
was 2 cm. Bottom subplot: The estimate of lateral shift, i.e., $P(x_{ee}|\hat{x}_{ee})$, for an optimal observer that combines the prior knowledge with the evidence. (C) The lateral deviation of the cursor at the end of the reach as a function of the imposed lateral shift for a typical participant. The dotted horizontal line at 0 indicates full compensation for the observed error. The dashed diagonal line indicates complete neglect of the observed error. The solid line is the Bayesian model, with the level of uncertainty fitted to the data. (D) The inferred prior knowledge for each participant and condition. B shows the true distribution. a.u., arbitrary units (From Kording and Wolpert[21] with permission.)

exceeds that in the other, then the estimate of end-effector location should depend more on the reliable sensor.

Because the two sensors simultaneously report the state of the robot, the network also knows which pairs of values tend to occur together, and which pairs almost never do. You can represent this information as a joint probability distribution $P(\hat{x}_{ee}, \hat{\theta})$. Finally, the true location of the end effector is likely to be somewhere near the center of the workspace, and rarely near the edges. You can represent this "prior" knowledge as the probability distribution $P(x_{ee})$. **Bayesian theory** suggests that an optimal estimate of end-effector location should take into account all of the above probabilities:

$$P(x_{ee}|\hat{x}_{ee}, \hat{\theta}) = \frac{P(x_{ee})P(\hat{x}_{ee}|x_{ee})P(\hat{\theta}|\theta)}{P(\hat{x}_{ee}, \hat{\theta})}.$$

Does the CNS employ such probabilistic approaches to estimate hand location during a reach? Recently, Konrad Kording and Daniel Wolpert[21] designed an elegant experiment to test this idea. They first trained participants with a visual display that provided feedback for only one hand location during the reach. At that location, visual feedback was 1.0 cm to the right of the actual hand location. This manipulation meant that after training, $P(x_{ee})$ should have been centered 1.0 cm to the right. Next, the experimenters manipulated the amount of noise in the visual feedback (figure 15.5A). For example, when the visual display had a large noise, $P(\hat{x}_{ee}|x_{ee})$ had a large variance around its mean. Kording and Wolpert predicted that the participants would respond to the noisy visual feedback based on both the prior distribution that they had learned during the initial training and the visual information they received:

$$P(x_{ee}|\hat{x}_{ee}) = \frac{P(x_{ee})P(\hat{x}_{ee}|x_{ee})}{P(\hat{x}_{ee})}.$$

For example, when the feedback suggested that the hand was 2.0 cm to the right, but it was very noisy, then the participants were more likely to rely on the prior knowledge, which predicted a smaller displacement than observed. If the feedback was noise free, participants tended to discount the prior knowledge and responded instead to the observed displacement (figure 15.5B). Thus, as the visual information became less certain (because of noise), the participants relied more on their prior knowledge (figure 15.5C), and as the visual information became clear, they relied more on what they observed. The experimenters estimated the distribution of the prior knowledge (figure 15.5D), which was centered on the training displacement of 1.0 cm.

In summary, it seems that the CNS combines visual and proprioceptive information from the arm to produce an estimate of end-effector location. The computation that produces this estimate takes into account the noise in the sensors. It also takes into account prior experience. The neural mechanisms for this computation remain an open question.

References

1. Deneve, S, Latham, PE, and Pouget, A (2001) Efficient computation and cue integration with noisy population codes. Nat Neurosci 4, 826–831.
2. Sperry, RW (1943) Effect of 180 degree rotation of the retinal field on visuomotor coordination. J Exp Zool 92, 263–279.
3. Sugita, Y (1996) Global plasticity in adult visual cortex following reversal of visual input. Nature 380, 523–526.
4. Aksay, E, Baker, R, Seung, HS, and Tank, DW (2000) Anatomy and discharge properties of pre-motor neurons in the goldfish medulla that have eye-position signals during fixations. J Neurophysiol 84, 1035–1049.
5. Zipser, D, Kehoe, B, Littlewort, G, and Fuster, J (1993) A spiking network model of short-term active memory. J Neurosci 13, 3406–3420.
6. Durstewitz, D, Seamans, JK, and Sejnowski, TJ (2000) Neurocomputational models of working memory. Nat Neurosci 3 Suppl, 1184–1191.
7. Renart, A, Song, P, and Wang, XJ (2003) Robust spatial working memory through homeostatic synaptic scaling in heterogeneous cortical networks. Neuron 38, 473–485.
8. Shadmehr, R, and Wise, SP (2004). In The New Cognitive Neurosciences, 3rd ed, ed Gazzaniga M, (MIT Press, Cambridge MA).
9. Sekiyama, K, Miyauchi, S, Imaruoka, T, et al. (2000) Body image as a visuomotor transformation device revealed in adaptation to reversed vision. Nature 407, 374–377.
10. Martin, TA, Keating, JG, Goodkin, HP, et al. (1996) Throwing while looking through prisms. II. Specificity and storage of multiple gaze-throw calibrations. Brain 119, 1199–1211.
11. Hallett, M, Pascual-Leone, A, and Topka, H (1996). In The Acquisition of Motor Behavior in Vertebrates, eds Bloedel, JR, Ebner, TJ, and Wise, SP (MIT Press, Cambridge MA), pp 289–301.
12. Kurata, K, and Hoshi, E (1999) Reacquisition deficits in prism adaptation after muscimol microinjection into the ventral premotor cortex of monkeys. J Neurophysiol 81, 1927–1938.
13. Welch, RB, Choe, CS, and Heinrich, DR (1974) Evidence for a three-component model of prism adaptation. J Exp Psychol 103, 700–705.
14. Harris, CS (1965) Perceptual adaptation to inverted, reversed, and displaced vision. Psychol Rev 72, 419–444.
15. Norris, SA, Greger, BE, Martin, TA, and Thach, WT (2001) Prism adaptation of reaching is dependent on the type of visual feedback of hand and target position. Brain Res 905, 207–219.
16. Clower, DM, and Boussaoud, D (2000) Selective use of perceptual recalibration versus visuomotor skill acquisition. J Neurophysiol 84, 2703–2708.
17. Kitazawa, S, and Yin, PB (2002) Prism adaptation with delayed visual error signals in the monkey. Exp Brain Res 144, 258–261.
18. Held, R, and Bauer, JA (1974) Development of sensorially-guided reaching in infant monkeys. Brain Res 71, 265–271.
19. Bullock, D, Grossberg, S, and Guenther, FH (1993) A self-organizing neural model of motor equivalent reaching and tool use by a multijoint arm. J Cog Neurosci 5, 408–435.

20. Haruno, M, Wolpert, DM, and Kawato, M (2001) Mosaic model for sensorimotor learning and control. Neural Comput 13, 2201–2220.
21. Kording, KP, and Wolpert, DM (2004) Bayesian integration in sensorimotor learning. Nature 427, 244–247.
22. Martin, TA, Keating, JG, Goodkin, HP, et al. (1996) Throwing while looking through prisms. I. Focal olivocerebellar lesions impair adaptation. Brain 119, 1183–1198.

Reading List

The two papers by Tom Thach and his colleagues,[10,22] provide a good starting point for exploration of this literature.

16 Aligning Vision and Proprioception II: Mechanisms and Generalization

Overview: This chapter examines some of the neural systems involved in adapting and learning the alignments between visual and proprioceptive information and some of the consequences of their mechanisms.

16.1 Neural Systems Involved in Adapting Alignments Between Proprioception and Vision

Current evidence points to three regions of the CNS as crucially involved in adapting to misalignments in visual and proprioceptive information: PMv, the cerebellum, and PPC.

16.1.1 PMv Deactivation Prevents Adaptation

Recall the experiment of Kurata and Hoshi, described in section 15.4. In their experiment, small amounts of muscimol were injected into PMv or PMd as monkeys adapted to the prisms. An **agonist** of GABA, muscimol inactivates the injected area. In effect, it causes a temporary lesion. The experimenters observed that their small injections did not significantly affect the reaction time or movement time of the monkey's reaching movements, nor did they change the accuracy of the reaching movements in the baseline (no prism) condition. Interestingly, muscimol injection into the left PMv affected only adaptation to the prism that shifted the visual image to the left. These results suggest that the PMv contributes to the neural network that performs the alignment of x_{dv} with $\Delta\hat{\theta}$, but other regions of the CNS may also contribute to this computation. Note that this view accords with the findings presented in section 14.1, which indicate that cells in PMv encode the direction and distance of the target from the end effector with activity fields that move with the hand,[1] and remain invariant to changes in hand orientation[2] and which hand a monkey uses to reach.[3] But why did inactivation of the left PMv affect only adaptation to the leftward-shifting prisms?

It turns out that the representation of space to the right of the fixation point plays a crucial role in adapting to a left-shifting prism. To understand this idea, reexamine the virtual robot with a prism in front of its camera (figure 15.4). At the end of the reach in figure 15.4B, the camera sensed the gripper's location. The left-shifting prism caused the gripper to miss to the left, which placed the real target to the right of the gripper at the movement's termination. Now imagine yourself as the reaching agent. You will see the end effector (your hand) and the target in their *viewed* location. When you look at the end effector, you must represent space to the right of your hand, and therefore to the right of your fixation point, to note the location of the target. If you could not represent space to the right of your fixation point, then you could not accurately register the error in the movement, which should slow adaptation. Is there any part of the CNS that specifically represents space contralateral to the fixation point?

16.1.2 Spatial Neglect

David Gaffan and Julia Hornak[4] performed a study that illustrates the role of posterior parts of the cerebral cortex in representing space contralateral to the fixation point. Their data also indicate the importance of communication between these posterior cortical areas and the frontal cortex. In their experiment, they studied the ability of monkeys to reach to an arbitrarily designated "correct" stimulus on a video screen that displayed five objectlike stimuli horizontally. For example, a scarlet letter A might serve as the correct target among five stimuli on the screen. By touching the correct stimulus, the monkey received a reward. On each trial, the correct stimulus appeared in a different location.

Gaffan and Hornak damaged the brain in various ways and measured visual **neglect**. In humans, neglect usually occurs after damage to the right hemisphere. Such patients tend to ignore information on the left side of space. For example, when asked to copy a drawing of a symmetrical object such as a butterfly, they might not draw the left half. When these patients look at the correct drawing, they tend not to fixate its left parts.

To evaluate neglect in monkeys, Gaffan and Hornak measured the distribution of errors that the monkeys made among five locations on the screen. Normal monkeys picked the correct stimulus about 90% of the time, with their errors distributed equally among the five locations. Gaffan and Hornak asked: If one side of the brain was damaged, say the right hemisphere or the right optic tract, did the monkey have a higher tendency to pick a stimulus on the right side of the screen, as if it neglected the left side (see figure 16.1)? If so, then the errors would tend to cluster toward stimuli on the right.

Figure 16.1A shows the distribution of errors that they observed after the lesions. Most lesions did not yield any difficulties with representing stimuli in contralateral space. For example, Gaffan and Hornak sectioned the optic tract on one side of the brain, and that had no effect. They also sectioned the commissures that connect the two cerebral hemispheres, the corpus callosum and the anterior commissure, without effect. Likewise, lesions of the PPC on one side caused no deficit. However, when

16.1. Neural Systems Involved in Adapting Alignments

Gaffan and Hornak combined unilateral optic tract section with section of the anterior commissure and corpus callosum, they found that the monkeys rarely responded to stimuli contralateral to the hemisphere with the optic-tract lesion. The same kind of neglect occurred after disconnecting the posterior cortical areas (much of the visual cortex) from the more rostral parts of the brain (including frontal cortex) in one hemisphere. Gaffan and Hornak concluded that the cortex of each hemisphere maintains a fixation-centered representation of the visible half-world contralateral to the current fixation point. That is, the left hemisphere maintains a representation of the space to the right of fixation, and the right hemisphere maintains a representation of the space to the left of fixation. When a hemisphere does not receive visual information from that half of the world, the monkey tends to ignore it.

These data deserve detailed consideration. To give them their due, recognize that even the most experienced experts resort to shortcuts to describe and understand neglect. You should not feel frustrated if you get somewhat lost during repeated left–right inversions as the discussion swings from contralateral to ipsilateral, contralesional to ipsilesional, and so forth. As shorthand for keeping these notions straight, think of "good" sides and "bad" sides. The bad side, of course, suffers the brunt of the damage. When Gaffan and Hornak sectioned the right optic tract, the "bad" right hemisphere could not receive information from any part of the retina to the right of the fovea, called the right *hemiretina*. The right hemiretina normally receives light from the part of the visual field to the left of the fixation point, called the left *hemifield*. Therefore, consider the left hemifield "bad" and the right hemifield "good." When the monkeys fixated a point, this lesion prevented the right hemisphere from "viewing" the left hemifield.

Note that the experimenters did not require the monkeys to maintain fixation on any particular point. They could move their eyes around, and presumably did so. By moving their eyes to the very left part of the screen, the monkeys could put the right part of the screen in the "good" hemifield. The "good" hemifield sends information to the "good" left hemisphere, and the left hemisphere projects—via the forebrain commissures—to the "bad" right hemisphere. When monkeys returned to a central fixation point, near the center of the screen, the left part of the screen returned to the "bad" visual hemifield. Through the commissural connections, however, the "bad" right hemisphere could construct a representation of space to the left of the fixation point (i.e., the "bad" visual hemifield). The results of Gaffan and Hornak show that the corpus callosum or anterior commissure must have transferred the *memory* of what had been on the left part of the screen to the "bad" right hemisphere from the "good" hemisphere, once the eyes reoriented and the left part of the screen returned to the "bad" hemifield.

A recent imaging experiment illustrates this point in humans. Pieter Medendorp, Doug Crawford, and their colleagues[5] asked participants to fixate a central point and flashed a target on either the left or the right side of the fixation point ("delayed pointing" task in figure 16.1B). After a

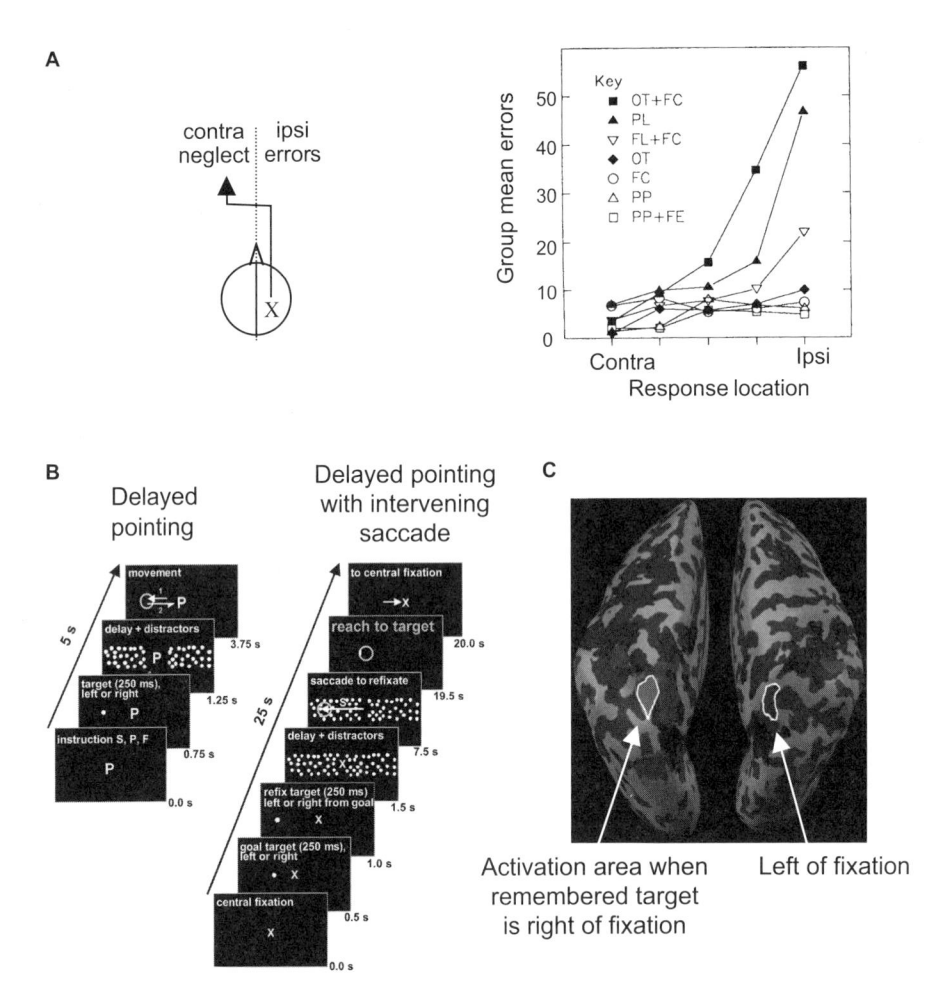

298   Chapter 16. Aligning Vision and Proprioception II

Figure 16.1
Representation of contralateral visual space in humans and monkeys. (*A*) Spatial neglect in monkeys. Left: In the plot, ipsilateral (ipsi) and contralateral (contra) are defined relative to the hemisphere lesioned (X). The experimenters predicted that for left-side neglect (contra), the monkey should choose targets to the right (ipsi), in error. Right: Errors as a function of screen location. Lesions: PP + FE, posterior parietal cortex plus frontal eye field (FEF) ablation; PP, posterior parietal cortex ablation; FC, forebrain commissurotomy, which involves transection of both the corpus callosum and the anterior commissure; OT, optic tract section; FL + FC, frontal lobe removal plus forebrain commissurotomy; PL, parietal leucotomy, the unilateral section of the white matter between the intraparietal sulcus and the lateral ventricle, thus disconnecting posterior (visual) cortex and frontal cortex; OT + FC, optic tract section plus forebrain commissurotomy. (*B*) Visual displays for studying fixation-centered updating of visual representations due to intervening eye movements. The participant fixates a central location, where a letter indicates the task to perform (S, saccade to a target; P, point to a target; or F, continue fixating the center of the screen). For the "delayed-pointing" task, the target

delay period, the participants reached to the remembered location of the target. They found that if the target was to the left of fixation, the right PPC was activated, and vice versa (figure 16.1C). In a subsequent experiment ("delayed pointing with intervening saccade" task in figure 16.1B), the participants fixated a point when the movement target was flashed to one side, say the right side of fixation, thus activating the left PPC. However, before reaching, a second fixation point appeared, so that now the remembered target fell on the left side of the fixation point. As the saccade moved the fovea from one fixation point to another, the remembered location of the target moved from the right side of the fixation point to the left side. Coincident with this, the activity in PPC shifted from the left hemisphere to the right. Therefore, despite the fact that no target was visible, movements of the eyes caused a remapping of the remembered reach target with respect to the fixation point. This finding illustrates that reach targets are encoded as a vector with respect to the fixation point.

With this finding in mind, you can interpret the Gaffan and Hornak result to mean that cutting the forebrain commissures prevents the right hemisphere from receiving this remapped memory, thus causing a left-side neglect. Gaffan and Hornak also found that disconnecting the posterior visual areas from the frontal lobe caused a contralateral neglect, which suggests that a representation of contralateral space in fixation-centered coordinates requires the interaction of posterior and frontal cortical areas. Recall that left PMv inactivation blocks adaptation for leftward light shifts, but not for rightward shifts.[6] The property of representing space contralateral to the fixation point might account for this specificity of the PMv-inactivation effect.

16.1.3 Effect of Prism Adaptation on Neglect

Yves Rossetti and his colleagues[7] wondered whether prism adaptation could ameliorate neglect. Their patients had right hemisphere damage and neglected their left visual field. When told to point straight ahead with blindfolds on, they pointed about 30° to the right of midline (figure 16.2A). When asked to copy a picture, they tended to ignore the left part of the picture (figure 16.2C and D).

Figure 16.1 (continued)
appears and, after a delay, the participant points to the remembered location of the target. In the "delayed-pointing-with-intervening-saccade" task, the fixation point changes locations before the pointing takes place. (C) The PPC region that shows activity (i.e., changes in blood flow) switches hemispheres when eye movements place the remembered target to the left of the fixation point, as opposed to when they place it to the right of that point. The "activated" regions appear as enclosed areas on an "inflated" map of the cortex. (A from Gaffan and Hornak,[4] B and C from Medendorp et al.[5] with permission.)

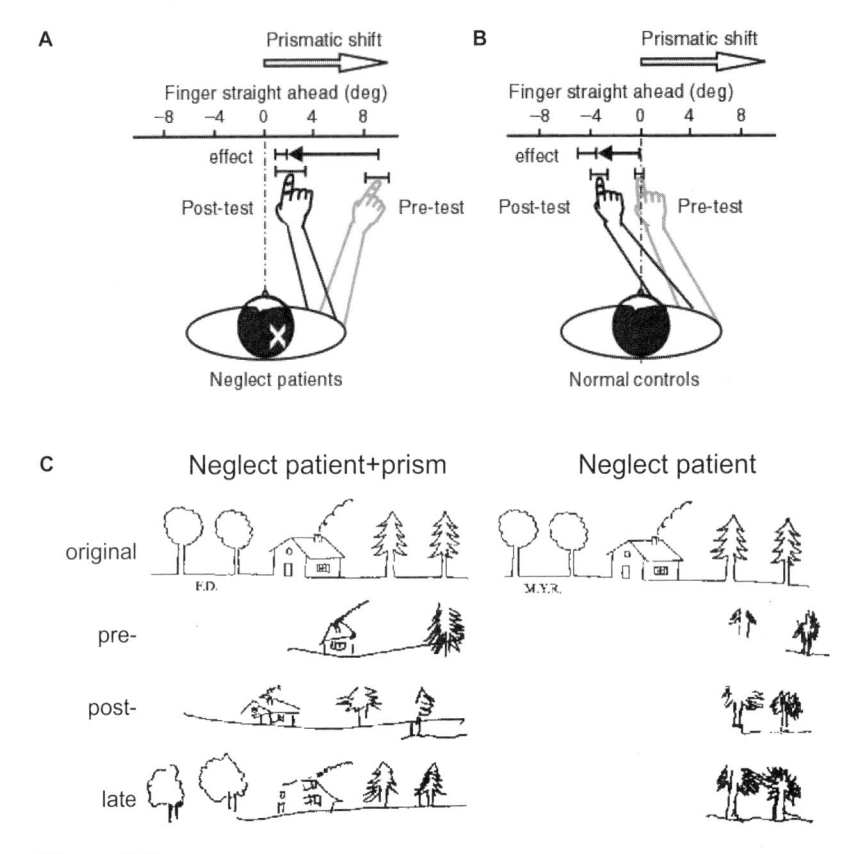

Figure 16.2

Prism adaptation in neglect patients. (A) The experimenters asked blindfolded neglect patients to point straight ahead in a pre-test condition (before treatment with prism glasses) and in a post-test condition (after making 50 reaching movements while looking through right-shifting prisms). In the pre-test condition, the patients' sense of their midline was 30° or so to their right, reflecting left-side neglect. After wearing the prisms, the neglect patients had a large aftereffect, which placed their sense of midline closer to its actual location. (B) As in A, but for healthy participants. Their aftereffects were smaller than those for neglect patients. (C and D) The experimenters gave some neglect patients prism treatment, and others acted as patient controls. Before treatment, each participant attempted to copy the templates on the top line. Examples of performance from each group appear beneath the template. Immediately after prism adaptation, patients drew more of the figures. Two hours after prism adaptation, the improvement in the patients' condition remained evident. (From Rossetti et al.[7] with permission.)

In this experiment, some of these patients wore prisms that shifted the visual field ~10° to the right. A series of targets of reaching movements appeared 10° to the right or left of the body midline, and the experimenters asked the participants to point rapidly to each target. When they reached, their hand ended up to the right of the target. Therefore, when they looked at their hand, the target appeared to the left of fixation. From the experiment of Kurata and Hoshi (section 16.1.1), you would guess that adapting to these errors required the right PMv. Therefore, in adapting to the prism, the patients were engaging the side of the cortex that had been damaged. After adapting to the prism (~5 min), the participants were blindfolded and asked to point straight ahead. Following the adaptation, both the patients and a control group had "straight-ahead" pointing movements that had shifted to the left (figure 16.2A and B). To account for this, consider that a rightward prism realigned the location map so that the estimate of hand location in fixation-centered coordinates shifted to the right. A misestimation of hand location to the right resulted in movements that missed the target to the left. Without vision, the patients imagined a target straight ahead, and their reach pointed to the left of the midline. Because neglect patients tended to point to the right of their midline, this aftereffect of prism adaptation improved the patients' performance.

In a second experiment, the experimenters investigated whether prism adaptation affected other spatial abilities. Before the 5-min adaptation session with the right-shifting prisms, they tested the patients on a battery of tasks, then repeated the tasks immediately after adaptation and again two hours later. The participants tried to copy the figures shown on the top line of figure 16.2C and D. Before wearing prisms, both the group of neglect patients with prism treatment and a control group of neglect patients copied only the items on the right side of the template. The patients with prism treatment showed improvement once they removed their prisms. Remarkably, this improvement increased two hours after the treatment. More recent experiments have shown that the improvement lasts 4 days after treatment. These results indicate that the process of prism adaptation, long considered a relatively passive process involving only recalibration of kinematic mappings, can also affect higher levels of spatial representation.

16.1.4 Damage to the Cerebellum Prevents Adaptation

Examination of patients with damage to the cerebellum and its pathways suggests that in addition to PMv, the cerebellum plays a crucial role in prism adaptation. Recall that, like other motor areas, PMv has reciprocal connections with parts of the motor thalamus that receive inputs from the cerebellum. It also sends signals to the cerebellum via the basilar pontine and inferior olivary nuclei (ION).

Mark Hallett and his colleagues[9] first showed that patients with cerebellar damage have deficits in adapting to prisms. Similarly, in monkeys,

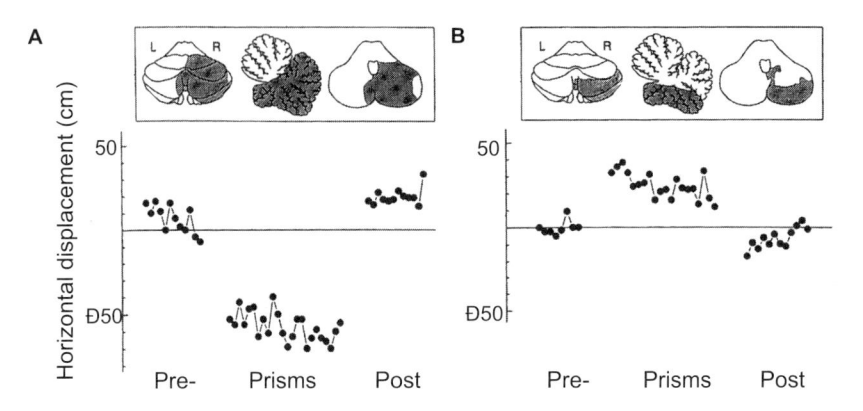

Figure 16.3
Inability to adapt to prisms in two patients (*A* and *B*) with damage to the ventral and lateral parts of cerebellar cortex (top). The plots (bottom) show throwing errors on consecutive trials, before (pre-), during (prisms), and after (post) wearing prisms. Format as in figure 15.3A. (From Martin et al.[12] with permission.)

lesions of the posterior lobe of the cerebellum cause impairments in prism adaptation.[10] In accord with these findings, Mistuo Kawato and his colleagues demonstrated that there were changes in cerebellar activations as people learned how to move a video cursor with a computer mouse.[11]

Tom Thach and his colleagues[12,13] confirmed and extended these results by examining a large group of patients with damage to the cerebellum and its pathways. As described in chapter 15, these experimenters asked the patients to throw clay balls at a target before, during, and after wearing wedge prisms. Many of the patient groups failed to adapt to the prisms, as illustrated in figure 16.3. Note that when the patients wore the prisms, they made throwing errors of over 50 cm. Importantly, after the prisms came off, many patients showed no aftereffects: They had not adapted, so no "de-adaptation" occurred.

The work of Thach's group[12] pointed to the importance of climbing-fiber inputs from the contralateral ION and mossy-fiber inputs from the contralateral basilar pontine nuclei. (Chapters 23 and 24 present the idea that climbing fibers convey an error signal to the cell bodies of Purkinje cells, whereas mossy fibers excite cells that make excitatory synapses on Purkinje cells.) Damage to the ION and the basilar pontine nuclei resulted in abnormal adaptation when patients threw with the limb contralateral to the lesion. The other arm in the same patients showed normal adaptation to the prisms. According to these data, diseases of the ION resulted in spared performance, as indicated by a normal scatter of thrown balls, but impaired adaptation. Patients with damage to the output system of the cerebellum (e.g., the dentate nucleus), or its chief targets in the thalamus, showed normal prism adaptation, although they had other motor disabilities that caused poor throwing performance.

16.1.5 Adaptation Activates PPC

In addition to the PMv and the cerebellum, the PPC probably plays an important role in the adaptation of kinematic maps. Dottie Clower and her colleagues[14] found that a part of PPC showed altered blood flow when human participants adapted to prisms. As in typical prism-adaptation experiments, their participants reached to a target under visual control and adapted to the 17° prismatic shifts in five to ten reaches. During the neuroimaging experiments, the participants underwent continuous adaptation due to repeated reversals of the prisms. In a control task, the target shifted before participants could complete a reaching movement. To a first approximation, this procedure controlled for the errors in reaching during prism adaptation and the participants' reactions (emotional and otherwise) to those errors.

A region of PPC contralateral to the reaching limb increased its blood flow during adaptation. (The precise area, in terms of maps such as that shown for monkeys in figure 6.3, was uncertain, although Clower et al. suggested, indirectly, that it corresponded to VIP.) This region might play a role in realigning visual and proprioceptive inputs, although it could simply record the misalignment without contributing to adaptation. On the latter view, this part of PPC might play a role in perceiving misreaching.

16.2 Generalization of Adaptation to Altered Visual Feedback

According to section 16.1, the maps that align visual and proprioceptive estimates of location and displacement (i.e., networks 1 and 4 in figure 14.1) probably depend on PMv, cerebellum, and PPC. However, the specific way that each area may take part in adaptation of these maps remains unknown. Psychophysical experiments can explore the mechanisms of adaptation, however, even in the absence of such knowledge. Typically, participants in these experiments make a small number of movements of a particular kind and, after adaptation, make some test movements. These test movements may be to parts of the workspace that had received no prior training. Does the training generalize to these movements? If so, can the pattern of generalization tell us something about how the CNS encodes the maps?

For example, in the location map $\hat{\theta} \leftrightarrow \hat{x}_{ee}$, you might guess that the encoding of end-effector location in proprioceptive coordinates depends on neurons that increase their discharge as a muscle, or group of muscles, increases in length. In section 10.3, you saw examples of this kind of broad encoding of limb configuration in the spinal cord, somatosensory cortex, M1, and PPC. Neurons in those locations did not seem to have a sharply defined preferred limb location, as might be described by a narrow Gaussian function. Rather, their discharge varied gradually and often linearly with limb configuration. How might this kind of proprioceptive representation influence generalization?

16.2.1 Representations Affect Generalization

To ask how a given kind of representation might affect generalization, reconsider the imaginary robot and camera setup in figure 15.1A. Imagine that the **controller** encodes the joint angles of the robot's arm with neuronlike computational elements that have a sharply tuned Gaussian activity field. This means that when the robot arm has a particular configuration $\theta^{(1)}$, neurons that have their activity field centered on that configuration have maximal activity, while most others remain silent. You might also have neurons that encode gripper location in camera-centered coordinates with Gaussian activity fields. You would observe that a small subset of these neurons discharges when the gripper appears at $x_{ee}^{(1)}$ and that others remain mostly silent.

To align the joint- and camera-centered information, you could strengthen the connections between neurons that had the most activity at any given time. Those neurons would correspond to alignments of joint-centered activity fields centered at $\theta^{(1)}$ and camera-centered activity fields centered at $x_{ee}^{(1)}$. The neurons that have an activity field centered close to $\theta^{(1)}$, but not exactly at that location, will show some (but a lesser amount of) activity. Accordingly, you could strengthen their connections to the camera-centered neurons by a smaller amount. In this scenario, the change in the connection strengths involves neurons that have their activity field centered at $\theta^{(1)}$, as well as those that have their peak activity at neighboring regions. The neighborhood where the weight changes took place depends on the choices that you made about how cells encode joint space. If you choose sharply tuned activity fields, that neighborhood will be small. If you choose a large activity field—for example, with neurons that linearly encode joint angles throughout their extent—then you will make the neighborhood large.

The size of the neighborhood where changes take place affects the generalization of learning and adaptation by the networks in the controller. For sharply tuned elements, the change in connection strengths for prism adaptation will affect only a small neighborhood near the data pair $\{\theta^{(1)}, x_{ee}^{(1)}\}$ for which the controller gained experience. With broadly tuned elements, the same data pair affects weights for a very large number of neurons because many of these neurons have some activity for any particular gripper location. The latter scenario leads to a situation where experience at one restricted set of locations affects the robot's movements over most (or perhaps all) of the workspace.

Thus, one way to examine how the location and displacement maps in figure 14.1 encode information is to measure the patterns of generalization for exposure to altered visual feedback. That is, to inquire about representation, it is useful to ask how training for some restricted places and states affects other movements. Section 21.3.2 will formalize this idea and provide mathematical tools to advance the argument, but for now the argument rests on intuition.

16.2.2 Realignment of the Location Map

Philipp Vetter, Susan Goodbody, and Daniel Wolpert[15] examined how adaptation to a visual displacement of the hand at a single location affected movements to other locations. In their setup, the participant could not view his or her hand because a mirror obscured it (figure 16.4A). However, the experimenters projected images on the mirror so that when the participants looked at it, they saw computer-generated images that represented their finger and the target. In this way, the participants performed unconstrained three-dimensional arm movements while the experimenters manipulated the location of the virtual finger and the target.

The experiment tested how displacement of the image of the finger from its actual location near a single target affected reaching movements to other targets. Participants assumed a starting posture with the finger close to the midline. In a *baseline* phase of the experiment, they pointed to targets distributed throughout the workspace with continuous, accurate feedback about their finger location. The situation differed little from unperturbed, direct vision of the hand. Next, in a *pre-exposure* phase, participants pointed to the same targets without visual feedback. Figure 16.4B shows their pre-exposure performance. Participants appeared to overshoot the targets slightly when they could not see their finger.

In the *exposure* phase, the participants repeatedly pointed to a single target. Figure 16.4A labels this as the "exposure target," which a box in figure 16.4C identifies. Participants could not see their virtual finger as they reached to this target, except for a 3-cm region surrounding the target location. When their virtual finger came within 3 cm of the target, the experimenters altered the visual feedback so that the image representing the finger appeared 6 cm to the left of its actual location along the x-axis. Therefore, to see their finger at the target, participants had to move their finger 6 cm to the right of the target.

After a few practice reaches, the participants quickly learned to reach so that they positioned their virtual finger at the target. After the exposure phase, participants reached to targets throughout the workspace, without visual feedback. Vetter and his colleagues observed that for the exposure target, the finger continued to move about 5 cm to the right. However, movements to other targets also changed, compared to the pre-exposure condition (figure 16.4C). The experiment demonstrated that when the visual feedback about finger location changed in a small part of space, there were global consequences.

To help understand this result, consider this experiment in the context of the computational framework outlined in figure 14.1. In the pre-exposure phase, the CNS received visual feedback about finger location. The virtual finger appeared in its expected location, which reinforced the location and displacement maps—$\hat{\theta} \leftrightarrow \hat{x}_{ee}$ and $x_{dv} \stackrel{\hat{\theta}}{\leftrightarrow} \Delta\hat{\theta}$, respectively—for unperturbed vision. Early in the exposure trials, the CNS recorded the location of the virtual finger, which did not correspond to the expected location. The only time the CNS received visual feedback about finger

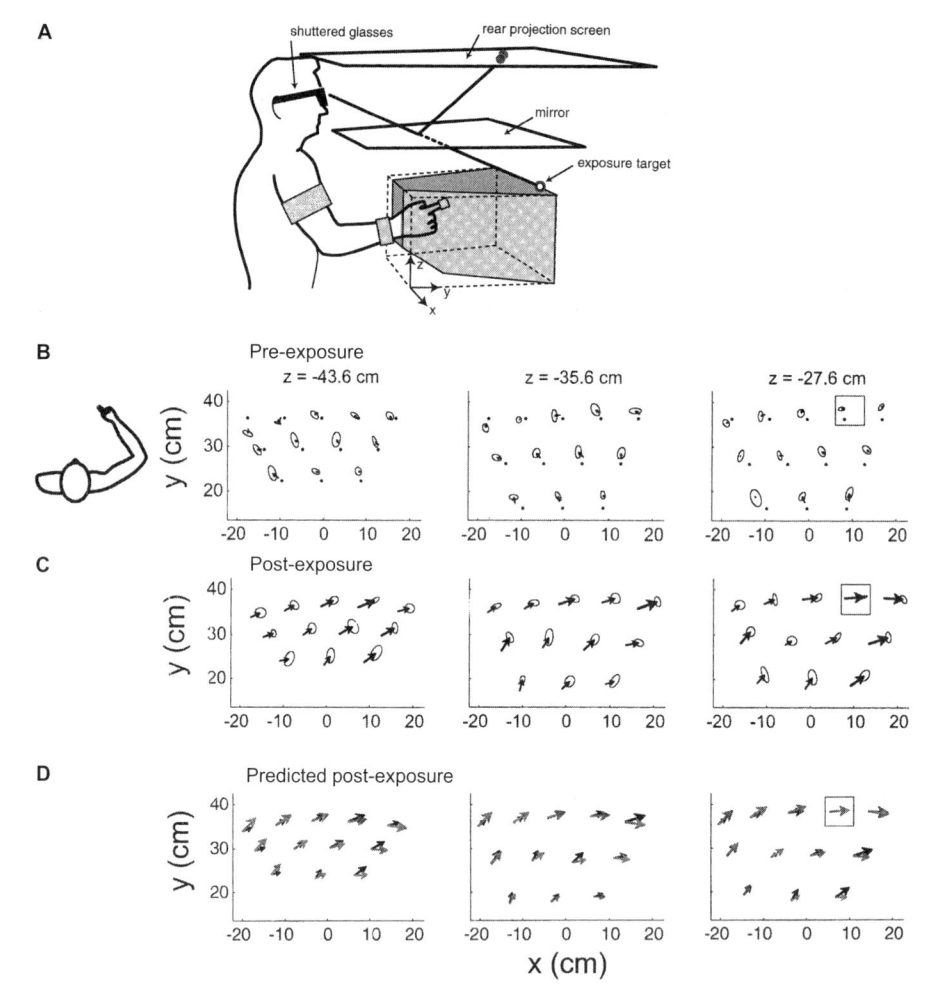

Figure 16.4
Modification of visual feedback at one location has global consequences. (*A*) The participant sees images that correspond to a virtual finger and a target. Shuttered glasses alternately blank the view from each eye in synchrony with the display, allowing each eye to have the appropriate planar view. Therefore, participants perceive a three-dimensional scene. The shaded area shows the workspace in which targets appeared. The drawing shows the arm at approximately its starting location. (*B*) Movements in the pre-exposure trials (without visual feedback). View from above the arm. Plots show two-dimensional slices along the horizontal plane. Each ellipse represents the 95% confidence limit for the endpoint of reaching movements for each target. (*C*) Each arrow represents the change in endpoint from the pre-exposure to the post-exposure trials. The square marks the exposure target. (*D*) Comparison of the predicted change in pointing based on spherical coordinates with an origin between the eyes (gray vectors) and the actual changes (black vectors). (From Vetter et al.[15] with permission.)

location, proprioceptive signals indicated that the finger occupied a location to the left of the predicted one. That is, in the space around the exposure target, the CNS "observed" that this particular fixation-centered location of the finger corresponded to a new proprioceptive-centered representation: $(\hat{\theta} + \varepsilon) \leftrightarrow \hat{x}_{ee}$. A test for generalization of adaptation examined how the alignment changed (or failed to change) for other parts of space. Accordingly, Vetter et al. asked the participants to reach to other targets without visual feedback, and observed that all movements appeared to terminate a few centimeters to the right of the target (figure 16.4C).

To explain this finding, note that each reaching movement started from a common location. However, participants could not see their finger at the starting location, and therefore—according to the computational framework in figure 14.1—must have used proprioceptive information to estimate finger location in fixation-centered coordinates. Before the exposure trials, proprioceptive information accurately predicted finger location. However, due to the altered feedback in the exposure trials, the alignment between proprioception and vision changed. If the starting location was near enough to the exposure target, this would have affected the mapping for that location. Accordingly, when the participants' virtual finger was at the starting location, the proprioceptive signals predicted a finger location approximately 5 cm to the right of its actual location. After calculating a difference vector, this error had global consequences; the realignment affected all movements, and they all missed to the right.

A prediction that follows from this hypothesis is that if participants point with their left hands to the location of the right finger, they will point a few centimeters to the right of its actual location. That experiment has not been done, but Vetter et al. did note that the pattern of errors in figure 16.4C appeared to rotate as a function of distance from the exposed target. They estimated that a spherical coordinate system centered directly between the eyes best fit this pattern of rotations (figure 16.4D). A fixation-centered representation of finger location produces a vector in spherical coordinates with its origin near a point midway between the eyes.

16.2.3 Realignment of the Displacement Map: Generalization to Other Arm Configurations

In the setup shown in figure 16.4, visual feedback occurred only in a small region of space centered on a single target. According to the model in figure 14.1, alteration of this feedback resulted in a realignment of the location map $\hat{\theta} \leftrightarrow \hat{x}_{ee}$. This realignment occurred because finger location observed in fixation-centered coordinates corresponded to a new and different arm configuration in proprioceptive coordinates. But did the altered visual feedback also affect the displacement map $x_{dv} \overset{\hat{\theta}}{\leftrightarrow} \Delta\hat{\theta}$? Perhaps, but that effect was probably small.

To understand this conclusion, recall that the displacement map is related to the location map through a Jacobian. When the location map changes, the Jacobian changes. However, the Jacobian changes fairly slowly as a function of arm configuration. For example, assuming a stiff wrist, in order to move your finger to the right, you must extend your elbow and shoulder. If you now move your arm a couple of centimeters to the left, the same extension of the elbow and shoulder will still produce a mostly rightward movement of the finger (though it will also have a small component away from you). This fact suggests that the displacement map is not very sensitive to small changes in location of the arm. Therefore, the 6-cm change in visual feedback in figure 16.4 can be safely assumed to have affected only the location map $\hat{\theta} \leftrightarrow \hat{x}_{ee}$.

To adapt the displacement map, you might show the finger moving in a direction other than the one in which it actually moves. Pierre Baraduc and Daniel Wolpert did that experiment.[16] They introduced a distortion between observed displacement of the finger and its actual displacement during a reaching movement. They asked participants to reach with the tip of their finger from a fixed starting point to a single target. The experimenters projected computer-generated images on a screen and participants viewed these images through a mirror. The participants could not see their hand (figure 16.5A). Instead, they saw a cursor that represented their finger location and a sphere that represented the target of movement, ~15 cm away. In pre-adaptation trials, the cursor location coincided with actual finger location, which reinforced the alignments in the $\hat{\theta} \leftrightarrow \hat{x}_{ee}$ and $x_{dv} \overset{\theta}{\leftrightarrow} \Delta\hat{\theta}$ maps.

As the adaptation trials began, the cursor moved along the x-axis in proportion to the finger's displacement along the y-axis. For example, as the finger moved directly along the y-axis, the cursor also moved along the y-axis, but it also traveled by a small, negative amount along the x-axis. A finger displacement of $[\Delta x\ \Delta y] = [0\ 15]$ cm (i.e., directly away) appeared as a cursor displacement of $[\Delta x - a\Delta y\ \Delta y]$, where a was gradually increased from 0 to 0.67. When a reached its final value of 0.67 (after 20 trials), the cursor appeared at $[-10\ 15]$ cm (i.e., away and to the left), when in fact the finger's location was equal to $[0\ 15]$ cm. Therefore, to view the virtual finger arriving at the target, $[0\ 15]$, the participants had to move their finger to $[+10\ 15]$ cm, away and to the right.

In the pre-adaptation trials, a desired displacement of the end effector in fixation-centered coordinates by amount x_{dv} required a change in arm configuration by amount $\Delta\hat{\theta}$. With the altered visual feedback, the same x_{dv} now required a different displacement in **joint coordinates**: $x_{dv} \overset{\hat{\theta}}{\leftrightarrow} (\Delta\hat{\theta} + \delta)$. For example, for the starting arm configuration labeled 2 in figure 16.5A, the participants had to learn to extend their elbows more to bring the virtual finger to the target. Therefore, the altered visual feedback required modification of the displacement map.

The location map changed in a more complicated way during the adaptation trials. During those trials, visual feedback at the start of the movement did not change, and therefore at this location the $\hat{\theta} \leftrightarrow \hat{x}_{ee}$ map did not change. However, at the end of the movement, the location map required the association of a new set of joint angles with a given finger

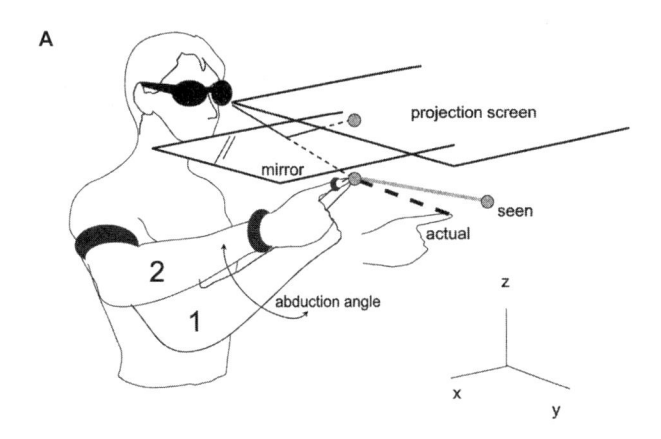

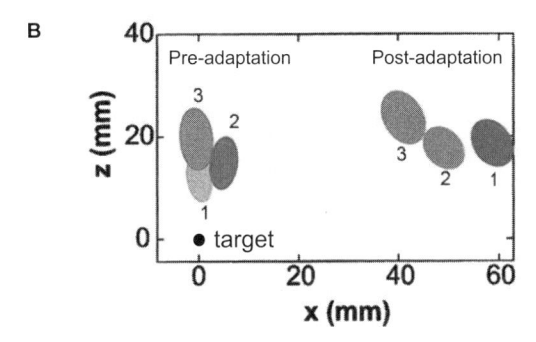

16.2. Generalization of Adaptation to Altered Visual Feedback

Figure 16.5
Alteration of visual feedback for reaching along a single direction of movement and its dependence on arm posture. (*A*) Experimental setup. Infrared markers placed on the arm and fingertip indicated arm configuration. A computer generated images representing the fingertip and target and projected them on a screen that was viewed through a mirror. The experimenters required participants to start each movement with a specific abduction angle of the elbow. The figure illustrates posture 1 (adducted) and posture 2 (30° abduction of the elbow). Posture 3 involved a 60° abduction. (*B*) Each 95% confidence ellipse represents the distribution of finger locations at the end of movements without visual feedback. Before the change in initial posture, the reaches terminated above the target (which was at coordinate 0, 0). Adaptation most affected reaches that began with the training posture (posture 1), and had less effect on reaches for the other postures. (From Baraduc and Wolpert[16] with permission.)

location. Therefore, at the target location $[-10\ 15]$ cm, the location map had to have changed: $(\hat{\theta} + \varepsilon) \leftrightarrow \hat{x}_{ee}$.

In the post-adaptation trials, the experimenters withheld visual feedback from the participants as they reached from the same starting location to the same target. To make this movement, they had to estimate end-effector location at the start of the movement, compare this location with the target location, compute a difference vector in fixation-centered coordinates, and transform that vector to a set of joint rotations. At the starting point, the $\hat{\theta} \leftrightarrow \hat{x}_{ee}$ map had not changed, and therefore the participant (presumably) estimated finger location x_{ee} accurately. According to the present model, by comparing x_{ee} with target location x_t, the participants' CNS calculated a desired difference vector x_{dv} for the finger. During adaptation trials, the participants learned that x_{dv} needed to be associated with a new set of joint rotations $\Delta\hat{\theta} + \delta$ (i.e., network 4 in figure 14.1, adapted). The arm's joints then rotated from their starting configuration $\hat{\theta}$ by $\Delta\hat{\theta} + \delta$, with the result that the actual finger location moved to the right of the target as the cursor hit the target.

To study generalization, you need to ask how the change in the displacement map affected other movements. Baraduc and Wolpert[16] asked how local changes in the displacement map affected reaches that started in the same finger location but had different initial arm configurations.

In the pre-adaptation trials, when the participants received veridical visual feedback, the experimenters constrained initial arm posture to one of three initial configurations. Figure 16.5A shows posture 1 and posture 2, the latter a 30° abduction of posture 1. The figure does not illustrate posture 3, which was a 60° abduction of posture 1, essentially putting the entire arm in the horizontal plane. Consider these postures as $\hat{\theta}^{(1)}, \hat{\theta}^{(2)}$, and $\hat{\theta}^{(3)}$. The desired finger displacement x_{dv} did not differ as a function of initial posture. However, arm posture affected the association of that hand displacement with joint rotations. Accordingly, to describe that dependence for the three different postures, you can write $x_{dv} \stackrel{\hat{\theta}^{(i)}}{\leftrightarrow} \Delta\hat{\theta}^{(i)}$.

In some of the pre-adaptation trials, the experimenters withheld visual feedback and found that the finger landed slightly above the target, regardless of the arm's initial posture (figure 16.5B, left). During the adaptation trials, participants had experienced only posture 1, and the altered visual feedback changed the displacement map at posture 1, as follows: $x_{dv} \stackrel{\hat{\theta}^{(1)}}{\leftrightarrow} (\Delta\hat{\theta}^{(1)} + \delta)$. The experimenters examined how this remapping generalized to the other two postures.

Again the experimenters withheld visual feedback. For posture 1—the one used during adaptation trials—movements landed about 6 cm to the right of the target (figure 16.5B, right). Recall that the altered feedback displaced hand location at the target by 15 cm. Thus, the error in reaching amounts to less than half of what you might expect if the participants fully adapted to the altered visual feedback. More interesting, however, was the finding that when participants began reaches from posture 2 or 3—postures not used in the adaptation trials—the movements still missed the target to the right (by ~5 cm and ~4 cm, respectively). Thus, the more the initial posture differed from that used during adaptation trials, the smaller the effect (figure 16.5B).

Therefore, the change δ that occurred in the displacement map for the adaptation trials, posture $\hat{\theta}^{(1)}$, generalized to postures $\hat{\theta}^{(2)}$ and $\hat{\theta}^{(3)}$. This finding indicates that the neural representation of the map from hand location to joint displacement does not "sharply" or "narrowly" depend on $\hat{\theta}$. If it did, then a change made to the map at $\hat{\theta}^{(1)}$ would be localized to a small region of arm configurations near that posture. It would not affect $\hat{\theta}^{(2)}$ and $\hat{\theta}^{(3)}$. The fact that it does affect the mappings for those "untrained" postures means that the neural system represents the map $x_{dv} \stackrel{\hat{\theta}}{\leftrightarrow} \Delta\hat{\theta}$ as a relatively "broad" function of arm configuration $\hat{\theta}$. A change in the map that occurs at a particular $\hat{\theta}$ affects other postures, even those far away.

16.2.4 Realignment of the Displacement Map: Generalization to Other Movement Directions

In the experiment depicted in figure 16.5, Baraduc and Wolpert[16] altered visual feedback near a particular direction of reaching movement and asked how this alteration affected movements in the same direction but beginning with other arm configurations. The results revealed a broad pattern of generalization across location space. You might also ask how adaptation of the displacement map for a single direction affects movements in other directions. This experiment has not yet been done, but Claude Ghez and his colleagues[17] performed an experiment that comes close. The participants viewed not their finger but a cursor that moved on a video monitor.

The participants studied by Ghez and his colleagues sat facing a computer monitor. They controlled a cursor displayed on the monitor by moving a handheld sensor across the surface of a digitizing tablet. Without vision of their hand or much of their arm, the participants made a rapid, uncorrected movement to the target and back to center (termed an *out-and-back movement*). The experimenters focused on the initiation of each movement and measured the angle of cursor motion with respect to direction of the target. Helen Cunningham and her colleagues[18–21] pioneered this motor-learning paradigm, often called "rotation experiments." Her experiments varied the relationship between joint-angle changes and end-effector displacements. For example, in one condition, movements of the participant's hand to the left caused a movement of a cursor to the left, but in a "rotated" condition, movements of the hand to the left caused the cursor to move downward (a −90° rotation). (You might recall that the monkey experiment illustrated in figure 14.5 used a similar experimental design.) To move the cursor to a target, the CNS had to learn to map the desired cursor displacement to a new set of joint rotations: $x_{dv} \stackrel{\hat{\theta}}{\leftrightarrow} (\Delta\hat{\theta} + \delta)$. Unlike prism-adaptation experiments, nothing distorts visual inputs in this experimental paradigm. Note that experimenters can induce any number of transforms between joint rotations and cursor movements,[22] but angular deviations (rotations) remain the most common.

In the *baseline* movements of the experiments performed by Ghez and his colleagues,[17] the cursor moved in the same direction as the hand. Viewing the motion of the cursor reinforces the map $x_{dv} \overset{\hat{\theta}}{\leftrightarrow} \Delta\hat{\theta}$, which describes how cursor movements relate to joint rotations. You can assume that the participants had learned this mapping in their previous experience with computer "mousing." Viewing the cursor could also modify the location map $\hat{\theta} \leftrightarrow \hat{x}_{ee}$, but the location of the cursor need not correspond to any particular arm configuration. As mentioned earlier, if you pick up your computer mouse and put it down somewhere else, the cursor does not move much, but your arm configuration can change a lot. Evidence suggests that the CNS attenuates proprioceptive inputs during such rotation conditions.[23]

In some cases during the baseline trials, the cursor disappeared and the movement took place without visual feedback. Next, in the *training* trials, different groups of participants made movements to either a single target or to eight targets arranged in a circle around the initial cursor location. In both cases, targets appeared ~4 cm from the origin, a rather small movement. In this training condition, participants always had visual feedback, but the cursor's movement relative to the hand's movement on the tablet differed by a counterclockwise rotation of 30°. It took about 20 trials for participants to adapt in the single-target condition, but more than 60 movements to adapt in the eight-target condition.

For movements of only ~4 cm, the rotation has quite a small effect on the location map $\hat{\theta} \leftrightarrow \hat{x}_{ee}$, even with intact proprioceptive feedback. Rather, rotation experiments mainly affect the displacement map. In the baseline trials, in order to displace the cursor in a particular direction by amount $x_{dv}^{(1)}$, participants have to change joint angles by amount $\Delta\hat{\theta}^{(1)}$. In the training trials, participants learned that the same cursor displacement required joint-angle changes of $\Delta\hat{\theta}^{(1)} + \delta$. Therefore, after adaptation the mapping changed to $x_{dv}^{(1)} \overset{\hat{\theta}^{(1)}}{\leftrightarrow} (\Delta\hat{\theta}^{(1)} + \delta)$. A study of generalization examines how this change affects the displacement map for other movement directions. That is, how does experience with a given rotation condition for $x_{dv}^{(1)}$—the *training target*—affect movements that require cursor displacements in direction $x_{dv}^{(2)}$ or $x_{dv}^{(3)}$?

To answer this question, Ghez and his colleagues[17] asked participants to make movements to other targets, without visual feedback from the cursor. They found that for the training target, participants moved their hands 30° away from the target, in the clockwise direction. This makes sense because the altered visual feedback had imposed a 30° counterclockwise rotation. However, for targets that the participants had not experienced during training trials, they moved more accurately, especially if the target differed by more than 45° or so from the training target. That is, participants showed little generalization to directions more than 45° distant from the training target in radial coordinates. Figure 16.6B shows the shape of this generalization as a function of the angular distance from the training target. This small generalization to neighboring directions might explain why the learning rate in the eight-target condition was so much slower than in the one-target condition (figure 16.6A).

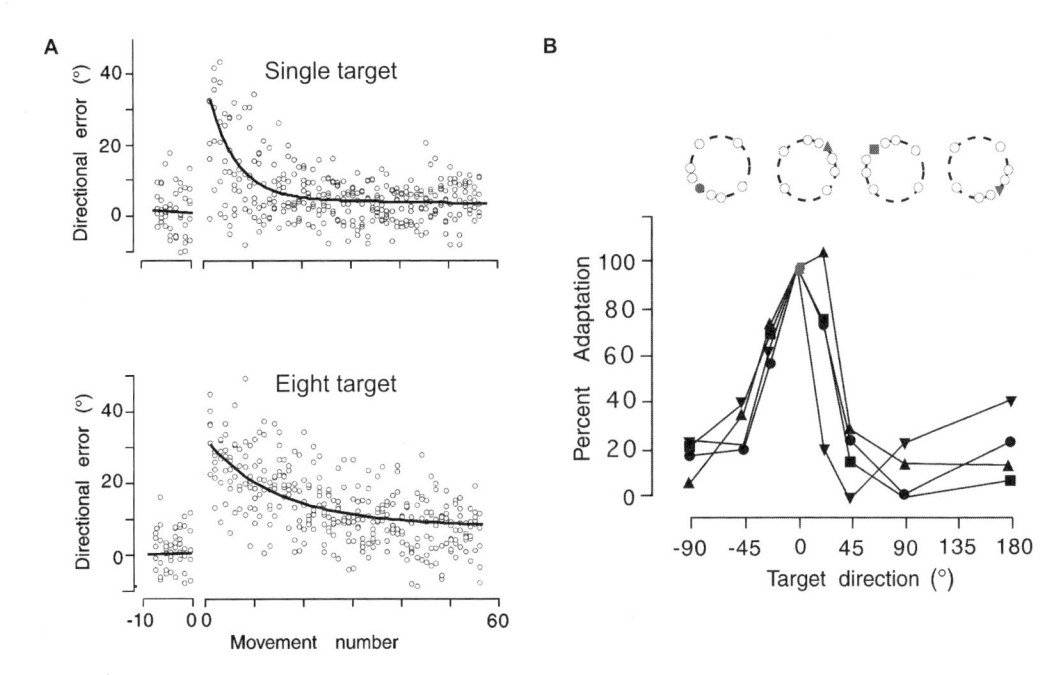

16.2. Generalization of Adaptation to Altered Visual Feedback

Figure 16.6
Visual-feedback rotation. (A) Participants moved a cursor to a target and, in the experimental condition, the cursor's displacement deviated by 30° with respect to hand displacement. Angular deviations of this kind are often called rotations. Training consisted of either a single target location or eight targets arranged in a circle around the origin. The plots show the angular difference between the initial movement direction and a straight path to the target for consecutive movements. (B) Generalization to neighboring directions after adapting to a single direction. The gray symbols on top show the four different training directions for four different days. (From Krakauer et al.[17] with permission.)

To explain this limited pattern of generalization, recall that the difference vector x_{dv} represents the location of the cursor with respect to the target. This vector results from the subtraction $x_{t'} - x_{ee}$, each represented in fixation-centered coordinates (see chapter 12). This computation provides a hint that neurons representing the difference vector might have Gaussian-like activity fields centered near a **preferred direction** or preferred vector. The finding that when a particular cursor difference $x_{dv}^{(1)}$ requires a particular joint rotation $\Delta\hat{\theta}^{(1)} + \delta$, adaptation of the displacement map generalizes only to nearby directions, indicates that the adaptation mechanism limits changes to neurons representing $x_{dv}^{(1)}$ or nearby vectors. These neurons should have their greatest activity when the CNS computes that vector or one close to its parameters.

You might expect that these neurons have localized activity fields and that the adaptation of the displacement map will affect only their connections with neurons that represent a given pattern of joint rotations. This property would explain why adapting to one direction of cursor displacement does not generalize to very different movement directions.

Studies on monkeys yield similar curves for both behavioral generalization and neuronal tuning for the direction of a reaching movement.[24] (Note, however, that movement amplitude has been ignored here, so you cannot distinguish between Gaussian-like representations of x_{dv} and representations that might encode direction, but not amplitude, as a preferred direction).

According to figure 14.1, the displacement map aligns the difference vector with joint rotations. The idea presented in the previous paragraph suggests that the larger the rotation, the more difficult it might be for the CNS to bring these two parameters into alignment. In support of this view, Sylvie Abeele and Otmar Bock[25] found that previous training on one rotation facilitated adaptation to subsequent ones, whether larger or smaller in degree. However, if the misalignment exceeded 80° or so, much less facilitation occurred. Abeele and Bock estimated 120° as the approximate limit for "easy" alignment of the displacement map. This finding explains why people (and monkeys) found rotations of approximately 90°–135° to be the most difficult in these experiments.[18] Abeele and Bock suggested that for rotations greater than 120° or so, the participants called up a different displacement map and adapted behavior from that starting point.

In another rotation experiment, Ethan Buch, Sereniti Young, and José Contreras-Vidal[26] studied adaptation in elderly versus college-age participants. They found that elderly participants had trouble adapting to relatively large rotations ($\sim 90°$). However, if they first adapted their displacement maps in smaller steps ($\sim 10°$), they then adapted to 90° rotations and showed aftereffects. These studies support the idea that kinematic mappings become more difficult to learn for large rotations (and become increasingly difficult with advancing age).

16.2.5 Realignment of the Displacement Map: Neural Basis

The cerebellum appears to play a role in adapting the displacement map, especially for relatively small rotations. Evidence shows that inactivation of the deep cerebellar nuclei in monkeys causes a deficit in adapting to rotations.[27] Cerebellar inactivation also affects the ability of monkeys to track a visual target. As chapter 23 sets out, the cerebellum plays an important role in predicting stimuli. Accordingly, movements made with an intact cerebellum correlate with the target's speed. After inactivation of the cerebellum, monkeys cannot accurately predict the future location of a target, and their movements correlate better with the difference between current end-effector location and target location.[28] Cerebellar inactivation appears to disrupt the prediction of target location without affecting the difference vector or the system that uses the difference vector to compute the next set of desired joint rotations.

Very little neurophysiological work has studied the changes in neuronal activity during adaptation of displacement maps. The most extensive analysis of neuronal activity during the adaptation of displacement maps has been for rotation experiments,[22,29] although those experiments

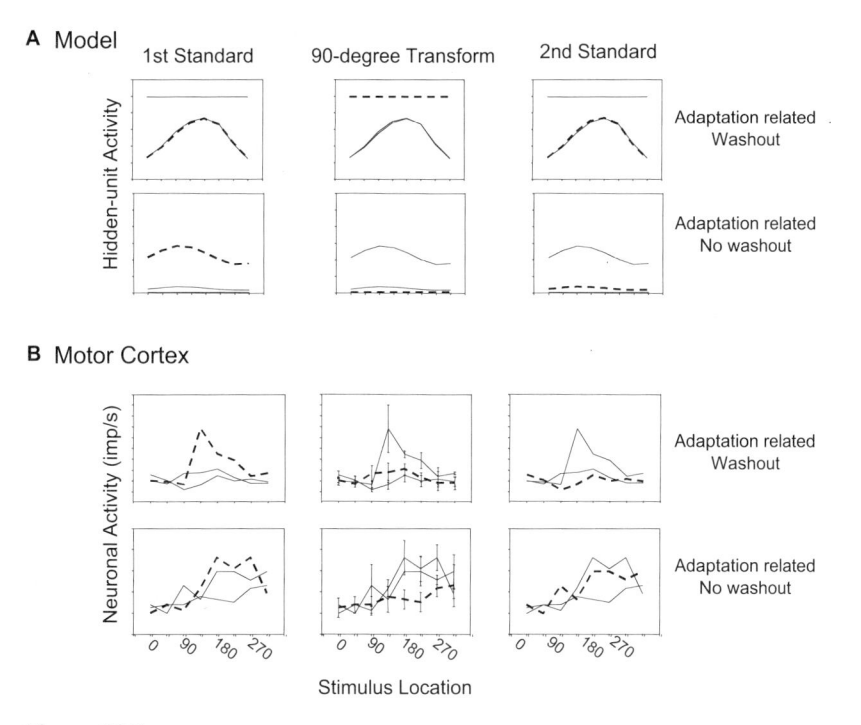

16.2. Generalization of Adaptation to Altered Visual Feedback

Figure 16.7
Neuronal activity during visual-feedback rotation. (*A*) Each row shows the activity of a single hidden unit of a model neural network, and the data are the same in all three columns. Highlighted in each column, by a thicker, broken line, are the tuning curves for the mapping stated at the top of the column. Thus, reading from left to right in the top row, the hidden unit was well tuned to ~180° once the standard mapping had been first learned (left column). Then, after the same network mastered the nonstandard mapping, involving a 90° counterclockwise rotation, the cell became completely untuned (middle column), but returned to its original tuning during the retraining to a standard mapping (right column). The adaptation-related change washed out. The hidden unit in the second row shows a different property. Like the unit in the top row, the cell lost its tuning after the transition from standard mapping to a 90° counterclockwise remapping. However, when the network was retrained on the standard mapping, the unit remained weakly tuned and much more closely resembled its tuning for transformational mapping. Thus, in this unit, the adaptation-related change did not wash out. (*B*) Neuronal activity from the monkey motor cortex in approximately the format of *A*. The same general phenomena occur in cortex as in the model. (From Moody and Wise[29] with permission.)

also involved other transforms. As a population, cells in PMd, SMA, and M1 all showed dramatic changes in their properties during adaptation of displacement maps. This plasticity took the form of changes in the cells' preferred directions, often large ones, and changes in both the depth and the width of tuning. Approximately half of the sample of neurons in these areas showed significant changes in activity during adaptation and, as a population, this activity change appeared to lag the adaptation as measured behaviorally. This finding may be relevant to the time course of consolidation, but it remains poorly understood.

Another property of cortical cells during adaptation of the rotation map resembles a finding that chapter 20 deals with when discussing the adaptation of the dynamics map. In studies of dynamics adaptation in monkeys,[30] some M1, PMd, PMv, and SMA cells show changes in preferred direction or modulation amplitude during adaptation, and these changes do not always wash out (adaptation-related, no washout in figure 16.7). Other cells do not change those properties during adaptation, but do upon washout (figure 20.8). Neuronal plasticity during adaptation of the displacement map has the same characteristics, at least for neurons in M1, PMd, and SMA.[22,29] Sohie Lee and Wise[29] showed that artificial neural networks—trained to compute the same adaptations and washout—do so as well (figure 16.7A). This result is not surprising. There is no reason to suppose that a neural network will compute a given input–output function the same way at two different times in its history, especially when, in the meantime, its weights have changed to compute a different mapping.

References

1. Graziano, MS, Hu, XT, and Gross, CG (1997) Visuospatial properties of ventral premotor cortex. J Neurophysiol 77, 2268–2292.
2. Kakei, S, Hoffman, DS, and Strick, PL (2001) Direction of action is represented in the ventral premotor cortex. Nat Neurosci 4, 1020–1025.
3. Hoshi, E, and Tanji, J (2000) Integration of target and body-part information in the premotor cortex when planning action. Nature 408, 466–470.
4. Gaffan, D, and Hornak, J (1997) Visual neglect in the monkey—Representation and disconnection. Brain 120, 1647–1657.
5. Medendorp, WP, Goltz, HC, Vilis, T, and Crawford, JD (2003) Gaze-centered updating of visual space in human parietal cortex. J Neurosci 23, 6209–6214.
6. Kurata, K, and Hoshi, E (1999) Reacquisition deficits in prism adaptation after muscimol microinjection into the ventral premotor cortex of monkeys. J Neurophysiol 81, 1927–1938.
7. Rossetti, Y, Rode, G, Pisella, L, et al. (1998) Prism adaptation to a rightward optical deviation rehabilitates left hemispatial neglect. Nature 395, 166–169.
8. Pisella, L, Rode, G, Farne, A, et al. (2002) Dissociated long lasting improvements of straight-ahead pointing and line bisection tasks in two hemineglect patients. Neuropsychologia 40, 327–334.

9. Weiner, MJ, Hallett, M, and Funkenstein, HH (1983) Adaptation to lateral displacement of vision in patients with lesions of the central nervous system. Neurology 33, 766–772.
10. Baizer, JS, Kralj-Hans, I, and Glickstein, M (1999) Cerebellar lesions and prism adaptation in macaque monkeys. J Neurophysiol 81, 1960–1965.
11. Imamizu, H, Miyauchi, S, Tamada, T, et al. (2000) Human cerebellar activity reflecting an acquired internal model of a new tool. Nature 403, 192–195.
12. Martin, TA, Keating, JG, Goodkin, HP, et al. (1996) Throwing while looking through prisms. I. Focal olivocerebellar lesions impair adaptation. Brain 119, 1183–1198.
13. Martin, TA, Keating, JG, Goodkin, HP, et al. (1996) Throwing while looking through prisms. II. Specificity and storage of multiple gaze-throw calibrations. Brain 119, 1199–1211.
14. Clower, DM, Hoffman, JM, Votaw, JR, et al. (1996) Role of posterior parietal cortex in the recalibration of visually guided reaching. Nature 383, 618–621.
15. Vetter, P, Goodbody, SJ, and Wolpert, DM (1999) Evidence for an eye-centered spherical representation of the visuomotor map. J Neurophysiol 81, 935–939.
16. Baraduc, P, and Wolpert, DM (2002) Adaptation to a visuomotor shift depends on the starting posture. J Neurophysiol 88, 973–981.
17. Krakauer, JW, Pine, ZM, Ghilardi, MF, and Ghez, C (2000) Learning of visuomotor transformations for vectorial planning of reaching trajectories. J Neurosci 20, 8916–8924.
18. Cunningham, HA (1989) Aiming error under transformed spatial mappings suggests a structure for visual-motor maps. J Exp Psychol 15, 493–506.
19. Cunningham, HA, and Vardi, I (1990) A vector-sum process produces curved aiming paths under rotated visual-motor mappings. Biol Cybern 64, 117–128.
20. Cunningham, HA, and Welch, RB (1994) Multiple concurrent visual-motor mappings: Implications for models of adaptation. J Exp Psychol 20, 987–999.
21. Cunningham, HA, and Pavel, M (1995). In Pictorial Communication in Virtual and Real Environments, eds Ellis, SR, Kaiser, MK, and Grunwald, AJ (Taylor and Francis, London), pp 281–294.
22. Wise, SP, Moody, SL, Blomstrom, KJ, and Mitz, AR (1998) Changes in motor cortical activity during visuomotor adaptation. Exp Brain Res 121, 285–299.
23. Jones, KE, Wessberg, J, and Vallbo, Å (2001) Proprioceptive feedback is reduced during adaptation to a visuomotor transformation: Preliminary findings. NeuroReport 12, 4029–4033.
24. Paz, R, Boraud, T, Natan, C, et al. (2003) Preparatory activity in motor cortex reflects learning of local visuomotor skills. Nat Neurosci 6, 882–890.
25. Abeele, S, and Bock, O (2001) Sensorimotor adaptation to rotated visual input: Different mechanisms for small versus large rotations. Exp Brain Res 140, 407–410.
26. Buch, ER, Young, S, and Contreras-Vidal, JL (2003) Visuomotor adaptation in normal aging. Learn Mem 10, 55–63.
27. Robertson, EM, and Miall, RC (1999) Visuomotor adaptation during inactivation of the dentate nucleus. NeuroReport 10, 1029–1034.

28. Miall, RC, Weir, DJ, and Stein, JF (1987) Visuo-motor tracking during reversible inactivation of the cerebellum. Exp Brain Res 65, 455–464.
29. Moody, SL, and Wise, SP (2001) Connectionist contributions to population coding in the motor cortex. Prog Brain Res 130, 245–266.
30. Li, CSR, Padoa-Schioppa, C, and Bizzi, E (2001) Neuronal correlates of motor performance and motor learning in the primary motor cortex of monkeys adapting to an external force field. Neuron 30, 593–607.
31. Ghez, C, Krakauer, J, Sainburg, RL, and Ghilardi, M-F (2000). In The New Cognitive Neurosciences, ed Gazzaniga, M (MIT Press, Cambridge, MA), pp 501–514.

Reading List

The review by Ghez et al.[31] presents an overview of the rotation experiments reported by him and his colleagues.

17 Remapping, Predictive Updating, and Autopilot Control

Overview: Reaching and pointing movements involve continuous monitoring of target and end-effector locations in fixation-centered coordinates with the goal of reducing the difference vector to 0. Your CNS recomputes the kinematic maps that estimate target and end-effector location as the eyes, targets, and end effector move. Because this remapping depends on a copy of motor commands to the eyes, the head, and the arm, your CNS can update these estimates predictively. Systems that predict consequences of motor commands in sensory coordinates are called **forward models**. *Forward models may also underlie your ability to imagine movements.*

17.1 Remapping Target Location

In the imaginary robot described in figure 15.1A, the camera never moved. Your eyes move, of course; they change orientation within the orbits at a very fast pace. Your eyes also move when your head changes orientation or you travel from one place to another. Because the **kinematic** computations for reaching involve vision-based coordinates, eye movements pose a fundamental problem. Imagine that the target in figure 15.1A fell outside the camera's field of view. The robotic **controller** would be lost. Without a target activating some pixels in the camera's field of view, the controller could not compute a difference vector and could not move the gripper to the target. To overcome this problem, you might place the camera on a swivel mechanism. If the angular excursion of the swivel mechanism had some limit (e.g., 45°), you might use two of these mechanisms, one on top of the other, to cover more territory. With the swivels, the robot's camera gains a large field of view from a relatively focal camera lens. This benefit, however, comes at a cost: the location of the target no longer depends solely on which camera pixels the target activates. The target's location in camera-centered space will vary as a function of the camera's orientation. Accordingly, the robotic controller must keep track of the camera's orientation, as well as of which pixels the target and gripper activate.

To illustrate these points, imagine yourself reading this book at a coffee shop. As you read, you decide to reach to the right for a pencil

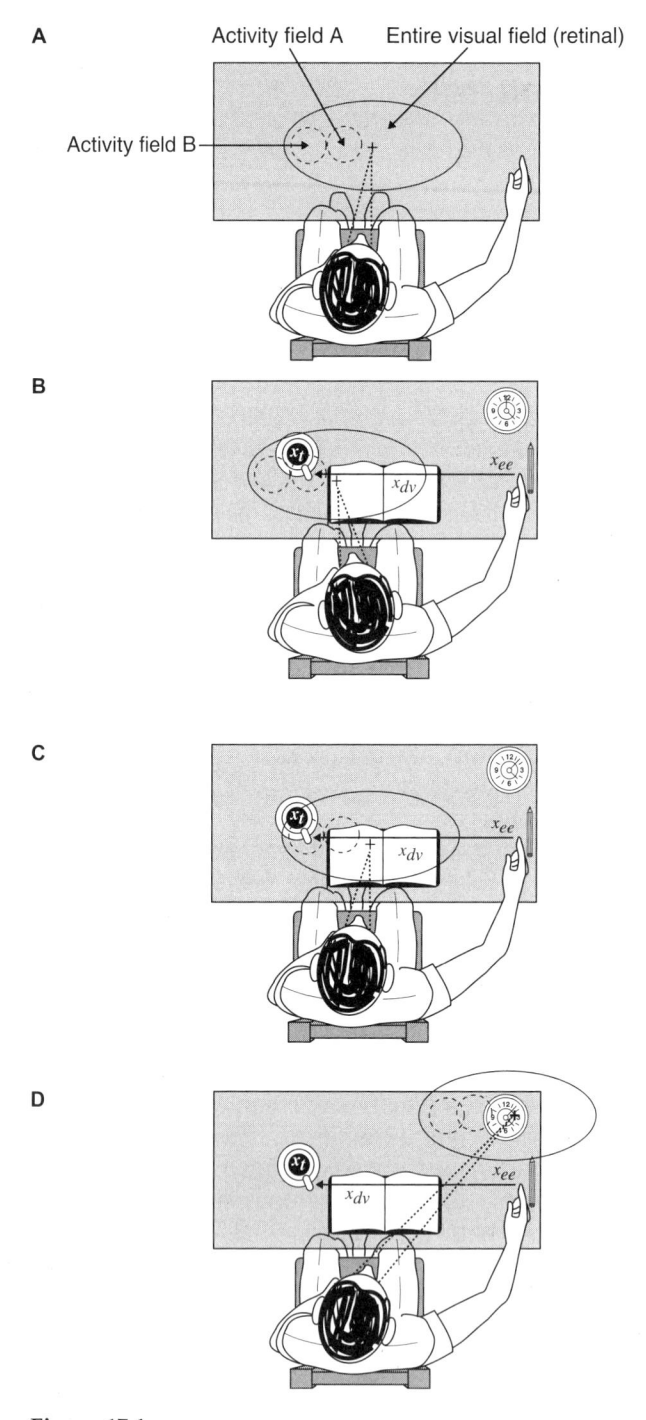

320 Chapter 17. Remapping, Predictive Updating, and Autopilot Control

Figure 17.1
Updating target location due to reorientation of the eye. (*A*) The dashed circles show the activity fields of two cells, A and B. The plus sign (+) shows the fixation point, and the ellipse shows the limits of the entire visual field (not to scale).

because you want to make a note (figure 17.1B). Because you continue to fixate a point in the book (+), the pencil lies in your peripheral field of view, to the right. But you know where it is, so you can reach for it with reasonable accuracy. For some unknown reason, before your hand gets to the pencil, you change your mind and decide you need some coffee—now! This new goal establishes a new difference vector, from your current hand location x_{ee} to the handle of the coffee cup, the new x_t (figure 17.1B). You continue to read, so your fixation point moves along the page from left to right (figure 17.1C). But wait: You remember that you need to be somewhere, so you check the clock (figure 17.1D). Both your head and your eyes orient toward the clock, but you can still reach fairly accurately to the coffee cup. Note that you did not look at the pencil initially, you did not look at the handle of the coffee cup when you changed your mind, and now—before you reach for the coffee cup—you look even farther away from the target as you check the time. (It might be a good idea to make a saccade to the handle just before your hand gets there, though.)

When your eyes and head moved, the location of your hand (the end effector) and the handle of the coffee cup (the target) changed in *retinotopic* coordinates (i.e., the part of the retina on which the image of the target and the end effector fell (figure 17.1B versus 17.1C). Although eye movements resulted in a change in the location of x_t and x_{ee} in these retinotopic coordinates, the difference vector x_{dv} remained approximately the same. According to the present model, changes in the fixation point resulted in a recalculation of the difference vector because of the changes in the retinotopic location of the target and the end effector, but the result was the same. After all, you got the coffee. Note, however, that as you made a saccade to the clock at the far right (figure 17.1D), the handle of the coffee cup went out of your visual field entirely. You could not see it, but you could continue to represent its location in fixation-centered coordinates. Fixation-centered coordinates depend on the orientation of the fovea, but do not depend on the visual field, unlike *retinotopic* coordinates (at least as usually construed).

This fixation-centered frame corresponds to the extrinsic, vision-based coordinates considered in chapter 11 (section 11.2 and figure 11.3). For the robot, fixation-centered coordinates can describe the location of

Figure 17.1 (continued)
(*B*) The person reads a book while considering whether to reach for a cup of coffee (left) or a pencil (right). A clock (upper right) shows the time. As the person fixates the beginning of a line of text, the handle of the coffee cup falls within the activity field of cell A. (*C*) As the fixation point shifts to the right part of the line, so does the visual field as a whole, as well as both cells' activity fields. Note that the target x_t—the handle of the coffee cup—has shifted out of cell A's activity field and into that of cell B. (*D*) Another eye movement and head movement shift the target out of the visual field entirely. Updating of the target's representation in fixation-centered coordinates maintains the motor plan unchanged, as indicated by the difference vector x_{dv}. (Drawn by George Nichols.)

the target and gripper beyond the camera's field of view. The same holds for your vision.

This chapter adds two concepts to the model sketched in figure 14.1: adaptive remapping and **efference copy**. First, movements of the eyes, the hand, and the target require updating the kinematic maps to reflect the changes in fixation-centered location of both x_t and x_{ee}. Second, this updating relies largely on a *copy* of motor commands to the eyes and the arm, and to a lesser extent on proprioceptive feedback. Further, the updating of these variables is predictive.

As noted above, according to the present model, when the CNS updates $\hat{x}_t$ and $\hat{x}_{ee}$ in fixation-centered coordinates, it should recalculate its estimate of x_{dv}. Return to the example of reaching for the pencil while reading this book. Imagine that the pencil starts rolling left as reaching begins. The updating mechanism ensures that the end effector reaches the target: Reaching movement terminates when joint rotations null the difference vector. If the target jumps or drifts during the movement, the difference vector changes as a continuous function to reflect the new direction and amplitude of joint rotations needed to reduce the difference vector to 0. Thus, the kinematic maps update not only when the eyes move, but also when either the target or the end effector moves. You saw neurophysiological evidence for this in section 13.1.2 for changes in the hand's initial location for PMd and M1. Similarly, cells in PMd and the parietal reach region (PRR) rapidly change delay-period activity to reflect a shift of a target into a cell's preferred direction or some kind of Gaussian activity field.[1-3] Similar data have been obtained for movement-related activity in M1.[4-6] The motor system thus acts something like an autopilot, guiding the end effector to the target, and PPC has been especially implicated in this computation.[7-9]

17.1.1 Remapping due to Intervening Eye Movements

Rene Duhamel, Carol Colby, and Mickey Goldberg[10] obtained experimental evidence that the CNS updated the remembered location of a stimulus—in fixation-centered coordinates—as a monkey reoriented its eyes. These investigators recorded the activity of cells in PPC and focused on the control of eye movements, but their data probably apply to reaching and pointing movements as well. They trained monkeys to fixate a light spot and make a saccade to it whenever it jumped from one location to another. Figure 17.2A illustrates that in addition to the fixation point (•), sometimes another visual stimulus appeared on the screen (∗). As the monkey looked at the fixation point, neurons in area LIP responded to the onset of the other visual stimulus (∗) at a latency of 70 msec, provided that the second stimulus appeared in the cell's activity field. Figure 17.2A shows the approximate location of one cell's activity field, above and to the left of the fixation point, by the dotted circle.

The activity field, being retinotopic, moved when the fixation point changed. When the fixation point jumped to a new location, the cell discharged after the saccade brought the stimulus into the cell's activity field,

17.1. Remapping Target Location

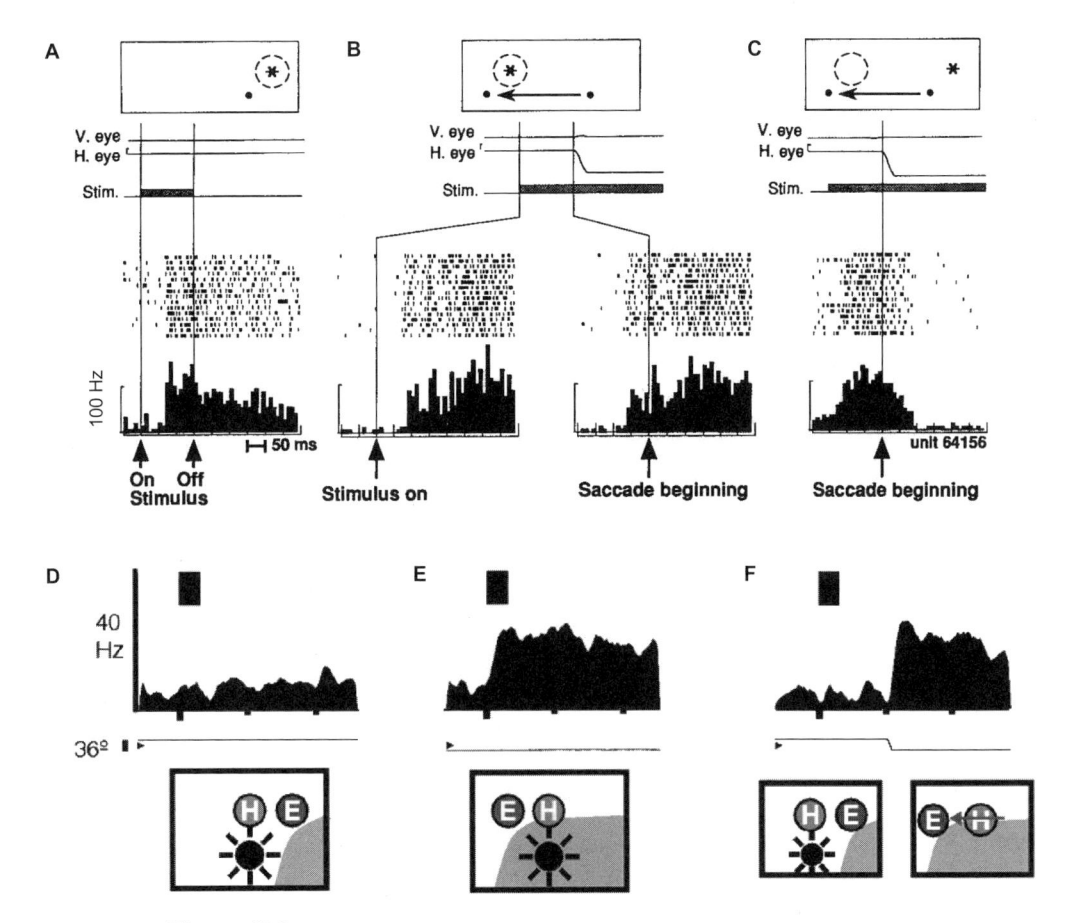

Figure 17.2

Effect of eye movement on the memory of a visual stimulus. (A–C) Remapping in an LIP neuron. In each panel, the filled circle represents the fixation point, the asterisk indicates the location of the visual stimulus, and the dashed circle indicates the activity field of an LIP cell. (A) Discharge to the onset of a visual stimulus in the cell's activity field. Abbreviations: H. eye, horizontal eye orientation; Stim, stimulus; V. eye, vertical eye orientation. (B) Discharge after a saccade brings the stimulus into the cell's activity field. (C) Response after a saccade takes the cell's activity field away from the stimulus. (D–F) Remapping in a PRR neuron. The monkey fixated a point (dark gray, marked E) while touching a button (light gray, marked H) on a vertical panel. A light appeared for 300 msec (dark bar on top) to instruct a reach target. (D) The target fell outside the neuron's activity field (gray region in the lower drawings), and the cell showed no significant modulation in activity. Note that a trigger (go) signal came along later. (E) The target fell inside the neuron's activity field, and the cell responded with sustained, delay-period activity. (F) The monkey fixated to the right when a reach target appeared outside the cell's activity field. During the delay period, a cue instructed the monkey to saccade to the left. Immediately after completion of the saccade, discharge in the neuron increased to resemble that in E. (A–C from Duhamel et al.,[10] D–F from Batista et al.[11] with permission.)

above and to the left of the fixation point (figure 17.2B). The most interesting observation occurred in a condition in which the experimenters presented a stimulus shortly before the saccade but removed it before the saccade ended. In this condition (figure 17.2C), they turned the second stimulus off at least 150 msec before the saccade. The cell discharged robustly after the saccade brought the location of the vanished stimulus into the cell's activity field. This finding suggests that when the eyes moved, the CNS had updated the memory of the stimulus location in fixation-centered coordinates.

A more recent experiment investigated this kind of remapping in the context of making reaching movements to remembered targets. Aaron Batista, Chris Buneo, Larry Snyder, and Richard Andersen[11] trained a monkey to fixate one location and reach to a target just outside the cell's activity field. For example, figure 17.2D shows that after the target appeared briefly outside the cell's activity field (the shaded area), the neuron did not respond. Not surprisingly, if the target appeared inside the cell's activity field, the cell discharged vigorously and maintained its discharge during the delay period (figure 17.2E). Now consider what happened when the monkey reoriented its gaze to the right and the reach target appeared, briefly, outside the cell's activity field. The cell did not discharge at first. However, if the monkey later made a saccade to the left, bringing the location of the vanished stimulus into the cell's activity field, the neuron showed delay-period activity for the reaching movement. This activity began almost immediately after the saccade (figure 17.2F). Therefore, in the period before the reach, this cell appeared to encode the target's location in a way that compensated for the intervening eye movement and the disappearance of the stimulus. The CNS maintained the memory of target location in fixation-centered coordinates and remapped those coordinates as the eyes moved.

The mechanism for remapping and predictive updating remains unclear. Evidence presented in section 16.1B, from the work of Gaffan and Hornak, suggests that cortical regions caudal to PPC must interact with frontal areas for such updating to occur, and a model has recently addressed updating for saccadic eye movements.[12] Despite the uncertainty about its mechanisms, the neurophysiological evidence indicates that predictive updating follows eye movements in both LIP and PRR. The next section presents psychophysical evidence for similar updating with head and trunk movements.

17.1.2 Remapping due to Intervening Head and Trunk Movements

Consider again your effort to reach for a coffee cup while reading this book. When you change your eyes' orientation to look at the clock (figure 17.1D), your CNS had to update the fixation-centered estimate for the location of the coffee-cup handle and your hand. Note that, as illustrated in the figure, you not only moved your eyes but also rotated your head (and maybe your trunk) at the same time. In theory, to have maintained accurate reaching, your CNS needed to remap the remembered locations of the target and the hand in fixation-centered coordinates as your eyes, head, and trunk rotated.

To investigate this question experimentally, Gabriel Gauthier and his colleagues[13] asked participants to sit in a chair and reach to the remembered location of a target that appeared briefly. A light, mounted on a table that did not move, served as the target. The chair, however, rotated (figure 17.3A). The participant fixated a light that was mounted on the rotating platform. The experimental setup ensured that the eyes remained fixed in the head and that the head remained fixed relative to the shoulder girdle. The chair began rotating ~30 msec before the reaching movement started, and both the amplitude and the direction of chair rotation varied randomly from trial to trial (figure 17.3B). Therefore, as the chair began rotating, target location but not hand location changed in fixation-centered coordinates.

The experimenters tested reaching movements in darkness. This meant that in order for the hand to reach the target, the location of the target needed to be remapped in fixation-centered coordinates to compensate for rotation of the trunk. Remarkably, after a few practice trials, reaching movements under these conditions had almost the same accuracy as without rotation.

From the perspective of a Cartesian coordinate system centered on the stationary table, the reaching movements looked about the same with or without chair rotation (figure 17.3C). However, from the perspective of the participant's shoulder, for example, the movements differed dramatically (figure 17.3D). If the chair rotated clockwise, the reach involved shoulder flexion. If the chair rotated counterclockwise, the reach involved a large shoulder extension. Regardless of the rotations, the reaches remained accurate.

17.2 Predictive Remapping of Target and End-Effector Locations with Efference Copy

Because reaching movements appear to be planned in fixation-centered coordinates, the fact that people can make accurate reaches despite body rotations (figure 17.3) and eye movements implies that the CNS re-estimates target $\hat{x}_t$ and end-effector $\hat{x}_{ee}$ locations. These computations, in turn, result—at least in theory—in a recalculation of the desired difference vector x_{dv}. As a reach progresses, the CNS must update $\hat{x}_t$, $\hat{x}_{ee}$, and x_{dv} continuously, even as x_{dv} approaches 0.

In principle, you can imagine two different kinds of mechanisms that might drive this reestimation of target and end-effector locations. One would depend primarily on proprioceptive and vestibular feedback. When some external agent moves your body, as in figure 17.3, this sensory information plays the largest role. Another process depends on motor commands, also known as **corollary discharge** or efference copy. For example, if you plan to make a saccade, your CNS might use that plan to reestimate the location of the target of a reaching movement in

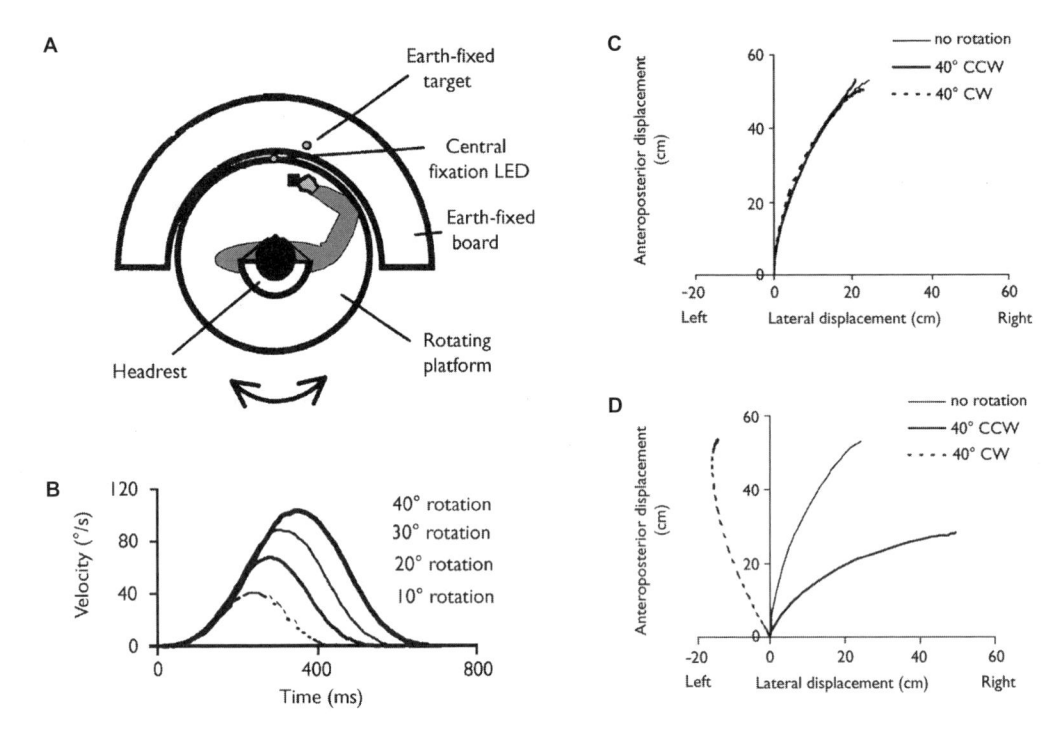

Figure 17.3
Reaching movements while the body and head rotated. (*A*) Experimental setup. Reaching movements made in the dark to the remembered location of a target 57 cm away. The participant fixated a light-emitting diode (LED). (*B*) Rotational velocity of the chair. Typically, a reaching movement began at around the time that the chair began its rotation and lasted about 700 msec. (*C*) Mean hand trajectories produced by a participant in the control condition, in which no rotation of the chair occurred, and for 40° clockwise (CW) and counterclockwise (CCW) rotations. The trajectories were represented in an earth-fixed frame (i.e., a Cartesian coordinate system centered on the table). (*D*) The same trajectories plotted in a participant-based frame of reference. (From Bresciani et al.[13] with permission.)

fixation-centered coordinates. Current evidence suggests that the CNS uses efference copy to update target and end-effector locations when it can do so.

17.2.1 Remapping Locations Before a Planned Eye Movement

In the experiment described in section 17.1.1, Duhamel et al.[10] made another important finding. They showed that the CNS updates the estimated location of a visual stimulus based on efference copy and that PPC activity before a planned saccade reflects this updating.[14] Consider the approximate location of a cell's activity field, as shown by the dotted circle in figure 17.4A. Figure 17.4B shows that this cell discharged when a stimulus fell in its activity field after the saccade was completed. If the cell's

17.2. Predictive Remapping of Target and End-Effector Locations

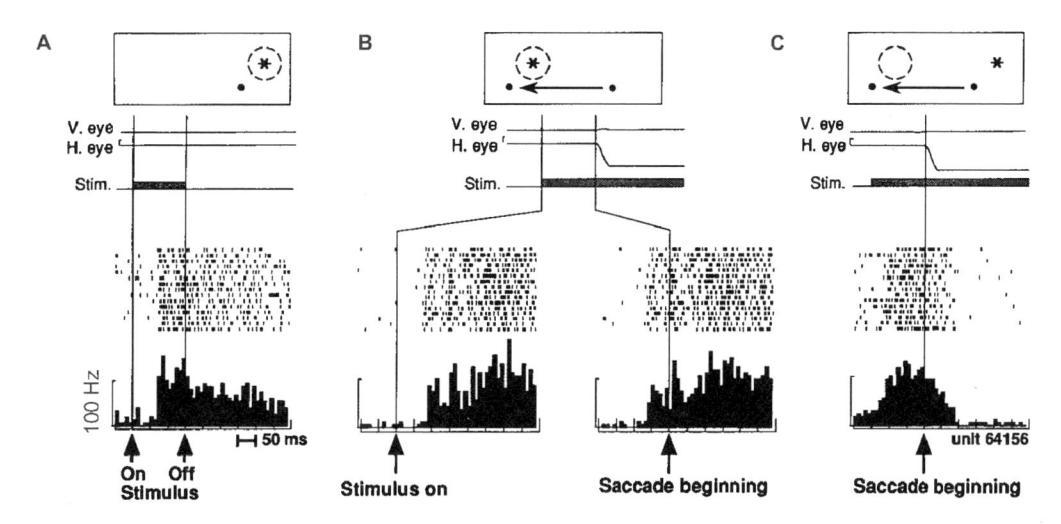

Figure 17.4

Effect of a planned saccade on adaptive remapping. Format as in figure 17.2A. (*A*) Activity during the onset and offset of a visual stimulus (thick bar to the right of "Stim"). (*B*) The fixation point jumped, and the cell's discharge increased before a saccade that brought the stimulus into the cell's activity field. (*C*) The saccade took the visual stimulus outside the cell's receptive field. (From Duhamel et al.[10] with permission.)

discharge resulted only from a passive response to the stimulus, then it should increase activity ~70 msec after the saccade's completion, when the saccade brought the stimulus into the cell's activity field. However, Duhamel et al. found that the cell's discharge increased 80 msec *before* the beginning of the saccade, 150 msec *before* the stimulus came into the cell's activity field. Many LIP cells showed this predictive property.

In an important control condition (figure 17.4C), the stimulus appeared in the cell's activity field (above and to the right of the fixation point) at first, but as the fixation point jumped, the stimulus disappeared. In this condition, the cell's discharge returned to baseline levels, as would be expected if the cell's activity field moved away retinotopically. This decline, however, occurred faster than if the stimulus simply disappeared (figure 17.4A, stimulus off), presumably because the saccade took the stimulus outside the cell's activity field, and the CNS updated its location relative to the fixation point. Carol Colby and her colleagues have shown that this predictive remapping occurs prominently in PPC and in FEF, but less commonly in more posterior visual areas.[15]

17.2.2 Remapping Locations with Efference Copy

Re-estimation of stimulus location—when it occurs ahead of a planned eye movement—must rely on a copy of the neural commands sent to the eyes. Marc Sommer and Bob Wurtz[16] confirmed this idea experimentally. They hypothesized that the cerebral cortex receives a copy of oculomotor commands in order to perform the updating function and that this information arrives via the thalamus. First, they recorded from the dorsomedial nucleus of the thalamus and found that its cells increased their activity about 140 msec before saccade onset (figure 17.5A). Because this activity began before movement, it could not have resulted from sensory feedback. Then the experimenters used a task called the *double-jump saccade task* to examine predictive updating. In this task, monkeys had to make successive saccades to two briefly flashed targets. The two targets, T1 and T2, appeared one at a time and disappeared while the monkey maintained fixation, long before either saccadic eye movement. Therefore, the monkey saw neither target when the saccades took place, and had to guide movements from memory.

To make the first saccade (S1), the monkey had to remember T1's location in fixation-centered coordinates. To make the second saccade (S2), the monkey had to remember the location of T2 and to recompute its location in fixation-centered coordinates to take into account the change in eye orientation caused by S1.

The experimenters used a GABA agonist, muscimol, to inhibit neurons in the mediodorsal nucleus of the thalamus. This inactivation should prevent the relay of the oculomotor efference copy to the cortex. After the injection, the monkey had little trouble making the first saccade, but consistently made large errors for the second saccade (figure 17.5C). S2 ended up to the right of T2, as predicted if the inactivation prevented a re-estimation of the location of T2. The monkey generated its second saccade as if the first one had not occurred.

17.2.3 Role of PPC in Updating of Target Location

Return again to the scenario presented earlier in this chapter: reaching for a coffee cup while reading (figure 17.1). To plan this movement, you estimate target location $\hat{x}_t$, end-effector location $\hat{x}_{ee}$, and the resulting difference vector x_{dv}. As you begin your reach, you decide to check the time (figure 17.1D), and so you reorient your gaze from the book to the clock. Because $\hat{x}_t$ and $\hat{x}_{ee}$ represent target and end-effector locations in fixation-centered coordinates, your eye movements result in a re-estimation of these variables. From the results summarized in figures 17.4 and 17.5, it seems that your CNS uses a copy of its oculomotor commands to remap these variables. Now imagine that someone bumps the table, causing a little shift in the location of the coffee cup. Therefore, target location changes as you make your reaching movement. In fact, the table is about to tip over, but you deftly snatch the handle of the cup as it slides away and rescue your coffee. In order to reach accurately, your CNS has to recompute $\hat{x}_t$, compare it against the current estimate of end-effector location $\hat{x}_{ee}$, and calculate a new difference vector x_{dv}.

Following some earlier leads,[17] Michel Desmurget and his colleagues[8] hypothesized that this kind of updating depends on the PPC. To test their idea, they did an experiment in which participants could see a fixation point and a reach target, but not their arm or hand. After the tar-

17.2. Predictive Remapping of Target and End-Effector Locations

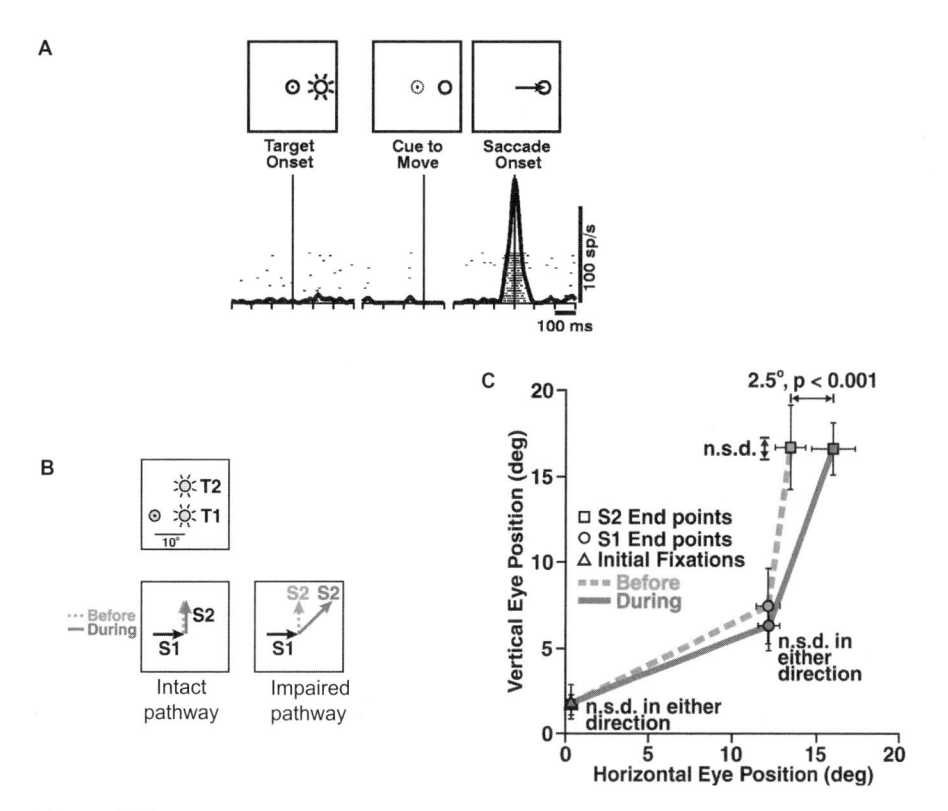

Figure 17.5

Effect of disrupting the pathway that carries an efference copy for saccades to the cortex. (A) The activity of a neuron in the dorsomedial nucleus of the thalamus. The monkey fixated the center location as a peripheral target appeared. The fixation spot then disappeared as a trigger (cue to move) stimulus for the saccade (arrow). The discharge of this neuron preceded saccade onset. (B) Injection of muscimol inactivated the mediodorsal nucleus as the monkey performed a double-jump saccade task. After the monkey fixated the spot depicted as a dot inside a circle, two targets appeared sequentially (T1 and T2); both disappeared before the first saccade (upper drawing). The monkey then made sequential saccades (S1 and S2) to the targets, in the order they had appeared. If the CNS remapped T2's location in fixation-centered coordinates after the first saccade, then S2 should be accurate (lower left). If not, then S2 should have an error to the right (lower right). (C) Means (and SDs) of eye orientation, in the format of B, lower right. Abbreviation: n.s.d., not significantly different. (From Sommer and Wurtz[16] with permission.)

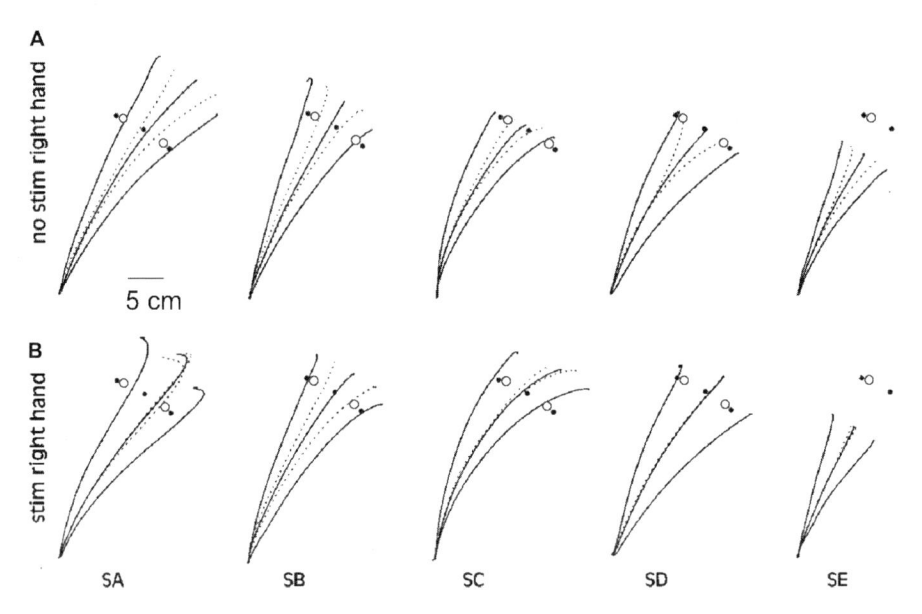

Figure 17.6

Disruptive stimulation of the PPC prevented online control of reaching movements. Five participants (left to right) reached to targets with their right hand. In some trials, the target was displaced by 7.5° as the participants saccaded to the original target before the end of the reaching movement. The experimenters applied single transcranial magnetic stimulation (TMS) pulses to the region of PPC just after movement onset. Solid lines show the mean trajectories toward stationary targets; dashed lines show the mean paths directed at displaced targets. Filled circles the show locations of stationary targets; unfilled circles show the locations of the displaced targets. Except for the second participant (SB), TMS prevented autopilot adjustments of trajectory. (From Desmurget et al.[8] with permission.)

get appeared, the participants made a saccade to fixate it and then started their reaching movement. On most trials, the target remained stationary and participants reached in an approximately straight line to the target (figure 17.6A). On some trials, however, as the participants made a saccade to the target, it jumped to a new location. The participants had no trouble compensating for the target's jump, although they did not notice it. Their hands smoothly veered to the right or left, corresponding to the shift in target location (figure 17.6A). The reach appeared to be on "autopilot" control: Reach was accurate without the person's noticing either the change in target location or the deviation of his or her reach trajectory.[18]

To assess the role of the PPC in this online modification of trajectory, Desmurget and his colleagues[8] disrupted its function with a single **transcranial magnetic stimulation** (TMS) pulse. The pulse occurred as hand movements to a target began. On trials in which the target did not jump, the brief disruption of the PPC caused a change in the endpoint of the reach, but these errors had no consistent pattern: The participants' hands ended up a few centimeters away from those for the nonstimulated, control condition. On trials in which the target jumped, most of the participants had hand movements that disregarded the shift in the target location (participants SA, SC, SD, and SE in figure 17.6B). They reached along a path much like that for a stationary target. Without the TMS pulse, of course, the participants reached accurately to the shifted target, as if on autopilot control. Stimulation of M1 and its vicinity produced a small twitch in the wrist but did not interfere with the autopilot. Therefore, it appears that the PPC participates not only in planning for the movement (i.e., the computation of a difference vector) but also in the online monitoring of target and end-effector location and the updating of these vectors.

In a subsequent experiment, Helene Grea and her colleagues[9] studied a patient who had suffered bilateral lesions of the PPC. This patient was interesting because she did not exhibit spatial neglect. She could reach to objects normally if the images of those objects fell in or near her fovea. However, she was impaired when the target jumped during the reach. The experimenters asked the PPC patient and a group of normal participants to reach for a cylinder (figure 17.7A). The cylinder usually remained stationary at a central location, but occasionally, during the onset of the reaching movement, it moved either left or right. The control participants had no trouble compensating for this shift in target location. The hand path changed midflight and veered smoothly toward the second location of the target (figure 17.7B). However, the patient with damage to her PPC could not make this autopilot correction. Her arm completed the reach toward the initial location of the target, and then made a second movement toward the current location of the target (figure 17.7C).

In summary, it appears that when you reach for a target, your CNS continuously updates its estimate of target location $\hat{x}_t$ and the difference vector x_{dv}. Because the CNS computes vector $\hat{x}_t$ in fixation-centered coordinates, this vector changes with eye, head, or trunk movements. Vector $\hat{x}_t$ also changes when the target moves. As movement progresses, your CNS monitors the progress of the end effector toward its goal and makes adjustments if the target changes or something perturbs the limb. Importantly, the mechanism for updating $\hat{x}_t$ appears to depend less on proprioceptive feedback and more on efference copy. What about $\hat{x}_{ee}$?

17.3 Remapping End-Effector Location

17.3.1 Updating End-Effector Location During Reach

Previously, you imagined reaching for a coffee cup while looking somewhere else. In that scenario, you did not look at your hand, but you could have. Typically, however, when you reach for something, you fixate the target of reach, not your hand. How could you test the hypothesis that as the reach progresses, your CNS continually updates its estimate of end-effector location in fixation-centered coordinates?

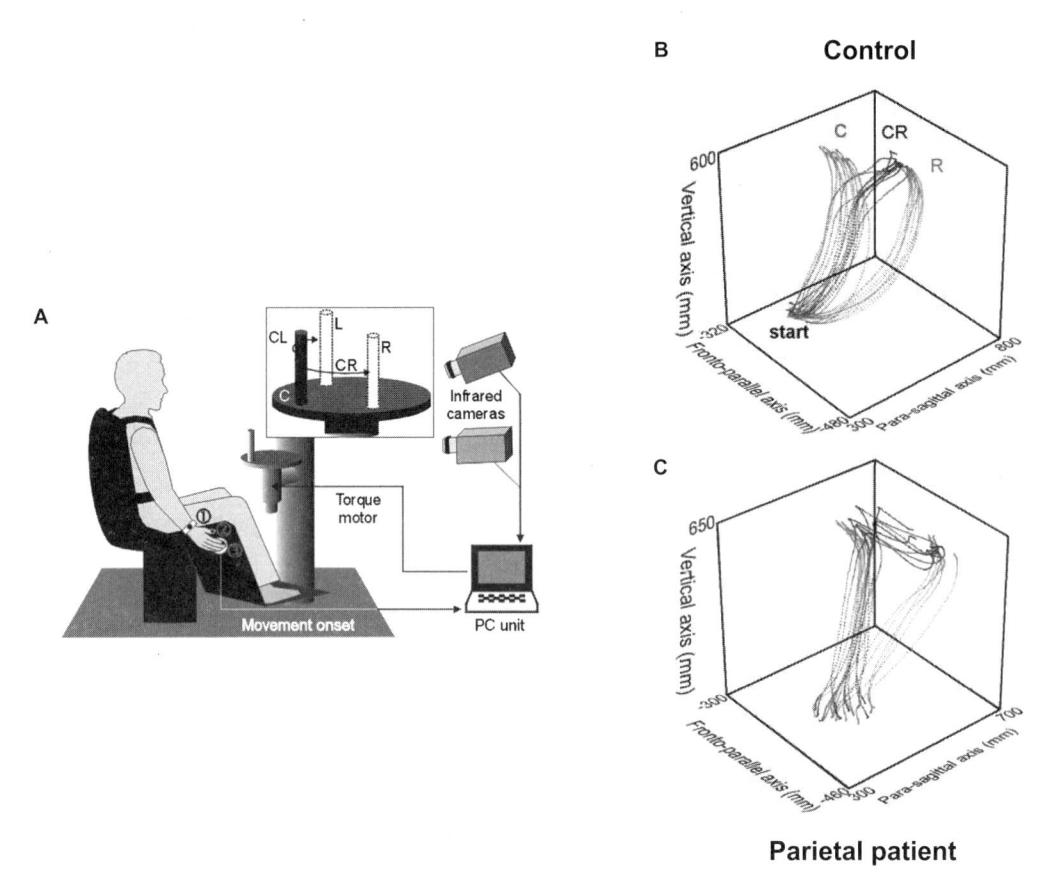

Figure 17.7
A patient with bilateral PPC lesions in an autopilot reaching task. (*A*) Participants sat in a chair facing a cylinder, the reach target. On some trials, the cylinder remained stationary at either the center (C), left (L), or right (R) location. On others, it shifted instantaneously from the center to the left (CL) or to the right (CR) as the reach started. (*B*) Hand paths of a control participant reaching to stationary targets at C and R, and to a displaced cylinder CR. (*C*) Hand paths of a patient with damage to the PPC for a stationary cylinder at R and a cylinder displaced from C to R (CR). (From Grea et al.[9] with permission.)

Greg Ariff, Opher Donchin, Thrishantha Nanayakkara, and Shadmehr[19] tested this hypothesis by asking participants to try to look at their hand's location during movement, but without vision of the hand. The experimental setup is shown in figure 17.8A. Participants held the handle of a robotic **manipulandum** and looked at their hand while the experimenters monitored eye orientation. A trial began with the participant fixating a light indicating the handle's location. A target briefly appeared at a random location, and after a "go" signal, the participant reached to the target. As the hand began to move, the handle light disappeared, removing visual feedback regarding hand location. A field of random dots appeared on the screen, preventing a smooth-pursuit eye movement.

17.3. Remapping End-Effector Location

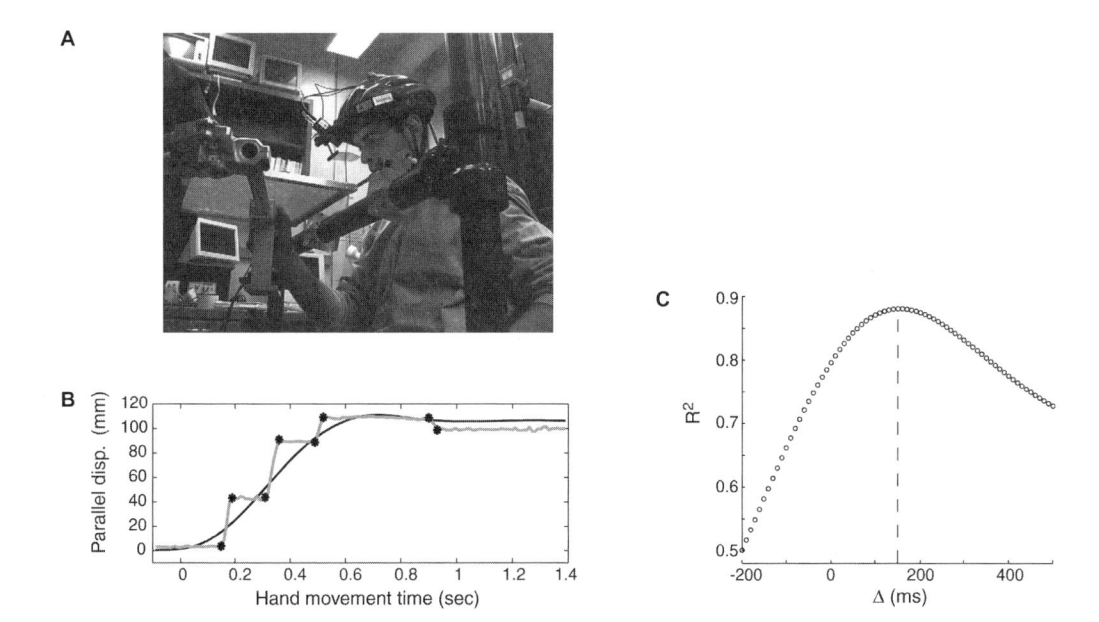

Figure 17.8
Estimating hand location during reaching movements. (A) Participants held the handle of a robotic manipulandum while making reaching movements. They were instructed to look at their unseen hand as they reached to a remembered target location. The participants could not see their arms in the actual experiment (unlike in the photograph). (B) Hand location (solid line) and eye orientation (gray line) for a typical movement, along one axis. (C) Eye orientation at each saccade endpoint correlated with hand location 150 msec later. (From Ariff et al.[19] with permission.)

In the absence of smooth eye movements, the participants attempted, by prior instruction, to make saccadic eye movements to the location of their unseen hand as the reach progressed.

As the movement unfolded, the participants' fixation points jumped ahead of hand location by about 150 msec (figure 17.8B). For a saccade at time t, the endpoint of the saccade $i(t)$ best correlated with hand location at ee ($t + 150$ msec) (figure 17.8C). This result suggests that the CNS updates $\hat{x}_{ee}$ as a reaching movement progresses. Because the estimate of hand location leads the actual hand location, the updating mechanism probably relies on efference copy and not solely on proprioception. In section 17.3.3, this mechanism gets a name: forward model. For now, you only need to know that you can use motor commands, in the form of efference copy, to update your estimation of end-effector location $\hat{x}_{ee}$.

You saw earlier that efference copy also contributes to remapping target location when your eyes reorient. Therefore, the updating of both target location $\hat{x}_t$ due to eye movements and of end-effector location $\hat{x}_{ee}$ due to reaching movements appears to involve predictive mechanisms that rely on efference copy.

17.3.2 Updating End-Effector Location After Perturbation

What happens to your estimate of end-effector location if a perturbation unexpectedly makes your hand jump as you reach for a target? Does your CNS use the proprioceptive feedback to recompute current end-effector location despite perturbations?

Ariff and his colleagues[19] examined this question with the setup shown in figure 17.8A. As before, participants held the handle of a robotic arm and made reaching movements to a remembered target without vision of their hands. The experimenters instructed the participants to look at their hand during the reaching movement. However, the robot occasionally perturbed the hand by pushing on it in a random direction with a force pulse of 50-msec duration. Figure 17.8B shows that participants look ahead of the hand during the reach, predicting end-effector location based on efference copy. However, for perturbations the CNS could not predict what would happen. It could only use proprioception to detect a change in end-effector location.

Figure 17.9A shows two sample trajectories of a perturbed reaching movement. The labels e1, e2, and e3 indicate eye orientation at the end of the first, second, and third saccades. Similarly, h1, h2, and h3 label hand locations at the time of each saccade. The rightmost trial also has a fourth data point: e4 and h4. In an unperturbed reaching movement, both hand location and eye orientation follow a straight line to the target, with the fixation point leading the hand. However, when the robot perturbs your hand, your eyes do not follow this straight path. Rather, your saccades deviate in the direction of the perturbation. Therefore, if you assume that the fixation point in this task reflects an estimate of end-effector location, then it cannot depend solely on efference copy for the reaching movement. Rather, $\hat{x}_{ee}$ must also depend on proprioception from the arm. However, proprioceptive information requires time to reach the CNS, especially the cortex. Therefore, you might expect the fixation point to lag the end-effector location. Interestingly, despite the fact that the perturbation displaced the end effector in a random direction and amplitude, the fixation point still led the hand. On average, a post-perturbation saccade at time t resulted in an eye orientation $i(t)$ that best correlated with ee $(t + 150$ msec). Figure 17.9B shows the relationship between eye orientation and hand location for this time lead.

The timing of saccades after a perturbation also followed an interesting pattern. In an unperturbed reaching movement, saccade timing was described by a broad distribution (figure 17.9C). However, when the hand was perturbed, saccades were inhibited about 150 msec after the onset of the perturbation. This interval is long enough for proprioceptive feedback to arrive in the brain, including the cerebral cortex. This signal indicates that the hand will not be at a predicted location in the near future and appears to cancel the planned saccade to that location. This pattern of post-perturbation saccade inhibition occurred regardless of when the perturbation occurred during the reaching movement. About 200 msec after the onset of the perturbation, saccade probability rose sharply in response

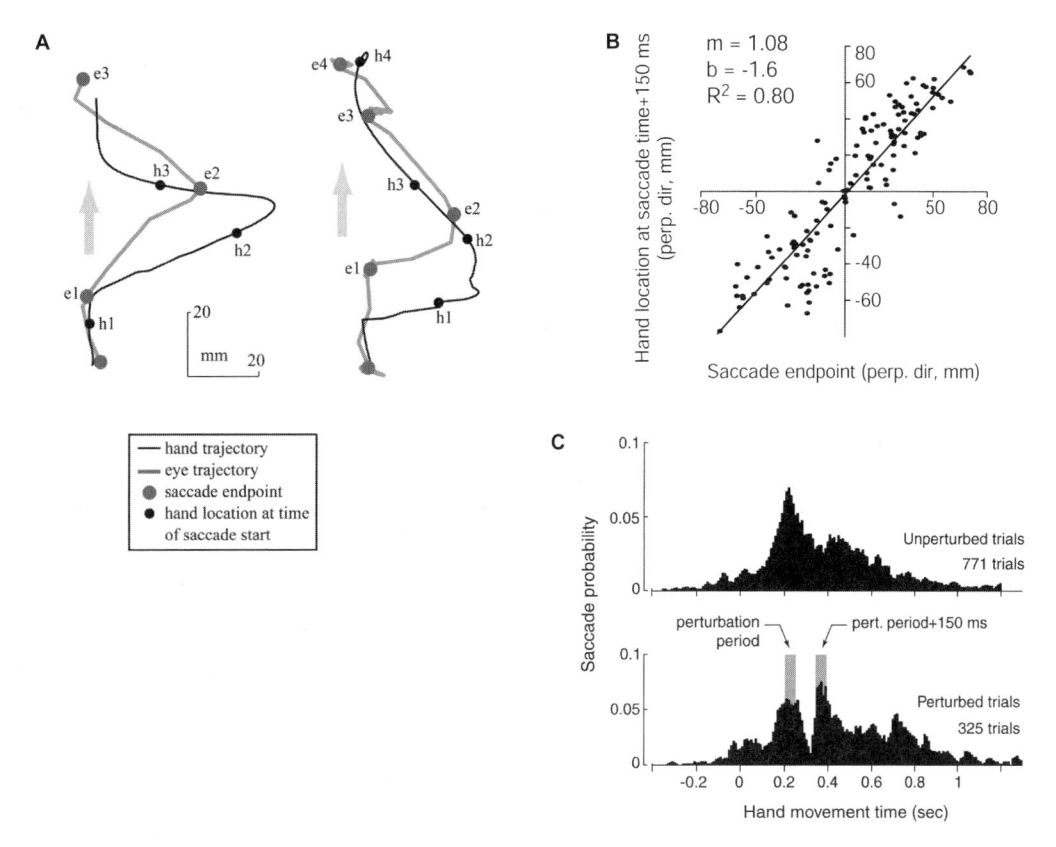

17.3. Remapping End-Effector Location

Figure 17.9
Eye movements during perturbed reaching movements. Experimental setup as in figure 17.8A. As a participant reached to the remembered location of a target, the robot perturbed the hand with a 50-msec force pulse. The direction of the force vector was perpendicular to the direction of motion. It varied randomly to push the hand either clockwise or counterclockwise. (*A*) Fixation point and hand trajectory for two typical trials. The labels e1, e2, and e3 indicate eye orientation at the end of the first, second, and third saccade, respectively. Labels h1, h2, and h3 correspond to hand locations at the same times. Abbreviations: pert., perturbation; perp., perpendicular. (*B*) Post-perturbation saccades resulted in fixation points corresponding to hand location 150 msec later. (*C*) Saccade probability for unperturbed (top) and perturbed reaching movements (bottom). (From Ariff et al.[19] with permission.)

to the arm perturbation. Figure 17.9B plots the locations of these post-perturbation saccades. The saccades continued to lead hand location by about 150 msec.

However, the post-perturbation saccades no longer predicted hand location accurately if the perturbation lasted much longer than 50 msec. For example, if a force that resisted the motion of the hand followed the force pulse, then the saccades tended to overshoot hand location. If a force that assisted the motion of the hand followed the force pulse, then the saccades tended to undershoot the hand. These observations suggest that the ability of the CNS to estimate hand location during reaching movements depends not only on proprioceptive feedback and efference copy, but also on some kind of internal model of how the arm should behave. This kind of internal model, called a forward model, apparently allows the CNS to predict behavior of the hand in the near future.

17.3.3 Forward Models

As presented earlier in this chapter, when you move your eyes, your CNS appears to re-estimate target location well before the saccade ends. When your hand moves, your CNS updates its estimate of hand location well before proprioceptive feedback returns to it. In both situations, it appears that the estimate of hand or target location relies on a copy of the planned motor commands to the arm and the eyes. In the case of reaching movements, however, knowledge of motor commands alone cannot specify its future location. You also need to know the arm's configuration before the CNS generates motor commands. Furthermore, even if you know the arm's configuration and the commands sent to its muscles, you still cannot predict the arm's future configuration, and therefore the location of the end effector, unless you know something about its physical dynamics.

Dynamics refers to the relationship between force and motion. If you know something's location and how much force you plan to impose on it, and if you also know something about its dynamics (for example, its mass), then you can predict its location in the near future. You can call that "something" that the CNS knows about its dynamics a forward model. A forward model is a kind of internal model that describes how a physical object, such as your arm, should change state (location or velocity), given the action of a force upon it (a motor command to the muscles). It therefore predicts the sensory consequences of a motor command. Indeed, Randy Flanagan and Roland Johansson[20] have recently shown that when participants observe someone else performing a task, the observer's eye movements also predict the performer's hand location by an average of 150 msec.

Forward models compute estimates of a system's dynamical behavior. Chapter 20 begins a detailed presentation on the dynamics of the arm and how forces produce a change in its state. For now, your intuition tells you that if you want to predict where your hand will be in the future, you need to know its mass and other variables that play an important part in

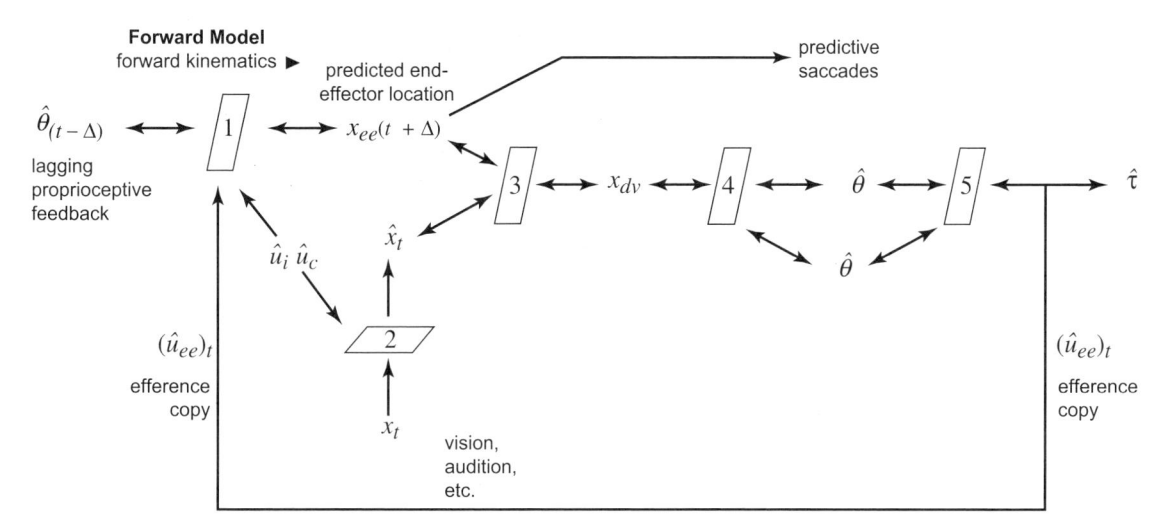

17.3. Remapping End-Effector Location

Figure 17.10
Predictive remapping with a forward model. $\hat{u}_i$ represents a copy of the motor command sent to the eye (i); $\hat{u}_c$ represents a copy of the motor command sent to head (c); and $\hat{u}_{ee}$ represents a copy of the motor command sent to the arm. Proprioceptive input arrives at the CNS after a delay Δ. A forward model uses sensory feedback and efference copy to predict end-effector location at some future time. (Drawn by George Nichols.)

dynamics. If you have a forward model that computes a good approximation of the arm's dynamics, then you can accurately predict future locations of your end effector.

The forward-model theory accounts for the oculomotor behavior seen in figure 17.8B. As the CNS sent commands u_{ee} to the arm muscles to produce force, a copy of these commands $\hat{u}_{ee}$ went to the neural system that estimates end-effector location $\hat{x}_{ee}$. This system also receives proprioceptive information $\hat{\theta}$ from the arm, but only after a delay. The proprioceptive information $\hat{\theta}(t - \Delta)$ tells you your arm's configuration a little while ago. The efference copy $\hat{u}_{ee}(t)$ tells you the commands that will act on the arm. By combining these two pieces of information with a forward model, you can predict where the end effector will be after the commands act on your limb. Therefore, given the proprioceptive information $\hat{\theta}(t - \Delta)$ and the efference copy $\hat{u}_{ee}(t)$, a forward model of arm dynamics predicts end-effector location for the near future $\hat{x}_{ee}(t + \Delta)$. Figure 17.10 shows a schematic of this idea.

Forward models supplement peripheral and spinal mechanisms for limb stability (chapter 8). Spinal reflexes and the muscles' length–tension properties provide fast-acting feedback mechanisms to resist perturbations. If something hits your arm as you reach, as in figure 17.9A, reflexes modify the motor commands so that stretched muscles receive stronger activation signals and vice versa. But your CNS does much more to overcome potential impediments. Based on lagging proprioception and current efference copy, a forward model estimates end-effector location in the near future $\hat{x}_{ee}(t + \Delta)$. Through forward models, efference copy, and proprioceptive feedback, your CNS learns to approximate the dynamics of your limb and can make adjustments to ongoing movements if they go off track.

17.3.4 Damage to PPC Inhibits Motor Imagery

You can think of a forward model as a neural simulator. In the context of reaching movements, a forward model predicts consequences of the motor commands to your arm. Because your CNS provides mechanisms to plan motor commands but not execute them, perhaps the act of imagining a movement reflects the "running" of a forward model using the planned motor commands.

Angela Sirigu and her colleagues[21] examined the ability of patients with PPC damage to monitor imagined movements of their fingers or arms. In their first task, they compared the maximum speed of imagined movements with the participants' actual movements. The experimenters asked participants to *imagine* touching the tip of the thumb with the tip of each finger of the same hand, in time to the sound of a metronome. They slowly increased the speed of the metronome until the individual reported that the imagined movement could no longer keep up. The experimenters then measured how fast the participants could *actually* make the movement. They found that in healthy people, the maximum speed of the imagined movements agreed remarkably well with the maximum speed of the actual ones (within 2%). Patients with unilateral PPC damage could not accurately estimate their performance with the hand contralateral to the lesioned hemisphere. They could, however, estimate their maximum movement speed for the hand ipsilateral to the lesion. In contrast, a patient with a motor cortex lesion in the right hemisphere could accurately estimate maximum speed with both hands, despite the fact that the left hand moved much more slowly.

Next, Sirigu et al. had their participants imagine movements that varied in difficulty. In one such task, participants produced a sequence of hand postures that varied from easy to hard. In another task, they asked the participants to imagine reaching with a pen in order to place the tip inside a small or a large square. In both the hand-posture and the reaching tasks, participants received a trigger signal and imagined performing five consecutive cycles of the same movement. Upon completion, they signaled the experimenter. Afterward, they performed the actual movements. In healthy participants, the time to completion of the imagined and actual movements agreed closely. As the task became more difficult, both the imagined and the actual movements took longer to complete. Patients with motor cortex lesions had trouble making the actual movement with the contralateral arm, but the duration of imagined movements matched that of the actual ones. In contrast, in patients with right PPC lesions, the duration of imagined movements with the left hand or the left arm was significantly less than the duration of actual movements. Imagined and actual movements of the less affected right arm, however, showed a close correspondence. Therefore, damage to PPC affects the ability to imagine the duration of movements.

A number of investigators have reported that similar brain regions show blood-flow fluctuations in actual and imagined movements.[22] Indeed, in M1, real and imagined wrist movements cause the same changes in local blood flow.[23] These results, taken together with the results of Sirigu et al., suggest that the act of imagining a movement involves a forward model that relies on planned (but not executed) motor commands, that this forward model depends on the integrity of PPC, and that M1 and other motor areas reflect the output of that neural simulator.

References

1. Wise, SP, and Mauritz, K-H (1985) Set-related neuronal activity in the premotor cortex of rhesus monkeys: Effects of changes in motor set. Proc Roy Soc London B 223, 331–354.
2. Batista, AP, and Andersen, RA (2001) The parietal reach region codes the next planned movement in a sequential reach task. J Neurophysiol 85, 539–544.
3. Snyder, LH, Batista, AP, and Andersen, RA (1998) Change in motor plan, without a change in the spatial locus of attention, modulates activity in posterior parietal cortex. J Neurophysiol 79, 2814–2819.
4. Georgopoulos, AP, Kalaska, JF, and Massey, JT (1981) Spatial trajectories and reaction times of aimed movements: Effects of practice, uncertainty, and change in target location. J Neurophysiol 46, 725–743.
5. Port, NL, Lee, D, Dassonville, P, and Georgopoulos, AP (1997) Manual interception of moving targets. 1. Performance and movement initiation. Exp Brain Res 116, 406–420.
6. Lee, D, Port, NL, and Georgopoulos, AP (1997) Manual interception of moving targets. 2. On-line control of overlapping submovements. Exp Brain Res 116, 421–433.
7. Desmurget, M, Pelisson, D, Rossetti, Y, and Prablanc, C (1998) From eye to hand: Planning goal-directed movements. Neurosci Biobehav Rev 22, 761–788.
8. Desmurget, M, Epstein, CM, Turner, RS, et al. (1999) Role of the posterior parietal cortex in updating reaching movements to a visual target. Nat Neurosci 2, 563–567.
9. Grea, H, Pisella, L, Rossetti, Y, et al. (2002) A lesion of the posterior parietal cortex disrupts on-line adjustments during aiming movements. Neuropsychologia 40, 2471–2480.
10. Duhamel, J-R, Colby, CL, and Goldberg, ME (1992) The updating of the representation of visual space in parietal cortex by intended eye movements. Science 255, 90–92.
11. Batista, AP, Buneo, CA, Snyder, LH, and Andersen, RA (1999) Reach plans in eye-centered coordinates. Science 285, 257–260.
12. Mitchell, J, and Zipser, D (2001) A model of visual-spatial memory across saccades. Vision Res 41, 1575–1592.
13. Bresciani, JP, Blouin, J, Sarlegna, F, et al. (2002) On-line versus off-line vestibular-evoked control of goal-directed arm movements. NeuroReport 13, 1563–1566.

14. Nakamura, K, and Colby, CL (2000) Visual, saccade-related, and cognitive activation of single neurons in monkey extrastriate area V3A. J Neurophysiol 84, 677–692.

15. Nakamura, K, and Colby, CL (2002) Updating of the visual representation in monkey striate and extrastriate cortex during saccades. Proc Natl Acad Sci USA 99, 4026–4031.

16. Sommer, MA, and Wurtz, RH (2002) A pathway in primate brain for internal monitoring of movements. Science 296, 1480–1482.

17. Prablanc, C, and Martin, OJ (1992) Automatic control during hand reaching at undetected two-dimensional target displacements. J Neurophysiol 67, 455–469.

18. Day, BL, and Lyon, IN (2000) Voluntary modification of automatic arm movements evoked by motion of a visual target. Exp Brain Res 130, 159–168.

19. Ariff, G, Donchin, O, Nanayakkara, T, and Shadmehr, R (2002) A real-time state predictor in motor control: Study of saccadic eye movements during unseen reaching movements. J Neurosci 22, 7721–7729.

20. Flanagan, JR, and Johansson, RS (2003) Action plans used in action observation. Nature 424, 769–771.

21. Sirigu, A, Duhamel, J-R, Cohen, L, et al. (1996) The mental representation of hand movements after parietal cortex damage. Science 273, 1564–1568.

22. Deiber, M-P, Passingham, RE, Colebatch, JG, et al. (1991) Cortical areas and the selection of movement: A study with positron emission tomography. Exp Brain Res 84, 393–402.

23. Naito, E, Roland, PE, and Ehrsson, HH (2002) I feel my hand moving: A new role of the primary motor cortex in somatic perception of limb movement. Neuron 36, 979–988.

24. Gaffan, D, and Hornak, J (1997) Visual neglect in the monkey—Representation and disconnection. Brain 120, 1647–1657.

Reading List

The paper by Gaffan and Hornak,[24] on the representation of targets relative to the fixation point, sheds considerable light on the topic of remapping, although the authors wrote that paper before ideas about representing targets in fixation-centered coordinates became widely known.

18 Planning to Reach or Point I: Smoothness in Visual Coordinates

Overview: A reaching or pointing movement can entail an infinite number of trajectories from the end effector's starting location to the target. However, for most reaching and pointing movements, your CNS plans the movement so that the end effector moves along just one of these trajectories: an approximately straight path with a smooth, unimodal velocity profile. Given the choice between a trajectory that looks straight in visual coordinates and one that is straight in reality, your CNS generates a visually straight trajectory.

Imagine that as you stand in line for coffee, a friend comes over and extends his or her hand to shake yours. Your friend's hand serves as the target of this movement, but rather than fixating the target, you look your friend in the eye. (You have sound ethological reasons for doing so; otherwise your friend might consider you shifty-eyed and strange.) You see neither your hand nor that of your friend. According to the ideas presented here, as you fixate on your friend's face and eyes, your PPC estimates your hand's location as well as the target's location in fixation-centered coordinates, and, along with premotor cortex, computes a difference vector. This difference vector points from your hand to the target in fixation-centered coordinates. PMd, PMv, and M1 cortex can convert this hand difference vector into a joint-rotation vector; and M1, the cerebellum, and maybe SMA further transform this information into a force vector, which it sends through various routes to the spinal cord. (Some experts think that M1 plays only an indirect role in computing a force vector.[1])

Now imagine that as your friend keeps moving toward you, and you make an eye movement to track his or her face, your CNS communicates the planned forces to your spinal cord and brainstem. As it does so, it updates (i.e., remaps) the estimated location of your hand (the end effector) and your friend's hand (the reach target) in anticipation of the planned and ongoing changes in retinal orientation, resulting in a recalculation of a difference vector. This process continues until the difference between the estimated hand location and the target location reaches 0 (i.e., your CNS minimizes the difference vector and your hand reaches your friend's hand).

The difference vector has its origin at the current hand location and points to the target, but it remains a fixation-centered vector. This counterintuitive conclusion follows from the source of the difference vector. According to the present theory, the difference vector results from the subtraction of two fixation-centered vectors: one that represents current end-effector location and one that represents target location. The theory assumes that reaching with your hand, or moving some end effector, involves a control process in which your CNS monitors target and end-effector locations online and brings the end effector to the target.

Now imagine a different way that you might reach to a target. Consider the imaginary robot illustrated in figure 15.1A. To move the robot's gripper from one point to another, you could compute the location of the target in camera-centered coordinates and then estimate the joint angles θ_f that would place the gripper at the target. This computation involves **inverse kinematics**. For example, if you had a map that aligned hand location in camera-centered coordinates with joint angles of the robot (i.e., $x_{ee} \to \hat{\theta}$), you could estimate that when $\hat{\theta} = \theta_f$, the gripper should reach the target. With initial robot angles $\theta = \theta_i$, you could estimate a joint-rotation vector $\Delta\theta = \theta_f - \theta_i$. This mapping tells you that to reach the target, the robot's "shoulder" should rotate by about 90° as its "elbow" angle changes very little (~0°). Now, if you command your robot's joint controllers to flex the shoulder by 90° and keep the elbow angle constant, the gripper moves to the target in an arc (figure 18.1). Interestingly, if you (not your robot) were to reach to the same target from the same starting location, your hand would not move in an arc; it would move in a straight line, unless you specifically wanted to move along a curved trajectory. To move from point a to point b (figure 18.1) along a straight line, your shoulder and elbow joints have to rotate in a fairly complicated way. With your robot, the joints rotate in a simple way, but the gripper moved along a more complex (curved) path.

The robot's joint angles moved in a simple way because you used a control system that focused on minimizing a joint-angle difference vector. In this chapter, you will see that your CNS does not seem to use this

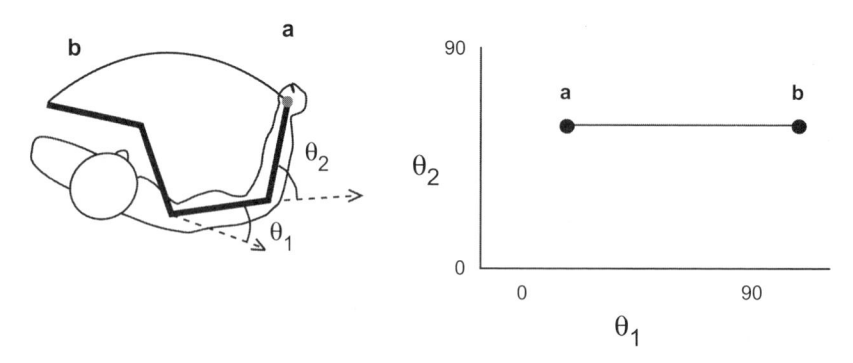

Figure 18.1
A movement from point a to point b can be accomplished by a 90° rotation of the shoulder joint (left), which produces a simple trajectory in joint-centered coordinates (right). (Based on Hogan et al.[2])

approach. Rather, it appears to minimize a hand difference vector, and it does so in a coordinate system closely tied to vision. You can imagine sound biological reasons for this state of affairs. Your visual system will record any error in reaching or pointing in a coordinate system commensurate with that of the target. Thus, the "choice" of fixation-centered coordinates has some advantage over body-centered coordinates. Section 18.1.5 takes up this topic.

18.1 Regularity in Reaching and Pointing

Although your CNS can move your hand through any one of an infinite number of paths to a target, under most unconstrained circumstances your hand (or some other end effector) moves with a smooth velocity and along a straight path.

18.1.1 Minimizing the Difference Vector

Pietro Morasso[3] first quantified the regularity of reaching movements. He performed a simple experiment in which participants held the handle of a manipulandum and reached to a target (figure 18.2A). Participants could see both their hand and the target.

Figure 18.2B shows examples of the hand's trajectory between two target pairs. Because the hand moved along an approximately straight line, the hand difference vector—a vector that points from the hand to the target—continuously became smaller. Now examine the movement in joint coordinates. For example, consider the movement between T2 and T5. The elbow joint initially flexed (i.e., the joint angle became larger), and then extended (figure 18.2C). The shoulder joint initially extended and then flexed (figure 18.2C). Imagine a difference vector that points from the current configuration of the arm in joint coordinates to its final configuration at the target. The elbow component of this joint difference vector first increased and then decreased, and the shoulder component first decreased and then increased. Figure 18.2D plots joint velocity, which represents joint displacement computed over a small period of time. Note how the joint displacements did not follow a unimodal trajectory. If your CNS wanted simply to move your elbow and shoulder to some final configuration, it could smoothly flex both your elbow and your shoulder, and your hand would arrive at a target. However, if that happened, your hand would not move along a straight path.

Morasso found that the x- and y-components of hand location changed monotonically until the hand reached the target (figure 18.2E). The hand difference vector always points from the current hand location to the target. Therefore, figure 18.2E suggests that both the x- and the y-components of the hand difference vector gradually decreased during the reach. The hand-velocity vector had a single peak (figure 18.2F), as compared to the multipeak velocities for the joint-velocity vector. The monotonic change in the components of the hand location vector during the

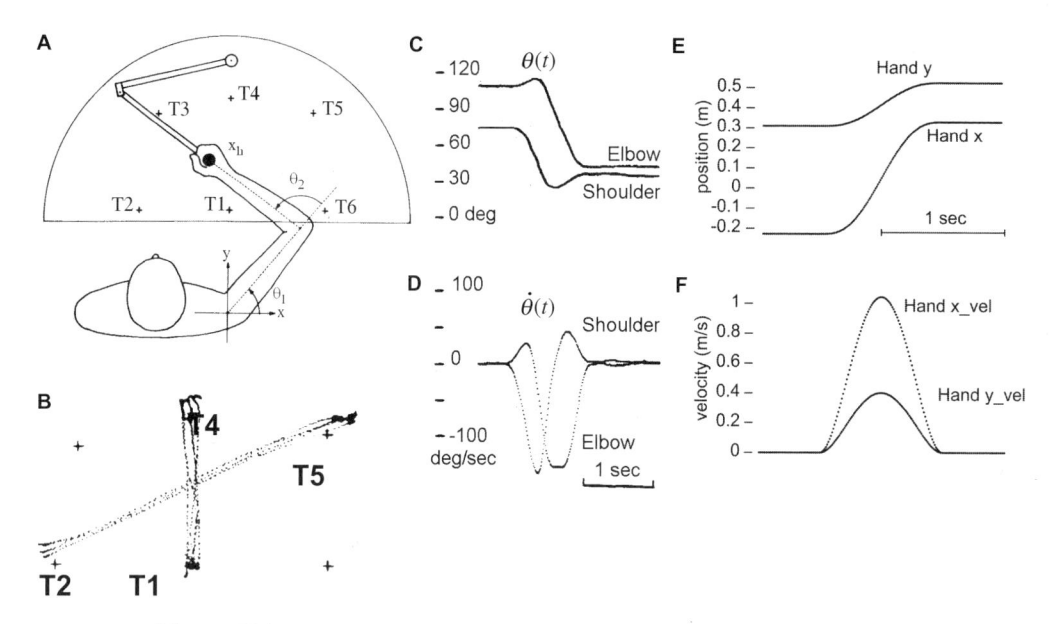

Figure 18.2
Straight trajectories for reaching and pointing movements. For point-to-point reaching movements performed in the horizontal plane, hand displacement follows an approximately straight line. This fact implies that each component of the difference vector changed monotonically as a function of time. (A) Targets (T1–T6) used for measuring reaching movements. (B) Typical reaches from T2 to T5 and from T1 to T4. (C) Joint angles during the reach. (D) Joint velocity. (E) Hand location (estimated). (F) Hand velocity (estimated). (A–D from Morasso[3] with permission, E–F computed from an estimate of hand location in B.)

reach seems consistent with a control scheme involving the gradual minimization of the hand difference vector, until the hand reaches the target.

18.1.2 Altering Perception of Path to a Target

The data presented in figure 18.2 expose a vulnerability in the present theory. Remember that the theory postulates that the CNS computes an end-effector difference vector in fixation-centered coordinates. Figure 18.2, however, plots end-effector locations in Cartesian coordinates. If something changes monotonically in Cartesian coordinates, does it also do so in fixation-centered coordinates? Morasso did not provide information about eye orientation for the data presented in figure 18.2, but, Daniel Wolpert and his colleagues performed two experiments that addressed this issue.

In their first experiment, Wolpert, Zoubin Ghahramani, and Michael Jordan[4] asked participants to sit in front of a digitizing table and hold a computer mouse. The participants viewed an image that corresponded to the location of their finger, though they could not see their hand (figure 18.3A). The participants needed to reach to a target at a distance of 30 cm.

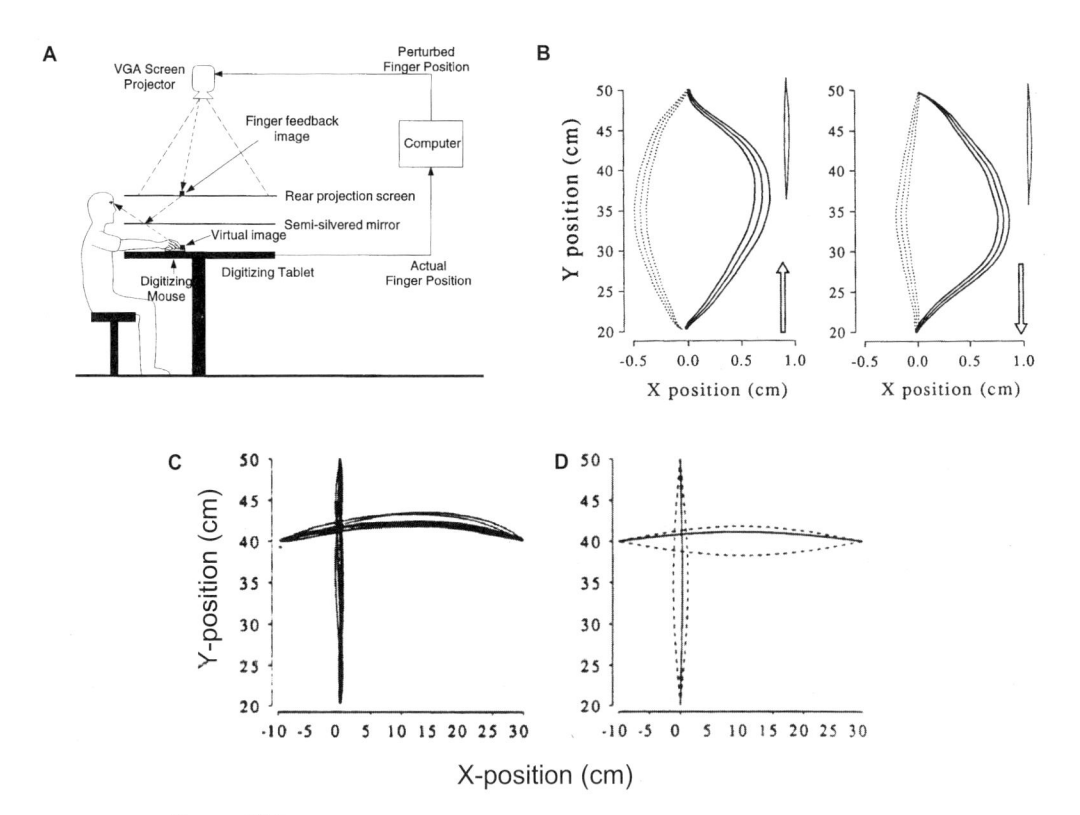

18.1. Regularity in Reaching and Pointing

Figure 18.3
Motor commands produce a reach that appears straight in visual coordinates. (*A*) Participants viewed a cursor that represented their hand's location while they reached to a target 30 cm away. In visually perturbed trials, the experimenters added a hemi-sinusoid to the hand-location feedback. Accordingly, the displayed hand location differed from the actual location during the movement but not at the target location. In this way, the experimenters altered the visually sensed trajectory of the hand without affecting feedback about end-point accuracy. (*B*) Dotted lines show the mean hand trajectories in the unperturbed condition (± 1 S.D.). Solid lines show hand trajectories in the visually perturbed condition. When the experimenters distorted the visual feedback, the participants produced a visually straight hand trajectory rather than an actually straight one. (*C*) The finger path for a single participant during trials without visual feedback. The axes are relative to the head orientation. (*D*) The dashed lines show the range of the paths used in the perception experiment. The solid lines show the paths that the participant estimated to have no curvature. (*A* and *B* from Wolpert et al.,[4] *C* and *D* from Wolpert et al.[5] with permission.)

During the movement, the computer displayed either the actual hand location or an altered version of it.

When the experimenters displayed an actual hand location, participants reached approximately straight to the target (the dotted lines in figure 18.3B). However, Wolpert and his colleagues wondered what would happen if they altered the displayed hand path between the starting location and the target but not at the target. Instead of displaying true hand location, the computer added a hemi-sinusoid in the negative x-direction. The sinusoid had a value of 0 at the starting and target locations. Thus, the experimenters distorted the visual feedback about hand location during the reach, but not at the beginning or the end of the movement. The visual distortion between the two end points remained so small that the participants did not report any awareness of the perturbation. So the question was: Given the choice between visually straight reaching movements and actually straight ones (in Cartesian coordinates), which choice does your CNS make? The answer is that the CNS chooses visually straight movements. After a few practice trials, people changed their hand trajectory so that the actual hand location (in Cartesian coordinates) moved in an arc to the right and the visually sensed hand location moved in an approximately straight line (figure 18.3B). This result supports the idea that the CNS uses a control process that monitors end-effector location in visual coordinates in order to minimize the hand difference vector.

In a second experiment, Wolpert and his colleagues[5] focused on the fact that although the hand's trajectory took approximately a straight line, it still showed some curvature (figure 18.3B). They wondered whether there was a relationship between visual perception of curvature and the curvature in a movement. Perhaps their participants' hand trajectories curved slightly because they (wrongly) perceived that small curvature as straight.

To test this idea, the experimenters asked participants to make reaching movements between targets. After about 40 movements with visual feedback, the participants performed 20 more movements in the dark. Figure 18.3C shows an example of trials without visual feedback. Note that the left–right movements have a fair amount of curvature, while the up–down movements seem straighter. Next, the same persons participated in a perceptual task in which they made no movements but observed a cursor traversing a 600-msec trajectory between the same targets. On each movement, the experimenters added a hemi-sinusoid of variable amplitude to the straight-line trajectory. In this way, the cursor path curved to one side or the other. The participants had to decide the direction in which the cursor had curved. In figure 18.3D, the dashed lines show the range of stimuli used in the study, and the solid lines show the trajectory for which the participants estimated 0 curvature. Note the correlation between the curvature perceived as straight in figure 18.3D and the curvature of actual arm movements in figure 18.3C.

Perhaps your CNS tries to make straight-line movements in visual coordinates, but the actual movements have a gentle curvature in Cartesian coordinates because your visual system distorts Cartesian space. This distortion makes a slightly curved movement in Cartesian space appear straight in visual coordinates.

18.1.3 Moving an End Effector

Although the curvature of movements in figure 18.3 was small, the results indicated that the CNS minimized a hand difference vector computed in visual coordinates. However, what happens if the coordinate system in which you view an end effector reflects joint-angle coordinates instead of Cartesian ones? Randy Flanagan and Ash Rao[6] designed an experiment to answer this question. They measured shoulder and elbow angles and programmed a video display so that as the shoulder angle increased, the cursor moved to the right. As the elbow angle increased, the cursor moved up.

Participants sat at a table with their upper and lower arms resting on an air sled. They could not see their hands, but could view a computer monitor that displayed a cursor. In the *hand-space* condition, the (x, y) location of the cursor corresponded to the Cartesian location of the hand. In the *joint-space* condition, the (x, y) location of the cursor corresponded to the shoulder and elbow angles of the arm (i.e., the limb's coordinates in joint space). Flanagan and Rao arbitrarily made the shoulder angle correspond to the ordinate and the elbow angle to the abscissa, in terms of the monitor's Cartesian coordinates.

Participants made consecutive movements among five targets. The experimenters observed that in both conditions, the cursor on the screen moved straight to the target. In the hand-space condition, it took only a few practice trials for this to occur. In the joint-space condition, however, it took about 400 trials, but practice resulted in straight cursor paths on the monitor.

In the hand-space condition, the straight cursor movements corresponded to curved motion of the arm in joint space but straight movements of the hand. In the joint-space condition, the straight cursor movements corresponded to curved motion of the hand but straight motion of the arm in joint space. Therefore, when participants moved a cursor to a target, they moved it straight to the target in visual coordinates, regardless of the hand's path in Cartesian coordinates and the changes in joint coordinates.

This experiment required a novel mapping between end-effector displacements and joint rotations (i.e., $x_{dv} \xrightarrow{\hat{\theta}} \Delta\theta$ in terms of figure 14.1 [network 4]). When cursor movements corresponded to motion of the hand in Cartesian coordinates—the hand-space condition—the mapping remained much like that used in controlling a cursor with a computer mouse. It took only a few practice trials to move straight to each target. When the cursor movements corresponded to joint rotations, however, x_{dv} mapped to $\Delta\theta$ differently. This new transformation took a couple of hours of practice to learn, but once participants learned this mapping, they moved the end effector straight to the target in visual coordinates.

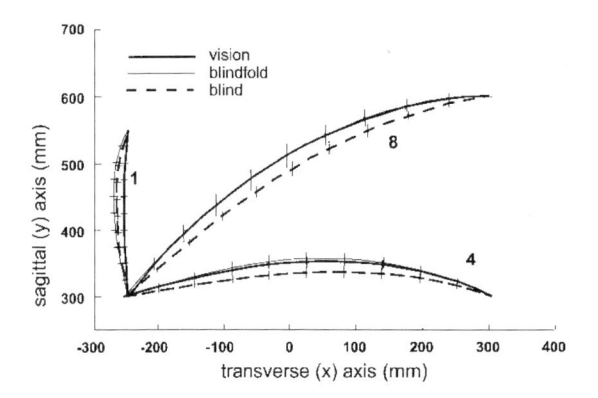

Figure 18.4
Reaching movements in visually impaired people. Solid lines show hand trajectories of participants with typical vision, who were either blindfolded (thin, solid line) or provided with sight of their hand during the reach (thick, solid line). Dashed lines show trajectories in the visually impaired (blind). (From Sergio and Scott[7] with permission.)

18.1.4 Reaching Movements in the Visually Impaired

You might wonder, with all of this attention to a visual, fixation-centered representation of hand location and the difference vector, how visually impaired people reach. For example, what about congenitally blind individuals who have had little or no experience of the world through their visual system? Lauren Sergio and Steve Scott[7] made the remarkable observation that some of these individuals make straighter movements than people with typical vision.

In their experiment, the control participants wore a blindfold and the experimenter moved their hand to the starting and target locations. Participants then made a few movements between the two points. Figure 18.4 shows that the control participants had nearly the same hand trajectory in the blindfolded condition as when they saw their hand. Thus, their CNS could still use proprioceptive feedback from the arm and efference copy to estimate hand location in fixation-centered coordinates (see section 17.3).

Next, Sergio and Scott examined reaching movements in individuals with congenital blindness. They found that their hand trajectories had slightly *less* curvature than those of the control participants (figure 18.4). Perhaps, as noted in the previous section, the movements of sighted individuals had a slight curvature because their vision distorted Cartesian space. Visually impaired individuals may make straighter movements because they do not experience that distortion. Beyond that supposition, the fact that hand trajectories did not differ dramatically for the two groups suggests that the CNS controls reaching movements in visual coordinates, even in congenitally blind people.

Figure 18.5
Reaching movements in an ancestral primate,[8] drawn by one of the fossil's discoverers, Douglas Boyer. Although these ideas remain somewhat controversial,[9,10] fossil evidence shows that the earliest primates had the ability to grasp distal tree limbs, probably to exploit their unique nutritional resources.

18.1.5 Why Vision-Based Maps?

You might wonder why vision should dominate the planning of movements. Recall from section 4.3.5 the idea that true primates originally evolved as specialized graspers. According to Jonathan Bloch and Douglas Boyer,[8] these animals grasped small tree branches, especially with their hindlimbs, and exploited a niche involving the nutrients found on the distal branches of the trees. Figure 18.5 shows one (controversial[9,10]) view of how ancestral primates reached, as envisioned by one of the paleontologists who discovered its fossil remains. Their feeding technique must have involved reaching accurately to an endpoint and using vision for deciding on the targets of reach. Note that the biological requirements of this animal do not include abstractions such as straight movements in a visual frame of reference, but such movements could have evolved under the pressure for highly accurate and rapid reaching. You can also appreciate the importance of predicting the consequences of a reaching movement on muscle activity that supports the body (see section 5.2.3). You might keep these ideas in mind as this chapter considers further abstractions and formalisms, such as the optimization of movement parameters (section 18.2). Your CNS did not evolve to optimize invariant movement parameters; it evolved to promote the transmission of your ancestors' genes to future generations. It was a tricky proposition to exploit the resources on the distal branches of trees. But, according to some current thinking about primate evolution,[8] your ancestors did exactly that, and this history may explain why your reaching and pointing movements are dominated by vision.

18.2 Description of Trajectory Smoothness: Minimum Jerk

Smoothness of movement seems to have some aesthetic appeal. In ballet, for example, movements appear beautiful partly because highly skilled dancers perform them smoothly. Indeed, ballet critics use the words "smooth" and "elegant" almost interchangeably in their description of performance. In a reaching movement, the hand not only tends to move in a straight line, but the individual components of a hand location vector appear to change smoothly, as in the x- and y-components of hand location in figure 18.2E. In contrast, joint angles may not change smoothly (figure 18.2C).

18.2.1 Minimum Jerk Trajectories

In the early 1980s, Tamar Flash and Neville Hogan[11] found a mathematical way to represent this smoothness property of hand trajectories. Their result went beyond the observation that the hand moves straight to the target. Rather, they found a way to describe the exact time course of the hand's trajectory during the reach, at the limits of the fastest movements. Many treatments of motor control begin with this description, and this approach seems natural enough. After all, what makes more sense in studying movement than to begin by describing—as accurately and in as much detail as possible—what needs explaining? The description of rapid reaching movements looks at the kinematics of the reach, which, because it begins and ends with a stationary end effector (a velocity of 0), must accelerate and decelerate in some way. It then minimizes the changes in acceleration and deceleration, a kinematic parameter called *jerk*.

Neville Hogan[12] first noted that smoothness can be quantified as a function of jerk, which is the time derivative of acceleration. For a system to move some end effector smoothly from one point to another, it needs to find the trajectory—the location as a function of time—that minimizes a cost. For any candidate trajectory $x(t)$, Hogan suggested that its smoothness can be quantified with a cost that is the sum of the squared jerk over the entire trajectory. For example, if the movement was to last 0.5 sec, then its cost would be

$$H(x(t)) = \frac{1}{2} \int_{t=0}^{0.5} \dddot{x}^2 \, dt.$$

This equation is called a *functional* because it assigns a cost (a scalar value) to each function. To find the smoothest possible movement, one needs to search among all possible trajectories to find the one that has the minimum cost. For example, assume that you wish to move some end effector

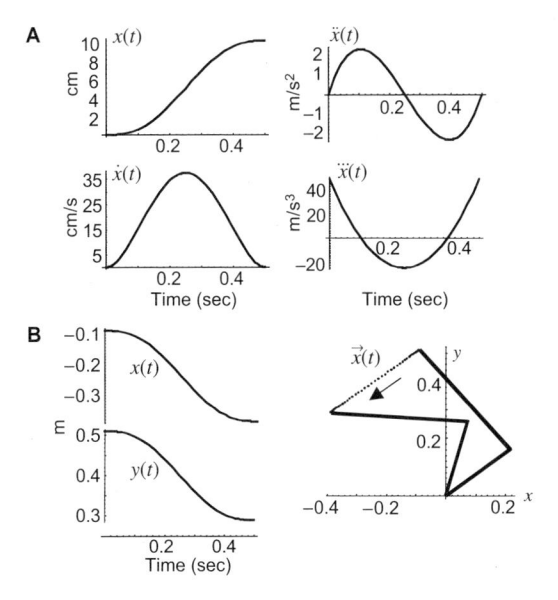

18.2. Description of Trajectory Smoothness: Minimum Jerk

Figure 18.6
Minimum-jerk trajectory. (A) The minimum-jerk scalar function $x(t)$, for moving 10 cm in 0.5 sec. (B) The end-effector location of a two-link arm as it moved from $(-0.09, 0.51)$ to $(-0.39, 0.29)$ in 0.5 sec. Left: Time course in x and y dimensions. Right: Cartesian plot; arrow shows movement direction.

by 10 cm in 0.5 sec. The boundary (beginning and ending) conditions are specified as

$$x(0) = \dot{x}(0) = \ddot{x}(0) = 0 \quad \text{and}$$
$$x(0.5) = 10, \quad \dot{x}(0.5) = \ddot{x}(0.5) = 0.$$

Figure 18.6A plots the trajectory $x(t)$ that minimizes the functional. The function $x(t)$ shown in this figure represents the minimum-jerk trajectory in one dimension. (To see the details of how to find the trajectory that minimizes this functional, see the web document *minimumjerk*.)

If you are interested in representing jerk of a system that has two or more dimensions, you simply add the squared jerk along each dimension. Tamar Flash and Neville Hogan[11] found that for end-effector locations specified as a vector of two or more dimensions, a minimum-jerk trajectory in two or three dimensions always corresponds to a straight line. Figure 18.6B exemplifies this relationship for a two-joint arm moving from an initial to a final location in 0.5 sec. Note how each component of location in Cartesian coordinates moves smoothly to its final value and the end-effector location moves along a straight line.

18.2.2 Why Not Minimum Snap, Crackle, or Pop?

The fourth, fifth, and sixth derivatives of end-effector location are called snap, crackle, and pop, respectively. How can you be sure that a minimum-jerk description provides the best description of your reaching movements? Why not minimum snap? Theory shows that the first derivative (speed) of each trajectory becomes progressively narrower and taller as you find the trajectory that minimizes jerk, snap, and crackle. Indeed, the ratio of peak speed to average speed increases for the trajectory as a system minimizes jerk, snap, crackle, and pop. Therefore, if you wish to minimize snap, the fourth derivative of location, you get a movement with a higher peak speed relative to its average speed than a trajectory that minimizes jerk. Psychophysical experiments reveal that your reaching movements most resemble minimum-jerk trajectories.[13]

References

1. Taylor, DM, Helms-Tillery, SI, and Schwartz, AB (2002) Direct cortical control of 3D neuroprosthetic devices. Science 296, 1829–1832.
2. Hogan, N, Bizzi, E, Mussa-Ivaldi, FA, and Flash, T (1987) Controlling multijoint motor behavior. Exercise Sport Sci Rev 15, 153–190.
3. Morasso, P (1981) Spatial control of arm movements. Exp Brain Res 42, 223–227.
4. Wolpert, DM, Ghahramani, Z, and Jordan, MI (1995) Are arm trajectories planned in kinematic or dynamic coordinates? An adaptation study. Exp Brain Res 103, 460–470.
5. Wolpert, DM, Ghahramani, Z, and Jordan, MI (1994) Perceptual distortion contributes to the curvature of human reaching movements. Exp Brain Res 98, 153–156.
6. Flanagan, JR, and Rao, AK (1995) Trajectory adaptation to a nonlinear visuomotor transformation: Evidence of motion planning in visually perceived space. J Neurophysiol 74, 2174–2178.
7. Sergio, LE, and Scott, SH (1998) Hand and joint paths during reaching movements with and without vision. Exp Brain Res 122, 157–164.
8. Bloch, JI, and Boyer, DM (2002) Grasping primate origins. Science 298, 1606–1610. Figure by Boyer from Sargis, EJ Science 298, 1564.
9. Kirk, EC, Cartmill, M, Kay, RF, and Lemelin, P (2003) Comment on "Grasping primate origins." Science 300, 741.
10. Ni, X, Wang, Y, Hu, Y, and Li, C (2004) A euprimate skull from the early Eocene of China. Nature 427, 65–68.
11. Flash, T, and Hogan, N (1985) The coordination of arm movements: An experimentally confirmed mathematical model. J Neurosci 5, 1688–1703.
12. Hogan, N (1984) Adaptive control of mechanical impedance by coactivation of antagonist muscles. IEEE Trans Auto Control AC-29, 681–690.
13. Richardson, MJ, and Flash, T (2002) Comparing smooth arm movements with the two-thirds power law and the related segmented-control hypothesis. J Neurosci 22, 8201–8211.

Reading List

Papers by Neville Hogan, Tamar Flash, and their colleagues[11-13] give the equations for minimum-jerk trajectories. The next chapter puts them to use in a model of movement planning.

19 Planning to Reach or Point II: A Next-State Planner

*Overview: Smooth hand trajectories may be an emergent property of a **feedback control** system that plans for a desired change in the limb's state based on an estimate of its current location and goal. Called a* next-state planner, *such a system allows the CNS to respond smoothly, as if on autopilot control, to unexpected changes in goals or perturbations to the limb. Evidence indicates that people carrying the gene for Huntington's disease, a disorder primarily of the basal ganglia, do not make these computations efficiently.*

Neville Hogan[1] noted that minimum-jerk trajectories do not explain "the cause of the behavior they describe but rather [provide] a distillation of its essence." He recognized that although minimum jerk provides a concise description of a wide variety of movements, it fails to address how the CNS might produce such movements or the advantages of doing so. This chapter considers those topics.

19.1 The Problem of Planning

Chapter 18 described the smoothness of reaching and pointing movements in visual coordinates: At their fastest, they have the least possible jerk. One idea about how your CNS controls these trajectories holds that when you decide to make a movement, your CNS computes a minimum-jerk trajectory, called a *desired trajectory*. Perhaps your CNS stores this trajectory somewhere and plays it out like a tape. At each instant of time, the tape provides the desired state of the limb—its location, velocity, and acceleration—and internal models transform these desired states into motor commands.

The notion of a desired trajectory entails two assumptions: (1) that your CNS plans the details of its movements far in advance and (2) that a mechanism keeps track of time. Both assumptions are problematic. For example, imagine that you want to stir a pot of soup with a spoon. As you begin stirring, it seems unlikely that your CNS would plan the number of times your hand would circle the pot and the entire trajectory of your hand during that motion. Or, to take another example, imagine that as you are reaching to a target, the target jumps halfway into the movement or something perturbs your hand. Your CNS would need to reevaluate the desired trajectory because such changes render the original plan obsolete.

Two groups of investigators—first Bruce Hoff and Michael Arbib, and later Stefan Schaal and his colleagues—noted these problems and developed models that do not depend on a precomputed desired trajectory. They suggest that the CNS produces movements by evaluating a goal in relation to the limb's current state, in real time, and then generates a small desired change in state. A smooth movement that minimizes jerk might be an emergent property of such an autonomous feedback system. A system like that does not plan the entire desired trajectory at or prior to the onset of the movement. Rather, it monitors the current state of the end effector and the target, and continuously formulates a desired change in end-effector state for the immediate future. The system acts as a *next-state planner*.

The next-state planner receives a high-level goal: bringing an end effector to the target. It iteratively accomplishes this goal by breaking the movement down into a sequence of small end-effector displacements: not in advance, but as the movement progresses. Put another way, it always plans the next state based on an evaluation of the current state with respect to the goal. This approach not only produces smooth trajectories when an end effector reaches to a stationary target, but it also explains why movements remain smooth when a target changes location during movement or some perturbation unexpectedly affects the arm. As you will see, this framework also does away with the notion of a timekeeper, something the desired-trajectory hypothesis requires. This general concept is often called "autopilot control" (see section 17.3).

While this approach clarifies a lot about how the CNS plans movements, it does not, at first, seem to explain the smoothness of end-effector trajectories. A theory put forth by Chris Harris and Daniel Wolpert suggests that two factors may combine to make smooth trajectories a good plan, and a next-state planner would produce them. First, neurons in the motor system produce a noisy output in which the standard deviation of the noise grows as the output increases. Second, the CNS needs to minimize variability at the end of movement in order to achieve goals. In a system that has these two characteristics, minimum-jerk movements work well because they do not require large motor commands and produce accurate endpoints. The next-state planner does not generate a motor command to reach the target in one go, so the commands at any given time remain small (and therefore relatively low in noise). But it always gets to the target, no matter what happens to the limb or the goal along the way. The final stages of a reaching or pointing movement always involve small motor-command signals, so the system can minimize noise-related errors. Section 19.4 explores these ideas in more detail.

19.2 Transforming a Displacement Vector into a Trajectory

The model depicted in figure 14.1 starts by computing an estimate of target location $\hat{x}_t$ and end-effector location $\hat{x}_{ee}$ in fixation-centered coordinates, and then computes a difference vector x_{dv}. How can your CNS gradually transform this vector into a change in joint angles $\Delta\theta$ so that your CNS produces a smooth movement to the target? And how can it specify the intention to move fast at one time but slow at another?

Dan Bullock and Steve Grossberg[2] proposed that in addition to the spatial variables x_{dv} and $\Delta\theta$, another variable describes intended movement speed. They considered this variable to be a scalar function of time $\gamma(t)$ and called it a "go" function. To derive some of the properties of this function, they noted that x_{dv} decreases monotonically during a movement. They proposed that in order to compute $\Delta\theta$ at any instant of time, $\gamma(t)$ should scale x_{dv} to produce a time-dependent difference vector, which is transformed into a joint-rotation vector. They proposed that $\gamma(t)$ should be a monotonically increasing function of time and should specify how fast you intend to make the movement.

To see the consequence of this approach, consider the behavior of the system over a small time interval $t = 0 \to \Delta t$. Initially, x_{dv} begins as a relatively large vector and $\gamma(t)$ begins small. Multiplying the two factors leads to the scaling of x_{dv}; hence the product begins as a relatively small vector. To map x_{dv} into $\Delta\theta$, you might use the Jacobian that relates joint rotations to changes in end-effector location:

$$x = f(\theta)$$

$$J = \frac{dx}{d\theta}.$$

Using this Jacobian, you can compute the joint rotations needed to move the end effector along a small displacement Δx:

$$\Delta x(t) = \gamma(t) x_{dv} \tag{19.1}$$
$$\Delta\theta(t) = J^{-1}\Delta x(t).$$

Note that the joint-rotation vector now corresponds to a velocity. In the final step, according to this model, your CNS transforms this velocity into a motor command and produces the forces that move the arm. As it computes these commands and transmits them to the spinal cord, it also sends an **efference copy** somewhere (perhaps to the PPC) to estimate current end-effector location $\hat{x}_{ee}$, which in turn produces a new x_{dv}. This series of computations produces a monotonically decreasing function $x_{dv}(t)$ multiplied by a monotonically increasing function $\gamma(t)$. This multiplication results in a function $\Delta x(t)$ that first increases and then decreases during movement. $|\Delta x(t)|$ is the speed profile for the movement. If $\gamma(t)$ is 0 at $t = 0$, $\Delta x(t)$ grows from 0 to some peak value, then declines to 0 as the vector x_{dv} becomes smaller. By manipulating $\gamma(t)$, this model can produce fast or slow movements.

Figure 19.1 shows an example of a movement generated by this model. For this simulation, $\gamma(t) = 0.1t^{1/4}$, where t has units of seconds. The end effector moves along a straight line to the target (figure 19.1A) with a unimodal speed profile (figure 19.1B). Figure 19.1C and 19.1D show the

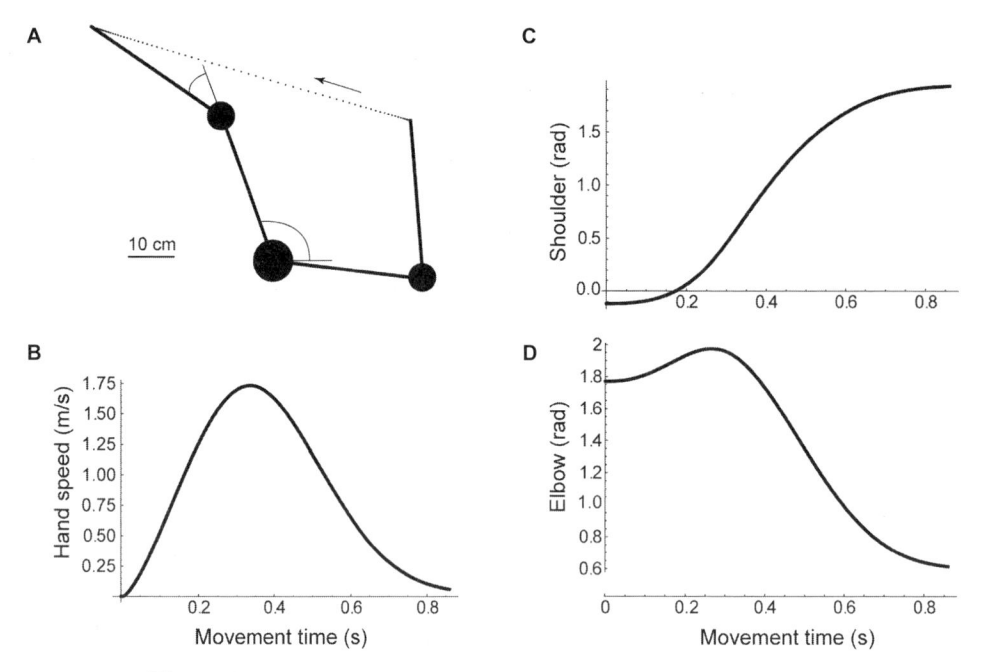

Figure 19.1
Transforming a difference vector into a trajectory via a "go" function. (*A*) Equation 19.1 simulated the motion of a two-joint arm. Dots indicate end-effector location at 10-msec intervals. The "go" function was $\gamma(t) = 0.1t^{1/4}$. (*B*) End-effector speed during the movement. (*C*) Shoulder angle. (*D*) Elbow angle.

joint rotations corresponding to this movement, as calculated through an iteration of equation 19.1.

Three important ideas emerge from this model. First, the model assumes that your CNS always computes an estimate of the current end-effector location. Second, it assumes that your CNS continuously estimates the vector that points from the current end-effector location to the target. And third, it scales the estimate by a time-dependent function $\gamma(t)$. At any time t, the resulting vector corresponds to the desired displacement for the end effector, $\Delta x(t)$. A transformation of $\Delta x(t)$ to joint coordinates, and a further transformation to forces, produces the motor commands that eventually move the limb. The so-called invariant properties of limb trajectory (i.e., its straightness in visual coordinates and minimum-jerk smoothness) represent emergent properties of both the feedback system and the "go" signal.

Scaling the difference vector x_{dv} by a time-dependent function $\gamma(t)$ has some limitations, however. Consider what should happen to $\gamma(t)$ if conditions change during the movement. For example, imagine that as you start reaching to pick up a pencil (figure 17.1B), someone accidentally bumps the table and the pencil begins rolling away from you. Target location $\hat{x}_t$ thus changes during the execution of the movement. Because the model continuously estimates x_{dv}, the change in target location affects the difference vector. However, does $\gamma(t)$ also change? Does it need to be reset, for example? Recall the discussion of autopilot control in section 17.3. When a target jumps during a movement, the movement adjusts smoothly, only slightly lengthening movement duration.[3] But by the time that happens, the go function would have reached a high level, which would create problems. Imagine further that somehow the pencil keeps rolling away at the same velocity as your reaching movement so that x_{dv} remains constant. The motor commands would get larger and larger. The "go" function of Bullock and Grossberg shows one way in which a smooth hand trajectory can emerge from a state-feedback system, but it relies on a feedback-independent function of time and an exponential rise to infinity, neither of which seems realistic. The model nevertheless illustrates that a control system does not need to plan for the entire trajectory at the onset of a movement.

19.3 The Next-State Planner

Instead of planning an entire trajectory in advance of movement onset, a controller can just plan for what to do next. This approach requires an appropriate *control policy*, one with a measure of target and end-effector locations. As the CNS executes its plan, and therefore the limb's state changes, the planner responds by updating the plan of what to do next. The planner has a policy (i.e., a set of equations that describe how to produce an output, your desired end-effector location in the near future), given some inputs (estimates of current limb configuration and target location). Through this approach, the planner becomes an autonomous system. If your *policy* involves smooth movements, the problem of planning becomes one of figuring out the autonomous system that will produce smooth movements regardless of the state of the limb or the location of the target.

Two theoretical developments have suggested a way to solve the problem of planning. First, Bruce Hoff and Michael Arbib showed how to derive the dynamics of the autonomous system with a policy of moving smoothly. Second, Stephan Schaal and his colleagues showed that whatever your policy may be, you can produce an autonomous system that implements that policy. That is, whether you intend to move to a target smoothly, in a zigzag, or more variably, a policy can describe those movements by setting the static parameters of an autonomous system. These parameters describe your intention, at an abstract level, of how to move. The next-state planner, then, computes the details of the plan during the actual movement.

19.3.1 The Hoff–Arbib Model

Recall that the Bullock–Grossberg model had a feedback-independent, but time-dependent, "go" function $\gamma(t)$ that scaled motor output. Bruce Hoff and Michael Arbib[4] developed a model that produces smooth movements and also relies on feedback, like the Bullock–Grossberg model, but to some extent does away with a time-dependent function that scales x_{dv}. In dispensing with a "go" signal, Hoff and Arbib considered a system having a current estimate of x_{dv}, as well as a current estimate of end-effector location, velocity, and acceleration $[\hat{x}_{ee}(t), \hat{\dot{x}}_{ee}(t), \hat{\ddot{x}}_{ee}(t)]$. This system would produce the desired state of the end effector at some time in the future: $[x_d(t+\Delta), \dot{x}_d(t+\Delta), \ddot{x}_d(t+\Delta)]$. Their model computed this desired state so that the trajectory of desired end-effector states would follow a smooth transition to the target. That is, they forced the transition of the states from one time to another to follow a criterion described by the minimum-jerk function.

Hoff and Arbib began by computing the solution to the minimum-jerk functional (see section 18.2.1) for arbitrary initial and final limb states. For example, if the initial state at $t = 0$ is something other than 0 velocity and 0 acceleration, then acceleration at time t becomes a function of these initial conditions:

$$\ddot{x}(t) = \ddot{x}_o(1 - 9\varepsilon + 18\varepsilon^2 - 10\varepsilon^3) + \dot{x}_o(-36\varepsilon + 96\varepsilon^2 - 60\varepsilon^3)/D$$
$$+ (\hat{x}_t - x_o)(60\varepsilon + 180\varepsilon^2 + 120\varepsilon^3)/D^2, \quad (19.2)$$

where at $t = 0$, $x = x_o$, $\dot{x} = \dot{x}_o$, and $\ddot{x} = \ddot{x}_o$. Variable D represents the desired duration of the movement and, therefore, corresponds to the difference between the final time of movement t_f and the original time of movement t_o such that $D = t_f - t_o$. The proportional elapsed time is $\varepsilon = (t - t_o)/D$. The time derivative of the function described in equation 19.2 is a description of how acceleration changes at any instant of time (i.e., jerk). If you find the jerk $\dddot{x}(t)$ and in its expression set $\varepsilon = 0$ and $D = t_f - t$, you have a rule with which you can optimally change the desired state of your system. You have

$$\Delta x_d(t) = \dot{x}_d(t)$$
$$\Delta \dot{x}_d(t) = \ddot{x}_d(t) \quad (19.3)$$
$$\Delta \ddot{x}_d(t) = \left(-\frac{60}{D^3}(x_t - \hat{x}_{ee}(t)) - \frac{36}{D^2}\dot{x}_d(t) - \frac{9}{D}\ddot{x}_d(t)\right).$$

Equation 19.3 describes a next-state planner that conforms to an imposed *policy* of minimum jerk. It specifies how much change should take place in the desired location, velocity, and acceleration of the end effector. It computes this change as a function of the current estimate for end-effector location $\hat{x}_{ee}$ and the estimated time D remaining to reach the target. You can express the vector that specifies desired change in end-effector location as $\Delta \vec{x}_d = [\Delta x_d, \Delta \dot{x}_d, \Delta \ddot{x}_d]$.

The Hoff–Arbib model not only generates a smooth end-effector trajectory to the target but also changes this desired trajectory in the face of perturbations either to the target or to the end effector. For example, you can use equation 19.3 to simulate behavior of the end effector when the target jumps to a new location during a reach. In figure 19.2A, the target

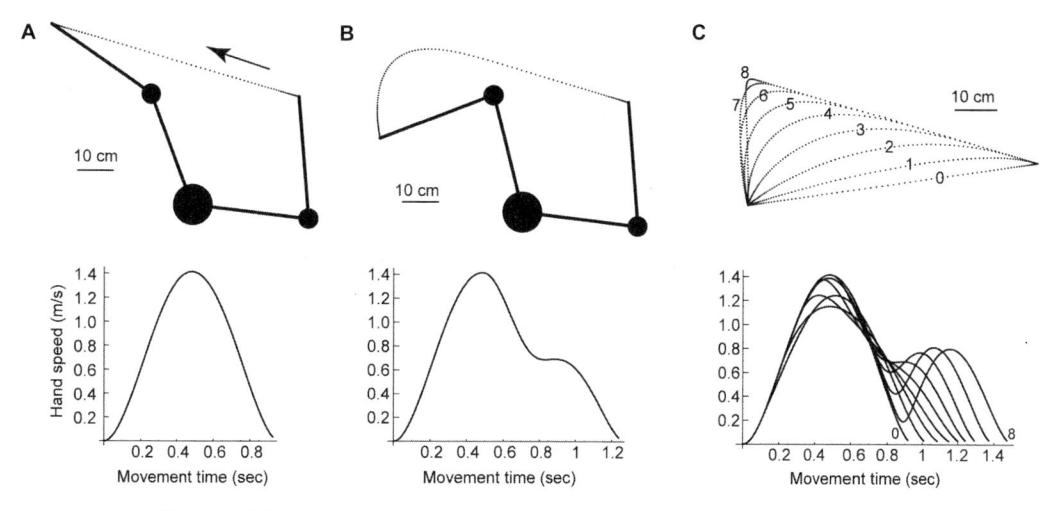

19.3. The Next-State Planner

Figure 19.2
Using feedback to generate a smooth desired trajectory. (A) Equation 19.3 was used to simulate desired end-effector location. Dots indicate end-effector location at 10-msec intervals, $D = 0.97$ sec. (B) At $t = 500$ msec, the target jumped to a new location. Desired end-effector trajectory changed smoothly to reflect the change in target location, as expected for autopilot control. At the time of target jump, time remaining D increased by 200 msec. (C) Target jumped at $t = 0$ msec, $t = 100$ msec, ..., $t = 800$ msec.

remains stationary and the desired trajectory for the end effector has a smooth speed profile. In figure 19.2B, at 500 msec into the desired trajectory, the target jumps to a new location. The system responds by modifying the desired trajectory, producing a second peak in the speed trace. The end-effector trajectory depends on when this perturbation takes place (figure 19.2C). If the target jumps early in the movement, the end effector simply moves toward the new location of the target. If it takes place late in the movement, the model generates two movements in sequence. In between these extremes, a family of trajectories guides the end effector smoothly to the new target location.

Notice that a change in the target location often increases the duration of the movement. Bruce Hoff[5] noted that this change in movement time correlates with the distance that the target moved. In equation 19.3, D (time remaining to target) is an input to the system. You have to provide this input by estimating the time remaining to the target when a change occurs in target location. Therefore, the system needs a timekeeper. Hoff asked whether this change in D could be optimally determined by another estimator, one that predicts the optimum length of time it takes to perform that movement.

To incorporate a movement-duration estimator, Hoff introduced a cost function that took into account both jerk and time. The faster the movement, the smaller the time cost but the higher the jerk cost. This led to a function that estimated time remaining to target (D) as a function of the current state of the end effector, the distance to target $x_t - \hat{x}_{ee}$, and a constant r that described the trade-off between moving fast versus moving smoothly. For example, Hoff found that for a movement that started from static initial conditions, movement time should be

$$D = (60(x_t - \hat{x}_{ee}))^{1/3} r^{1/6}.$$

In this formulation, r "weighs" the relative cost of a movement's smoothness versus the cost of its duration. If you imagine that you can "afford" jerkier movements for larger movement targets, then this relationship leads to Fitts's Law: a systematic trade-off between movement duration and target size.

$$D = a + b \log_2(2x_{dv}/S),$$

where S is the size or width of the target (i.e., the accuracy requirement), and a and b are regression coefficients. If you wish to move very accurately, the movement will take systematically longer and vice versa. Hoff showed that you can measure movement duration for one target, estimate r from that movement, and then predict movement durations for other movements, including perturbed movements. His approach produced movement durations and end-effector trajectories that matched psychophysical data in a number of experiments.

The Hoff–Arbib model combines optimization with control and shows that a "large-scale goal" of bringing the end effector to the target can be accomplished through a sequence of small goals $\Delta \vec{x}_d(t)$. The system that plans for the next state of the limb, the next-state planner, does not plan very far in advance. When the target appears, the planner does not compute a desired end-effector trajectory for the entire movement, nor does it use a signal, such as the "go" signal of Bullock and Grossberg's model, that is an explicit function of time. Rather, the planner keeps track of the current state of the end effector, the target's location, and the amount of time needed to reach the target, and then estimates the desired next state for the end effector. The ability to change movement plans in response to perturbations of the end effector or displacements of the target emerges from the properties of this next-state planner.

19.3.2 Planning for Arbitrary Trajectories

With the Hoff–Arbib model, you can abandon the notion of planning an entire reaching or pointing movement from beginning to end. Instead, you can rely on an autonomous feedback-control system to plan a movement through a next-state planner. In each instant, the next-state planner evaluates the goal of the task as well as an estimate of the current state of the limb, and then generates a desired change in that state for the immediate future. However, what if you do not want simply to reach straight to a target, but desire to move your hand in a zigzag path or take some other route to the target? To make such movements, you would need a different next-state planner.

19.3. The Next-State Planner

Auke Jan Ijspeert, Jun Nakanishi, and Stefan Schaal[6] considered this problem and proposed a way to represent most trajectories in terms of a parameterized, autonomous system. In their model, the next-state planner consists of a set of nonlinear differential equations. The parameters of these equations describe the "landscape" over which the system brings the limb to the goal. Therefore, as the system learns to plan a movement, it learns the parameters of the autonomous system.

For example, imagine that you have a robot (figure 15.1A) and you want to guide its gripper to a target. However, in this case you do not want it to go straight to the target; rather, you want the gripper to take a zigzag path along the way. You need to set the parameters of a next-state planner. At any instant, it evaluates the end effector's location with respect to the target and generates a small desired change in state. When the changes in state accumulate over a long period, the gripper moves in a zigzag pattern.

In this model, the next-state planner has two inputs, x_t and $\hat{x}_{ee}$ (target location and estimated end-effector location), and one output, $\dot{x}_d$ (desired change in end-effector location). It has four internal states, labeled $z, v, x, \tilde{x}$, and a few time constants a_z, b_z, and so on. A set of differential equations describes the dynamics of this system:

$$\dot{z} = a_z(b_z(x_t - x_d) - z)$$
$$\dot{v} = a_v(b_v(x_t - x_d) - v)$$
$$\dot{x} = v(1 + a_{px}(\hat{x}_{ee} - x_d))^{-1} \quad (19.4)$$
$$\tilde{x} = (x - x_o)/(x_t - x_o)$$
$$\dot{x}_d = z + \frac{\sum_{i=1}^{n} w_i g_i(\tilde{x})}{\sum_{i=1}^{n} g_i(\tilde{x})} v + a_{py}(\hat{x}_{ee} - x_d).$$

In this formulation, x_o represents end-effector location at the start of the movement, and $\tilde{x}$ represents the fraction of the distance progressed to the target. A set of nonlinear **basis functions** g_i encodes the distance $\tilde{x}$:

$$g_i = \exp\left(-\frac{1}{2\sigma_i^2}(\tilde{x} - c_i)^2\right).$$

Each basis has a "center" at c_i. The parameters of this system are the weights w_i and the time constants a_z, b_z, and so on. If the time constants are all positive, the system will settle at $(z, v, x, x_d) \to (0, 0, x_t, x_t)$. That is, the desired end-effector location will go to the target, without doubt. However, by manipulating the time constants and the weights of the basis functions, you can dictate how the system brings the end effector to the target. The task of learning to plan a movement becomes one of learning the parameters of this dynamical system.

For example, consider a movement around a single joint, where $x_t = 1$, $\hat{x}_{ee} = 0$. You want to complete the movement in approximately 1 sec. The desired movement time dictates the time constants of equation 19.4. In this case, you set the time constants $a_z = a_v = 12$ and $b_z = b_v =$

$a_z/4$. You do not know for sure how you want to move to the target (i.e., whether smoothly or in a zigzag), so you simply set the weights w_i to be some random numbers. Further, assume that your planned movements have no error (i.e., $\hat{x}_{ee} = x_d$). The resulting movement (figure 19.3A) gets to the target in approximately the right time but not with a smooth trajectory: The velocity profile has a number of peaks. If you want to plan a smooth movement (e.g., one with a minimum-jerk profile), you need to change the weights in the model so that the next-state planner produces such a movement through its own dynamics. Once you find the "right" weights (the web document *schaalmodel* provides the procedure for finding them), the end effector moves smoothly to the target (figure 19.3B). By design, the next-state planner easily handles changes in the target location during the movement or perturbations to the limb. Accordingly, figure 19.3B displays behavior of the system when the target jumps at various times in the movement.

Now consider a reaching movement that takes a zigzag path. To plan the zigzag, you set the weights of your dynamical system so that its "landscape" produces the desired behavior. In this case, you set the weights so that the end effector moves slightly toward the target, then moves away from the target, and then moves to the target again. Figure 19.3C shows that the system can produce such a behavior. More interestingly, you can ask the system to display its desired trajectory when the target jumps. You find that the trajectory has a single zigzag if the target jumps at the very start of the movement (at time 0) but a double zigzag if it jumps at the middle of the movement (at time 0.5). If the target jumps at around 1 sec, when the movement to the first target is nearly complete, you will see a smaller second zigzag.

People may not plan their movements using exactly this kind of mechanism, but it illuminates certain principles. The formalism introduced by Schaal and his colleagues suggests that movement planning can be represented as the static parameters of an autonomous system. When someone switches the target of the movement, the autonomous system generalizes from a movement that its weights have been trained on to another movement. The pattern of generalization depends on the basis functions g_i.

Figure 19.4 embeds this model in a control system. In this figure, the displacement map transforms the change in desired end-effector state $\dot{x}_d$ into a desired change in joint state $\dot{\theta}_d$. Mathematically, the transformation involves an estimate of the inverse of the Jacobian, $J^{-1}(\hat{\theta})$. $\dot{\theta}_d$ is then transformed into motor commands (torques) using an internal model of inverse dynamics. Chapters 20–22 explain limb dynamics in some detail. For now, you need only recognize that these motor commands eventually reach the limb, causing it to move. Because of delays in the system, the next-state planner relies on efference copy and a model of the forward dynamics of the limb to estimate current end-effector location. This aspect of the model captures the remapping function presented in figure 17.10.

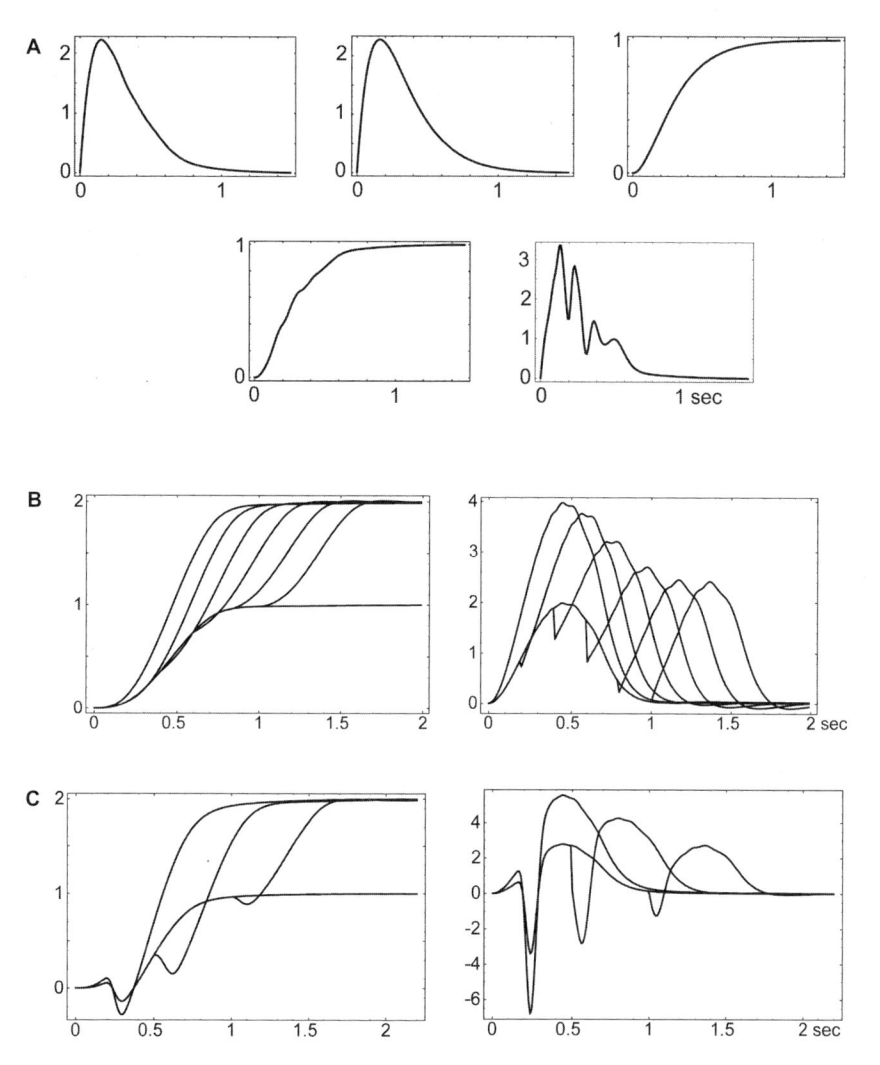

Figure 19.3

Generating arbitrary trajectories with a next-state planner. (A) Equation 19.4 was simulated with 25 basis functions having centers uniformly distributed over $\tilde{x} : 0 \to 1$ and random weights w_i, sampled over an interval from -1 to 1. Top: From left to right, trajectories of z, v, and x. Bottom: From left to right, trajectories of x_d and $\dot{x}_d$. (B) Weights w_i were found that produced a minimum-jerk trajectory x_d. At 0.2-sec intervals into the movement (including at time 0), the target jumped from location 1 to location 2 for x_d (left) and $\dot{x}_d$ (right) in both unperturbed and perturbed conditions. (C) Weights w_i were found that produced a zigzag trajectory to the target. Format as in B.

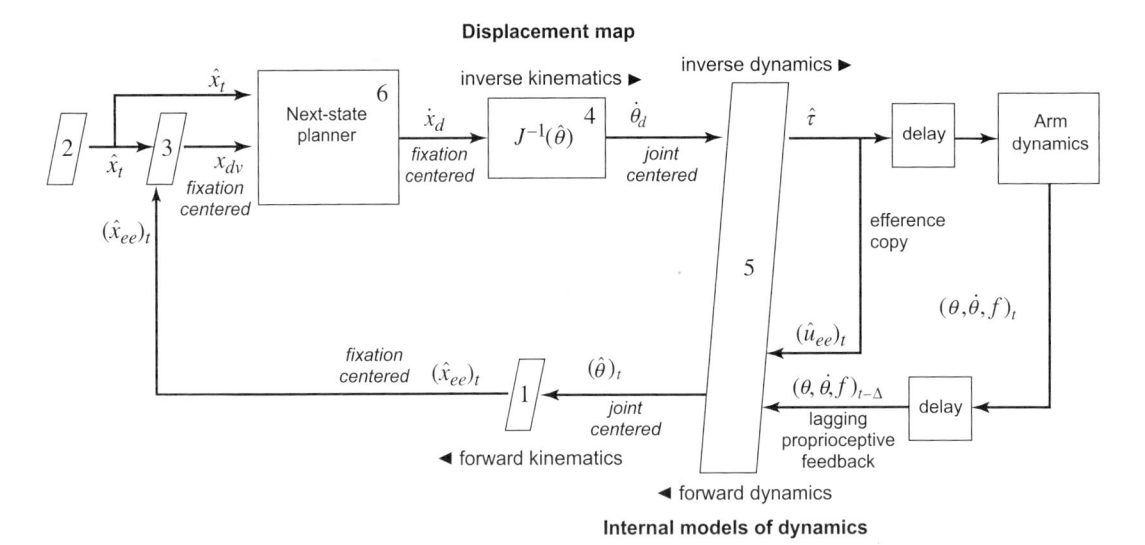

Figure 19.4

A system for planning and control of reaching. Schematic representation of the next-state planner in conjunction with the system that executes the plan using internal models of dynamics. $\hat{x}_t$ represents estimated target location; $\hat{x}_{ee}$, the estimated end-effector location; x_{dv}, the difference vector; $\dot{x}_d$, the desired change in hand state, and $\dot{\theta}_d$, the desired change in joint angles; τ, the torque needed to accomplish the plan. Abbreviations: f, force; J, the Jacobian; θ, joint angles; Δ, time delay. (Drawn by George Nichols.)

19.3.3 Learning to Plan a Movement

The next-state planner implements a *control policy*, expressed through the selection of the static parameters of an autonomous system. This approach suggests that in order to learn how to perform a task, you first need to learn a control policy in terms of the parameters of the next-state planner. Emo Todorov, Shadmehr, and Emilio Bizzi[7] studied this possibility by asking right-handed participants to hold a paddle in their left hand and hit a ball accurately. A *training* group began by watching a video monitor that displayed the movements of their paddle superimposed on the movements of a teacher's paddle. A *control* group had no such experience. The experimenters sought to teach the training group the trajectory that an expert teacher performed. Although the training group received 50% less actual practice than the control group, they nevertheless improved performance to about the same level as controls.

19.4 Minimizing the Effects of Signal-Dependent Noise

If you set the weights of the next-state planner appropriately, it can produce state transitions that smoothly bring the end effector to the target.

The resulting movement minimizes jerk. Why should the nervous system want to optimize such a cost function?

Chris Harris and Daniel Wolpert[8] provided a theory that answers this question. They argued that if your goal is to bring your end effector to the target in some desired amount of time D, then a more natural cost is something that measures the error in end-effector location x_{ee} with respect to the target x_t at time D. To minimize this error, it seems reasonable to start the movement with high acceleration and then slow down near the end, in order to home in on the target. That is, given that you have D time to get to the target, you should get close to the target quickly, and then spend most of the remaining time slowly fine-tuning end-effector location. The result of this high initial acceleration will not be a smooth movement because end-effector speed would rise quickly but fall slowly.

How, then, can you account for the finding that reaching movements have a symmetrical speed profile? Wolpert and his colleagues[8,9] suggested that the reason might have something to do with noise. Neurons that convey motor commands to the muscles have a certain amount of noise in their signal. Perhaps this noise results from the fact that a neuron's response to an input varies from one trial to the next. In their experiment, participants maintained a target force by flexing their thumb muscles. The experimenters then measured the standard deviation of the resulting force. They found that it grew roughly linearly as a function of mean force output (figure 19.5). The growth in force variability did not result from properties of the muscles: Electrical stimulation of the thumb flexor muscle did not cause increasing variance with stimulation magnitude. The signal-dependent noise came from the neural, motor-command signals.[9]

In exploring the significance of this signal-dependent noise, Harris and Wolpert modeled a two-joint arm with pairs of shoulder and elbow muscles. They gave each muscle a motor neuron that had a noise property. The motor neuron's output equaled the sum of its input and a noise term that had 0 mean and a standard deviation that grew linearly as a function of the input. The ratio of standard deviation to the mean is known as the coefficient of variation, which Harris and Wolpert set to about 15%. The larger the input to the motor neuron, the larger its mean output and the larger its standard deviation around this mean.

Because the modeled motor neurons had noise, two identical inputs to the modeled arm did not produce identical movements. In fact, the larger the input, the larger the variance in the movement. Therefore, if you tried to move this model limb with a high initial acceleration, and then you evaluated $x_t - x_{ee}$ at the end of the movement (i.e., at $t = D$), you would have the *endpoint error*. If you repeated this measure over a number of trials, you would notice a large variance.

Harris and Wolpert suggested that your CNS learns to generate motor commands so that—across repeated movements to the same target—it minimizes endpoint error. Given a desired movement duration and a target location, they began by assuming some desired hand trajectory and then worked back through the equations of motion and estimated the

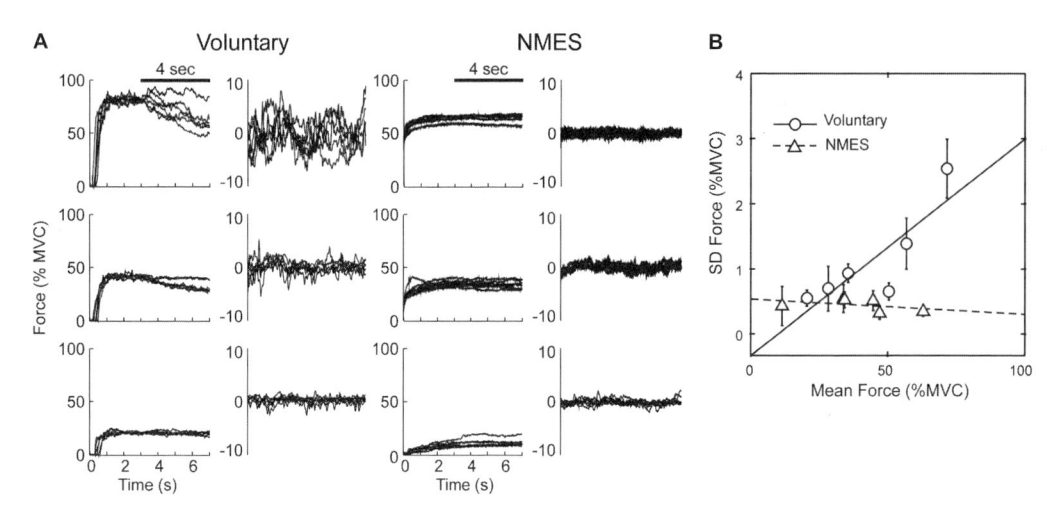

Figure 19.5

The standard deviation of noise grows with mean force in an isometric task. Participants produced a given force with their thumb flexors. In one condition (labeled "voluntary"), the participants generated the force, whereas in another condition (labeled "NMES") the experimenters stimulated their thumb flexors artificially to produce force. To guide force production, the participants viewed a cursor that displayed thumb force, but the experimenters analyzed the data during a 4-sec period in which this feedback had disappeared. (*A*) Force produced by a typical participant. For each two-column panel (voluntary and NMES), the period without visual feedback is marked by the horizontal bar at the top, labeled "4 sec." These data are expanded to make up the right half of each panel. (*B*) When participants generated force, noise (measured as the standard deviation) increased linearly with force magnitude (with a slope of ~1). Abbreviations: NMES, neuromuscular electrical stimulation; MVC, maximum voluntary contraction. (From Jones et al.[9] with permission.)

pattern of inputs to the motor neurons. Harris and Wolpert then produced that pattern a number of times and recorded the variance in endpoint error. They searched for the desired hand trajectory that produced the smallest variance and found that in the optimal trajectory, the end effector moved with a smooth, bell-shaped speed profile along an essentially straight line to the target (figure 19.6). Therefore, perhaps end-effector trajectories have smooth accelerations and decelerations because such a pattern minimizes endpoint error in the presence of signal-dependent noise.

19.5 Online Correction of Self-Generated and Imposed Errors in Huntington's Disease

The approach outlined in figure 19.4 combines optimization with control so that at any given instant, the next-state planner produces a desired

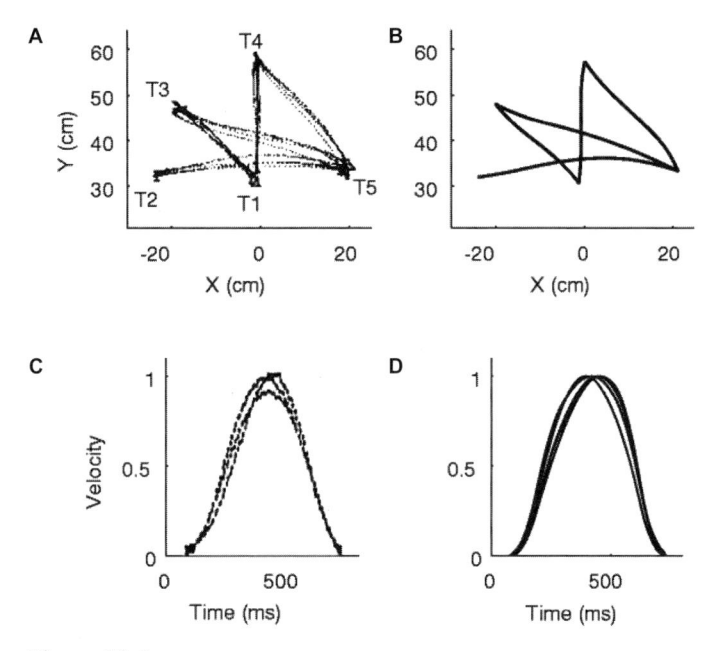

Figure 19.6
Comparison of observed end-effector trajectories and those predicted in a system with signal-dependent noise. (*A*) Observed end-effector paths for a set of point-to-point movements. (*B*) Optimal end-effector paths for a simulation of a two-joint, four-muscle arm, in which the inputs to the muscles contained signal-dependent noise. (*C*) Observed end-effector speed for movements between T1 and T3 in A, normalized to have a maximum of 1. (*D*) Normalized end-effector speed in all the movements shown in part B. (From Harris and Wolpert[8] with permission.)

state for the limb that takes into account both the target's location and the current state of the limb. Despite the fact that the control policy remains unchanged (i.e., parameters of the next-state planner remain static throughout the movement), the system responds smoothly to unexpected events (e.g., the target jumping midway through the movement or the end effector hitting some obstacle). However, even if nothing like those perturbations occurs, you should expect some amount of unexpected behavior because of the noise that exists in the system. If the next-state planner or the forward model functions poorly, then the feedback system of figure 19.4 cannot deal with these errors effectively.

This section presents the idea that the basal ganglia needs to function properly for error corrections of this kind (i.e., normal autopilot control requires healthy basal ganglia). To investigate this idea, Maurice Smith, Jason Brandt, and Shadmehr[10] examined the movements of patients with Huntington's disease (HD). They also studied people called *asymptomatic gene carriers* (AGCs), who had the gene for Huntington's disease but had not developed symptoms. But before we present the results, it might be useful to consider the nature of HD and its causes.

19.5.1 Why Do Striatal Cells Die in Huntington's Disease?

Significant, progressive atrophy of the basal ganglia occurs in HD,[11] and its hallmark motor sign involves rapid, irregular, involuntary movements. The gene that causes HD has an abnormal lengthening of a particular sequence of nucleotides which codes for the amino acid glutamine. Within the protein encoded by this gene, called *huntingtin*, the long stretches of glutamine appear to compromise both exocytosis and endocytosis, which affect synaptic plasticity. Thus, some of the initial problems in HD might reflect disruption of the cells' plasticity mechanisms. In addition, the long run of glutamine in the huntingtin protein weakens its interaction with another protein, Hip1, to the extent that the latter disassociates from a huntingtin–Hip1 complex. Hip1 then runs riot in the cell, interacting with other proteins, some of which degrade proteins and others of which compromise the function of mitochondria, eventually causing cell death.[12] The effects of this mutation appear relatively late in life, when aging processes have compromised the ability of cells to withstand the insult.

For some reason, striatal cells have a special susceptibility to the effects of this mutation, and in HD the striatum degenerates dramatically. The earliest stages of the disease begin with degeneration of the striatal *patches*, or *striosomes*.[13] It has been established that the patches provide the outputs from the striatum to the dopaminergic cells of the midbrain (see also section 23.4.4). In addition to degeneration in the patches, the dorsal and lateral aspects of the caudate and the putamen seem to be affected earlier than more ventral and medial parts, which in turn degenerate before the ventral striatum. In late stages of the disease, the degeneration becomes more widespread and includes the cerebral cortex in addition to the striatum. Nevertheless, most experts treat HD as a basal ganglia disorder, and especially in its early stages and for ACGs, this assumption seems reasonable. HD does, however, affect signal processing in the cerebral cortex.

19.5.2 Huntington's Disease and Cortical Function

Many experts view the autopilot mechanism outlined in chapter 17 and above in this chapter as a function of the PPC. This section presents the idea that this mechanism does not work properly in HD and in ACGs, which seems to point to basal ganglia rather than to the PPC. However, as emphasized in section 6.2.1, the cortex and basal ganglia work together in cortical–basal ganglionic modules, and the PPC participates in such networks. The PPC sends massive projections to the striatum and the basal ganglia sends outputs to the PPC. Therefore, you should not be surprised that striatal dysfunctions in HD and AGCs include problems with autopilot control.

Recall from section 8.8.2 that somatosensory evoked potentials are diminished in patients with HD, as are long-loop reflexes. These findings indicate that proprioceptive signals have a smaller amplitude in HD patients, at least at the level of the cortex. Proprioceptive feedback plays a critical role in the autopilot mechanism and in the computations of the

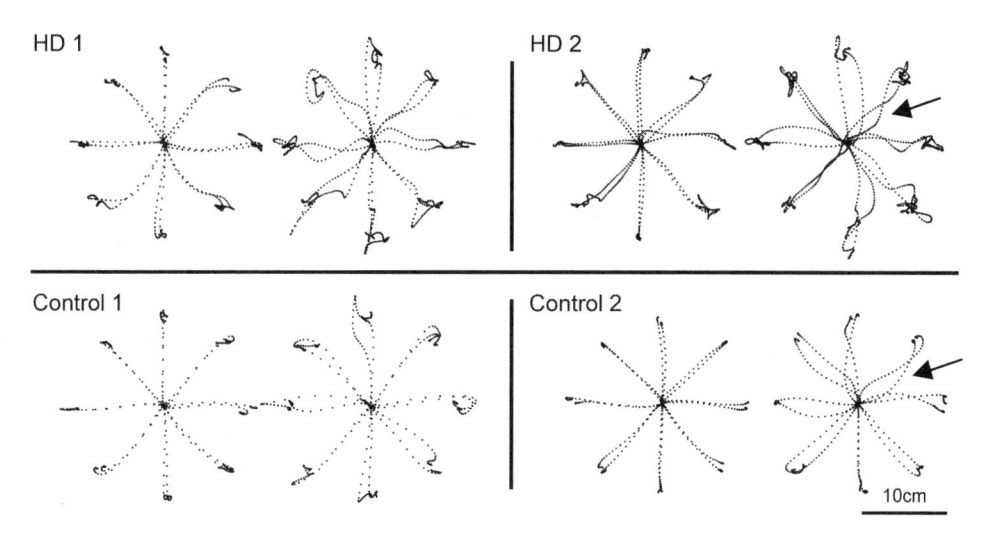

Figure 19.7
Best and worst movements of patients with the gene for Huntington's disease (HD) and control participants. Both groups exhibited small initial errors in some movements (arrows), but movements in the HD patients had numerous sharp, jerky changes in direction as the hand neared the target. (From Smith et al.[10] with permission.)

forward model because they depend on it to predict the limb's state in the near future. Among other relevant findings, the early component of the somatosensory evoked potential deceases in correlation with the number of glutamine repeats in the huntingtin protein,[14] the electrical sources of the somatosensory evoked potential are reduced in HD,[15] and corticostriatal pathways are the first to show abnormalities in mice that have the *huntingtin* gene.[16]

19.5.3 Adjustments to Noise in Huntington's Disease

Smith et al.[10] examined the feedback-control process in HD by asking patients to reach to a set of targets (figure 19.7). They noticed that in HD patients, many movements had abrupt changes in direction as the hand approached the target. In some trials, movements began in slightly the wrong direction and failed to stop smoothly at the target. However, sometimes movements appeared quite normal from beginning to end. When Smith et al. examined movements of control participants, they also saw movements that began in slightly the wrong direction. Unlike the HD patients, however, control participants smoothly corrected movements that began with small errors.

To understand these differences better, Smith et al. measured movement jerk by comparing the raw squared jerk profiles in the two groups. Early in movement, movement jerk in the HD patients matched that in the control group. Later, ~300 msec after movement onset, it began to exceed that of the controls (figure 19.8A). Interestingly, that was about the time

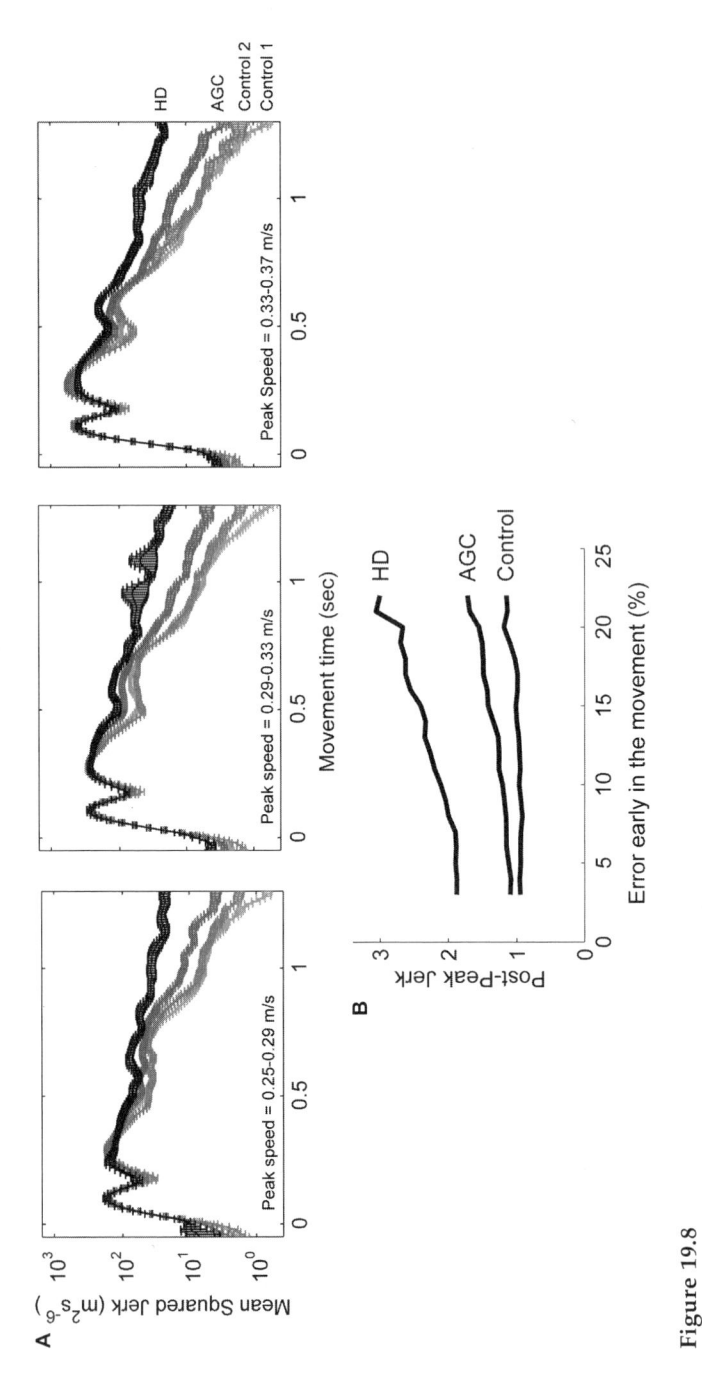

Figure 19.8

A deficit in error feedback control in HD. (*A*) The mean squared jerk ± SE on a logarithmic scale, as a function of movement time in each speed range, in three participant groups. The second peak in the squared jerk profile corresponded approximately to the peak in the speed profile. Note that groups had quite similar jerk profiles until 300–400 msec, but diverge after this point. Abbreviation: AGC, asymptomatic gene carriers for HD. (*B*) Average movement jerk near the end of movement (i.e., after peak speed), as a function of the amount of error early in the movement. (From Smith et al.[10] with permission.)

the hand reached peak speed. Near the end of movement, HD patients showed several times the amount of jerk as controls did. In ACGs, movement jerk fell between the values for controls and HD patients, but significantly exceeded control levels.

Why might the movements of HD patients begin to become irregular 300 msec into their course and not before? Smith et al. concluded that control participants smoothly correct the errors that occur early in a movement, but in HD patients, the feedback process that subserves these corrections malfunctions. To quantify error early in the movement, Smith et al. assumed that regardless of the intended speed of movement, hand location at peak hand speed should be near the center of the movement path, a prediction of the minimum-jerk model. They estimated the error, therefore, as the difference between hand location at peak speed and the midpoint of the movement. The experimenters found that the error early in movement did not differ between the groups. However, compared with the control group, in HD patients and in ACGs a given error early in the movement led to a higher jerk in the remainder of the movement (figure 19.8B). Therefore, in HD, the motor control system appeared to respond poorly to small, self-generated errors. This finding suggested that damage to the basal ganglia in HD somehow reduced the integrity of the feedback-control system, perhaps due to some dysfunction of the next-state planner.

How could a problem in the next-state planner produce an inappropriate motor command only when there is substantial error in the movement and not when the movement is error free? One possibility is that the error-dependent increase in jerkiness results from sudden, midmovement changes in the control policy (i.e., changes in the parameters of the next-state planner). Whereas in a healthy individual, a static control policy suffices for responding to self-generated errors during a reach, it is possible that in HD, the control policy changes in the middle of movement, producing corrective "submovements." Even in healthy individuals, however, there must be a threshold of error beyond which the initial control policy is abandoned. Perhaps this threshold is much lower in HD.

The same general result for HD patients and AGCs was also observed when perturbations were imposed on the hand during movement. Similar results have been obtained in a different motor task by Kevin Novak, Lee Miller, and Jim Houk.[17,18] As with the results on control participants by Smith et al., Novak and his colleagues found that corrective submovements began at about the time of peak velocity, near 300 msec after movement onset. They also suggested that the basal ganglia played an important role in these corrective aspects of the movement. Basal ganglia disease appears to lead to an inability to correct for errors that result from the noisiness of motor commands.

19.6 Transforming Plans into Trajectories: The Problem of Redundancy

In the next-state planner, the transformation of a large-scale plan into a sequence of small, near-term plans preserves the coordinate system of the large-scale plan. Both x_{dv} and $\dot{x}_d(t)$ represent vectors in visual, fixation-centered coordinates. How do you transform a desired change in end-effector location in fixation-centered coordinates into joint rotations? Many changes in arm configuration can lead to the same displacements of the end effector.

You have only three **degrees of freedom** to describe end-effector location in fixation-centered coordinates because you need to deal only with the three dimensions of real space. Nevertheless, your arm configurations at the shoulder and elbow occupy at least a five-dimensional space. (If you include your wrist, they occupy a seven-dimensional space.) Because of this high dimensionality, the task of reaching or pointing involves many combinations of joint angles that can produce the same wrist or finger location. By analogy, many combinations of joint rotations correspond to the same end-effector displacement. Specifying what your end effector should do does not determine what your arm should do.

To make the problem a bit more explicit, imagine a professor pointing to a spot on the blackboard. When the professor wants to point to some other location, the tip of the pointer moves roughly parallel to the surface of the blackboard with a smooth, straight trajectory. For pointing movements of this sort x_{dv} is a two-dimensional displacement. The movement retains its low dimensionality when transformed into a time sequence of end-effector displacements $\dot{x}_d(t)$. Imagine, for example, the professor walking parallel to the blackboard as he or she points to a series of locations along a horizontal line. The professor could do this with many different joint angles, and if some obstacle made him or her move away from the board, the trajectory of the end effector could continue unchanged by simply extending the arm more. The next section considers two ways to solve this problem, often called the problem of kinematic *redundancy*.

19.6.1 The Stiffness Approach

Imagine that you attach a pair of rubber bands to each joint of your robot's arm, configured in an antagonistic architecture. After the rubber bands are attached, the arm comes to rest at some unique configuration which depends on the stiffness and length of each rubber band. Sandro Mussa-Ivaldi, Pietro Morasso, and Roberto Zaccaria,[19] and Mussa-Ivaldi and Neville Hogan,[20] constructed such a system and observed that despite the limb's kinematic redundancy, if they pulled on the end effector, the joint angles changed in a consistent and reproducible way.

This happened because each rubber band acted like a spring, and springlike systems minimize potential energy. The equilibrium point of the arm represents the minimal potential energy when no one pulls on the arm's end effector. When someone does pull on the end effector, the system still has a minimal potential energy for that new location, although the energy level exceeds that of the equilibrium point at rest. At each point in the trajectory, some minimal potential energy exists, and the limb should move along that path regardless of its redundancy.

Of course, your arm also has stiffness because muscles and the associated reflexes act like springs. For example, recall from section 8.2 that if someone displaces your hand by Δx, the springlike arm muscles produce a restoring force. If you represent this restoring force with vector f, you have

$$f = K_{ee}\Delta x.$$

In the above expression, K_{ee} is a matrix that specifies stiffness of the arm as measured at the end effector (in this case, your hand). The force at the hand corresponds to a vector of joint torques specified by the arm's Jacobian matrix:

$$\tau = J^T f.$$

The muscles that produce these torques have a stiffness as measured at the joints. If you label this joint stiffness with matrix K_θ, torque at the joints is related to joint displacement $\Delta \theta$ by

$$\tau = K_\theta \Delta \theta.$$

This expression leads you to a relationship between end-effector displacement and joint displacement:

$$\Delta \theta = K_\theta^{-1} J^T K_{ee} \Delta x. \tag{19.5}$$

The stiffness of muscles produces a potential energy "landscape" that can act as a minimization principle with which to solve the problem of redundancy. To displace the end effector in your redundant limb, your joints should move in a pattern specified by the stiffness of the arm and the energy "landscape" it creates.

19.6.2 Trade-Offs Between Controlled and Uncontrolled Variables

In the model of figure 19.4, the controlled variable is the end effector's location with respect to the target. In a redundant system, this controlled variable has fewer degrees of freedom than planned joint states or the resulting motor commands. This fact implies that as long as the end effector gets to the target, the system may not care about the precise configuration of the arm. For example, assume that there is noise in the transformation $\dot{x}_d \to \dot{\theta}_d$ so that the system introduces some errors in calculating the forces necessary to make the movement. The system, however, does not attend to these errors equally. The only relevant errors involve those affecting end-effector location with respect to the target. The system allows other errors to remain uncorrected.

Emo Todorov and Michael Jordan[21] proposed that in a redundant system, the fluctuations in redundant joint-angle dimensions can be ignored unless they have an impact on the task-relevant variables. As long as those fluctuations have no relevance to the crucial variables, the redundant degrees of freedom remain largely uncontrolled.

Todorov and Jordan found that *control laws* for redundant tasks do not enforce a specific trajectory; instead, they obey a *minimal intervention principle*, which states that only those deviations from the average behavior that interfere with the task-level goals should be corrected. As long as fluctuations among the redundant degrees of freedom do not interfere with task performance, an optimal control strategy leaves them uncorrected. The advantage of doing so results from the fact that correcting (and *acting* in general) is expensive: It increases both signal-dependent noise and energy consumption.

But what are the relevant variables that the CNS needs to control? The answer to that question depends on the task. In a complex task, such as tying your shoelaces, the relevant variables describe the configuration of the shoelaces, regardless of your hand movements. In a reaching or pointing task, it makes sense for the CNS to care about endpoint error because your goal involves reaching the target (as opposed to, say, making a movement along a particular trajectory). The theory of Todorov and Jordan explains trajectory smoothness as the outcome of minimizing endpoint error and energy consumption in the face of sensory and motor noise (the latter being signal-dependent). As Novak, Miller, and Houk[18] note: "A biologically plausible optimization criterion is the minimization of the occurrence and amplitude of corrective submovements.... We postulate that ... other criteria ... are instead secondary...."

References

1. Hogan, N (1984) An organizing principle for a class of voluntary movements. J Neurosci 4, 2745–2754.
2. Bullock, D, and Grossberg, S (1988). In Dynamic Patterns in Complex Systems, eds Kelso, JAS, Mandell, AJ, and Shlesinger, MF (World Scientific Publishers, Singapore).
3. Pelisson, D, Prablanc, C, Goodale, MA, and Jeannerod, M (1986) Visual control of reaching movements without vision of the limb. Exp Brain Res 62, 303–311.
4. Hoff, B, and Arbib, MA (1993) Models of trajectory formation and temporal interaction of reach and grasp. J Mot Behav 25, 175–192.
5. Hoff, B (1992) A Computational Description of the Organization of Human Reaching and Prehension, doctoral dissertation, University of Southern California.
6. Ijspeert, AJ, Nakanishi, J, and Schaal, S (2002) Movement imitation with nonlinear dynamical systems in humanoid robots. In Proc IEEE International Conference on Robotics and Automation (IEEE, New York) pp 1398–1403.
7. Todorov, E, Shadmehr, R, and Bizzi, E (1997) Augmented feedback presented in a virtual environment accelerates learning of a difficult motor task. J Mot Behav 29, 147–158.
8. Harris, CM, and Wolpert, DM (1998) Signal-dependent noise determines motor planning. Nature 394, 780–784.
9. Jones, KE, Hamilton, C, and Wolpert, DM (2002) Sources of signal-dependent noise during isometric force production. J Neurophysiol 88, 1533–1544.
10. Smith, MA, Brandt, J, and Shadmehr, R (2000) The motor dysfunction in Huntington's disease begins as a disorder in error feedback control. Nature 403, 544–549.

11. Ross, CA, Margolis, RL, Rosenblatt, A, et al. (1997) Huntington disease and the related disorder, dentatorubral-pallidoluysian atrophy (DRPLA). Medicine 76, 305–338.

12. Mattson, MP (2002) Accomplices to neuronal death. Nature 415, 377–379.

13. Hedreen, JC, and Folstein, SE (1995) Early loss of neostriatal striosome neurons in Huntington's disease. J Neuropathol Exp Neurol 54, 105–120.

14. Beniczky, S, Keri, S, Antal, A, et al. (2002) Somatosensory evoked potentials correlate with genetics in Huntington's disease. NeuroReport 13, 2295–2298.

15. Seiss, E, Praamstra, P, Hesse, CW, and Rickards, H (2003) Proprioceptive sensory function in Parkinson's disease and Huntington's disease: Evidence from proprioception-related EEG potentials. Exp Brain Res 148, 308–319.

16. Cepeda, C, Hurst, RS, Calvert, CR, et al. (2003) Transient and progressive electrophysiological alterations in the corticostriatal pathway in a mouse model of Huntington's disease. J Neurosci 23, 961–969.

17. Novak, KE, Miller, LE, and Houk, JC (2003) Features of motor performance that drive adaptation in rapid hand movements. Exp Brain Res 148, 388–400.

18. Novak, KE, Miller, LE, and Houk, JC (2002) The use of overlapping submovements in the control of rapid hand movements. Exp Brain Res 144, 351–364.

19. Mussa-Ivaldi, FA, Morasso, P, and Zaccaria, R (1988) Kinematic networks: A distributed model for representing and regularizing motor redundancy. Biol Cybern 60, 1–16.

20. Mussa-Ivaldi, FA, and Hogan, N (1989) Solving kinematic redundancy with impedance control: A class of integrable pseudoinverses. In Proc IEEE International Conference on Robotics and Automation (IEEE, New York) pp 283–288.

21. Todorov, E, and Jordan, MI (2002) Optimal feedback control as a theory of motor coordination. Nat Neurosci 5, 1226–1235.

Reading List

If you have an interest in a detailed mathematical description of trajectory computations, especially the approach to resolving redundancy known as the Moore–Penrose pseudoinverse, you might consult two papers: Mussa-Ivaldi et al.[19] and Mussa-Ivaldi and Hogan.[20]

IV Prediction, Decisions, and Flexibility

20 Predicting Force I: Internal Models of Dynamics

*Overview: In planning a reaching or pointing movement, your CNS relies on an **internal model** that predicts the forces needed to reach the target. This internal model maps desired limb states—for example, the limb's configuration and the rate at which that configuration changes—to forces.*

Imagine that your friend just opened a can of soda and put it on a table in front of you. Your friend happens to be someone who likes practical jokes; she gave you an empty can. As you lift the can, your hand jerks up faster than you intended or expected. Fortunately, your motor system reacts quickly and stops the motion of the hand before the can chips a tooth. The fact that you expected a full can made a big difference in your movement. Surprisingly, miscalculating the weight of an object by a small amount affects reaching a lot. This fact implies that control of reaching involves more than computing a kinematic plan that brings the hand to the target. The CNS must also compute something about the requisite forces.

If you hold the can and intend to lift it to your mouth, the current end-effector location (the opening at the top of the can), the difference vector that points from this location to your mouth, and the transformation of that vector into desired hand and joint velocities compose the movement's **kinematics**. The problem of **dynamics** addresses the transformation of these intended movements into forces. Dynamics specifies how the motion of a system depends on the forces that act on it. A given displacement of a can full of liquid requires somewhat larger lift (and grip) forces than an empty can. Presumably, because you have lifted many cans full of liquid, your CNS has learned something about the forces needed to lift cans filled to various degrees of capacity. The expectation of a full can results in "larger" commands sent by the CNS to the spinal cord, which eventually reach your arm muscles.

Picking up a can provides yet another example of predictive motor commands. Just as computing target location, a hand difference vector, and a joint-rotation vector depended on internal models (sections 15.8 and 17.3.3), computing forces and **torques** also depends on an internal model. That is, when you want your hand to lift a can to your mouth, your CNS apparently predicts the forces and torques necessary to accomplish the desired limb displacement: $\Delta\hat{\theta} \xrightarrow{\theta} \Delta\hat{\tau}$. Like figure 14.1, figure 19.4

represents this final stage of computation as network 5 in a hypothetical framework for the neural control of reaching and pointing movements. This network aligns joint rotations and torque, and thus can be called a *dynamics map*; it is an internal model of dynamics. Because a given joint rotation requires different forces for full cans, as opposed to empty ones, this map must depend on context.

Now take the case of a server approaching your table with a tray full of food. As he or she moves the tray away from his or her body to rest it on your table, the **moment arm** associated with the tray increases so that as it reaches your table, the tray imposes a significant torque on the waiter's trunk and leg bones. As the waiter's arms extend the tray, his or her leg muscles probably become activated to compensate for the added torque. In fact, this occurs in anticipation of the effect that the movement will have on postural stability (see section 5.2.3).[1]

These examples might give you the impression that the problem of computing dynamics occurs only for reaching with an object in your hand. However, the same principles apply to any fast movement. For example, take the case of flexing your elbow with your arm in the horizontal plane. If your CNS simply activates the biceps muscle, the elbow will flex, but its motion will impose a torque, called an ***interaction torque***, on your shoulder. If you do not want that torque to move your upper arm while you move your lower arm around the elbow, you will need to activate shoulder muscles to negate the interaction torque. Therefore, activating a muscle will produce a force that may rotate a joint, and the rotation of that joint puts forces on other joints, which in turn causes further motion. For your body to remain stable, your CNS must "know" about these interaction forces. That is, your CNS must know physics. Internal models embody your CNS's **implicit** knowledge of physics.

As one way to think of internal models, you might envision them as a sort of memory of what has worked in a particular situation. That way, when you reach and point, you do not have to perform a lot of calculations beforehand. Your motor plan, especially at relatively high levels, need not specify how much to contract each muscle in order to arrive at the correct joint angles. Instead, your CNS recognizes the state of current inputs and motor commands (in the form of **efference copy**), and uses that information to predict the state of your limb at some future time. How does your CNS represent internal models of dynamics?

20.1 Internal Models of Dynamics

Dynamics describes how forces that act on a mechanical system produce a change in its location and velocity. For example, imagine resting your elbow on a table. If you produce some torque τ with your biceps, your elbow flexes. Dynamics describes the relationship between the torque τ and the resulting trajectory of the forearm. If we use θ to represent the elbow angle, m to represent the mass of the forearm, l to represent the location of the forearm's center of mass relative to the origin, and b to represent

20.1. Internal Models of Dynamics

the viscosity of the elbow joint, then the following equation describes the dynamics of the arm reasonably well:

$$\ddot{\theta} = \frac{1}{ml^2}(\tau - b\dot{\theta} - mgl\cos\theta).$$

This equation describes how forces acting on the system produce a change in its state (acceleration). In part because force causes the change in state, you can think of this equation as describing **forward dynamics**. (See the web document *dynamics* if you want to see where these equations come from and want to know how to derive them for arbitrary systems, including systems that resemble the human arm.) If you want to move your elbow in a particular way (i.e., you intend to change its state so that it moves along some desired trajectory $\theta_d(t)$), then you need to solve this equation for τ:

$$\tau = ml^2\ddot{\theta}_d + b\dot{\theta}_d + mgl\cos\theta_d.$$

This equation describes **inverse dynamics**, the forces (or torques) needed to make a particular movement. The notation $\Delta\hat{\theta} \xrightarrow{\theta} \Delta\hat{\tau}$ signifies an internal model of inverse dynamics for the system illustrated in figure 19.4 (see also figure 14.1). For simplicity, the remainder of this chapter ignores the distinction between forces and torques and concentrates on forces; hence the notation $\Delta\hat{\theta} \xrightarrow{\theta} \Delta\hat{f}$ (or $\Delta\hat{F}$).

Consider again the imaginary robot in figure 15.1A. If you wanted to move the robot's gripper to a new location, you might compute target and end-effector locations in camera-centered coordinates, compute a difference vector, and plan a desired next state. The problem remains as to how to transform this plan into commands to the robot's motors. You might estimate the dynamics of the system, then invert that model to form a model of inverse dynamics, and use that model to compute the robot's motor forces. However, most robots do not work this way. Most commercial robots assume that a spring and a damper lie between the current configuration of the robot's arm and its configuration at a target location. To implement this imaginary spring–damper system, the robot's controller uses a fast-**feedback** control system that measures the current state of the robot's end effector (i.e., the gripper's location and velocity), compares it against a desired state, and produces torques proportional to the difference. The performance of such a system depends on three factors: the accuracy of the sensors, the delay in the feedback loop, and the speed with which the motors can produce force.

Optical encoders on most robots can measure joint angles to a small fraction of a degree. The delay in the sensors and feedback pathways rarely amounts to 1 msec, and a reasonable torque motor responds to a step change in its input with a time constant of about 2 msec. If you want your robot to make a movement that takes about 0.5 sec, such a system would have little trouble relying on its feedback to move its arm to the target location. However, the larger the arm—and therefore the larger the arm's **inertia**—the more difficulty a spring–damper system has: It may slightly overshoot the target before settling at it. Nevertheless, a robotic arm the size of a human arm would have little trouble making accurate movements with this kind of feedback controller. Indeed, for such a system it hardly matters whether the gripper holds a full can of soda or an empty can.

However, for your CNS it does matter whether the soda can is full or empty. Your CNS relies on transducers (sensory neurons), communication pathways (axons), and force production systems (muscle cells) that produce a cumulative delay nearly two orders of magnitude larger than the spring–damper robot described above (see chapter 3). Chapter 8 describes two feedback pathways, one that depends on length–tension properties of muscles and another that relies on spinal and long-loop reflexes. These feedback systems maintain the stability of your limb, but they do not suffice for control of rapid movements of limbs as large and as heavy as yours. To overcome these fundamental physical and physiological limitations, your CNS has evolved neural structures that learn to predict the relationship between state changes in your limb and motor commands to your muscles. That is, your CNS has acquired the ability to learn internal models of dynamics.

Why don't commercial robots use a similar approach? In part, because of their high degree of specialization. Commercial robots typically do one thing well, and neither predictions nor adaptability has a lot of importance in these systems. Beyond that, however, an accurate prediction of dynamics requires a surprising amount of computational power. For example, the equations that represent dynamics of a primatelike arm with five degrees of freedom involve the multiplication of thousands of trigonometric functions. In turn, these equations depend on estimates of several physical parameters of the arm, including mass, centers of mass, moments of inertia, and link lengths. If the robot held different tools in its gripper, these parameters would change, and the controller would have to adapt to each tool.

Your physiology puts you at a disadvantage in regard to robots: You have much slower sensorimotor feedback pathways and the other limitations enumerated in chapter 3. However, your CNS has the advantage in terms of computation. You display that advantage whenever you effortlessly adapt your reaching movements to the rapidly changing and only partially predictable dynamics of the world.

20.1.1 Internal Models of Dynamics: Reaching with Robots

To approach the question of how your CNS represents the dynamics of reaching and pointing movements, Shadmehr and Sandro Mussa-Ivaldi,[2] working in Emilio Bizzi's laboratory, developed an experimental paradigm in which participants held the handle of a robotic arm and reached to visual targets (figure 20.1). By giving people real objects to reach with, the experimenters could approximate the dynamics of an arbitrary system with the robot's motors. In this way, they could produce a wide variety of dynamics and quantify how the CNS changed its motor commands

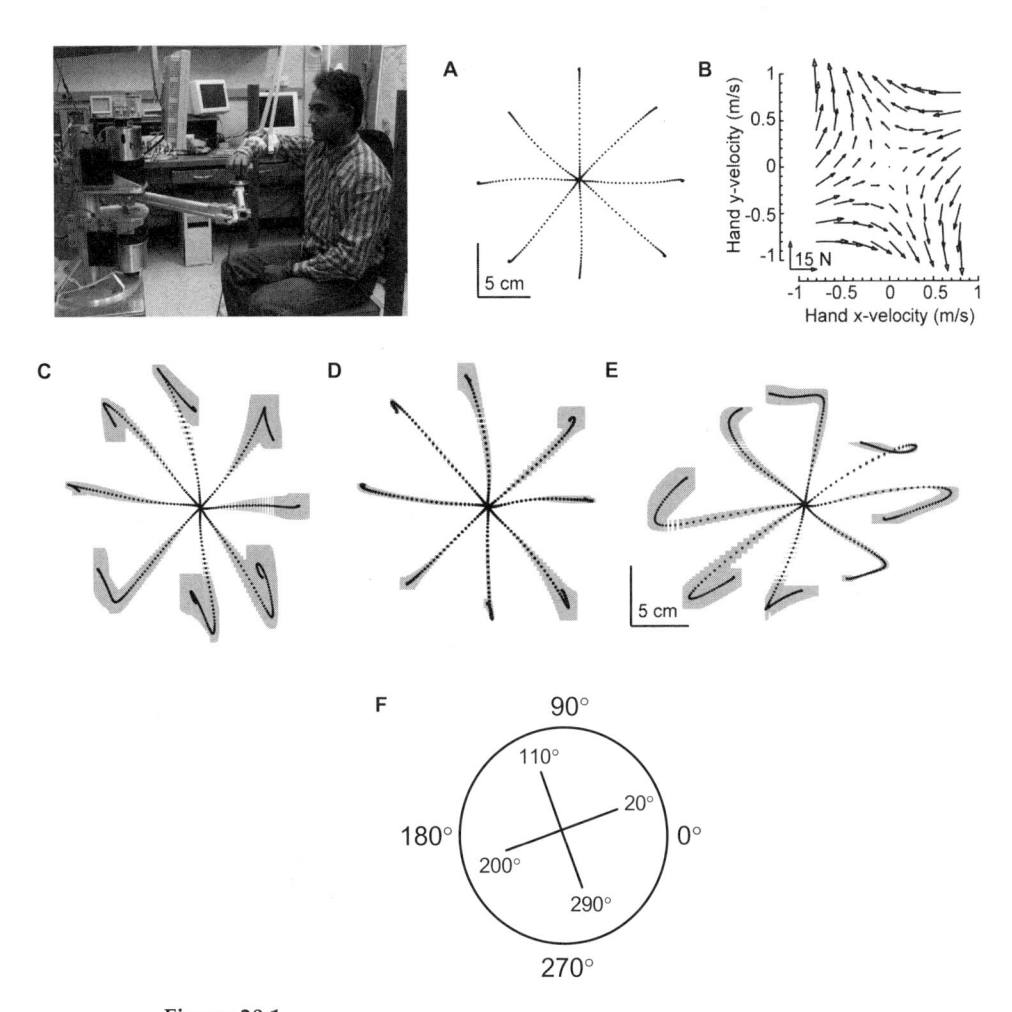

20.1. Internal Models of Dynamics

Figure 20.1
Adaptation of reaching movements to force fields. Top, left: Participants held the handle of the robot and reach to a target. (*A*) Hand trajectory (with the dots 10 msec apart) for typical movements to eight targets. (*B*) The force field produced by the robot. The figure plots forces as a function of hand velocity. (*C*) Average hand trajectories for movements during the initial 200 trials in the force field. (*D*) Average hand trajectories for movements 600–800. (*E*) Average hand trajectories during catch trials. (*F*) Selected polar coordinates. (From Shadmehr and Mussa-Ivaldi[2] with permission.)

because of motor learning. Shadmehr and Mussa-Ivaldi tested the hypothesis that the CNS computes motor commands based on an internal model of limb dynamics.

The participants saw a target on a video monitor and reached to it while holding the robot's handle. With the robot's motors off (the *null-field* condition), participants made straight movements (figure 20.1A). After these baseline measurements, the experimenters engaged the robot's motors and programmed it to produce a force field that depended on the velocity of the hand (figure 20.1B). When it was stationary, the robot imposed no force on the participants' hands. However, when their hands moved, the robot's imposed forces resulted in skewed trajectories (figure 20.1C). After a few hundred movements, participants adapted to the robot's forces and again made smooth and nearly straight movements to the target (figure 20.1D).

Notice that this force field tended to resist the motion of the hand for some directions of movement (~20° and ~200°), to assist the hand in other directions (~110° and ~290°), and, for other directions, it displayed more complex influences. Shadmehr and Mussa-Ivaldi hoped that by using this force field, they could alter the dynamics of reaching in a way that the participants had never experienced, which would therefore require gradual adaptation.

Somewhat like the situation in which you mistook an empty soda can for a full one, Shadmehr and Mussa-Ivaldi occasionally turned off the forces as the participants trained on reaching in the force field. They called such trials *catch trials*. Very early in training, the participants moved their hands straight to the targets in catch trials. With further training in the field, as trajectories in the force-field trials became straighter, the trajectories in catch trials became more skewed (figure 20.1E). At the end of training, hand trajectories in catch trials roughly resembled a mirror image of the trajectories in force-field trials, sometimes termed "fielded trials" in laboratory jargon. The trajectories in the catch trials reflect the **aftereffects** of adaptation.

Improvement in performance occurred because, with training, the participants changed the motor commands to their arms. Participants could improve by coactivating antagonist muscle groups. This strategy would stiffen the limb, making it more resistant to perturbations imposed by the robot (section 8.3). However, in a catch trial, this strategy would not produce any aftereffects because catch trials do not contain a perturbation. The fact that in a catch trial the hand's trajectory mirrors the early force-field trials shows that the motor commands to the muscles changed, but not by simply increasing limb stiffness. Rather, the CNS better predicted the forces imposed by the robot.

20.1.2 Internal Models of Dynamics: Reaching While Rotating

Most robotic arms have a base that is bolted to the ground. Your arm, however, moves along with your body, and the movements of your body affect the dynamics of reaching. Jim Lackner and his colleagues have

Box 20.1
Coriolis forces

> What are Coriolis forces? Imagine that sometime in the future, humans build a space station big enough to require artificial gravity. They build a very large, airtight hollow sphere and spin it around its polar axis. They then build homes on the inside of the sphere, locating them around the equatorial region of the space station. They put the houses there because spinning the sphere produces centrifugal forces that push on any mass placed on the inner surface, and those forces are largest at the equator. This force acts like gravity.
>
> Now imagine that a family on this space station has just moved into their new equatorial house and decides to play volleyball in their back yard. The server holds the ball and throws it straight up to hit it. Remarkably, the ball does not come down toward the server, but lands behind him or her. What happened resulted from the Coriolis force. The motion of the ball in the spinning sphere resulted in a force perpendicular to the ball's velocity, much like a curl field illustrated in figure 20.5A. To play volleyball, the server will have to learn to compensate for this dynamics. (The idea of people living in a hollow sphere was based on a story by Isaac Asimov, "For the Birds," in *Winds of Change and Other Stories* [Ballantine Books, New York, 1983, pp. 62–73].)

demonstrated that even when you do not hold anything in your hand, reaching dynamics change as you rotate your body, and your CNS adapts in order to predict these dynamics.

Jim Lackner and Paul Dizio[3] produced force fields that differed from those imposed by a robot. Their forces affected reaching not through contact with the participants' hands, but through inertial forces that acted on the entire limb. Their participants sat in the center of a cylindrical room that slowly accelerated until it reached an angular velocity of 10 revolutions per minute (rpm) or 60°/sec. This rotation imposed force on every part of the body, except those at the precise center of the room. One component of this force is an "outward" force that grows linearly with the distance of the body part from the center of the room, linearly with the mass of the body part, and as the square of the angular velocity. Drawing the limbs toward the body minimizes this force.

When any part of the body moves in a rotating room, that part will experience a Coriolis force (see box 20.1). This force is approximately the cross product of the linear velocity of the moving part and its angular velocity. For example, in the setup shown in figure 20.2A, the room rotates counterclockwise, so the angular velocity vector points out of the page toward you. If you perform a center-out reaching movement, your hand will experience forces that push it to the right. This force is the cross product between the hand-velocity vector and the angular-velocity vector (see appendix C). After the outward reach is complete, as your hand returns toward the center, it will experience a leftward force. Coriolis

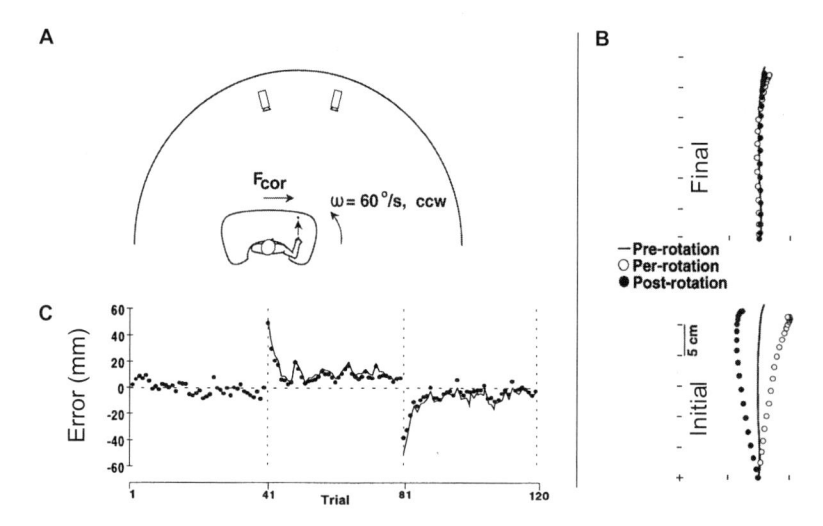

Figure 20.2
Reaching in a rotating room. (A) Participants sat near the center of a room that rotated at constant angular velocity of 60°/sec and reached to a target 35 cm away. (B) Top view of average hand trajectories for fast reaches. The figure shows the initial (bottom) and final (top) reaching movements in the prerotation period, the per-rotation period (i.e., during rotation), and the postrotation period. Note both the endpoint error and the perpendicular displacement from a straight path to the target. Time moves up in each part of the figure. (C) Movement errors as averaged across participants. Filled circles indicate endpoint errors; lines show the maximum perpendicular displacement from a straight path to the target. Trials 0–40 came from the prerotation period; trials 41–80, from the per- (during) rotation period; and trials 81–120, from the postrotation period. (From Lackner and Dizio[3] with permission.)

forces are nil when your hand does not move. The faster your hand goes, the larger the hand-velocity vector, and the larger the Coriolis forces.

The participants pointed to a visual target in a darkened room. The target was illuminated briefly, and then the light was turned off before the end of movement. Before the room began rotating, participants made straight movements to the target (figure 20.2B). The room accelerated so slowly that the participants often failed to detect its rotation. However, the rotation produced Coriolis forces on the moving hand, deflecting the reaching movement to the right (figure 20.2B). These deflections vanished, for the most part, after about eight trials. After 40 trials, the room slowly returned to zero angular velocity and reaching resumed. Despite the fact that no Coriolis forces perturbed the hand, its trajectory deflected leftward. It took approximately the same number of trials for these aftereffects to wash out as for the initial adaptation.

These results suggest that during a rotation of your body, your CNS learns to alter the patterns of muscle activation to compensate for forces perpendicular to the direction of hand motion. When the rotation stops, the internal model that predicted these forces has changed and now predicts inappropriate dynamics. The results illustrate that adaptation of the internal models of dynamics takes place not only when your hand comes into contact with a novel object (e.g., the robotic arm described in the previous section), but also when forces act on your limb because of the motion of your body.

Reaching in a rotating room might appear strange, but you experience Coriolis forces frequently. For example, your CNS compensates for such forces when you reach to a target that requires rotation of your trunk. Suppose that you rapidly turn your trunk counterclockwise as you reach with your right hand to a target on your left. The trunk rotation produces Coriolis forces on the arm, pushing it in a direction perpendicular to the hand-velocity vector (in this case, resulting in a clockwise force pattern). As the rotational velocity of the trunk increases, the forces become larger. Remarkably, your CNS accounts for these forces in the motor commands to your limb, producing nearly straight movements for both slow and fast rotations.[4]

20.1.3 Internal Models of Dynamics as Transformations from Limb States to Forces

The ability to compensate for varied dynamics of reaching movements results from an internal model of dynamics (i.e., a neural system that predicts force as a function of a given desired state of the limb, given a current state). When the CNS experiences novel dynamics, the internal model is "wrong" in that it cannot predict the forces that are imposed on the hand. After a period of training, it adapts to and correctly predicts forces that produce the planned movement.

What is an internal model? You need to recognize the difference between "model" in the sense used in "internal model" and "model" as used in other ways. The term "model" in "internal model" applies to a broad class of computations that involve learning about the world and/or the effectors, such as the physics of an arm used for reaching and pointing. The term "model" as in "the model illustrated in figure 19.4" often refers to anything that attempts to capture some aspect of a biological system with a nonbiological method, be it physical, mathematical, or computational (see boxes 11.2 and 11.3). Various models, in the general sense of that term, make use of internal models, including inverse and forward models.[5-7]

Consider again the example of a person learning to reach in a force field. Shadmehr and Mussa-Ivaldi[2] derived the equations of motion of a humanlike arm holding the handle of a robot. (The derivations are available in the web document *dynamics*.) These equations describe how the mass of the limb responded to force input from the muscles. To represent the error feedback system of the muscles and the spinal **reflexes**, Shadmehr and Mussa-Ivaldi augmented the equations by including a simple, low-gain spring–damper element that stabilized the limb near a desired state. The desired state traveled along a trajectory straight to the target. To produce a movement, Shadmehr and Mussa-Ivaldi assumed that the joint

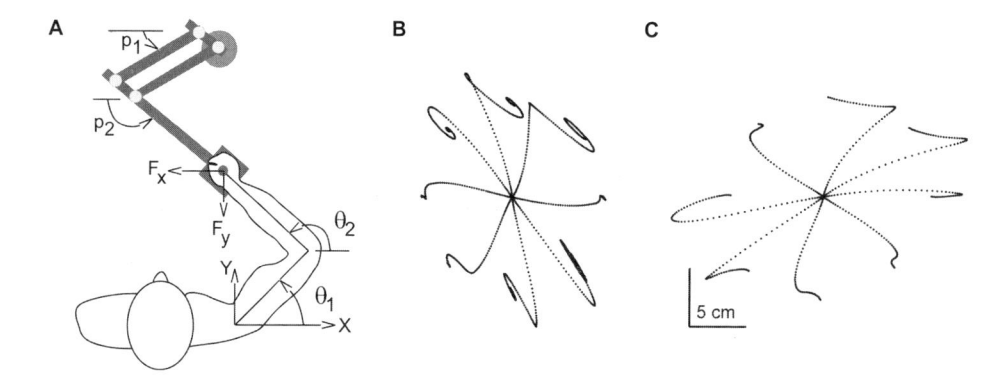

Figure 20.3
Simulated trajectories for a humanlike arm during force-field training. (A) Schematic with angles p for the robot and angles θ for the humanlike arm. F_x and F_y refer to force in the x and y dimensions, respectively. (B) Trajectories for an early stage of training for the force field shown in figure 20.1B. (C) Catch trials after the controller for the humanlike arm had acquired an accurate **internal model** of the field.

torques depended on knowledge of the inverse dynamics of the limb (i.e., a map—an internal model of the arm's physical dynamics—that transformed the desired sensory state of the limb into torques to compensate for the arm's inertial dynamics). Equations 20.1 and 20.2 describe the internal model:

$$I_r(p)\ddot{p} + G_r(p,\dot{p})\dot{p} = \tau_r - J_r^T(p)F$$
$$I_s(\theta)\ddot{\theta} + G_s(\theta,\dot{\theta})\dot{\theta} = C(\hat{\theta},\dot{\hat{\theta}},\ddot{\hat{\theta}}) + J_s^T(\theta)F \tag{20.1}$$

$$C(\theta,\dot{\theta},\theta_d,\dot{\theta}_d,\ddot{\theta}_d) \equiv \hat{I}_s(\theta_d)\ddot{\theta}_d + \hat{G}_s(\theta_d,\dot{\theta}_d)\dot{\theta}_d - K(\theta - \theta_d)$$
$$- V(\dot{\theta} - \dot{\theta}_d) - J_s^T(\hat{\theta})\hat{F}(\dot{\theta}_d). \tag{20.2}$$

Equation 20.1 describes the coupled dynamics of the four-link system shown in figure 20.3A. In equation 20.1, I represents the inertia matrix of the arm: the top equation for the robot r and the bottom one for the participant s. G represents the centripetal/Coriolis matrix. The symbol τ_r stands for the torque vector produced by the robot; F, for the force at the robot's handle; J, for the Jacobian matrix; and C, for the motor controller. Equation 20.2 defines C. It has three components: (1) an internal model of the dynamics of the participant's arm (i.e., $\hat{I}$ and $\hat{G}$), (2) a feedback control system that has stiffness K and viscosity V, and (3) an internal model of the forces imposed by the robot. Shadmehr and Mussa-Ivaldi further assumed that the internal models depended on a desired limb state $\vec{\theta}_d$, and that the CNS had acquired an accurate model of its own arm's dynamics. With training, the CNS learned to predict the velocity-dependent, robot-imposed forces $\hat{F}$.

In this system, the ability to compensate for varied dynamics of reaching movements results from a system that predicts force as a function of a given desired state of the limb. When the CNS experiences novel dynamics, the internal model is "wrong" in that it cannot predict the forces that are imposed on the hand. After a period of training, it adapts and correctly predicts forces that produce the planned movement.

The robot had only a small amount of inertia, and therefore when the motors were not engaged, only small forces acted on the hand. When the motors were engaged, forces depended on the hand's velocity and significantly displaced the hand during movements. You can think of the internal model at this early stage of training as naïve (i.e., $\hat{F} \approx 0$). The internal model predicted what you might think of as the "normal" forces encountered during reaching and pointing movements, such as the interaction torque imposed by movements of the lower arm on the upper arm. However, it could not predict the forces imposed by the robot. Therefore, when the motors were engaged, simulations of movements based on naïve equations 20.1 and 20.2 did not result in straight movement to the targets (figure 20.3B). Note, however, the smaller movement errors for targets at 0° or 180°, and the larger errors for targets at 90° or 270°. The model tells you that this difference does not result from the adaptation process; rather, it depends on the inertial and stiffness properties of the arm (see section 8.9). Because these mechanical properties are not uniform, a given error in prediction of force produces different errors in hand trajectory for different directions of movement.

Next, the internal model adapted so that it could accurately predict the robot-imposed forces (i.e., predicted force matched actual force $\hat{F} \approx F$). Figure 20.3C shows a simulation of catch trials (i.e., movements with no robot-imposed forces, near the end of training). This simulation resembles the hand trajectories recorded in people, as illustrated in figure 20.1E. The internal model learned to map the desired limb velocity to forces (i.e., $\dot{\theta}_d \to \hat{F}$). You can think of this mapping as a state–force model because it maps the desired state of the limb onto the forces needed to produce that desired state.

20.1.4 Internal Models as Maps of States and Not of Time

You might consider an alternative hypothesis: The motor control system did not learn this state–force mapping, but rather a time–force mapping. Imagine a tape that plays out as a function of time for each movement direction. This tape may record an average of forces that the system sensed during previous movements in that direction. Mathematically, direction and time compose the inputs to this alternative model, which has force as its output. According to this hypothesis, your CNS engages in "rote learning" or memorization of the temporal sequence of forces, not their associations with limb states. As long as you test participants on movements that they trained on earlier, such a mechanism would adapt to the force field and produce catch trials similar to those observed in the state–force model.

Michael Conditt, Francesca Gandolfo, and Sandro Mussa-Ivaldi[8] found a way to test this hypothesis. They trained participants to reach to a

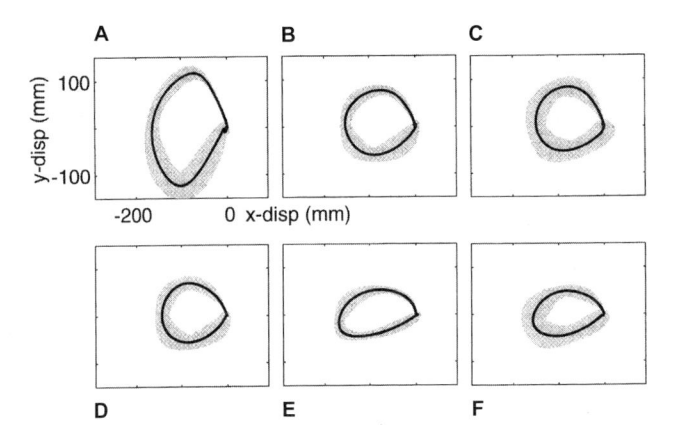

Figure 20.4
Generalization of force adaptation. (A) Circle trajectories made in a force field similar to figure 20.1B during initial training trials. (B) Trajectories recorded in the late stages of circle training (after about 700 movements). (C) Circle trajectories recorded from participants who adapted to straight reaching movements. (D) Circles drawn by naïve participants in the null field. (E) Catch trials during circle-drawing adaptation trials. (F) Catch trials from participants who adapted to straight reaching movements. (From Conditt et al.[8] with permission.)

small number of targets in a force field, then later asked them to draw a circle in the same field. They reasoned that if the CNS learned through rote memorization, like a tape recording of the forces encountered in reaching to each target, then the training for brief reaching movements should contribute little toward the execution of longer, circular movements. On the other hand, if the CNS learned an association between limb states and forces, then the order of states that it had experienced should not matter. As long as the circle involved limb states the CNS had experienced during various reaching movements, the participants should draw a reasonably geometrical circle.

Figure 20.1B shows the force field Conditt and his colleagues used. When naïve participants drew circles in this field, the robot distorted their hand trajectories in the direction of the destabilizing force, as shown in figure 20.4A. When these participants practiced their circle drawings, performance improved so that they eventually drew an accurate circle (figure 20.4B). However, other participants who had practiced short, straight reaching movements and had no practice with circles drew a circle equally well, as shown in figure 20.4C. Therefore, training in straight reaching transferred to drawing of circles. This finding suggested that adaptation to altered limb dynamics during reaching resulted in an internal model that could improve performance for other kinds of movements.

Importantly, Conditt and his colleagues observed that if participants trained on straight reaching movements but were given catch trials for circular movements, they showed aftereffects (figure 20.4E). These aftereffects resembled those of participants who had trained on circular movements (figure 20.4F).

The findings of Conditt and his colleagues suggested that the internal model of dynamics did not predict forces explicitly as a function of time. Rather, in performing the reaching movements the CNS had learned to associate the sensory states of the limb—specifically its location and velocity—with forces. The particular order in which the CNS experienced those sensory states did not matter. The critical factor was the region of the *state space*—the limb's location and velocity—that the reaching movements had visited.

Recall, however, that the force field the robot produced, varied as a function of hand velocity. Therefore, you could argue that the CNS learned to associate states to forces, rather than to some input that included time, because the force field imposed by the robot was not time-dependent. Conditt and Mussa-Ivaldi[9] pursued this issue by investigating whether participants could adapt to force fields that explicitly depended on time. They produced a "wavelike" force field that always pushed the hand to the left and varied only as a function of time. The amplitude of the force was 0 at movement initiation, reached a maximum value at ~160 msec, then declined to 0 at 333 msec and remained at 0. The force did not depend on either the location or the velocity of the hand, but only on movement time (i.e., the time into the movement). They observed that participants readily adapted to these time-dependent forces, and hand trajectories developed aftereffects that mirrored the pattern recorded in the early trials of adaptation.

If the internal model of dynamics maps limb states to forces, then how can you learn a force field that explicitly depends on movement time? First, note that adapting to time-dependent forces was more difficult than adapting to state-dependent ones. Conditt and Mussa-Ivaldi suggested that participants could build a *state-dependent* internal model that approximated the *time-dependent* force field. In the experiment, participants rarely experienced the same limb state at different movement times because targets required movements in eight directions. Because each limb state was therefore associated with a time, the participants could have learned to compensate for time-dependent forces by associating limb state, specifically its velocity, with force. If so, then the internal model that adapted to the time-dependent forces in straight reaching movements should not generalize to the same time-dependent forces for circular movements. Experimental results confirmed this prediction. Therefore, when a time-varying force alters the dynamics of the limb, the CNS adapts by erroneously building an association between limb states and forces.

20.2 Correlates of Adapting to Altered Dynamics

Participants in these experiments had little trouble reaching directly and accurately to the targets when the robot's motors produced no active force. However, when the robot's motors did produce forces, they pushed the

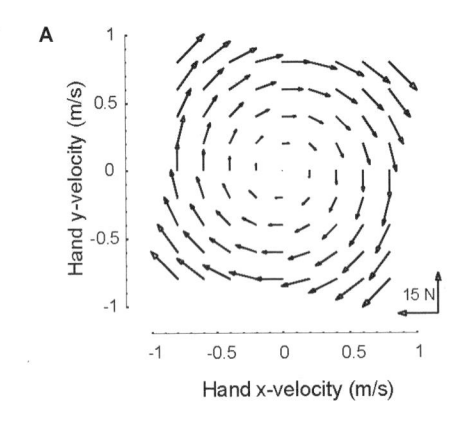

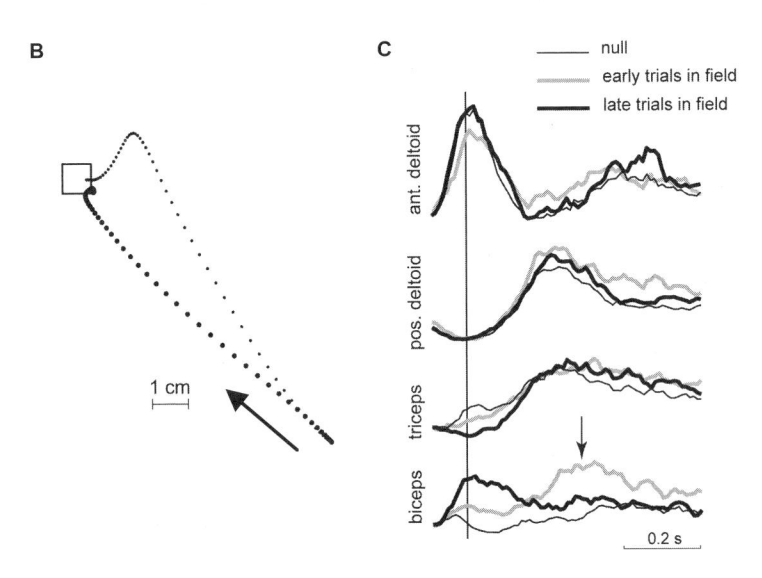

Figure 20.5

Adapting to the dynamics of a reaching movement. (*A*) A curl, velocity-dependent force field. The robot's motors produced a pattern of forces that depended on the velocity of the participant's hand. For each part of the workspace, the robot produced a force perpendicular to the hand's velocity vector in proportion to the hand's speed. (*B*) Sample hand trajectories for two 10-cm reaching movements to one of eight targets. The square marks the target; see figure 20.1A for the pattern of targets. Early in training (small dots), the force field shown in A resulted in displacement of the hand to the right of the target. With further training (big dots), the hand trajectory became straighter. (*C*) EMG activity from four arm muscles, aligned on movement onset (vertical line). In the null field (thin solid line), in which the robot's motors produced no active force, the movement began with activation of a shoulder flexor muscle (the anterior deltoid) and terminated with increased activity in the shoulder and elbow extensors (posterior deltoid and triceps). Early in training with the force field shown in *A* (gray line)—when the robot limb off course. For example, in some experiments the motors produced a force that always pushed the hand perpendicular to its velocity:

$$F = \begin{bmatrix} 0 & a \\ -a & 0 \end{bmatrix} \begin{bmatrix} \dot{x} \\ \dot{y} \end{bmatrix}.$$

Figure 20.5A illustrates the force pattern produced by this expression, called a curl force field or simply a *curl field*. This force field resembles those experienced for reaching during trunk or whole-body rotation (section 20.1.2). Coriolis forces push in a direction perpendicular to the hand's velocity vector and proportional to its speed, similar to a curl field. Figure 20.5B shows a typical hand trajectory for a movement in this field. The participant reached to the target box, but the robot perturbed the hand's trajectory. With some practice, the movement became straight. The next section considers how the motor command changed to restore straight trajectories.

20.2.1 Error-Feedback-Driven Changes in EMG During Adaptation

Kurt Thoroughman and Shadmehr[10] measured the activity of arm muscles in an experiment involving curl fields. In the null field, the particular movement shown in figure 20.5B began with activation of the anterior deltoid (see figure 3.2), a shoulder flexor muscle, and terminated with activation of the posterior deltoid and triceps, shoulder and elbow extensor muscles, respectively (figure 20.5C). In contrast, the biceps showed little modulation of its activity. When the experimenters introduced the force field, the motor commands began as before, but as the hand progressed toward the target, the robot's forces displaced it from a straight trajectory. The displacement of the hand stretched the biceps and activated **short-loop reflex** and **long-loop reflex** mechanisms (chapter 8). These mechanisms increased activity in the biceps as it stretched unexpectedly. However, this reflex occurred only after a substantial delay (arrow in figure 20.5C), too late to prevent the displacement of the hand. Nevertheless, this reflex reduced the magnitude of that displacement by producing a force that countered its effect. As the participant practiced this movement, the increased biceps activity shifted to earlier in the movement, so that eventually that muscle became active as early as the anterior deltoid muscle (i.e., it increased its activity just before the movement started).

Figure 20.5 (continued)
displaced the participant's hand from a straight path to the target—the movement stretched the biceps beyond its expected length, resulting in reflex-initiated increase in EMG activity (arrow). With further training (thick solid line), the increased EMG activity shifted to earlier in the movement as internal models in the CNS came to predict the forces imposed by the robot. Abbreviations: ant., anterior; pos., posterior. (From Thoroughman and Shadmehr[10] with permission.)

Figure 20.6
Composite EMG activity for targets in eight directions. The progression of plots from top to bottom shows EMG traces from early to late in training. The dashed lines in the top and bottom plots represent the magnitude of force created by the robot, which remained consistent throughout training. Early in training, the peak of the EMG lagged the imposed force, but with training it preceded the imposed force. The best fit to a fourth-order polynomial function estimated the timing of the peak, marked by the line that crosses the EMG traces from top to bottom. The dotted portions of this line represent 3-min breaks between training sets. (From Thoroughman and Shadmehr[10] with permission.)

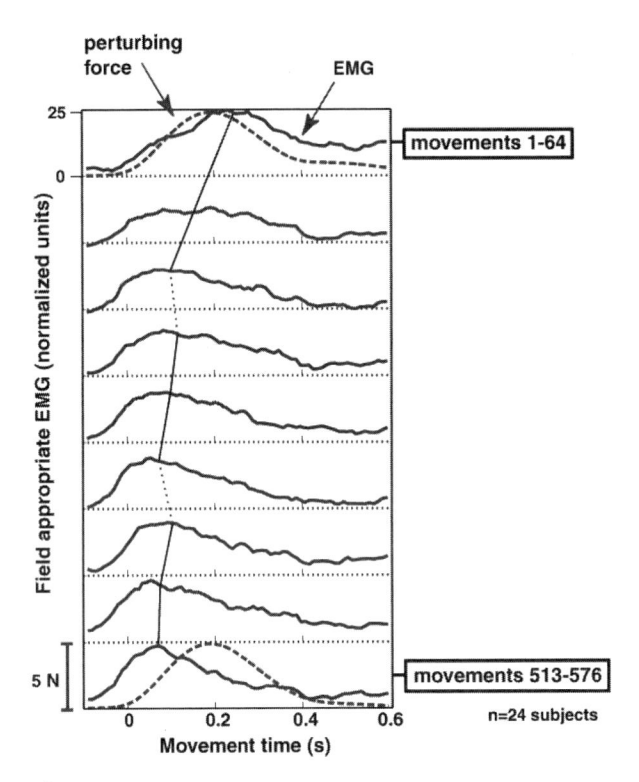

Figure 20.6 shows the transition in the activity of the muscles from the early training trials to the later trials. Because the participants made movements to targets in different directions, Thoroughman and Shadmehr combined the activity in the various muscles to estimate the component of EMG that produced a force that *opposed* the force produced by the robot. You can see that in the early trials, this composite EMG lagged the perturbing force. However, with training, the EMG peaked earlier in the movement until, at the end of the training, it "predicted" the robot-imposed force.

Figure 20.6 illustrates two key ideas: First, when your CNS miscalculates forces needed for a particular displacement of your hand, the hand no longer travels in a straight line to the target. Spinal and other feedback mechanisms, including feedback mechanisms that estimate current hand location with respect to the target and plan for the next state of the limb, react to the unexpected movement of the arm. This reaction produces muscle activity patterns that counter the effects of the miscalculation, maintaining your limb's stability so that your hand eventually reaches the target. However, the feedback process occurs too slowly—and its effects happen too late—to prevent the forces from significantly perturbing your hand's trajectory. Second, with training, your CNS adapts its computation of expected forces. It learns to produce the muscle activity that feedback pathways produced in earlier trials. For example, if you imagine that part of the feedback-driven change in motor commands resulted from local stretch reflexes, then the change caused by those reflexes can be transmitted to the CNS to serve as a "teacher" for adaptation of the internal model. With input from this "critic," the CNS can change its prediction and, perhaps, do better the next time. (This topic comes up again when we discuss learning mechanisms in chapters 23 and 24, but for now, note the intriguing similarity of the shift in biceps activation time shown in figure 20.6 to that illustrated for a dopaminergic neuron in figure 23.6A.)

20.2.2 Adaptation Causes Rotation in Muscle Preferred Direction

The equations of motion described in equations 20.1 and 20.2 produce the trajectories illustrated in figure 20.3. These equations include a torque C, a vector specifying the torque on each joint but not muscle-activation patterns. How could you incorporate a muscle model, such as that developed in chapter 7 (figure 7.2), into the equations? Thoroughman and Shadmehr[10] considered a simplified model of antagonist muscle pairs at the shoulder and the elbow: biceps, triceps, and anterior and posterior deltoids (see figure 3.2). This produced a four-dimensional muscle-force vector F_m. A Jacobian matrix J_m that represented the moment arms (as a function of arm configuration) transformed this vector into a two-dimensional torque vector C:

$$C = J_m^T F_m$$
$$F_m \equiv [f_b, f_t, f_a, f_p]^T.$$

Subscripts in the equation represent the force generated by biceps (b), triceps (t), anterior deltoid (a), and posterior deltoid (p), respectively. To solve for the force contribution of each muscle, Thoroughman and Shadmehr[10] assumed that antagonist muscles showed reciprocal activation. That is, the more one muscle was activated, the less its antagonist was activated:

$$f_b f_t = k_{bt}$$
$$f_a f_p = k_{ap}.$$

This assumption allowed the experimenters to translate a joint-torque vector C into muscle-produced forces and to quantify how muscle forces changed due to adaptation. Because muscle force depends on the direc-

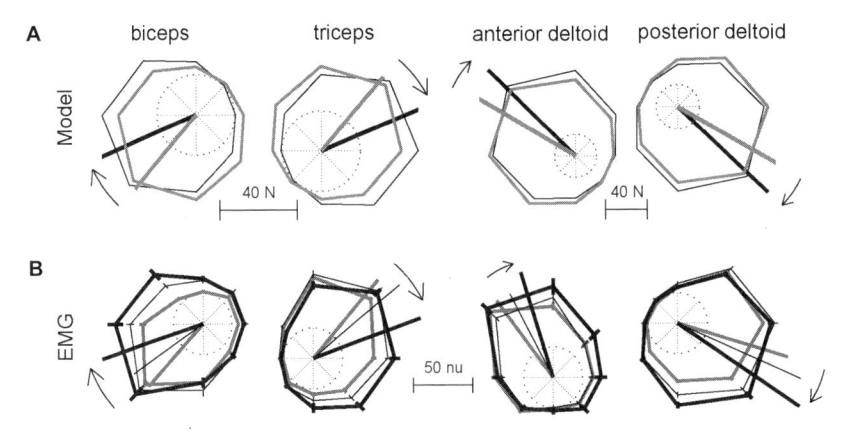

Figure 20.7
Rotation in EMG tuning functions with adaptation to a curl force field (see figure 20.5A). (*A*) Each polar plot shows the force predicted—for each muscle and for each movement direction—by the model described in equations 20.1 and 20.2 for movements in the null field (gray) and in the curl field (black). The straight lines show each muscle's preferred direction for both conditions (null field and curl field), computed as a vector sum. The arrows show the directional change in preferred direction. (*B*) Tuning functions of EMG activity in the null field (gray), early in adaptation to the force field (thin black line), and late in adaptation (thick black line). Otherwise, format as in A. Abbreviation: nu, normalized units. (From Thoroughman and Shadmehr[10] with permission.)

tion of motion, the experimenters averaged force for each muscle over the initial component of a movement (0–150 msec) for a given direction and then plotted the average for all directions. A spatial tuning curve resulted that described that muscle's preferred direction of motion in the null field (figure 20.7A).

Recall the concept of preferred direction described for cells in M1, as presented in section 14.2. The concept of directional tuning for muscles follows the same line of analysis. For muscles, however, you can intuitively understand why a muscle has a preferred direction. It produces forces on limb segments. Accordingly, it will move the limb in some direction which depends on the limb's configuration and the forces imposed on the limb. For example, consider the spatial tuning for biceps with your hand located directly in front of you. Figure 20.7 shows the amount of activation expected in biceps as a function of direction of reaching in the null field and in the curl field. If you think of activity in each direction as a vector, then the sum of these vectors specifies a resultant vector that describes the activation for that muscle. This vector has a magnitude and a direction, the latter corresponding to the muscle's preferred direction.

The model described in equations 20.1 and 20.2 predicted that with adaptation, both the direction and the magnitude of this vector should change. You can think of this change as the rotation of the muscle's preferred direction, accompanied by some variation in the vector's length. For example, for a clockwise curl field (figure 20.5A), the model predicted that the preferred direction should rotate clockwise by ~27° for elbow muscles (biceps and triceps) and ~18° for shoulder muscles (anterior and posterior deltoid). The vectors should also increase in size because, for straight-movement trajectories, the muscles must produce a force perpendicular to the direction of movement to overcome the robot-imposed forces. EMG measurements confirmed these predictions (figure 20.7B).

20.2.3 Adaptation Causes Rotation in Neuronal Preferred Direction

The change in the muscle activations and the resulting rotations in their preferred directions should come as no surprise. The CNS has to change its commands to the muscles to produce forces that counter those imposed by the robot. Rather, the results and analysis of Thoroughman and Shadmehr[10] have heuristic value for two reasons. First, they provide a compact way for predicting and quantitatively testing the learning-related EMG change that should occur with force adaptation. Second, in monkeys, when preferred directions of muscles change, so do the those of M1 cells.[11] Accordingly, you can predict that the rotation in EMG preferred directions should accompany a similar change in M1 cells. Note that this statement makes no assertions about causality. It says only that if M1 participates in the computation of the internal model for dynamics, and if EMG preferred directions change, then so should at least some preferred directions in M1 cells.

Chiang-Shan Ray Li, Camillo Padoa-Schioppa, and Emilio Bizzi[12] trained a monkey to reach to an array of eight targets while holding the handle of a robot. As the monkey held the hand at a central target, a peripheral target would appear. After a delay period, the central target would disappear, serving as the "go," or trigger, signal, and the monkey then reached to the target. The experimenters divided each experimental session into three periods: (1) a *baseline* period, in which the monkey performed the task in the null field; (2) a *force-adaptation* period, in which the robot imposed a clockwise curl field; and (3) a *washout* period with the null field again.

In the baseline period, the monkey made essentially straight reaching movements. When the robot imposed the force field, it perturbed the monkey's movement trajectories, but after a few trials, trajectories returned to something resembling those in the baseline period. In the washout period, aftereffects appeared at first, but the trajectories quickly returned to their original form. Over several days of training, adaptation rates improved. In early training sessions, it took the monkey ~200 movements to reach a performance plateau; after 25 sessions it took only ~30 trials.

Bizzi and his colleagues recorded the activity of M1 neurons and muscles. They compared neuronal and EMG activity during trials with similar hand trajectories. During the force-adaptation period, the preferred

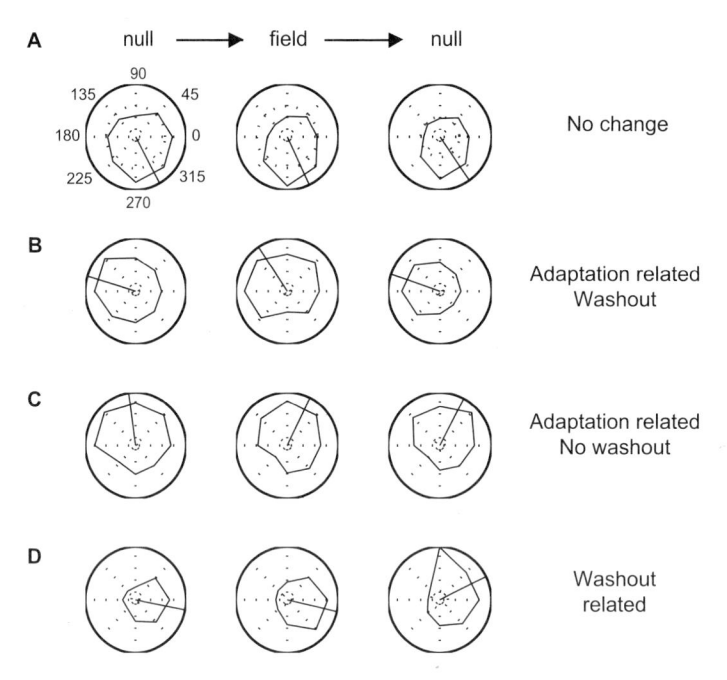

Figure 20.8
Spatial tuning functions of four cells in M1 during training in a clockwise curl force field. In each plot, the circle (dashed line) indicates the cell's activity prior to target appearance. Otherwise, format as in figure 20.7A. (*A*) Example of a cell unchanged by force adaptation. (*B*) A cell affected by adaptation that returned to its preadaptation directional tuning after washout. (*C*) A cell affected by adaptation, as in B, but which retained its adapted directional properties after washout. (*D*) A cell unaffected by adaptation, but which showed changes in its activity rate and preferred direction during washout. (From Li et al.[12] with permission.)

direction of muscles rotated in a clockwise direction, as expected for a clockwise curl field. In the ensuing washout period, the preferred direction of muscles rotated back to near baseline levels. On average, the muscles' preferred directions rotated by 19° in the force-field period (as compared to the baseline) and shifted back by 22° in the washout period (as compared to the force-field period).

The experimenters averaged M1 activity from 200 msec before movement onset until the end of the movement. Figure 20.8 shows this activity for four M1 cells, averaged for each movement direction. Bizzi and his colleagues computed preferred direction during each period (baseline, force-adaptation, and washout) by summing the activity vectors for each of the eight directions of movement. As a population, they found a 16° clockwise rotation in the preferred direction of M1 cells and a 19% increase in activity rates. In the washout period, the preferred directions rotated in the opposite direction (with respect to the force-field period) by an average of 14°, so that preferred directions returned, *on average*, to the baseline condition. Therefore, at a population level, the average change in M1's preferred direction closely resembled that in muscles.

Recall, however, that the monkey's adaptation rate gradually changed from about 200 trials to 30 trials or less. Although the changes in muscles and M1 neurons grossly resemble each other and show something about adapting an internal model, they do not speak to forming a new internal model (i.e., skill acquisition; see sections 4.3 and 15.4). Somehow, the CNS must have retained some changes even after the washout phase, but the average M1 activity does not reflect that development.

Bizzi and his colleagues used preferred direction to classify changes in M1 activity during force-field adaptation. According to the model presented in figure 19.4, they studied adaptation of the dynamics map (network 5 in the figure). Their results resemble those of Wise and his colleagues,[13] who studied changes in motor cortex activity as monkeys adapted their displacement maps (network 4 in figure 19.4). Recall from section 16.2.5 that after washout, some cells do not return to their original baseline properties. Yet different cells change their properties during washout but not adaptation trials. Studies of M1 activity during the adaptation of dynamics[12] and displacement[13] maps yield qualitatively similar results.

Some M1 neurons (34%) had a preferred direction that showed no change during adaptation and washout. Bizzi and his colleagues called these "kinematics" cells, but figure 20.8A refers to them in a terminology closer to the observation: no change. The preferred direction of other neurons (22%) did change during adaptation to the force field, but returned to baseline levels after washout. Figure 20.8 refers to these as *adaptation-related, washout* cells. They clearly reflected some aspect of movement dynamics, most likely adaptation of the dynamics map, but they could not contribute importantly to skill acquisition. (This conclusion follows from the idea that the skill remains—the monkey learns to adapt more and more quickly—although the cells have washed out their adaptation-related changes in activity.) The change in the preferred direction of these cells resembled that in the muscles: The cells showed a rotation in their preferred direction that was the same direction as the force field, and then a counterrotation of equal magnitude during washout.

However, in some other cells (20%), preferred direction changed in the force-adaptation period, and maintained that change in the washout period. These experimenters called these cells "memory, type I" cells; figure 20.8 labels them "adaptation related, no washout." This terminology provides a more descriptive name for what these cells do: They adapt to the imposed force field and maintain that adaptation throughout the washout period. Finally, another group of cells (21%) changed their preferred direction not in the force-adaptation period but after washout. The experimenters called these "memory, type II" cells (figure 20.8 labels them "washout related"). The change in preferred direction of washout-related cells occurred in the direction opposite to the force field. Indeed, in the washout period, the change in adaptation-related cells that did not show washout was about equal and opposite to the change in washout-related cells. Therefore, the population as a whole showed no net change in preferred direction in the washout period, but changes in over 40% of M1 neurons (figure 20.8C and 20.8D) outlasted the adaptation period.

If we make the reasonable assumption that M1 provides an important motor command signal, then these results suggest that in order to move the arm along similar trajectories in the baseline and washout periods, the neuronal population in M1 must activate muscles in approximately the same way. However, because this activation reflects summed output of M1 neurons, that sum may be unchanged although the activity of individual neurons differs markedly from the baseline period. Indeed, the properties of cells at the end of the washout period of one day's recording session should correspond to those for the baseline period the next day. The washout-related cells (figure 20.8D) appear to mask the effects of the adaptation-related cells that do not show washout (figure 20.8C). In this way, the sum of M1 output could remain unchanged for the purpose of adaptation, but could also retain some influence of the monkey's experience for the purpose of skill acquisition. This idea remains speculative and, at best, the details of this mechanism remain poorly understood. An improved understanding of the neuronal mechanisms of motor adaptation and skill acquisition represent important challenges for future work. Nevertheless, and although many gaps in knowledge remain, these results suggest that M1 plays a role in representing the internal model of limb dynamics, which maps limb states to forces.

20.2.4 Adaptation to Altered Dynamics Depends on the Cerebellum

Recall that when people wear prism glasses, accurate reaching movements require an adaptation of the kinematics maps that align location and displacement maps in visual and proprioceptive coordinates (see chapter 15). Adaptation of these internal models of kinematics depends on the integrity of the cerebellum: Patients with cerebellar damage show little or no aftereffects in the catch trials (see section 16.1.4). Maurice Smith[14] observed that cerebellar damage also prevents adaptation of internal models of dynamics. Working in Shadmehr's lab, Smith examined reaching movements of cerebellar patients with global degeneration in the cerebellar hemispheres and deep nuclei. The experiment was divided into sets of ~100 trials, each involving reaching to a target while holding the handle of a robotic arm. The patients and control participants performed the first four sets in the null field (the baseline condition); the next three sets in a clockwise curl field, and one to four additional sets in a counterclockwise curl field.

Smith found that the movements made by cerebellar patients in catch trials did not differ from their movements in the baseline condition. This finding indicates that they did not adapt to the force fields much, if at all; accordingly, they showed no aftereffects. In contrast, patients with Huntington's disease (see also section 19.5) did adapt, and had after-

20.2. Correlates of Adapting to Altered Dynamics

effects that resembled those of healthy, age-matched participants. Therefore, adaptation of the internal model of dynamics appears to depend on the integrity of the cerebellum, but not on that of the striatum. This basic result has also been reported based on an independent study by Tim Ebner and his colleagues.[15]

References

1. Cordo, PJ, and Nashner, LM (1982) Properties of postural adjustments associated with rapid arm movements. J Neurophysiol 47, 287–302.
2. Shadmehr, R, and Mussa-Ivaldi, FA (1994) Adaptive representation of dynamics during learning of a motor task. J Neurosci 14, 3208–3224.
3. Lackner, JR, and Dizio, P (1994) Rapid adaptation to Coriolis force perturbations of arm trajectory. J Neurophysiol 72, 299–313.
4. Pigeon, P, Bortolami, SB, Dizio, P, and Lackner, JR (2003) Coordinated turn-and-reach movements. I. Anticipatory compensation for self-generated Coriolis and interaction torques. J Neurophysiol 89, 276–289.
5. Kawato, M, and Gomi, H (1992) A computational model of four regions of the cerebellum based on feedback-error learning. Biol Cybern 68, 95–103.
6. Wolpert, DM, Ghahramani, Z, and Jordan, MI (1995) An internal model for sensorimotor integration. Science 269, 1880–1882.
7. Haruno, M, Wolpert, DM, and Kawato, M (2001) Mosaic model for sensorimotor learning and control. Neural Comput 13, 2201–2220.
8. Conditt, MA, Gandolfo, F, and Mussa-Ivaldi, FA (1997) The motor system does not learn the dynamics of the arm by rote memorization of past experience. J Neurophysiol 78, 554–560.
9. Conditt, MA, and Mussa-Ivaldi, FA (1999) Central representation of time during motor learning. Proc Natl Acad Sci USA 96, 11625–11630.
10. Thoroughman, KA, and Shadmehr, R (1999) Electromyographic correlates of learning an internal model of reaching movements. J Neurosci 19, 8573–8588.
11. Sergio, LE, and Kalaska, JF (1997) Systematic changes in directional tuning of motor cortex cell activity with hand location in the workspace during generation of static isometric forces in constant spatial directions. J Neurophysiol 78, 1170–1174.
12. Li, CS, Padoa-Schioppa, C, and Bizzi, E (2001) Neuronal correlates of motor performance and motor learning in the primary motor cortex of monkeys adapting to an external force field. Neuron 30, 593–607.
13. Wise, SP, Moody, SL, Blomstrom, KJ, and Mitz, AR (1998) Changes in motor cortical activity during visuomotor adaptation. Exp Brain Res 121, 285–299.
14. Smith, MA (2001) Error Correction, the Basal Ganglia, and the Cerebellum, doctoral dissertation, Johns Hopkins University.
15. Maschke M, Gomez CM, Ebner TJ, and Konczak J (2004) Hereditary cerebellar ataxia progressively impairs force adaptation during goal-directed arm movements. J Neurophysiol 91, 230–238.

Reading List

Maurice Smith's thesis[14] is available via the Reza Shadmehr home page.

21 Predicting Force II: Representation and Generalization

Overview: In computing an internal model of dynamics, your CNS maps limb states to forces. The patterns of generalization for this kind of learning suggest that in computing this map, your CNS represents limb states in intrinsic coordinates such as joint angles or muscle lengths.

21.1 The Coordinate System of the Internal Model of Dynamics

If a neural system computes a transformation from an input variable to an output variable, then the **adaptation** of that map should produce generalization patterns that reflect how neural elements in that map encode the input variable. Broad or coarse encoding of the input variable should lead to broad generalization. Localized encoding should lead to little generalization beyond the region of the input space where training took place. For example, if you train in a given force field with one configuration of your arm, should that training generalize to other arm configurations? As a concrete example, imagine swinging a new hammer. Does learning to swing the hammer at a nail over your head generalize to swinging it underhand at a knee-level nail? The answer depends on how cells code for arm configuration.

21.1.1 Encoding of Limb State in the Motor Cortex

M1 cells discharge as a function of both the direction of a reaching movement and the location of the arm. Cells have a preferred direction, but that tuning is not restricted to a small range of arm configurations. Rather, a typical cell remains tuned throughout much, if not all, of the workspace. The tuning does not remain constant, however. The preferred direction of a typical M1 cell will rotate approximately with the shoulder angle[1,2] and with other changes in arm configuration.[3-5] The reason for this rotation may be that many M1 cells reflect the force requirements of the task. For example, section 14.2 presented the work of Sergio and Kalaska, who studied changes in M1 activity as hand location varied in the workspace. On average, M1 cells showed a significant clockwise rotation of their preferred directions from a central hand location to locations to the right of center, and a significant counterclockwise rotation of their preferred directions for hand locations to the left of center (figure 14.8).

Shoulder and elbow muscles change their preferred directions in the same way. Recall that Bizzi and his colleagues showed that cells rotated their preferred directions in the curl field, also in a "musclelike" way (see section 20.2.3). Of course, you would like to know whether cerebellar Purkinje cells and cells in many other parts of the CNS have tuning functions that rotate with initial end-effector location, but because no one has yet reported the results of such experiments, this section concentrates on M1 neurons.

According to the available evidence, you can expect that a typical M1 cell will have direction tuning in the entire reachable workspace, but its preferred direction will rotate in a way similar to the preferred direction of arm muscles in a given task. Therefore, the preferred direction of a typical M1 cell changes as a function of the configuration of the arm (see figure 14.6). Now imagine that the force-field-related changes in preferred direction and the posture-related changes in preferred direction accumulate. Then training in one workspace should result in the rotation of preferred directions by a certain amount, and the change in the configuration of the arm to a new workspace should result in an additional rotation by an amount approximately equal to the rotation around the shoulder joint. In the new workspace, despite the fact that no prior training had taken place there, an effect of the training elsewhere should be observed (i.e., you should observe generalization).

Recognize, however, that not all M1 cells have "musclelike" tuning properties. A significant proportion of cells in M1 reflect parameters of reaching movements in **extrinsic coordinates**.[6–9] Perhaps the 34% of M1 cells that do not change during force-field adaptation[10] reflect extrinsic coordinate frames or some other aspect of movement kinematics. Assume, nevertheless, that M1 cells with more musclelike properties contribute most to the computations of internal models of the dynamics. The sensitivity of their tuning to the configuration of the arm and arm velocity should be reflected in the generalization patterns observed after adaptation.

21.1.2 Generalization from One Arm Configuration to Another

To see how generalization patterns relate to changes in M1's directional tuning, imagine placing your right arm in the horizontal plane and flexing your shoulder so that your right hand occupies a location in the "left" workspace. In this situation, your triceps muscle has a preferred direction of about 90° (see figure 20.1F for coordinates). Further, imagine that when you train on some force field with this arm configuration, you observe a 30° clockwise rotation in the preferred direction of your triceps. Now imagine M1 cells that also rotate their preferred direction by a similar amount. Adaptation to the force field in the left workspace should cause a musclelike M1 cell with a preferred direction of 180° to change by 30° to 150° (i.e., 180° − 30°). Now extend your shoulder by 90°, so that your hand occupies a location in the right workspace, as illustrated in figure 21.1A. You would expect your M1 cells to remain directionally tuned and

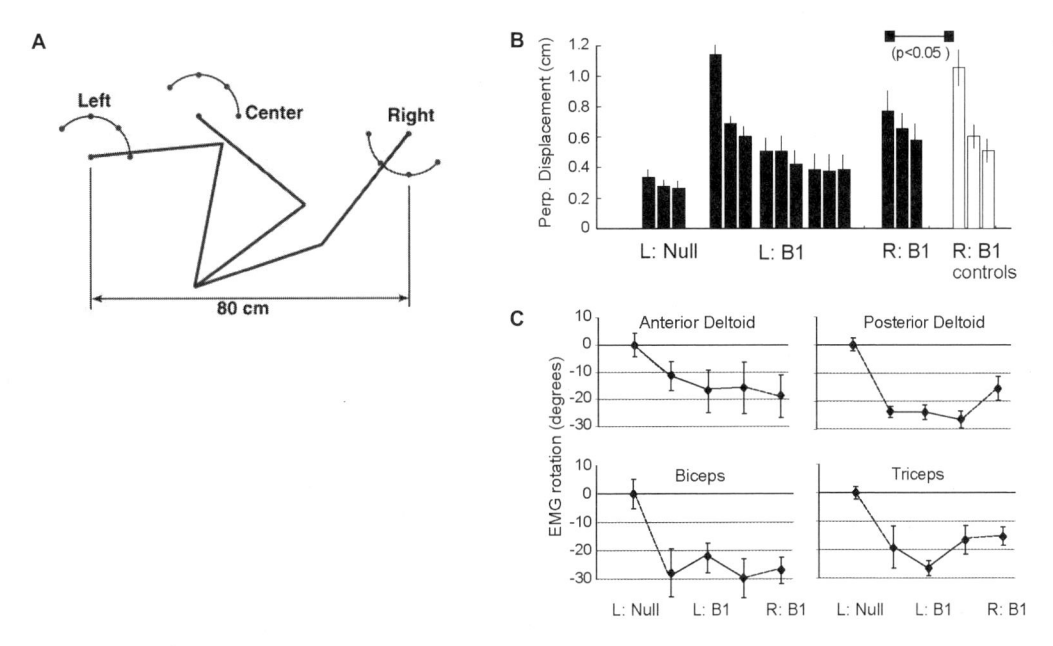

21.1. The Coordinate System of the Internal Model of Dynamics

Figure 21.1
Generalization from one arm configuration to another after training in a counter-clockwise curl force field. (A) A schematic of the right arm in the horizontal plane, viewed from above. Participants trained on reaching movements in a force field at the left and then moved their hands 80 cm to the right to test for generalization. (B) Movement errors in terms of displacements in hand trajectory perpendicular to the direction of the target. Field B1 is a counterclockwise curl field. Abbreviations: L, left; perp, perpendicular; R, right. (C) The change in muscle preferred directions with respect to the baseline condition (the null field) at the same arm configuration. (From Shadmehr and Moussavi[11] with permission.)

that, on average, the 90° clockwise rotation in the shoulder joint should cause the preferred direction of these cells to rotate by an average of ~90°. Accordingly, movement of the arm to the right workspace should bring the adapted M1 cell's preferred direction to 60° (i.e., 180° − 30° − 90°, a combination of the effects of training and relocating your hand). If you had not practiced movements in the force field, this cell would have had a preferred direction of 90°. This difference should affect behavior.

Shadmehr and Zahra Moussavi[11] used this idea to predict how adapting to a force field at one arm configuration should generalize to another. They considered force fields that produced a constant rotation in preferred directions of muscles regardless of the region of the workspace in which the task was performed. To a first approximation, these **torque** fields depend on joint velocity:

$$\tau = W\dot{\theta}.$$

To produce such a torque field, the robot must produce the following force field:

$$F = J_s(\theta)^{-T} W \dot{\theta},$$

where J_s is the Jacobian relating the participant's hand velocity to joint rotations. To avoid the need to measure a participant's joint velocities, you can replace $\dot{\theta}$ with $J_s(\theta)^{-1}\dot{x}$ and write

$$F = J_s(\theta)^{-T} W J_s(\theta)^{-1} \dot{x}.$$

If you define a matrix B to represent the above transformation from hand velocity to force, you have

$$F = B\dot{x}$$

$$B = J_s(\theta)^{-T} W J_s(\theta)^{-1}.$$

In general, the matrix B will be a function of θ. However, for certain values of matrix W, you can produce a skew-symmetric matrix B that is essentially independent of θ. This property means that the amount of change in muscle preferred directions should be independent of arm configuration for curl force fields. Shadmehr and Moussavi[11] predicted that adapting to a curl field at one arm configuration should generalize to all configurations.

They had participants put their hands in the left workspace and make about 600 movements in a counterclockwise curl field. Participants then moved their hands to the right workspace and made movements in the same force field (figure 21.1A). Shadmehr and Moussavi observed partial generalization: Participants with practice in the left workspace performed better in the right workspace than participants without such experience (figure 21.1B). They also observed that muscle preferred directions rotated counterclockwise when the participant trained in the left workspace, and maintained that rotation in the right workspace (figure 21.1C).

The results shown in figure 21.1 demonstrate that adaptation to a velocity-dependent force field results in broad generalization as a function of static location of the hand. If the neural elements that computed this internal model encoded limb location with narrow tuning functions, adaptation of movements at one arm configuration would not generalize to an arm configuration far away. The fact that force adaptation generalizes widely in spatial terms implies that the computational elements that map limb states to forces encode location very broadly.

21.1.3 Internal Model of Dynamics Represents Limb States in Intrinsic Coordinates

The results regarding large generalization in terms of limb location do not specify the coordinates in which the internal model codes for dynamics. Is the internal model of dynamics a map in extrinsic coordinates (e.g., hand velocity to hand forces $\dot{x} \xrightarrow{x} F$), or is it a map in intrinsic coordinates (e.g., joint velocity to joint torques $\dot{\theta} \xrightarrow{\theta} \tau$)? Because the force field was curl, you cannot tell the difference from the experiment related above.

To answer this question, Shadmehr and Moussavi[11] produced force fields that behaved differently in intrinsic and extrinsic coordinate systems. They considered a field described as

$$F = B\dot{x}$$

where $B = \begin{pmatrix} -11 & -11 \\ -11 & 11 \end{pmatrix}$ N.sec/m.

Figure 21.2 (top left) illustrates this force field. An internal model of this field in extrinsic coordinates is simply $\hat{F} = \hat{B}\dot{x}$. Imagine that you build an internal model of this field with your hand in the left workspace of figure 21.1A. When your arm moves to some other location, your expectation of force remains the same in Cartesian coordinates. For example, you would expect to encounter approximately the same forces during a movement toward a target at 90° regardless of your arm's configuration. This kind of internal model generalizes in extrinsic coordinates.

Contrast that kind of internal model with a different one. Although the robot imposes forces on the hand, this internal model associates joint velocities with joint torques (i.e., one learns that $\hat{\tau} = \hat{W}\dot{\theta}$, and therefore $\hat{F} = J_s(\theta)^T \hat{W}\dot{\theta}$). Generalization of this model to another arm configuration depends on joint angles. For example, if the internal model represents the field shown in the left column of figure 21.2 in intrinsic coordinates, then it will generalize this field to the right workspace, but the resulting forces will change. Note that in the right workspace, the intrinsic representation of the field correlates inversely with the extrinsic representation. If you experienced a force for a target at 270° that pushed your hand to the right in the left workspace, you should expect that the forces should push your hand to the left in the right workspace. Does your CNS generalize the force field in intrinsic or extrinsic coordinates?

Shadmehr and Moussavi trained two groups of participants in the force field shown in figure 21.2 (top left) in the left workspace. Following ~600 trials, they moved the participants' hands by 80 cm to the right workspace and tested reaching in either the extrinsic or the intrinsic representation of that field. Figure 21.3A shows the performance of participants who were tested in the intrinsic representation of the field. Figure 21.3B shows the performance in the extrinsic representation of the field. Shadmehr and Moussavi found that performance in the right workspace exceeded that in naïve controls when the field was represented in *intrinsic* coordinates, and was worse than controls when the field was represented in *extrinsic* coordinates. This finding means that adaptation resulted in generalization to the new arm configuration, as well as that generalization occurred in *intrinsic* coordinates. That is, the internal model maps $\dot{\theta} \xrightarrow{\theta} \tau$ and not $\dot{x} \xrightarrow{x} F$. Stated somewhat differently, the task is easy when the fields in the left and right workspaces require the same rotations in muscle preferred directions (figure 21.3A), but is hard when they differ (figure 21.3B).

21.1.4 Transfer of Learning Across Arm Configurations

It appears, then, that the internal model of arm dynamics represents limb states and force variables in intrinsic, jointlike or musclelike coordinates. In addition, it appears that these elements encode the static configuration of the limb very broadly, so that adaptation at one arm location generalizes

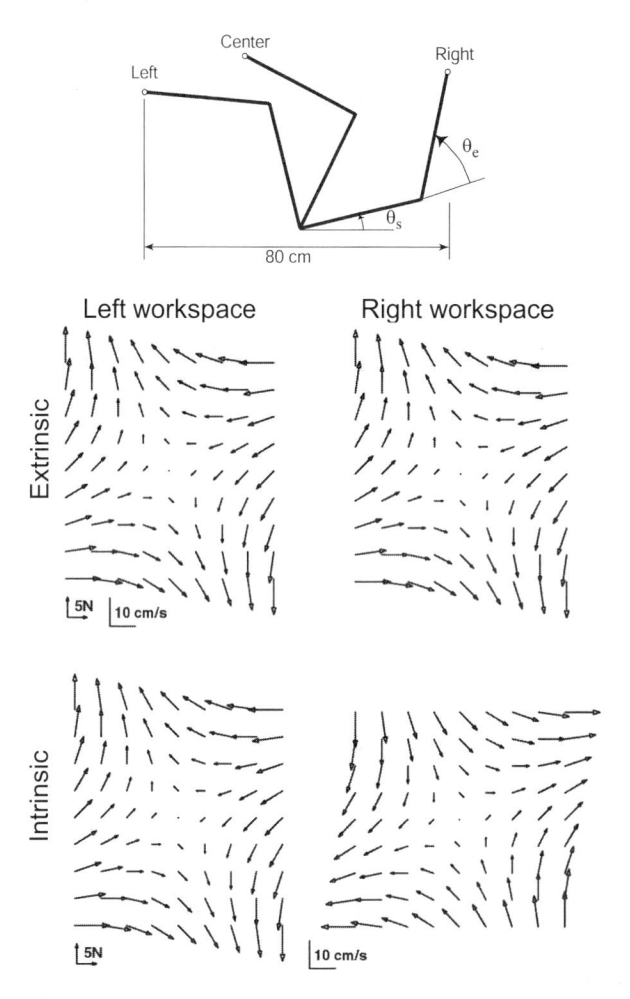

Figure 21.2

Generalization of a velocity-dependent force field in **intrinsic**, not **extrinsic**, coordinates. The vectors represent forces imposed by a robot on the hand. The labels at the top (left workspace and right workspace) refer to the arm configurations illustrated in figure 21.1A. Top: The test for an extrinsic representation assumed a mapping from hand velocity to forces on the hand. Accordingly, in this part of the experiment, the imposed forces did not differ for the left (acquisition) workspace versus the right (generalization) workspace. If the internal model used extrinsic representations, then the adaptations seen in the acquisition phase of the experiment should have generalized well (but see figure 21.3B). Bottom: For the same hand forces in the left (acquisition) workspace, the test for an intrinsic mapping assumed that the **internal model** approximated these forces in terms of a map from joint velocity to joint torques. Thus, for the right (generalization) workspace, the imposed forces differed in a predictable way. Because the internal model used intrinsic representations, the adaptations seen in the acquisition phase generalized well (as shown in figure 21.3A).

21.1. The Coordinate System of the Internal Model of Dynamics

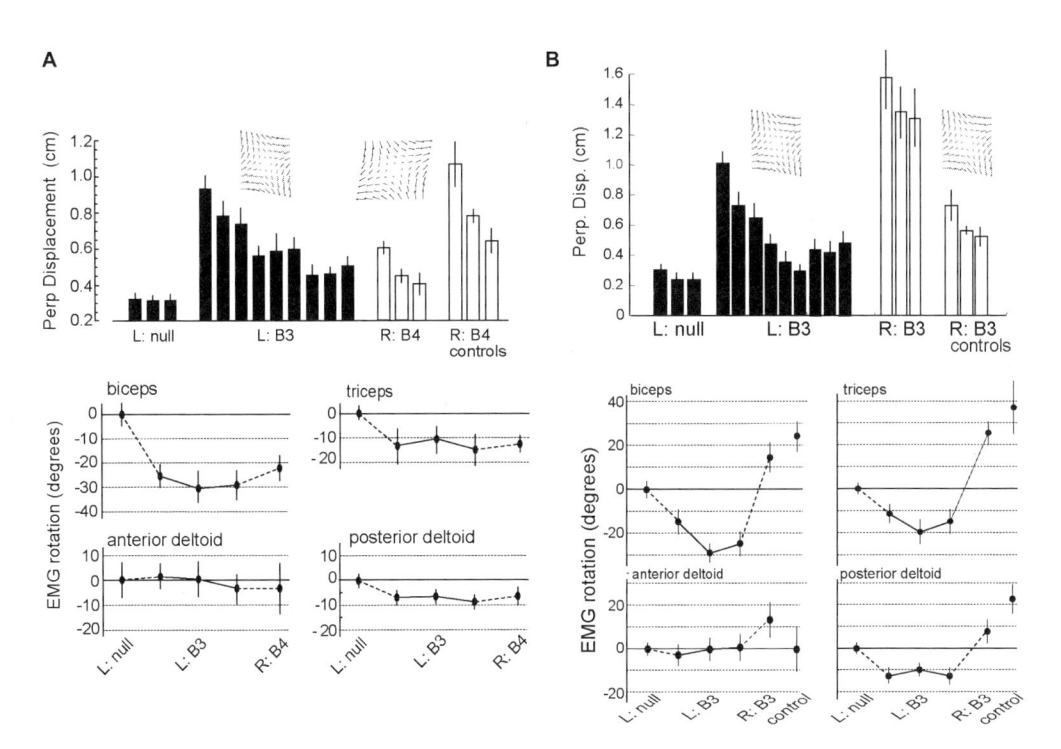

Figure 21.3

The internal model of dynamics represents limb states in intrinsic coordinates. Participants performed reaching movements in eight directions in the left and right workspaces, as shown in figure 21.1A. Training took place in the left workspace, and then participants were tested in the right workspace. Control participants were naïve to the force fields and trained only in the right workspace. The velocity-dependent forces experienced at each workspace are shown as a field (B3 or B4). The experimenters measured performance in terms of perpendicular displacement from a straight line to the target. The EMG measure at the bottom of *A* refers to a change in the preferred direction of the muscle activation function (see figure 20.7B). (*A*) The force field in the right workspace corresponded to an intrinsic representation of the field experienced in the left workspace. Performance exceeded that of controls. The fields in the left and right workspaces cause similar rotations in muscle preferred directions. (*B*) The force field at the right workspace corresponded to an extrinsic representation of the field experienced in the left workspace. Performance deteriorated compared to control participants. The fields in the left and right workspaces produced different rotations in muscle preferred directions. (From Shadmehr and Moussavi[11] with permission.)

to distant arm configurations. Nicole Malfait, Douglas Shiller, and David Ostry[12] have argued that this hypothesis implies that as far as the internal model of dynamics is concerned, joint velocities rather than hand velocities determine the similarity of one movement to another. Therefore, if you practice reaching movements in a force field in a single direction at one arm configuration, you should generalize to a similar movement in another arm configuration, provided that joint velocities are similar.

To test this idea, the experimenters trained participants on a single reach direction in the left workspace and then, in the right workspace, tested them in two ways: The test movements had either the same direction of hand displacements or the same joint rotations. For example, participants trained on reaching toward 0° in the left workspace, learning to overcome a curl field (labeled 1h in figure 21.4A) and then were tested for a target at 0° in the same curl field in the right workspace. These two movements involved the same hand displacements Δx but different joint displacements $\Delta \theta$. The hypothesis predicts that because your CNS builds an association with $\dot{\theta} \xrightarrow{\theta} \tau$ and not $\dot{x} \xrightarrow{x} F$, a 0° movement in the right workspace differs from the same movement, in terms of hand displacement, in the left workspace. Therefore, you might expect little or no generalization. In accord with this expectation, Malfait et al.[12] observed that despite training toward 0° in the left workspace, participants had a hard time reaching toward 0° in the right workspace (labeled 1h in figure 21.4B).

Next, they trained participants on a single reach direction in the left workspace and then tested them in the right workspace for movements that had similar joint rotations $\Delta \theta$. For example, in figure 21.4A, the part labeled 1j indicates the direction trained in the left workspace, and that labeled 1 indicates the direction tested in the right workspace. The experimenters observed that for two movements with similar joint rotations, training in the left workspace aided performance in the right workspace.

It appears from this result that your CNS represents the dynamics of reaching movements in terms of a map that associates state of the limb to forces in intrinsic coordinates (e.g., $\dot{\theta} \xrightarrow{\theta} \tau$).

21.2 Computing an Internal Model with a Population Code

If you adapt to a velocity-dependent field with one arm configuration, your CNS generalizes the map to other arm configurations, even for hand locations 80 cm apart. You saw similar results for adaptation of the displacement map $x_{dv} \xleftrightarrow{\hat{\theta}} \Delta \hat{\theta}$ in section 16.2.2. These results suggest that the neurons that compute either of these internal models encode static limb configuration with broad tuning functions. If they used narrow tuning functions, then a change made to the maps at $\hat{\theta}^{(1)}$ would not affect other arm configurations far away at $\hat{\theta}^{(2)}$ and $\hat{\theta}^{(3)}$.

As noted in chapter 10, many parts of the CNS—including the cerebellum, the parietal cortex, and the motor cortex—broadly encode limb configuration, with monotonic and approximately linear functions. This

21.2. Computing an Internal Model with a Population Code

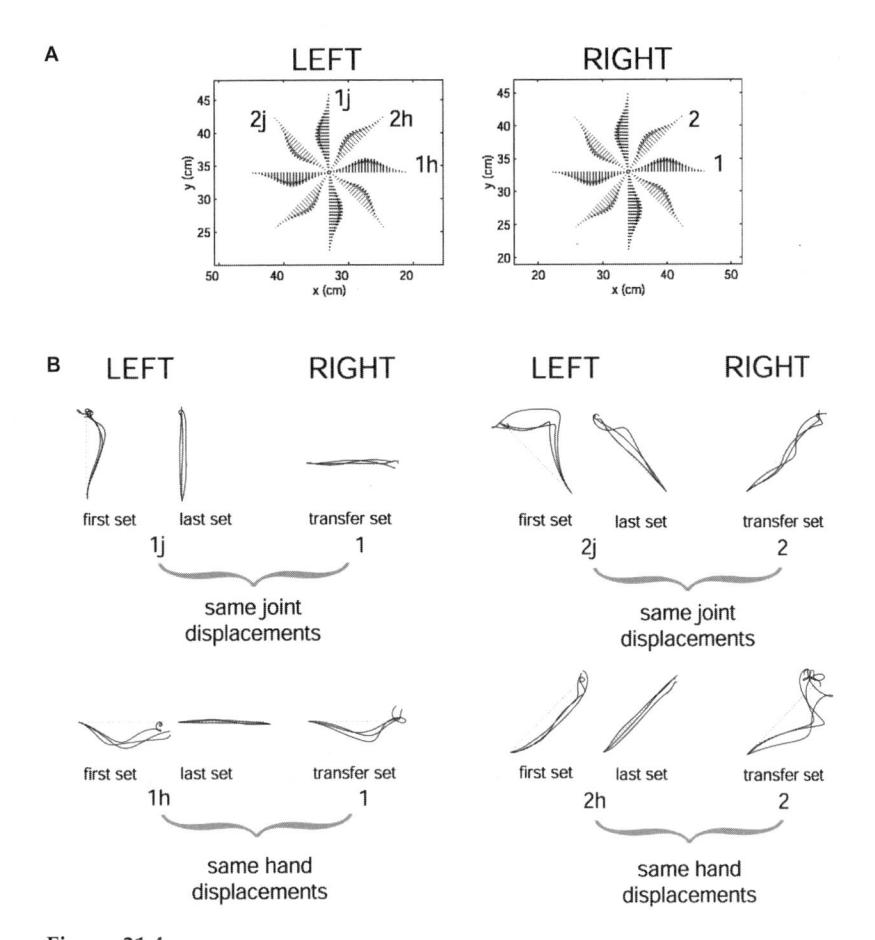

Figure 21.4
Training in a single direction at one arm configuration generalized to another arm configuration. (A) Participants trained on a single movement direction in the left workspace and the experimenters tested them in the right workspace. Directions of training and testing were the same in terms of either hand displacements or joint rotations. Movements labeled 1h and 1 had the same hand displacement. Movements labeled 1j and 1 had the same joint rotations. The figure displays the forces present for each movement. (B) Representative movements in the first set of training, the last set, and the transfer set. Training resulted in transfer only when the two movements had similar joint rotations. (From Malfait et al.[12] with permission.)

encoding scheme agrees with the broad pattern of generalization as a function of arm configuration.

Patterns of generalization provide a powerful tool for investigating how an adaptive system represents the maps that it learns. However, until this section, we have not presented a method for mathematically relating patterns of generalization to the representation of the information. To make this link, you need to go beyond a description of the input–output variables encoded by the internal models, such as sensory state of the arm and force, to consider how the transformation from input to output might take place. This section considers one way that the CNS might compute an internal model, and goes on to consider how generalization relates to representation.

Population coding remains one of the most widely used models of neural computation. The idea of using populations of neurons to code variables of interest has a long history,[13] and it has become a useful tool when combined with a simple decoding strategy called a population vector, which for cells in M1[14] (and other areas) corresponds to the direction of reaching movements (see section 14.2).

21.2.1 Encoding Identity Maps: Movement Direction in a Population

Section 14.2 presented the concept of a population vector. A typical cell in any of the frontal motor areas, in many parts of PPC, in the cerebellum, and in the basal ganglia discharges during a reaching movement, and its average discharge depends to a large degree on the movement's direction. The preferred direction of movement has the highest average discharge. You can represent the preferred direction of neuron i with the unit vector $\mathbf{w}_i$. If you confine reaching movements to the horizontal plane, $\mathbf{w}$ is a two-dimensional vector that points along some direction.

In a given trial n, movement might be in direction $\phi^{(n)}$, and neuron i might discharge by amount $r_i^{(n)}$. This value usually represents the average discharge over the entire movement period, but many other measures will do as well. If the monkey reaches in this direction five times, the cell will discharge at a slightly different mean rate each time. You might average the activity over repeated movements and get a discharge rate as a function of movement direction $g_i(\theta)$. This function describes the cell's directional tuning curve. On any given trial, noise adds to this average so that the discharge becomes

$$r_i^{(n)} = g_i(\phi) + \eta_i^{(n)}.$$

In this equation, the first term describes the tuning curve of the cell, and the second term describes the noise. Experiments show that in M1,[15] the tuning curve approximates a cosine-like function of movement direction with a half-width at the half-maximum value of approximately 56°. If you think of the target direction as the input to this equation, the noise term amounts to what you cannot account for on the basis of movement direction. Experiments suggest that this noise term in cortical neurons is often normally distributed with a variance that is proportional to the mean value of the tuning function.[16] If cells did not have a noise term and you could record from many cells simultaneously, you could normalize their discharge with respect to direction so that at their preferred direction, each cell had a maximum discharge of 1 unit. Then for any given movement, you could simply note that cell j happened to discharge maximally. Because cell j happens to have a preferred direction $\mathbf{w}_j$, you can estimate the movement direction $\hat{\phi}$ to be

$$\hat{\phi} = \mathbf{w}_j.$$

This approach amounts to *winner-take-all* coding. However, because the cells in your CNS have noise terms, your estimate would have a large variance from movement to movement, even though the actual direction of movement did not change. A population code provides a more robust approach.[14] For the population vector, as one example of a population code, each cell's discharge is weighted by its preferred-direction vector. The sum of these vectors produces the estimate of movement direction:

$$\hat{\phi}^{(n)} = \sum_i \mathbf{w}_i r_i^{(n)} = \sum_i \mathbf{w}_i g_i(\phi^{(n)}) + \mathbf{w}_i \eta_i^{(n)}.$$

This approach outperforms the winner-take-all computation in the sense that it produces a smaller variance in its estimate of movement direction from movement to movement. In fact, if the cells encode movement direction with exactly cosine functions, the estimate would be optimal in the sense that it would have the minimum variance.[17] Therefore, the success of population coding depends on computing the estimated movement direction with neurons that broadly encode state. Where this condition has been approximately met, experiments have generally demonstrated that a population code successfully estimates the input variable from noisy neuronal discharge.[18,19]

21.2.2 Population Coding with Basis Functions

In the above example, you estimated movement direction from the discharge rates of cells that each encode movement direction. This computation provides an example of an *identity mapping* (i.e., a mapping in which the output provides an estimate of the input variable). In general, a population code could map an input variable x into another variable y.[20,21] In this case, the tuning curves of the neurons that participate in this computation become the **basis functions** that represent the output. When linearly combined, a set of such functions can approximate almost any linear or nonlinear function. For example, Alex Pouget and Terry Sejnowski[22] suggested that neurons in the PPC might serve as basis functions for computing the location of a visual target with respect to the head. Cells in this region of the CNS typically have a discharge r that depends on both orientation of the eye x_{eye} in the orbit and location of the target on the retina x_r. As described in section 11.3, these cells have a preferred location on the retina and their discharge varies approximately linearly with the orientation of the eye.[23] The tuning function of a cell *i* can be labeled $g_i(x_{eye}, x_r)$. Using a weighted sum of these functions, you could estimate target location with respect to the head:

$$\hat{y} = \sum_i \mathbf{w}_i g_i(x_{eye}, x_r) + \mathbf{w}_i \eta_i.$$

Your CNS would have to learn the appropriate weighting $\mathbf{w}_i$ to form this map. However, Pouget and Sejnowski[22] pointed out that because the CNS constructs the map on the basis of tuning functions, it can use the same basis functions to form any other representation, for example, a shoulder-centered representation of the target or, as in the present model (figures 14.1 and 19.4), a fixation-centered representation. Note the significance of this idea: It demonstrates that population coding, a method that can compute identity maps, has applicability to the general problem of computing nonlinear maps. Whereas in the population code of an identity map, the weight vectors never change and point in the preferred direction of a cell, in computing nonlinear maps involving other functions and transforms, the weight vectors will change and will have no specific relationship to the tuning function. In this context, note that the preferred directions of cells in motor areas of the frontal cortex change during the learning of novel kinematic transforms[24] (section 16.2.5) and novel dynamics maps[25,26] (section 20.2.3), and when experimenters use neural signals to drive robots in their role as prosthetic devices.[27]

21.2.3 The Relationship Between Tuning Curves and Generalization

Now return to the problem of how your CNS might compute an internal model of dynamics. You can think of an internal model as a map that transforms an estimate of the state of the arm $\hat{\theta}, \hat{\dot{\theta}}$ (i.e., the configuration of the arm and its velocity) into an estimate of forces $\hat{\tau}$. Assume your CNS performs this computation with a population code. Each element *i* that participates in this computation encodes limb configuration and velocity with a function specified by g_i. Each element also has a preferred force vector $\mathbf{w}_i$. A movement comprises a time trajectory of limb states, and therefore the following equation describes the estimated forces along that trajectory:

$$\hat{\tau}^{(n)}(t) = \sum_i \mathbf{w}_i g_i(\theta^{(n)}(t), \dot{\theta}^{(n)}(t)) + \mathbf{w}_i \eta_i^{(n)}(t).$$

To relate tuning properties to the generalization of motor learning, this presentation makes a number of simplifications. First, assume that the time history of the states of the arm during a reaching movement can be reduced simply to its direction ϕ. Second, assume that you can represent the time-dependent activity of your computational element during this

21.2. Computing an Internal Model with a Population Code

trajectory by its average. Third, assume the absence of noise. For a given movement n, you now have

$$\hat{\tau}^{(n)} = \sum_i \mathbf{w}_i g_i(\phi^{(n)}).$$

If you have m elements in your computation (corresponding to the columns in the weight matrix W below) and you assume that force is a two-dimensional vector (corresponding to the rows in W), you can represent the sum as

$$\hat{\tau}^{(n)} = W\mathbf{g}(\phi^{(n)})$$

$$W = \begin{bmatrix} w_{11} & \cdots & w_{1m} \\ w_{21} & \cdots & w_{2m} \end{bmatrix}$$

$$\mathbf{g}(\phi^{(n)}) = [g_1(\phi^{(n)}) \quad \cdots \quad g_m(\phi^{(n)})]^T.$$

Now you want to change the weights W so that the internal model accurately estimates force τ as a function of movement direction ϕ. To do so, you can use a method called *gradient descent*, as described below. For movement n in a series of movements, assume that your estimate of torques does not match those actually obtained. Accordingly, you can compute the difference between actual force τ and the estimated force as

$$\tilde{\tau} = \tau - \hat{\tau} = \begin{bmatrix} \tilde{\tau}_1 \\ \tilde{\tau}_2 \end{bmatrix}.$$

To improve your estimate, you want to change W so that you begin to reduce "squared" error e:

$$e = \frac{1}{2}\tilde{\tau}^T\tilde{\tau}.$$

To do so, you need to measure the gradient of e with respect to W. After some algebra, you find

$$\frac{de}{dw_{ij}} = -\tilde{\tau}_i g_j.$$

This expression implies that after performing movement n, you can use the error for that movement $\tilde{\tau}^{(n)}$ to change the weights $W^{(n)}$ of the internal model. That change will be in the opposite direction from the gradient (because the gradient tells you the direction of change in weights that increases the error), and will be scaled by a small constant α:

$$w_{ij}^{(n+1)} = w_{ij}^{(n)} + \alpha \tilde{\tau}_i^{(n)} g_j(\phi^{(n)}).$$

Writing this in vector form, you have

$$W^{(n+1)} = W^{(n)} + \alpha \tilde{\tau}^{(n)} \mathbf{g}(\phi^{(n)})^T.$$

If you multiply both sides of this equation by $\mathbf{g}(\phi^{(n+1)})$, you have

$$W^{(n+1)}\mathbf{g}(\phi^{(n+1)}) = W^{(n)}\mathbf{g}(\phi^{(n+1)}) + \alpha \tilde{\tau}^{(n)} \mathbf{g}(\phi^{(n)})^T \mathbf{g}(\phi^{(n+1)}),$$

which is equivalent to

$$\hat{\tau}^{(n+1)}(\phi^{(n+1)}) - \hat{\tau}^{(n)}(\phi^{(n+1)}) = \underbrace{\alpha \mathbf{g}(\phi^{(n)})^T \mathbf{g}(\phi^{(n+1)})}_{\text{generalization function}} \tilde{\tau}^{(n)}. \tag{21.1}$$

Thus, the generalization function becomes

$$b(\phi^{(n)}, \phi^{(n+1)}) = \alpha \mathbf{g}(\phi^{(n)})^T \mathbf{g}(\phi^{(n+1)}) \tilde{\tau}^{(n)}.$$

Equation 21.1 says that the error in trial n times a generalization function completely describes the change in the force predictions of the internal model from movement n to movement $n+1$. That generalization function amounts to the projection of the basis functions in trial n upon the basis functions evaluated at trial $n+1$. Intuitively, you can see that that projection will be largest for two consecutive movements in the same direction (i.e., when $\phi^{(n)} = \phi^{(n+1)}$). The shape of the tuning functions will determine how the generalization falls off as a function of the distance between the two movements. For sharp (narrow) tuning functions, the generalization function falls off sharply. In practice, this means that with sharp tuning, the error experienced in movement n toward direction $\phi^{(n)}$ does not help (or hurt) the performance in the subsequent movement toward some other direction $\phi^{(n+1)}$, unless the two directions do not differ very much. However, for broad tuning functions, the generalization function will remain high even though the two consecutive movements differ widely in direction. In this case, what your CNS learns from the error in movement n significantly affects the subsequent movement in a wide spread of directions, perhaps in all directions.

21.3 Estimating Generalization Functions from Trial-to-Trial Changes in Movement

21.3.1 Problems in Measuring a Generalization Function

To measure generalization of an adaptive system, you might train that system with an input $\mathbf{x}_1$ and then test it with a new input $\mathbf{x}_2$. For example, you might provide a target at a given direction and then ask how training affected performance for some other movement direction. However, training and test trials do not really differ. In both, your CNS directs your arm to make a movement, it visits some arm states, experiences some errors, and adapts. The "training" trial has neither more nor less effect than the test trial on what you might think of as "error management."

To overcome this problem, experiments testing internal models and their adaptation have many more training trials than test trials. Nevertheless, the problem recurs when you try to measure how training in $\mathbf{x}_1$ affects another input $\mathbf{x}_3$. Because you may have tested the system previously with $\mathbf{x}_2$, you cannot know for sure how much that training affected performance for input $\mathbf{x}_3$. To get around this problem, you might have to recruit another participant, and in that untrained participant measure the effect of $\mathbf{x}_1$ on $\mathbf{x}_3$. Thus, measuring the generalization function has some practical difficulties. Experimenters often limit their designs to training and testing with one or two pairs of inputs, and conclusions come in terms of qualitative statements regarding the shape of the generalization (i.e., wide or narrow).

The second problem involves coordinate systems. Generalization depends on the distance between training and test inputs, in terms of the vectors encoded by the internal model. However, before you can measure that distance, you need to know the coordinate system.

The third problem is that the generalization function measured in one experiment may not be consistent with those inferred from another. For example, adaptation to one force field might result in a pattern of generalization that disagrees with that observed for another field. If the generalization function always had the same shape, you could argue that you have estimated the motor primitives, or basis functions, with which your CNS computes the internal model of dynamics.

Finally, even if you solve all of these problems, you still have the problem of interpretation: You hope that the inferred generalization function not only explains behavior but also reflects something known about the neurophysiology of the motor system. How can you approach these problems?

21.3.2 Measuring Generalization from Trial-to-Trial Changes

Kurt Thoroughman and Shadmehr,[28] and later Opher Donchin and Shadmehr,[29] proposed a method to approach these questions. They proposed that in order to estimate generalization, it may not be necessary to train in one set and test in another. Rather, all possible inputs (or movements) should be presented in a random sequence. From the trial-to-trial changes in performance, you can estimate how error in one movement affected the subsequent movement as a function of the distance of the two movements in your favorite coordinate system (for example, angular distance in directions of the two movements). If you picked the "wrong" coordinate system, you will see a rotation as well as a scaling of the error vector as a function of distance.

For example, imagine that you want to quantify how errors experienced in one movement direction generalize to all other directions. For now, assume that generalization changes the error vector only by scaling it up or down. Indeed, the specific way that equation 21.1 describes the internal model implies a scaling function: The generalization function results from projection of two vectors onto each other, and therefore is a scalar variable. As an alternative, the error experienced in one limb state might be generalized through both a scaling factor and a rotation to another state. Donchin et al.[30] considered this possibility but found that for reaching movements of about 10 cm, generalization of error from one movement direction to another did not involve significant rotation but only significant scaling.

If your experiment has eight directions of reaching movement, the generalization function $b(\phi^{(n)}, \phi^{(n+1)})$ of equation 21.1 is an 8×8 matrix. Element i, j of this matrix describes the fraction of error in movement

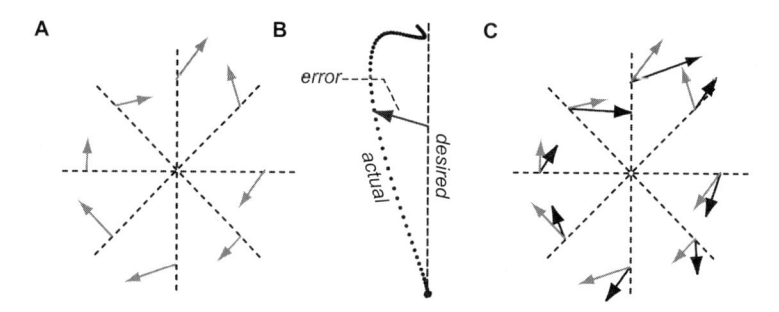

Figure 21.5
Effect of errors on an internal model. This internal model associates movement direction with force. (A) The state of the internal model just before the performance of movement n. (B) Movement n had an error (defined as the difference between the actual and desired trajectories). (C) The error updates the internal model for all directions. The figure shows the updated force estimates (black vectors) as well as the old ones (gray vectors). Note that the change for some directions exceeded that for other directions, with nearby and opposite directions showing the largest adjustments. This model assumes a bimodal pattern of generalization.

direction i (made in movement n) that generalizes to direction j (made in movement $n + 1$). To simplify things, assume that the angular distance $\Delta\phi$ between the two movements matters, not the specific direction of either (i.e., $b(\Delta\phi)$). This assumption reduces the generalization function to an 8×1 vector. If movement n had direction $\phi^{(n)}$, then each element of this vector indicates the fraction of the error in that movement that generalizes to movement $n + 1$ at angular distance $\Delta\phi$.

Note that although the error in movement n potentially affects the internal model for other movement directions, you can observe its effect for only one direction, the actual direction of movement $n + 1$. Therefore, you cannot observe the effects of the generalization to the other seven possible directions, although they exist and affect the trial-to-trial changes in the internal model.

Figure 21.5 illustrates this idea. Your CNS has an internal model that associates each movement direction with a force vector. Just before you perform movement n, your CNS evaluates the internal model along one direction, and applies the predicted force needed to produce the desired movement. During the movement, the CNS registers an error. That error affects the internal model for the direction in which it was just experienced and for all other directions as well. Generalization measures how this error affects directions other than the one most recently experienced.

Recall your goal: to estimate the generalization function $b(\Delta\phi)$, even though you can measure movements only one at a time. To estimate the generalization function, you need to be able to keep track of state changes that you cannot measure directly. However, you hope that the rules that govern these changes (i.e., the generalization function) remain constant from movement to movement. You can take the following approach: Let k

represent the states of the internal model that you would like to know (e.g., the eight possible directions of movement). Thus, k, an integer, takes a value from 1 to 8. After you complete movement n in direction $\phi^{(n)}$, your CNS updates your internal model for all directions, as illustrated in figure 21.5C. This computation depends on the error in movement n and the generalization function $b(\Delta\phi)$:

$$\hat{\tau}_k^{(n+1)} = \hat{\tau}_k^{(n)} + b(k - \phi^{(n)})\tilde{\tau}^{(n)} \qquad k = 1, \ldots, 8.$$

You have eight equations in the above expression. A force error $\tilde{\tau}^{(n)}$, experienced during movement in direction $\phi^{(n)}$, updates the internal model $\tilde{\tau}_k$ for all eight directions. This computation depends on the generalization function $b(\Delta\phi)$, which in turn depends on $\Delta\phi$, the distance between direction k for each of the eight possible movement directions in the experiment, and movement direction $\phi^{(n)}$, the movement you just made. You might expect, for example, that as the distance $k - \phi^{(n)}$ increases, the generalization function might decrease.

You should be able to estimate the generalization function b by fitting this system of equations to the trial-by-trial behavior during this experiment. However, you cannot do so directly because during the reaching movement, your movement does not express—at least not directly—the force predicted for that movement. You can only move your hand along some trajectory. Your CNS will guide a straight movement when the internal model computes a correct force estimate (i.e., when $\tilde{\tau}^{(n)} = \tau^{(n)}$). An incorrect force estimate will produce errors in the movement trajectory. Therefore, you need a way to relate errors in force estimation to errors in end-effector trajectory.

You can use the movements in the baseline condition (when the robot did not generate any active forces) as a measure of hand trajectory when your internal model produced accurate force predictions. To simplify matters, you might represent the error in hand trajectory when the robot does produce active forces as a vector that describes hand location at peak velocity compared to straight movements. Call that location-error vector $\mathbf{y}$. Next, you can relate this location error $\mathbf{y}$ to an error in force estimation. Perhaps the simplest model involves the assumption that $\mathbf{y}$ is related to force error $\tilde{\tau}$ via a *compliance matrix* D (compliance is the inverse of stiffness). You now have the following relations:

$$\begin{aligned}\mathbf{y}^{(n)} &= D\tilde{\tau}^{(n)} \\ \hat{\tau}_k^{(n+1)} &= \hat{\tau}_k^{(n)} + b(k - \phi^{(n)})\tilde{\tau}^{(n)} \qquad k = 1, \ldots, 8.\end{aligned} \qquad (21.2)$$

These equations describe a dynamical system. Our system has an output $\mathbf{y}$ and an internal state $\hat{\tau}$. To link the changes in the internal state to the changes in output, you can introduce a new variable $\mathbf{z}$ and define it as follows:

$$\mathbf{z}_k^{(n)} \equiv D\hat{\tau}_k^{(n)}.$$

Introduction of this new variable makes things simpler because with its substitution into equation 21.2, you arrive at a dynamical system that relates changes in its internal state to errors in movement:

$$\mathbf{y}^{(n)} = D\tau^{(n)} - \mathbf{z}_{k^{(n)}}^{(n)}$$
$$\mathbf{z}_k^{(n+1)} = \mathbf{z}_k^{(n)} + b(k - \phi^{(n)})\mathbf{y}^{(n)}. \qquad k = 1, \ldots, 8. \qquad (21.3)$$

You have nine equations in the above expressions. The first equation describes how error in hand location (at peak velocity) in movement n, represented by $\mathbf{y}^{(n)}$, relates to external forces $\tau^{(n)}$ and the internal model's estimate of these forces $\hat{\tau}_{k^{(n)}}^{(n)}$ for the movement direction in trial n. That is, this equation describes the errors you made during movement n in terms of the internal model's estimation (however erroneous) of the forces needed to make a straight movement to that target. The eight other equations describe how that error updates the internal model's force expectations for all directions of movement. Therefore, to estimate the generalization function b, you measure performance in a sequence of trials and measure errors $\mathbf{y}^{(n)}$. You then fit this sequence of movement errors to the system of equations in equation 21.3 and find the best fit for matrix D and vector b. These two variables have 12 unknown parameters, 4 in D and 8 in b. Donchin et al.[30] describe a procedure for fitting these equations to a sequence of movements. If correct, the model should describe all the trial-to-trial changes in performance that take place during adaptation and provide you with an estimate of the generalization function.

21.3.3 Estimating the Generalization Function in an Artificial System

Donchin and Shadmehr[29] tested this theory on an artificial adaptive system that learned to approximate dynamics using a fixed set of basis functions. They wanted to see whether the theory could extract the generalization function that resulted from these basis functions.

They simulated the equations of motion for a humanlike arm holding a robotic arm and reaching to a sequence of targets, as in equation 20.1. The motion of the robotic arm depended on the forces produced by its motor, forces acting on its handle by the humanlike arm, and forces resulting from its inertial dynamics. The motion of the humanlike arm depended on the forces acting on its hand by the robotic arm, forces due to its inertial dynamics, forces produced by a location-derivative feedback system that roughly represented the behavior of the muscles and the spinal reflex pathways, and forces predicted by an internal model of dynamics. With practice in the force field, the internal model learned to associate arm velocity with forces, effectively canceling the force field produced by the robot. This internal model was computed as a weighted sum of bases, each encoding a portion of the velocity space with a Gaussian function:

$$\hat{F} = \sum_i \mathbf{w}_i g_i(\dot{\mathbf{x}})$$
$$g_i(\dot{\mathbf{x}}) = \exp\left(\frac{-|\dot{\mathbf{x}} - \dot{\mathbf{x}}_i|^2}{2\sigma^2}\right).$$

Each basis function g_i had a preferred limb velocity $\dot{x}_i$. These "centers" were distributed so that they densely covered the velocity space relevant to reaching movements. The important variable was σ. It represented the width of the velocity coding (i.e., how broadly each basis encoded velocity). It also determined how errors experienced in one portion of the velocity space affected simulated movements in other parts. Therefore, unlike the simplified description used in the derivation of the generalization function, in which the basis functions encoded only direction, adaptation in this system relied on bases that encoded velocity.

The simulation performed a sequence of reaching movements in a curl force field (figure 20.5A), with occasional catch trials. During each reach, Donchin and Shadmehr sampled the trajectory of force errors and desired limb velocity at 10-msec intervals and used these variables in a gradient-descent algorithm to change the weights of the basis functions. To represent the performance for each movement, Donchin and Shadmehr stored hand displacement with respect to an unperturbed trajectory at maximum hand velocity for further analysis. This value measured the error experienced in each movement. They then fitted a sequence of these errors to the system of equation 21.3, to produce an estimate of limb compliance and a generalization function.

Figure 21.6 shows the performance of two different simulated adaptive systems of this type. One system (figure 21.6A and 21.6B) has narrow basis functions encoding velocity ($\sigma = 0.1$ m/sec), whereas the other system (figure 21.6C and 21.6D) has broad ones ($\sigma = 0.3$ m/sec). In both systems, performance improves with training. In particular, the errors in hand trajectory become small. Simulated catch trials reveal that errors during such movements mirror the errors in movements with the simulated force field.

Note the large variability in performance in figure 21.6A and 21.6C. For each direction, the internal model's errors produce a hand displacement that depends on the arm's inertia and stiffness. The direction-to-direction variability in these parameters causes the noisy appearance of figure 21.6A and 21.6C. Consideration of only the errors among this sequence for a single direction gets around this problem. For example, figure 21.6B shows errors for reaches toward a target at 180° for the narrow encoding of velocity space, and figure 21.6D shows the same data for high σ levels. Note that the adaptation curve for the broadly tuned system has slightly more noise. That is, performance has more "wiggles" between each catch trial in figure 21.6D than in figure 21.6B. The broadly tuned system does not, however, have any more inherent noise than the narrowly tuned one. Rather, between each pair of movements toward 180° the system made movements in other directions. Each of those movements affects the internal model toward 180°. Narrow encoding of limb velocity leads to smaller effects. That is, less (if any) generalization occurs for movements with a greater angular distance from 180°, and therefore performance for movements at 180° improves as if the system had not made these other movements. The "wiggles" in figure 21.6D result from the greater generalization associated with broader basis functions.

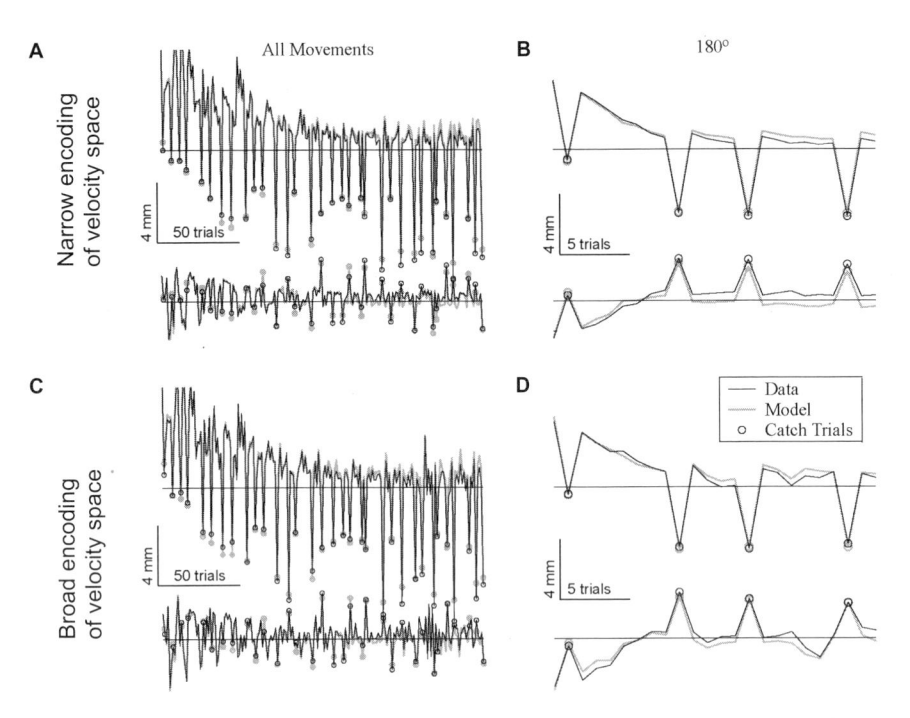

Figure 21.6
Performance of a simulated adaptive controller that learned an internal model of dynamics. In the simulation, the robot imposed a curl field on the humanlike arm. Each subfigure shows the error for each trial as a vector of hand displacement at maximum velocity with respect to an ideal trajectory. Each panel plots the vector as one component parallel to the target direction (bottom) and another component perpendicular to it (top). Movements were simulated toward eight targets, presented in random order. The black line shows the performance of the simulated controller. The gray line shows the fit of these data to equation 21.3. (*A*) Performance of an internal model that encoded velocity with narrow Gaussian bases $\sigma = 0.1$ m/sec. (*B*) Same data as in A, but plotted for only the movements toward 180°. (*C*) Internal model encoded velocity with wide Gaussian bases $\sigma = 0.3$ m/sec. (*D*) Same data as in C, but plotted for only the movements toward 180°. (Simulations by Opher Donchin and Shadmehr)

To estimate the generalization function, Opher Donchin, Joe Francis, and Shadmehr[30] fitted the sequence of movement errors to the system of equation 21.3. As the gray lines in figure 21.6 suggest, these simple linear equations account for almost all the variance in the simulated nonlinear arm–robot coupled system ($r^2 > 0.95$). The fit produced two parameters: a compliance matrix D and a generalization function $b(\Delta\phi)$. As expected, the estimate of arm compliance did not differ with changes in the width of the basis functions (figure 21.7A). Rather, the generalization function became wider (figure 21.7B). As the bases became broader, the errors spread further beyond the velocities that the system had experienced. Indeed, in theory you can derive the precise shape of the generalization function for

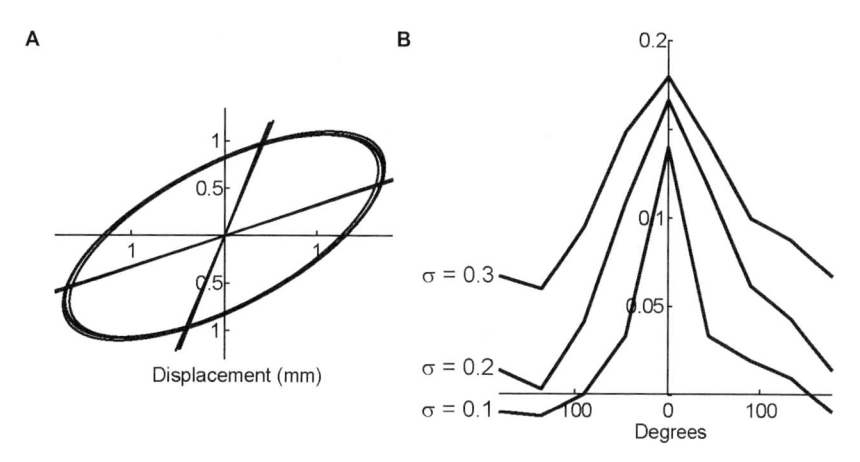

Figure 21.7
Estimate of compliance and generalization parameters for a simulated adaptive controller. The system described by equation 21.3 was fitted to the data in figure 21.6A and 21.6C to estimate the arm's compliance matrix D and generalization function b. (A) A graphical representation of matrix D. D was multiplied by a force vector of unit length that rotated around a circle, resulting in the illustrated hand displacement. The two lines represent the axes of the ellipse. The estimate of arm compliance was invariant to changes in the width of the basis functions. (B) The generalization function, $b(\Delta\phi)$, characterizing the spread of the error experienced in one direction on the other directions of movement. The x-axis of this figure gives the difference (in degrees) between the direction in which the system experienced an error and other movement directions. Broader encoding of velocity produced wider generalization. (Simulations by Opher Donchin and Shadmehr)

a given set of basis functions in this task. In effect, the algorithm provides a method for identifying the makeup of an adaptive system, an enterprise called *systems identification*.

21.3.4 Generalization as a Function of Direction in Force Adaptation

Donchin et al.[30] applied this system identification technique to estimate the generalization function of a group of participants during adaptation to a curl field. They fitted equation 21.3 to the sequence of errors and estimated a compliance matrix D and generalization function $b(\Delta\phi)$. Once they had estimated the parameters, Donchin et al. found that the equations accounted for about 80% of the variance in the data. They estimated that the compliance matrix (figure 21.8C) had a shape that was consistent with that of the simulation of the adaptive controller in the previous section (figure 21.7A). This means that the simulated adaptive controller had approximately the same inertial and stiffness parameters as the real human arm. However, the generalization function (figure 21.8D) differed from that in the simulation in an important way. It showed two peaks rather than

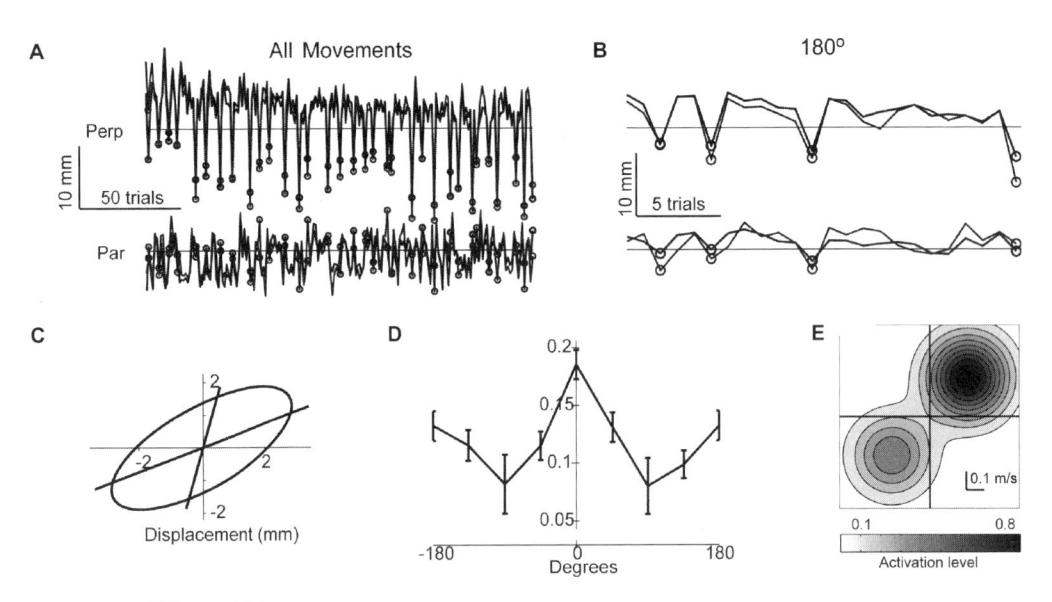

Figure 21.8

Estimate of the compliance and generalization parameters for the movements of people reaching in a curl force field. (*A*) Performance of participants and a model for a sequence of movements in random directions. The figure plots the error in each movement as components parallel (par) and perpendicular (perp) to target direction. Circles show catch trials. The actual Cartesian errors were fitted to the system of equation 21.3 to estimate arm compliance matrix *D* and generalization function *b*. (*B*) Same data as in A, but plotted only for the movements toward 180°. (*C*) A graphical representation of matrix *D*. Format as in figure 21.7A. (*D*) The generalization function, $b(\Delta\phi)$, in the format of figure 21.7B. Note the bimodal nature of the generalization function. (*E*) A basis function that encoded limb velocity and accounted for the bimodal generalization function. (From Donchin et al.[30] with permission.)

one. This bimodality means that the basis functions that represent the internal model in the CNS probably did not encode velocity with an isotropic, Gaussian-like function.

The shape of $b(\Delta\phi)$ remained consistently bimodal for various force fields. In one condition, Donchin et al.[32] considered a force field that randomly changed from trial to trial. In this task, the internal model could never develop accurate predictions of the force needed to make straight movements. However, the internal-model theory predicts that the rules that govern adaptation should apply even when the internal model cannot make long-term progress toward reducing error. An error in a given movement should continue to generalize to neighboring directions, and the effect should drop off as a function of the angular distance between movements. Analysis based on the random force field produced a compliance matrix and a generalization function that were consistent with those estimated for the curl field and for other patterns of imposed force. In particular, the generalization function remained bimodal.

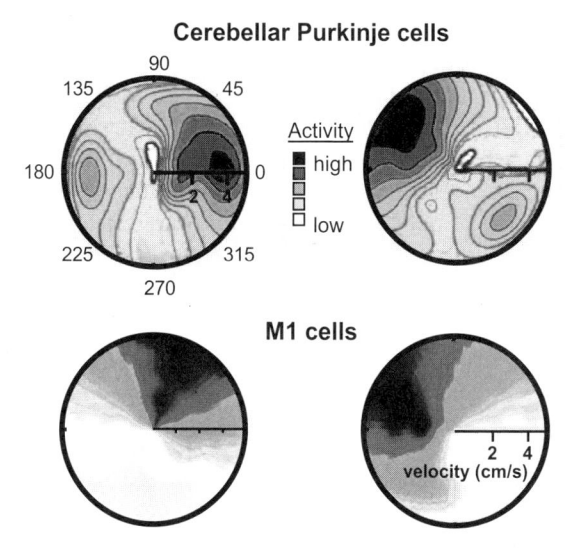

Figure 21.9
The encoding of limb velocity during reaching in the cerebellum (two cells) and the M1 cortex (two cells). The plots show the average rate of simple spikes during reaching movements, plotted as a function of average hand velocity during that movement. The radius of each circle corresponds to 5 cm/sec. Note the bimodal tuning function for the cerebellar Purkinje cells. (Modified from Coltz et al.[31] and Johnson and Ebner[33] with permission.)

To account for this bimodal generalization function, Donchin et al.[30] suggested that the bases might encode velocity space with tuning functions that also showed bimodality. For example, they considered the following encoding of velocity:

$$g_i(\dot{\mathbf{x}}) = \exp\left(\frac{-|\dot{\mathbf{x}} - \dot{\mathbf{x}}_i|^2}{2\sigma^2}\right) + \frac{1}{K}\exp\left(\frac{-|\dot{\mathbf{x}} + \dot{\mathbf{x}}_i|^2}{2\sigma^2}\right). \quad (21.4)$$

This function has a preferred velocity at $\dot{\mathbf{x}}_i$, the maximal velocity, and a second but smaller peak at $-\dot{\mathbf{x}}_i$. The constant K controls the relative size of the secondary peak. Donchin et al.[30] found that when $K = 1.7$ and $\sigma = 0.20$ m/sec, these basis functions produce a generalization function that approximates the pattern recorded in the experiments involving human participants. Figure 21.8E shows an example of this function.

21.3.5 Encoding of Limb Velocity in Cerebellum and Motor Cortex

The psychophysical results regarding generalization as a function of movement direction suggest that the basis functions that encode the internal model of arm dynamics represent limb velocity with broad but bimodal tuning functions. Do cells like that exist in your CNS?

You saw earlier that patients with cerebellar degeneration cannot adapt their reaching movements to force fields (section 20.2.4), so the cerebellum seems like a good place to look for such neurons. Tim Ebner and his colleagues[31-33] recorded from cerebellar Purkinje cells (see section 5.3) and M1 cells as a monkey moved its hand 5 cm at peak velocity of 2, 3, 4, or 5 cm/sec. Figure 21.9 shows the average discharge of two typical cells in each area as a function of hand velocity (direction and speed). In the Purkinje cells, the discharge reflected both a preferred speed and a direction of movement. Thus, Purkinje cells appeared to encode limb velocity, and they did so with a bimodal tuning function. For example, each cell fired maximally for a given direction and speed of reaching movement, the preferred direction. The discharge dropped off for movements perpendicular to that direction, then rose again for movements 180° from the preferred direction. At the preferred speed, the neuron showed bimodal tuning for movement direction. By contrast, M1 cells had a preferred direction and no bimodality. In these cells, as long as the hand moved in the preferred direction, discharge increased with speed of movement. Dan Moran and Andy Schwartz[7] have made a similar observation.

21.3.6 Combined Representation of Limb Configuration and Velocity via a Gain Field

The results thus far suggest that when you perform reaching movements in one part of the work space (e.g., with a flexed shoulder posture), adapting to velocity-dependent forces produces two kinds of generalization. First, training with one arm configuration results in aftereffects in another, even when the postures differ by as much as 80 cm in terms of hand displacement (section 21.1.2). Second, training for one direction of movement results in a bimodal pattern of generalization to other directions (section 21.3). Although the robot imposed forces on the hand, the patterns of generalization as a function of arm posture implied that the internal model encoded limb states and forces in an intrinsic coordinate system. The way this dynamics map (see figures 14.1 and 19.4) encodes arm configuration θ remains uncertain, but one way to account for broad generalization as a function of arm configuration is to assume that the basis functions have no sensitivity to limb configuration. According to this account, the basis function bimodally encodes joint angular velocity, as described in the following equation:

$$g_i(\dot{\theta}) = \exp\left(\frac{-|\dot{\theta} - \dot{\theta}_i|^2}{2\sigma^2}\right) + \frac{1}{K}\exp\left(\frac{-|\dot{\theta} + \dot{\theta}_i|^2}{2\sigma^2}\right).$$

Such basis functions would account for all the adaptation and generalization of dynamics described thus far. However, if the bases do not encode limb configuration, then they cannot learn to control the limb when the force pattern depends on that configuration. Eun-Jung Hwang, Opher Donchin, Maurice Smith, and Shadmehr[36] considered this question. In their experiments, forces imposed by the robot depended not only on limb velocity but also on limb configuration.

Participants made three parallel reaching movements from different starting locations to targets in the same direction, which led to similar joint velocities in different parts of the workspace. The experimenters covered the participants' arms with a semitransparent screen, upon which an overhead projector painted targets. The handle of the robot projected a spot of light on the screen so that participants had visual feedback regarding hand location. In these movements, opposite curl fields acted on the hand for the left and right targets (figure 21.10A). For movements to the middle target, the robot did not exert any active force on the participants' hands. The robot brought the hand to a random start position after the completion of each movement. Therefore, the force field depended both on velocity of the hand and on its location. The three parallel movements involved the same limb velocity but different forces. If the bases encoded only limb velocity and not limb configuration, then the participants should not be able to learn this task.

Hwang et al.[34] observed that when the distance between the three movements was small (0.5 cm), participants could not adapt within 550 trials. (Usually, for a single direction of movement, it takes only about 10 trials to adapt.) The participants made straight movements in catch trials and skewed movements in trials with the imposed force fields (figure 21.10B). When the three movements were separated by 12 cm, performance improved; participants showed clear signs of adaptation. Indeed, performance improved with the amount of separation (figure 21.10E).

Hwang et al.[34] also observed that for the center movement, when the robot imposed no active forces, the variance of the trajectory decreased as movement separation increased. This variance did not result from random noise, but reflected generalization from the two neighboring movements. Errors experienced during these movements generalized to the center movement. Because these errors had opposite directions, the generalization to the center movement had a mean value of 0 but a variance that depended on the distance of the two neighbors. The interference gradually decreased as the participants made movements farther apart from each other (figure 21.10D).

From this result, you can conclude that the basis functions must be sensitive to limb configuration. Because the force field is a nonlinear function of location and velocity $F = (x/d)B\dot{x}$, the basis functions could not encode location and velocity additively. That is, for each element i, the bases could not have the form $g_i = g_{i,1}(\dot{\theta}) + g_{i,2}(\theta)$. To approximate a force field that is a nonlinear function of arm configuration and velocity, the bases need to have the form $g_i = g_i(\theta, \dot{\theta})$. The bases might encode these two variables in a multiplicative fashion:

$$g_i = g_{i,1}(\theta)g_{i,2}(\dot{\theta}).$$

Hwang et al.[34] hypothesized that coding of arm configuration was linear, modulating the coding of velocity. A linear coding of arm configuration exemplifies very broad tuning, perhaps extending across the entire workspace. When you multiply this linear coding by a coding of velocity, gainfield coding results (see section 11.3).

Figure 21.10C shows an example of the output of one such basis function: a sample as the hand moves in one of eight directions from nine

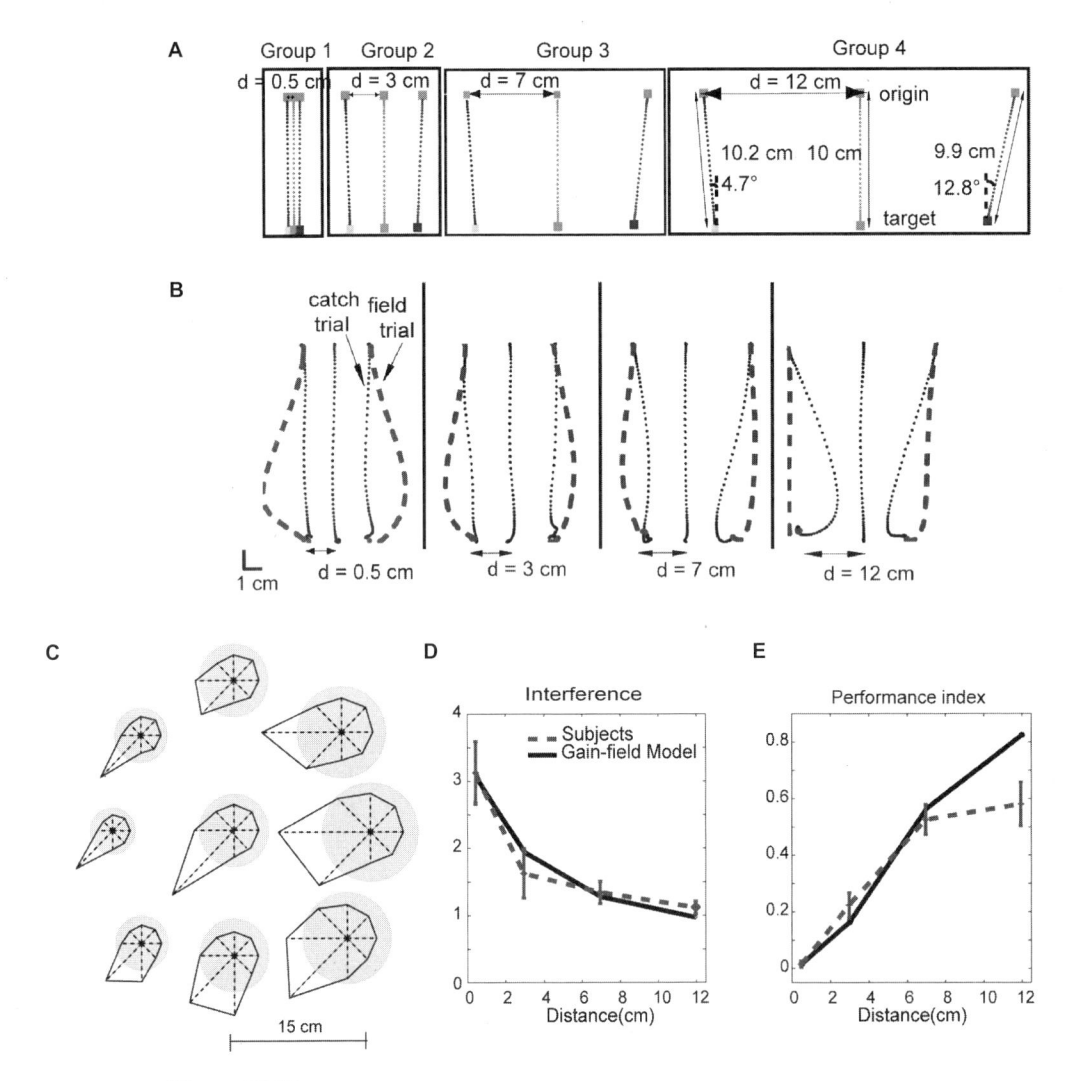

Figure 21.10
Adapting to a field in which forces depended on both the location and the velocity of the hand. (*A*) Participants performed a series of reaching movements from initial hand locations separated by distance *d*. The target layout ensured that the movements involved the same rotations in joint coordinates. Therefore, the movements involved similar joint velocities but different arm configurations. (*B*) For the left movement, the robot produced a counterclockwise curl field; for the right movement, it produced a clockwise curl field, and for the center movement, it produced no active forces at all. The figure illustrates typical trials during the final training set. As distance *d* increased, the task became more learnable. (*C*) The function described by a gain-field coding of limb configuration and velocity. The basis function encoded joint velocity and arm configuration in a multiplicative way. The basis function had a preferred velocity, and it encoded arm configuration linearly. The figure shows the output of the basis function, sampled as the hand

different starting locations. The output of the function depends on both limb configuration and velocity. The tuning with respect to direction is modulated as a function of the limb's initial configuration. Note that the preferred direction rotates as the initial configuration changes. This rotation results from the coding of velocity in a joint-based coordinate frame. Further, note that the depth of tuning also changes with arm configuration. This change in tuning depth results from the multiplicative nature of coding limb configuration and velocity. The similarity to the activity of M1 cells during isometric contractions (see figure 14.8) suggests that these ideas represent more than mere theoretical possibilities. Of course, given the crucial role of the cerebellum in force-field adaptation (see section 20.2.4), you probably would like to see similar data from neurons in that structure, but that experiment remains to be done.

Hwang et al.[34] used these basis functions to learn reaching movements in location- and velocity-dependent force fields. They found that these functions could account for the pattern of learning and interference recorded in their participants (figure 21.10D and 21.10E). They also found that the same basis functions produced the pattern of generalization across an 80-cm change in hand location.

In summary, the experiments and theory summarized here suggest that the internal model of dynamics (network 5 in figures 14.1 and 19.4) computes the map $\dot{\theta} \xrightarrow{\theta} \tau$ with basis functions that encode velocity with bimodal tuning, arm configuration linearly (constrained to yield a positive signal throughout the workspace), and their combination multiplicatively:

$$g_{i,1}(\theta) = k_1 \theta - k_2 \quad (g_{i,1} > 0) \quad \text{linear encoding of arm configuration}$$

$$g_{i,2}(\dot{\theta}) = \exp\left(\frac{-|\dot{\theta} - \dot{\theta}_i|^2}{2\sigma^2}\right) + \frac{1}{c}\exp\left(\frac{-|\dot{\theta} + \dot{\theta}_i|^2}{2\sigma^2}\right) \quad \text{bimodal encoding of joint angular velocity}$$

$$g_i(\theta, \dot{\theta}) = g_{i,1}(\theta) g_{i,2}(\dot{\theta}) \quad \text{multiplicative gain-field encoding}$$

(21.5)

Collectively, these properties appear to combine to produce many of the patterns of generalization that have been documented during adaptation to arm dynamics.

Figure 21.10 (continued)
moved in one of eight directions from nine different starting locations. The gray circle represents discharge before the start of the movement. Note that both the preferred direction of the basis function and the depth of tuning changed with limb configuration (compare with figure 14.8D). (D) Interference was measured as the variance of the error for the center movement. (E) A ratio of errors in catch trials and field trials made up the performance index. The x-axis refers to distance d. As the movements became more separated in the workspace, the task became learnable. (From Hwang et al.[34] with permission.)

21.3.7 Predicting Which Force Fields Should Be Easy or Hard to Learn

If internal models of dynamics use basis functions as basic building blocks for their representations, then the shape of the bases dictates the kinds of dynamics that the internal model can learn easily or only with difficulty. The bases describe a pattern of generalization that in turn specifies how error in one movement direction affects movements in neighboring directions. If the force field matches this generalization pattern, fast and easy learning results. If the force field does not match the generalization pattern, learning occurs slowly.

A broad coding of limb velocity does not match curl force fields very well. The curl field imposes opposite forces on the hand for movements in opposite directions. The pattern of generalization that comes from broad coding of velocity, however, scales the error experienced for one movement direction to all other directions. This means that with broad coding of velocity, every movement in a curl field interferes slightly with the next movement in the opposite direction. For example, every movement in the 270° direction causes some "unlearning" that results in poorer performance for the next movement toward 90°.

If you modify the curl fields slightly, you should be able to produce one that the internal model can learn more easily. For example, a broad generalization function implies that in an "easy" field, the direction of force should not change when the direction of movement changes. This means that unlike curl fields, forces in an easy field cannot be a linear function of velocity.

Figure 21.11A shows examples of fields in which forces depend on hand velocity. The broad generalization predicts that participants should learn field 1 more easily than field 2, and field 2 more easily than field 3. This is because in field 1, direction of force does not vary with direction of movement. The broader the generalization function, the easier it is to learn this field. In field 2, while some forces rotate as a function of direction, forces are equal for opposite directions of movement. This means that with broad generalization, an error experienced in one direction improves the internal model for the opposite direction. Finally, in field 3, forces not only change as a function of direction, but opposite directions have different forces. Here, broad generalization is counterproductive.

To measure performance of participants, you may compute a performance index as follows. During adaptation, you can expect that errors in fielded trials should gradually decrease and the error in catch trials should gradually increase. To quantify learning, you can use a ratio of the error in catch trials to the sum of displacements in catch trials and fielded trials. Early in training, when the participants made small errors in catch trials and large errors in fielded trials, the performance index was near 0. Late in training, errors in catch trials increased and errors in fielded trials decreased, so the index should approach 1. The performance of participants agreed with the prediction that they would learn field 1 more easily than field 2, and field 2 more easily than field 3 (figure 21.11B).

21.3. Estimating Generalization Functions

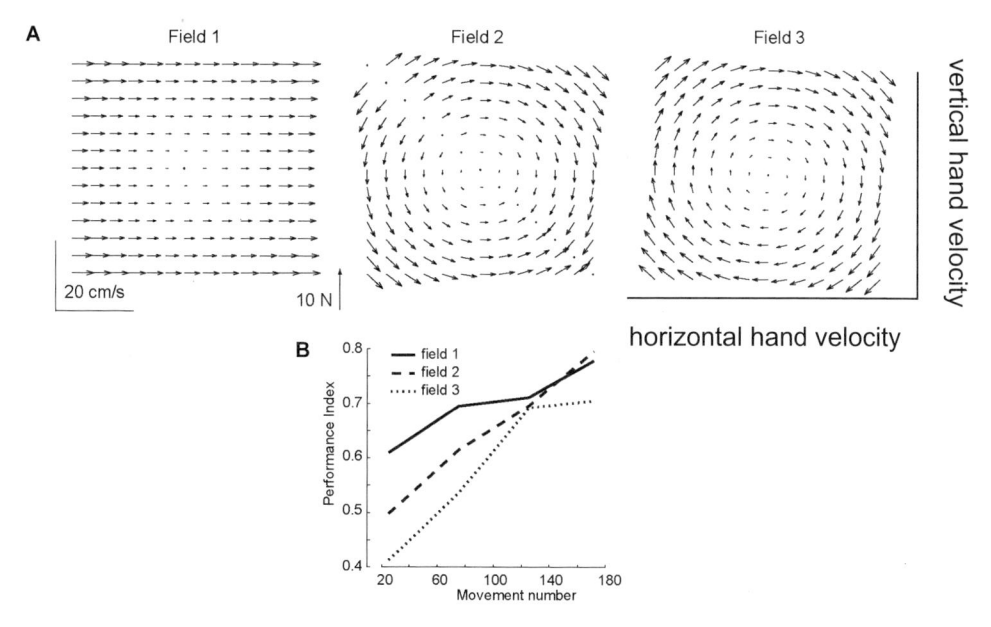

Figure 21.11
Easy and hard velocity-dependent force fields. (*A*) Three force fields that depend on hand velocity. Based on patterns of generalization, field 1 should be easier to learn than field 2, and field 2 easier to learn than field 3. (*B*) Performance of participants. (Opher Donchin and Shadmehr, unpublished data)

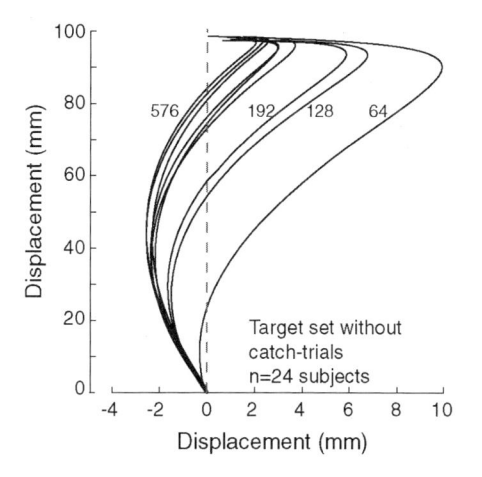

Figure 21.12
S-shaped reaching movements. Participants made reaching movements in a curl field in eight directions (no catch trials). The figure plots only movements at 90°. (From Thoroughman and Shadmehr[28] with permission.)

21.4 A Not-So-Invariant Desired Trajectory

According to the ideas presented here, the internal model of dynamics computes a mapping from a desired limb state to force. Chapter 19 presented the idea that a next-state planner could compute this desired limb state. The next-state planner, in turn, relies on a *control policy*. Up to this point, the presentation has assumed that when arm dynamics change, only the internal model adapts, while the control policy remains constant. Therefore, the underlying assumption has been that the desired trajectory never changes during force adaptation. However, what if the next-state planner changes its control policy?

For example, after about 500 movements in a curl field, reaches do not converge to a straight line. People overcompensate for the forces, and their reaching movements become slightly S-shaped (figure 21.12). One possible cause of this small deviation from straight trajectory involves the relatively broad tuning of the relevant basis functions.[28] However, it is also possible that the control policy guiding the next-state planner may change to produce S-shaped movements. Which of these possibilities causes the curvatures? This remains an open question.

References

1. Caminiti, R, Johnson, PB, Burnod, Y, et al. (1990) Shift of preferred directions of premotor cortical cells with arm movements performed across the workspace. Exp Brain Res 83, 228–232.
2. Caminiti, R, Johnson, PB, Galli, C, et al. (1991) Making arm movements within different parts of space: the premotor and motor cortical representation of a coordinate system for reaching to visual targets. J Neurosci 11, 1182–1197.
3. Ajemian, R, Bullock, D, and Grossberg, S (2001) A model of movement coordinates in the motor cortex: Posture-dependent changes in the gain and direction of single cell tuning curves. Cerebral Cortex 11, 1124–1135.
4. Scott, SH, and Kalaska, JF (1997) Reaching movements with similar hand paths but different arm orientation: I. Activity of individual cells in motor cortex. J Neurophysiol 77, 826–852.
5. Scott, SH, Sergio, LE, and Kalaska, JF (1997) Reaching movements with similar hand paths but different arm orientations. II. Activity of individual cells in dorsal premotor cortex and parietal area 5. J Neurophysiol 78, 2413–2426.
6. Moran, DW, and Schwartz, AB (1999) Motor cortical representation of speed and direction during reaching. J Neurophysiol 82, 2676–2692.
7. Moran, DW, and Schwartz, AB (1999) Motor cortical activity during drawing movements: Population representation during spiral tracing. J Neurophysiol 82, 2693–2704.
8. Schwartz, AB, and Moran, DW (1999) Motor cortical activity during drawing movements: Population representation during lemniscate tracing. J Neurophysiol 82, 2705–2718.
9. Kakei, S, Hoffman, DS, and Strick, PL (1999) Muscle and movement representations in the primary motor cortex. Science 285, 2136–2139.

10. Li, CS, Padoa-Schioppa, C, and Bizzi, E (2001) Neuronal correlates of motor performance and motor learning in the primary motor cortex of monkeys adapting to an external force field. Neuron 30, 593–607.

11. Shadmehr, R, and Moussavi, ZM (2000) Spatial generalization from learning dynamics of reaching movements. J Neurosci 20, 7807–7815.

12. Malfait, N, Shiller, DM, and Ostry, DJ (2002) Transfer of motor learning across arm configurations. J Neurosci 22, 9656–9660.

13. Humphrey, DR, Schmidt, EM, and Thompson, WD (1970) Predicting measures of motor performance from multiple cortical spike trains. Science 170, 758–762.

14. Georgopoulos, AP, Kalaska, JF, Caminiti, R, and Massey, JT (1982) On the relations between the direction of two-dimensional arm movements and cell discharge in primate motor cortex. J Neurosci 2, 1527–1537.

15. Amirikian, B, and Georgopoulos, AP (2000) Directional tuning profiles of motor cortical cells. Neurosci Res 36, 73–79.

16. Tolhurst, DJ, and Thompson, ID (1982) Organization of neurones preferring similar spatial frequencies in cat striate cortex. Exp Brain Res 48, 217–227.

17. Seung, HS, and Sompolinsky, H (1993) Simple models for reading neuronal population codes. Proc Natl Acad Sci USA 90, 10749–10753.

18. Georgopoulos, AP, Schwartz, AB, and Kettner, RE (1986) Neural population coding of movement direction. Science 233, 1416–1419.

19. Schwartz, AB (1994) Direct cortical representation of drawing. Science 265, 540–542.

20. Poggio, T (1990) A theory of how the brain might work. Cold Spring Harbor Symp Quant Biol 55, 899–910.

21. Pouget, A, Dayan, P, and Zemel, R (2000) Information processing with population codes. Nat Rev Neurosci 1, 125–132.

22. Pouget, A, and Sejnowski, TJ (1997) Spatial transformations in the parietal cortex using basis functions. J Cog Neurosci 9, 222–237.

23. Andersen, RA, Essick, GK, and Siegel, RM (1985) Encoding of spatial location by posterior parietal neurons. Science 230, 456–458.

24. Wise, SP, Moody, SL, Blomstrom, KJ, and Mitz, AR (1998) Changes in motor cortical activity during visuomotor adaptation. Exp Brain Res 121, 285–299.

25. Padoa-Schioppa, C, Li, CS, and Bizzi, E (2002) Neuronal correlates of kinematics-to-dynamics transformation in the supplementary motor area. Neuron 36, 751–765.

26. Li, CS, Padoa-Schioppa, C, and Bizzi, E (2001) Neuronal correlates of motor performance and motor learning in the primary motor cortex of monkeys adapting to an external force field. Neuron 30, 593–607.

27. Taylor, DM, Helms-Tillery, SI, and Schwartz, AB (2002) Direct cortical control of 3D neuroprosthetic devices. Science 296, 1829–1832.

28. Thoroughman, KA, and Shadmehr, R (2000) Learning of action through adaptive combination of motor primitives. Nature 407, 742–747.

29. Donchin, O, and Shadmehr, R (2002) Linking motor learning to function approximation: Learning in an unlearnable force field. Adv Neural Inf Proc Sys 14, 197–203.

30. Donchin, O, Francis, JT, and Shadmehr, R (2003) Quantifying generalization from trial-by-trial behavior of adaptive systems that learn with basis functions: Theory and experiments in human motor control. J Neurosci 23, 9032–9045.

31. Coltz, JD, Johnson, MTV, and Ebner, TJ (1999) Cerebellar Purkinje cell simple spike discharge encodes movement velocity in primates during visuomotor arm tracking. J Neurosci 19, 1782–1803.

32. Johnson, MTV, Coltz, JD, and Ebner, TJ (1999) Encoding of target direction and speed during visual instruction and arm tracking in dorsal premotor and primary motor cortical neurons. Eur J Neurosci 11, 4433–4445.

33. Johnson, MTV, and Ebner, TJ (2000) Processing of multiple kinematic signals in the cerebellum and motor cortices. Brain Res Rev 33, 155–168.

34. Hwang, E, Donchin, O, Smith, MA, and Shadmehr, R (2003) Gain-field encoding of limb position and velocity in the internal model of arm dynamics. Pub Lib Sci Biol 1, 209–220.

22 Predicting Force III: Consolidating a Motor Skill

Overview: Passage of time alters the representation of internal models. With sleep and with passage of time, the functional properties of motor skills change.

22.1 Consolidation

22.1.1 Retention of Internal Models of Dynamics in Amnesiacs

You can view the internal models of dynamics as skills that you instantiate based on context. When you learn to reach while holding a robot that pushes your hand around, you learn something specific about motor control for that machine. When you stop participating in the experiment involving robot-imposed forces and grab a cup of coffee, your CNS should not expect the cup to impose a curl force field on your hand. Experiments with amnesic patients show that your CNS can make this context-dependent selection of internal models **implicitly**.

Shadmehr, Jason Brandt, and Sue Corkin[1] found that when participants adapt to a force field, leave the lab, and then return the next day, their initial reaching movements with the robot exhibit **aftereffects** of the field that they had trained in the day before. Amnesic individuals claimed that they had not seen the robot before and could not describe the task in any way, despite 3 hours of training the previous day. They nevertheless showed the same aftereffects. One of these amnesiacs was the celebrated H.M., who underwent bilateral removal of large regions of his medial temporal lobe in the 1950s. The performance of H.M. has some noteworthy features. Normally, naïve participants sit quietly until the experimenters ask them to put their arm in a sling that supports their arm. They then grab the handle and move it to a center location. Despite his lack of **explicit** knowledge, when H.M. sat in front of the robot, unprompted and without saying a word, he grabbed the sling, put his arm in it, reached for the robot, brought it to a center position, and looked up to the monitor in a seamless sequence of movements. Therefore, despite his claim of never having seen the robot before, his CNS remembered how to do this task, much as he had nearly 40 years earlier when tested by Corkin[2] on various motor-learning tasks.

22.1.2 A Time-Dependent Change in the Internal Model's Properties

When you practice reaching movements while holding the handle of a robotic arm, the internal models that you acquire guide reaching specifically with that machine. The next time that you see the robot and prepare to move its handle, your CNS recalls the appropriate internal model and uses it to make reaching movements. For example, participants who completed training with the robot on day 1 and then returned on day 2 and moved the robot in a null field had aftereffects of the field that they had learned the previous day.[1] You might have awareness of this knowledge or not; it probably does not matter at all for your performance. Indeed, if you train in a field and make reaching movements in the same field 24 hours later, your performance exceeds that in a naïve condition (figure 22.1A). In this experiment, Shadmehr and Tom Brashers-Krug[3] trained participants in a clockwise curl field and tested them 24 hours later either on the same field or on the counterclockwise curl field. Performance had improved only for the field that the participants had previously adapted to (figure 22.1B). Furthermore, performance continued at the same high levels when the experimenters tested the same participants 5 months later. Therefore, training results in an **adaptation** that eventually becomes encoded into long-term memory (see table 4.2).

Although the neural processes that culminate in long-term motor memories remain unknown, for the most part a consistent feature of learning is that some process continues after practice has stopped. In general, memory appears to progress from a short-lived, fragile form to a long-

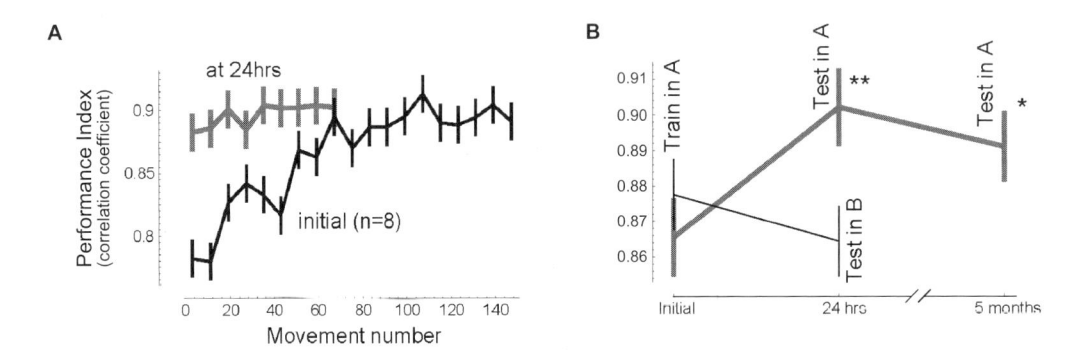

Figure 22.1

Retention of an internal model of **dynamics**. (*A*) Performance of naïve participants in a curl force field (black line) and in a test of retention 24 hr later. The performance index measured the correlation of movement trajectory with a baseline trajectory involving no active, robot-imposed forces. (*B*) Summary performance scores for a group of participants who trained in one field and were retested in the same field 24 hr and 5 mo later. The experimenters retested a second group on the opposite force field 24 hr after the initial training. (From Shadmehr and Brashers-Krug[3] with permission.)

lasting, more established form.[4,5] Phases of memory have different sensitivities to new experiences and susceptibilities to interference and injury.[6,7] The term **consolidation** refers to the progression of memory from a fragile form to a more resistant one.

Therefore, the passage of time alters the internal models in some way. Shadmehr and Brashers-Krug[3] found that the memory of the internal model of dynamics undergoes a change in its functional properties during the hours that follow completion of practice. The memory then becomes relatively resistant to disruption.

In this experiment, participants practiced in a clockwise curl field A, and sometime later (5 min to 5.5 hr) practiced reaching movements in a counterclockwise field B. The experimenters presented catch trials about once every six trials. Catch trials in field A had positive errors, and catch trials in field B had negative errors. Normally, naïve participants have initial catch-trial errors near 0, as shown for the control group by the first point in figure 22.2A. However, the participants in the experimental groups had just completed training in field A. Therefore, in field B, they had positive catch-trial errors initially (i.e., they expected field A). Interestingly, the size of these errors depended on the time since completion of training in field A. The longer the time delay, the smaller the average error in catch trials.

Participants who had trained in field A performed significantly worse in field B than did controls (figure 22.2B). This finding resulted from the expectation, developed by their CNS, of the forces that the robot had imposed in field A. The longer the delay between training in A and B, the better the performance in B.

A week later, the participants returned for a test of recall of field A. Shadmehr and Brashers-Krug[3] observed that the group that adapted to field B soon after field A showed little or no retention of field A (figure 22.2C). The longer the temporal interval between fields A and B, the better the performance during test of recall in field A. Some of these results have recently been extended: Claude Ghez and his colleagues[8,9] reported that in a task in which participants learned internal models of an inertial object, the internal model of inertial object 1 could be disrupted if practice was immediately followed by movements with inertial object 2. Kelly Goedert and Dan Willingham[10] reported a similar disruption of retention in a prism adaptation experiment when a shift to one side was followed by a shift to the opposite side. However, they did not observe that this retention was a function of the time difference between the two adaptation blocks. Why there might be a difference between the two protocols remains poorly understood.

The data for catch trials suggest that the rate of adaptation in field B did not differ among the groups; only the initial bias did. Because the initial bias appeared to be a function of time between the completion of practice in field A and the disruptive practice in field B, some aspect of the memory for field A appears to have faded with time. However, this must be only one aspect of the memory because another component persisted as much as 5 months (figure 22.1B). Taken together, the data suggest that

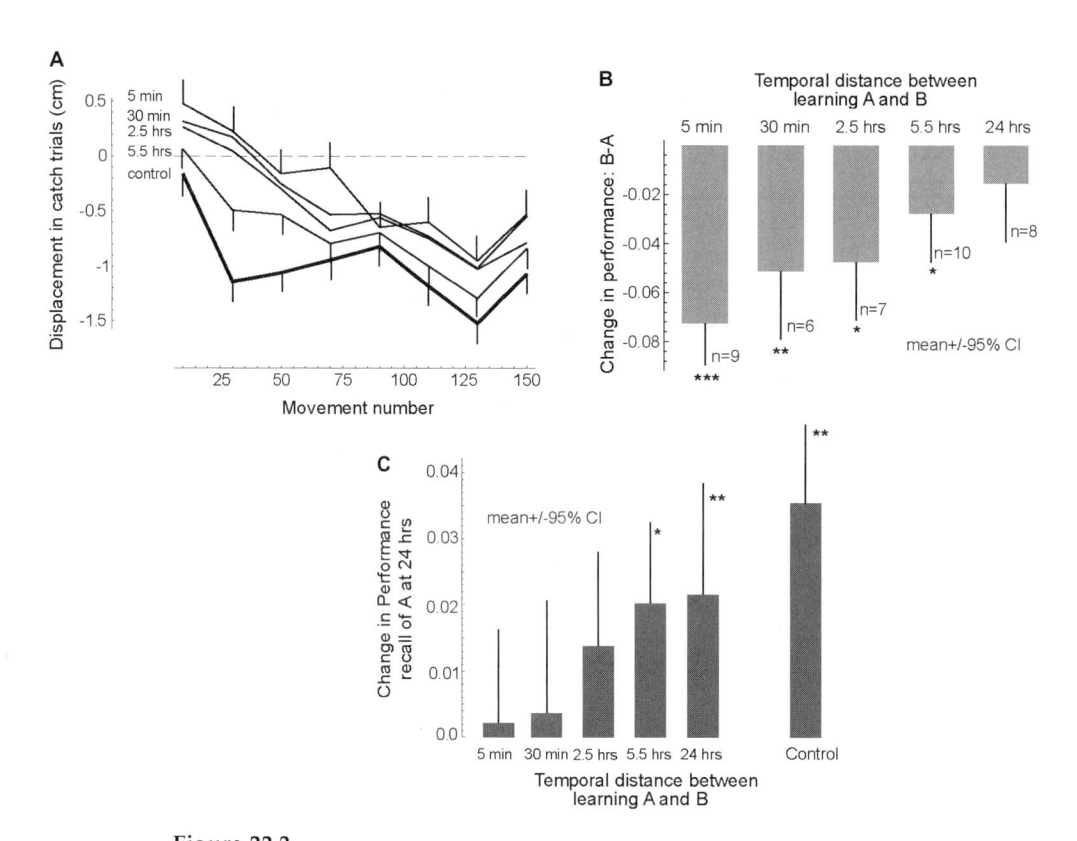

Figure 22.2

Functional properties of the internal models change as a function of time. (*A*) Errors in catch trials during adaptation in force field B after adaptation in field A. Error size corresponded to the distance from a straight line to the target 300 msec into the movement. Catch trials occurred once in every six movements. Because the size of errors depended on the direction of movement, the lines did not show a simple monotonic progression. However, all participants performed the same sequence of movements. (*B*) Performance in field B as a function of time since completion of training in field A. The experimenters measured performance as a correlation coefficient between the movement made in experimental conditions and that made in baseline, unperturbed movements. Abbreviation: Cl, control. (*C*) Performance in a test of recall of field A. Performance did not differ from that of naïve participants when training in B followed training in A at short intervals, but did differ after 5–6 hr. (From Shadmehr and Brashers-Krug[3] with permission.)

passage of time affects these motor memories, but that some aspects fade in 3–6 hours, whereas others persist indefinitely.

Thus, your CNS initially stores as least an aspect of motor memory in a labile form. Current thinking holds that this form involves neuronal discharge patterns generated through reverberating circuits. The firing pattern persists after completion of practice and leads to a change in synaptic weights, resulting in long-term memory. Long-term potentiation (LTP) provides a prominent example of synaptic plasticity: After inducing LTP, certain low-frequency stimuli can depotentiate the synapse,[11] effectively reducing the synapse's efficacy to near baseline levels. These stimuli, however, do so only within a narrow time frame: 20 minutes after induction of LTP, the low-frequency stimuli depotentiate the synapse by 70%; at 100 min, this value falls to 30%. Experimental evidence for LTP[12] and long-term depression (LTD)[13] has been obtained for M1 and the cerebellum. Most likely, these mechanisms occur ubiquitously.

These data accord with the findings that neuronal activity in M1, PMd, and SMA evolves as monkeys adopt new visuomotor skills, including changes in the preferred direction of neurons and in their depth of modulation.[14,15] Further, M1 excitability changes during skill acquisition.[16] M1 also exhibits substantial plasticity in response to pharmacological manipulation,[17] repetitive stimulation of the cortex,[18] alterations of the periphery,[19–21] changes in posture,[22] and various forms of motor experience.[23–25] Local rates or cerebral blood flow also changes in M1 during skill acquisition.[26–33] Plasticity of this type occurs in response to peripheral nerve damage, including amputation, vascular damage to the brain (as in stroke), and motor activity.[34–36]

A first-order model of motor memory consolidation might begin with Hebb's ideas regarding the initial representation of memory as a labile form of neuronal firing patterns, and synaptic plasticity as the means for representing long-term memory. According to this model, practice leads to recruitment of activity in neuronal circuits and establishes a reverberating pattern that persists after training ends. This pattern gradually decays, but it serves as the teacher for a slower but more resistant form of memory storage,[37,38] such as those that depend on synaptic plasticity. You might expect that the initial stage has a finite life and decays after completion of practice in task A. If you practice a sufficiently different skill while this neuronal firing pattern occurs, strong interference results and you both lose some of the "A skill" and have a harder time learning the "B skill." If sufficient time passes—at least a few hours—changes in synaptic efficacy gain stability and serve as a more permanent representation of the motor memory for the A skill. In an appropriate context, your CNS recalls, on cue, the internal model for the A skill.

No full-fledged theory accounts for consolidation, but this phenomenon is not unique to force adaptation. In instrumental learning (see section 4.4B), animals can learn to press one of two levers to obtain a reward. Disruption of protein synthesis in a particular part of the basal ganglia, the nucleus accumbens, blocks memory consolidation if it occurs immediately after learning. If 2–4 hours elapse until that treatment,

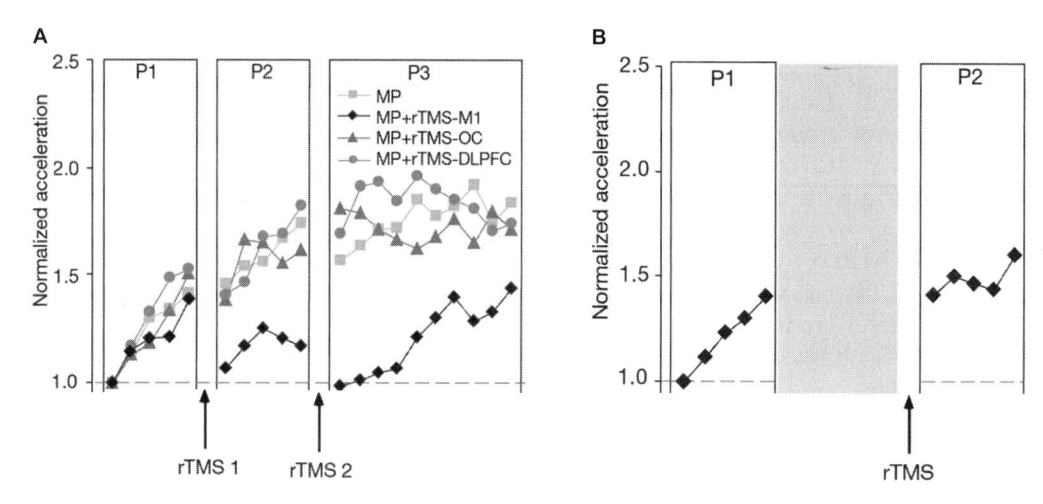

Figure 22.3
Stimulation of the motor cortex (mainly M1), after practice, had a time-dependent effect on retention. (A) Each symbol represents the (normalized) mean peak acceleration for each practice condition. Abbreviations: P1, practice 1; P2, practice 2; P3, practice 3; MP, movement potential; rTMS, repetitive transcranial magnetic stimulation; OC, occipital stimulation control; DLPFC, dorsolateral prefrontal cortex stimulation control. TMS over M1 canceled the retention and/or retrieval of the knowledge gained during P1 and P2. The ability to improve behavior by subsequent practice (P3) remained unimpaired, but the final improvement did not reach the level of the participants who had had no stimulation of M1. (B) TMS of M1 after a 6-hr consolidation period did not disrupt the retention and/or retrieval of the newly acquired motor memory. (From Muellbacher et al.[40] with permission.)

however, consolidation progresses normally.[39] Thus, a consolidation period on the order of hours may be a general feature of certain forms of learning.

22.1.3 Role of Motor Cortex in Time-Dependent Consolidation of Motor Memories

Results obtained by Wolf Muellbacher and this colleagues[40] suggest that perhaps some of the time-dependent changes in motor memories are crucially dependent on events in the motor cortex, perhaps even M1 cortex (figure 22.3). They studied an elementary motor task: **ballistic** pinching of the index finger and the thumb on the nondominant (left) hand to the 0.5 Hz beat of a metronome. Participants sat in front of a monitor. For every brisk pinch with an acceleration of more than 980 cm/sec^3, a green light indicated successful performance. A red light indicated unsuccessful performance. The session began with two 5-minute practice periods (practices 1 and 2), each followed by 15 minutes of rest, followed by another 10 minutes of practice (practice 3). Finger acceleration increased steadily across the sessions.

After completion of practice 1, the experimenters placed a **transcranial magnetic stimulator** (TMS) over the part of the skull nearest the M1 cortex. They stimulated this region at a low frequency during the 15-minute interval between practice 1 and practice 2. This caused performance in the subsequent session to suffer significantly. During practice 2, performance resembled that of naïve participants, as if consolidation had been entirely prevented. In comparison, TMS of the prefrontal cortex or the occipital cortex had no significant effect. When another TMS session followed session 2, performance in session 3 again did not differ from that of naïve participants. Therefore, TMS near M1 appeared to disrupt the memory in a way that canceled the effect of the immediately completed practice session.

In another group of participants, 6 hours of normal activities followed the initial practice session (5 min, practice 1). After the passage of those 6 hours, the experimenters stimulated motor cortex with TMS for 15 minutes. Immediately after completion of the stimulation, participants performed the task. Muellbacher et al.[40] observed that in this case, the stimulation had no significant effect on retention and/or retrieval. Therefore, stimulation of the motor cortex immediately after practice blocked consolidation of the motor skill, but after 6 hours the CNS had become resistant to the same stimulation. Thus, the motor cortex probably plays an important role in the early phases of consolidation.

In a related study, experimenters directly stimulated the motor cortex with TMS, using both anodal currents, which enhance cortical excitability, and cathodal current. Anodal (but not cathodal) stimulation improved the implicit learning of a movement sequence. Stimulation over either the premotor or the prefrontal cortex had no effect.[41] Taken together with the results of Muellbacher et al.,[40] these results suggest that increasing M1 excitability enhances the consolidation of motor memories, but disrupting its function blocks consolidation.

22.2 A Role for Time and Sleep in Consolidation of Motor Memories

The memory of the internal model of reaching does not appear to stop developing when the practice in the task ends. Rather, changes continue to occur in the hours after practice. In a different motor task, finger tapping, it appears that during the postpractice hours, and especially during sleep, the participants acquire distinct gains in their performance.

Avi Karni and his colleagues[42] asked participants to practice a finger-to-thumb opposition task with their nondominant hand (figure 22.4A). After a baseline measure of performance on two different sequences, each participant completed a few minutes of externally paced training on one of the sequences, and then returned the next day. The experimenters found that performance for the trained sequence had undergone further improvement overnight, without any intervening practice (figure 22.4A).

Stefan Fischer and his colleagues[43] discovered that sleep plays a critical role in this time-dependent improvement. They trained participants on the same finger-to-thumb task for three 5-minute blocks interrupted by

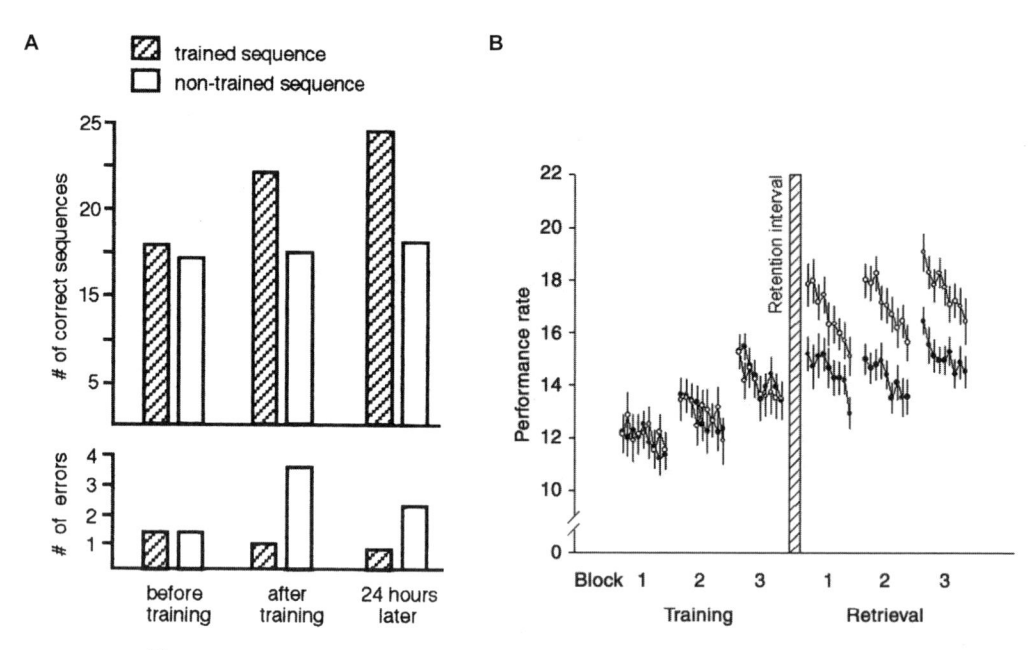

Figure 22.4
Sleep-improved performance of a sequence of finger-movement tasks. (*A*) Speed and accuracy of performance were measured over a 30-sec period for two sequences. For the trained sequence, the experimenters took the measure before a short period of training, immediately after training, and the next day. For the control sequence, no training took place. (*B*) An 8-hr retention interval separated performance blocks; during that time participants either slept (unfilled circles) or remained awake (filled circles). Data points present means ± SEM, adjusted to the first block of training and collapsed across both daytime and nighttime groups. (*A* from Karni et al.,[42] *B* and *C* from Fischer et al.[43] with permission.)

two 2-minute periods of rest. They defined performance as the number of correctly completed sequences per 30 sec, and performance gain as the change in average performance from training to testing after a retention interval.

One group trained at around 11 A.M. and had a retention interval (8 hours) during the day. During this period, some participants slept and others stayed awake. The experimenters found that both groups showed gains in performance, but those who slept had a greater gain. A second group trained at around 11 P.M. and then had a retention interval of the same duration during the night. Again, during this period some participants slept and others stayed awake. Figure 22.4B shows the average performance of the two groups. The results show that the mere passage of time resulted in small gains in performance, but sleep led to much greater improvement.

Recently, Matthew Walker, Tiffany Brakefield, Allan Hobson, and Robert Stikgold[44] found that practicing a sequence B after practicing sequence A affected the patterns of consolidation. They placed the left hand

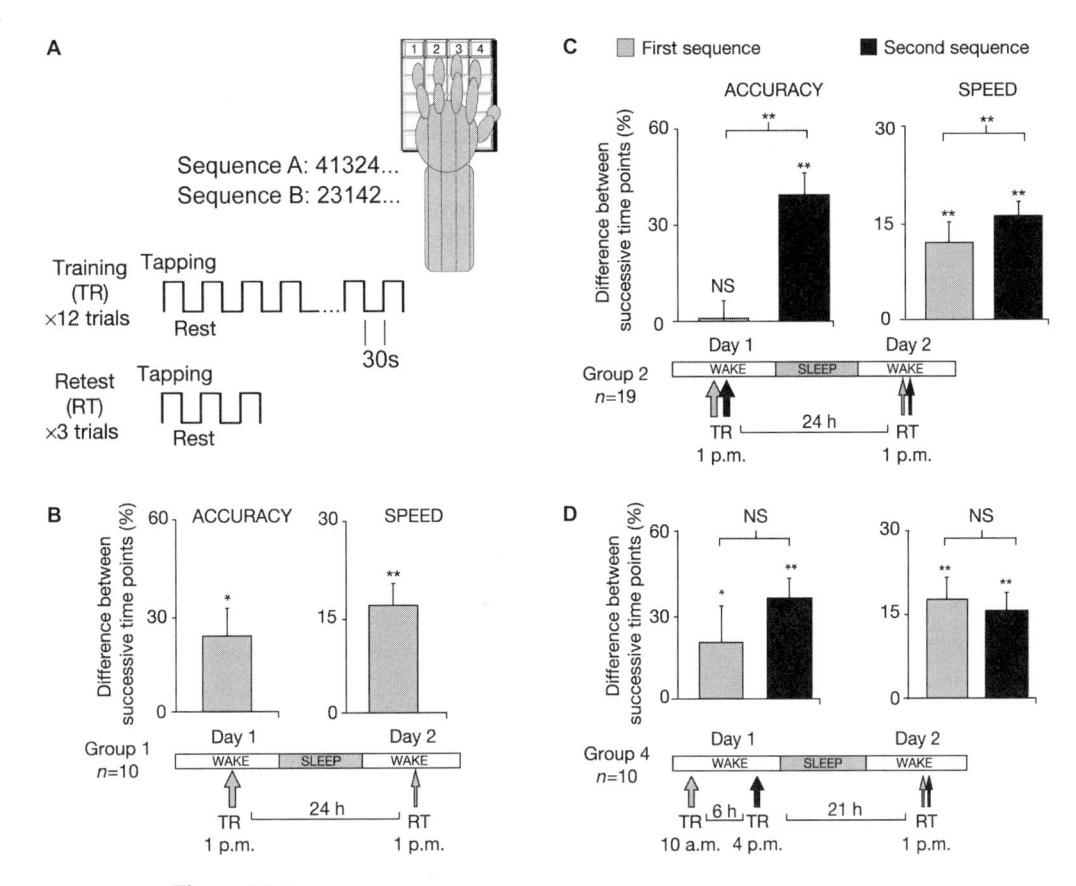

22.2. A Role for Time and Sleep in Consolidation of Motor Memories

Figure 22.5

Dissociable stages of memory consolidation in a finger tapping task. (*A*) Right handed participants trained on a sequence with their left hand, labeled sequence A and B. Performance measures were the number of complete sequences (speed), and the number of errors (accuracy). Retest values were compared to the final three trials of training. (*B*) Following training on a single sequence on day 1, overnight improvements were seen at the 24-hour retest on day 2. (*C*) Immediately after learning the first sequence on day 1, subjects trained on the second sequence. Overnight improvements in accuracy were noted for only the first sequence. Speed improved for both sequences. (*D*) Six waking hours after learning the first sequence on day 1, participants trained on the second sequence. Overnight improvements in both accuracy and speed were found for both sequences. (From Walker et al.[44] with permission.)

on a keyboard and had volunteers train on sequence A (figure 22.5A). They observed that 24 hours later, after a period of sleep, performance in terms of both accuracy and speed had improved (figure 22.5B). If sequence B was practiced minutes after A, then after a night's sleep following that practice there was no significant change in accuracy of sequence A (figure 22.5C). However, if sequence B was practiced either 6 hours after A (figure 22.5D) or 24 hours after A (data not shown), then after a night's sleep there was significant improvement in accuracy of sequence A. Therefore, at 5 minutes, the interfering sequence appeared to make the skill associated with sequence A ineligible for sleep-dependent improvements in performance. At 6 hours, this effect was reduced. The effect of the interfering task appeared to depend on the time that had passed since completion of training on the first task.

Interestingly, they observed that after A was practiced on day 1, if it was again briefly practiced on day 2, it once again became susceptible to interference. If on day 2 sequence B followed A, on day 3 accuracy of sequence A was significantly affected. This observation suggests that reactivation of consolidated memories might cause it to once again become fragile.

References

1. Shadmehr, R, Brandt, J, and Corkin, S (1998) Time dependent motor memory processes in amnesia. J Neurophysiol 80, 1590–1597.
2. Corkin, S (1968) Acquisition of motor skill after bilateral medial temporal-lobe excision. Neuropsychologia 6, 255–265.
3. Shadmehr, R, and Brashers-Krug, T (1997) Functional stages in the formation of human long-term motor memory. J Neurosci 17, 409–419.
4. Bailey, CH, and Kandel, ER (1995). In The Cognitive Neurosciences, ed Gazzaniga, M (MIT Press: Cambridge, MA), pp 19–36.
5. DeZazzo, J, and Tully, T (1995) Dissection of memory formation: from behavioral pharmacology to molecular genetics. Trends Neurosci 18, 212–218.
6. Tully, T, Preat, T, Goynton, SC, and Del Vecchio, M (1994) Genetic dissection of consolidated memory in *Drosophila*. Cell 79, 35–47.
7. Hammer, M, and Menzel, R (1995) Learning and memory in the honeybee. J Neurosci 15, 1617–1630.
8. Krakauer, JW, Ghilardi, MF, and Ghez, C (1999) Independent learning of internal models for kinematic and dynamic control of reaching. Nat Neurosci 2, 1026–1031.
9. Krakauer, JW, Pine, ZM, and Ghez, C (1999) Visuomotor transformations for reaching are learned in extrinsic coordinates. Soc Neurosci Abs 25, 2177.
10. Goedert, KM, and Willingham, DB (2002) Patterns of interference in sequence learning and prism adaptation inconsistent with the consolidation hypothesis. Learn Mem 9, 279–292.
11. Fujii, S, Saito, K, Miyakawa, H, et al. (1991) Reversal of long-term potentiation (depotentiation) induced by tetanus stimulation of the input to the CA1 neurons of guniea pig hippocampal slices. Brain Res 555, 112–122.

12. Sanes, JN, and Donoghue, JP (2000) Plasticity and primary motor cortex. Annu Rev Neurosci 23, 393–415.

13. Castro-Alamancos, MA, Donoghue, JP, and Connors, BW (1995) Different forms of synaptic plasticity in somatosensory and motor areas of the neocortex. J Neurosci 15, 5324–5333.

14. Wise, SP, Moody, SL, Blomstrom, KJ, and Mitz, AR (1998) Changes in motor cortical activity during visuomotor adaptation. Exp Brain Res 121, 285–299.

15. Li, CS, Padoa-Schioppa, C, and Bizzi, E (2001) Neuronal correlates of motor performance and motor learning in the primary motor cortex of monkeys adapting to an external force field. Neuron 30, 593–607.

16. Pascual-Leone, A, Dang, N, Cohen, LG, et al. (1995) Modulation of muscle responses evoked by transcranial magnetic stimulation during the acquisition of new fine motor skills. J Neurophysiol 74, 1037–1045.

17. Jacobs, KM, and Donoghue, JP (1991) Reshaping the cortical motor map by unmasking latent intracortical connections. Science 251, 944–945.

18. Nudo, RJ, Wise, BM, Sifuentes, F, and Milliken, GW (1996) Neural substrates for the effects of rehabilitative training on motor recovery after ischemic infarct. Science 272, 1791–1794.

19. Donoghue, JP, Suner, S, and Sanes, JN (1990) Dynamic organization of primary motor cortex output to target muscles in adult rats. II. Rapid reorganization following motor nerve lesions. Exp Brain Res 79, 492–503.

20. Kew, JJM, Ridding, MC, Rothwell, JC, et al. (1994) Reorganization of cortical blood flow and transcranial magnetic stimulation maps in human subjects after upper limb amputation. J Neurophysiol 72, 2517–2524.

21. Brasil-Neto, JP, Valls-Sole, J, Pascual-Leone, A, et al. (1993) Rapid modulation of human cortical motor outputs following ischaemic nerve block. Brain 116, 511–525.

22. Sanes, JN, Wang, J, and Donoghue, JP (1992) Immediate and delayed changes of rat motor cortical output representation with new forelimb configurations. Cerebral Cortex 2, 141–152.

23. Nudo, RJ, Milliken, GW, Jenkins, WM, and Merzenich, MM (1996) Use-dependent alterations of movement representations in primary motor cortex of adult squirrel monkeys. J Neurosci 16, 785–807.

24. Pascual-Leone, A, Wassermann, EM, Sadato, N, and Hallett, M (1995) The role of reading activity on the modulation of motor cortical outputs to the reading hand in Braille readers. Ann Neurol 38, 910–915.

25. Kleim, JA, Barbay, S, and Nudo, RJ (1998) Functional reorganization of the rat motor cortex following motor skill learning. J Neurophysiol 80, 3321–3325.

26. Hazeltine, E, Grafton, ST, and Ivry, R (1997) Attention and stimulus characteristics determine the locus of motor-sequence encoding—A PET study. Brain 120, 123–140.

27. Grafton, ST, Woods, RP, and Tyszka, M (1994) Functional imaging of procedural motor learning: Relating cerebral blood flow with individual subject performance. Human Brain Mapping 1, 221–234.

28. Schlaug, G, Knorr, U, and Seitz, RJ (1994) Inter-subject variability of cerebral activations in acquiring a motor skill: A study with positron emission tomography. Exp Brain Res 98, 523–534.

29. Karni, A, Meyer, G, Jezzard, P, et al. (1995) Functional MRI evidence for adult motor cortex plasticity during motor skill learning. Nature 377, 155–158.
30. Seitz, RJ, and Roland, PE (1992) Learning of sequential finger movements in man: A combined kinematic and positron emission tomographic (PET) study. Eur J Neurosci 4, 154–165.
31. Kawashima, R, Roland, PE, and O'Sullivan, BT (1995) Functional anatomy of reaching and visuomotor learning: a positron emission tomography study. Cerebral Cortex 5, 111–122.
32. Shadmehr, R, and Holcomb, HH (1997) Neural correlates of motor memory consolidation. Science 277, 821–825.
33. Seitz, RJ, Canavan, AGM, Yaguez, L, et al. (1994) Successive roles of the cerebellum and premotor cortices in trajectorial learning. NeuroReport 5, 2541–2544.
34. Friel, KM, and Nudo, RJ (1998) Recovery of motor function after focal cortical injury in primates: compensatory movement patterns used during rehabilitative training. Somatosens Motor Res 15, 173–189.
35. Nudo, RJ, Plautz, EJ, and Milliken, GW (1997) Adaptive plasticity in primate motor cortex as a consequence of behavioral experience and neuronal injury. Seminars Neurosci 9, 13–23.
36. Nudo, RJ, Wise, BM, Sifuentes, F, and Milliken, GW (1996) Neural substrates of the effects of rehabilitative training on motor recovery after ischemic infarct. Science 272, 1791–1794.
37. Alvarez, P, Zola-Morgan, S, and Squire, LR (1994) The animal model of human amnesia: Long-term memory impaired and short-term memory intact. Proc Natl Acad Sci USA 91, 5637–5641.
38. Alvarez, P, and Squire, LR (1994) Memory consolidation and the medial temporal lobe: a simple network model. Proc Natl Acad Sci USA 91, 7041–7045.
39. Hernandez, PJ, Sadeghian, K, and Kelley, AE (2002) Early consolidation of instrumental learning requires protein synthesis in the nucleus accumbens. Nat Neurosci 5, 1327–1331.
40. Muellbacher, W, Ziemann, U, Wissel, J, et al. (2002) Early consolidation in human primary motor cortex. Nature 415, 640–644.
41. Nitsche, MA, Schauenburg, A, Lang, N, et al. (2003) Facilitation of implicit motor learning by weak transcranial direct current stimulation of the primary motor cortex in the human. J Cog Neurosci 15, 619–626.
42. Karni, A, Meyer, G, Rey-Hipolito, C, et al. (1998) The acquisition of skilled motor performance: Fast and slow experience-driven changes in primary motor cortex. Proc Natl Acad Sci USA 95, 861–868.
43. Fischer, S, Hallschmid, M, Elsner, AL, and Born, J (2002) Sleep forms memory for finger skills. Proc Natl Acad Sci USA 99, 11987–11991.
44. Walker, MP, Brakefield, T, Hobson, JA, and Stickgold, R (2003) Dissociable stages of human memory consolidation and reconsolidation. Nature 425, 616–620.

23 Predicting Inputs and Correcting Errors I: Filtering and Teaching

Overview: The neural mechanisms for predicting inputs and correcting errors play a central role in motor learning. Although relatively little is known about the mechanisms of motor learning for reaching and pointing, more is known about those for Pavlovian and instrumental learning. These forms of learning depend on the cerebellum and basal ganglia, and they can serve as models for other forms of motor learning. The cerebellum and basal ganglia both function to correct errors, but of different kinds. The cerebellum functions to correct errors in the prediction of sensory signals, and perhaps of neural signals more generally. One consequence of these predictions is the production of motor commands that anticipate and meliorate the potentially damaging effects of predicted stimuli. Learning in the basal ganglia, on the other hand, is driven by an error in predicted biological value. Associating biological value with the state of the system aids in deciding what to do in a given context (e.g., the selection of control policies for performing an action).

In this book, you have come across several kinds of skill acquisition and motor adaptation relevant to reaching and pointing movements: prism adaptation, rotation adaptation, and adapting to altered limb dynamics or kinematics (table 23.1). In each case, adaptation was necessary because the maps that align various sensory or motor variables were perturbed or distorted in some way. In many cases, damage to the cerebellum appeared either to compromise the computation of the maps or to prevent their adaptation (sections 16.1.2, 16.2.5, and 20.2.4, above, and section 24.2.2, below). In contrast, damage to the basal ganglia in Huntington's disease (HD) did not appear to affect the adaptation to altered limb dynamics (figure 20.9). Rather, patients with HD and people carrying the gene for that disease (but not yet symptomatic) failed to adjust properly for errors in their ongoing reaching movements (section 19.5 and figure 19.8). Section 19.5 suggested that such a deficit could result from a disorder in the next-state planner (figure 19.4), the computational structure that implements a *control policy* that transforms a long-range goal, such as "bring the hand to the target," into a sequence of short-term desired "next-states" for the immediate future. In a subsequent chapter (chapter 25), you will read about the ability to learn sequences of movements. Evidence suggests that the basal ganglia likely plays a central role in this kind of skill acquisition,

Table 23.1
Evidence pointing to cortex, cerebellum, and basal ganglia as important in skill acquisition, adaptation, and switching

Perturbation	Error detected	Output adjusted	Structures Disease	Section
Wedge prisms	visual end-point error at <50 msec	$x_{ee} \leftrightarrow \hat{\theta}$, the location map; $x_{dv} \stackrel{\hat{\theta}}{\leftrightarrow} \Delta\hat{\theta}$, the displacement map, in fixation-centered coordinates	CB PMv PPC	16.1.2 16.1.1 16.1.3
Visual feedback rotation	unexpected end-effector displacements $x_{dv} \stackrel{\hat{\theta}}{\leftrightarrow} \Delta\hat{\theta}$	$x_{dv} \stackrel{\hat{\theta}}{\leftrightarrow} \Delta\hat{\theta}$, the displacement map, in fixation-centered coordinates	CB	16.2.5
Force field and interaction torques	undesired muscle stretch, etc.	$\Delta\hat{\theta} \stackrel{\hat{\theta}}{\leftrightarrow} \hat{\tau}$	CB	20.2.4 24.2.2
Predictive remapping		future state of limb (location, velocity)	PPC	17.1.6
Smooth trajectories	initial deviation based on proprioception	$\hat{\theta} \leftrightarrow \hat{x}_{ee} \leftrightarrow \Delta\vec{x}_d$, output of the next-state planner (in patients, the planner does not receive $\hat{x}_{ee}$ based on $\hat{\theta}$)	HD HD-AGC	19.5.3
Sequence, SRTT		next-state on path to goal (patients do not learn sequence)	HD	26.1.3
Skill switch		context-dependent switch to appropriate IM (patients do not switch off reflexes)	PD	6.1.4

which also involves putting together a series of next states in order to achieve some long-term goal.

The contributions of the cerebellum and basal ganglia to motor learning remain poorly understood, although almost all experts acknowledge that both play an important role. Two key ideas about their contributions involve adaptive filtering of neural signals and error correction, and this chapter addresses both topics. Adaptive filtering refers to the idea that your CNS needs to extract signals of interest from the tremendous amount of sensory information available from the periphery, including proprioceptive and visual signals. Some properties of cerebellar-like structures within the brainstem sensory nuclei provide hints about how the CNS might do this by using adaptive filters. The first part of this chapter, section 23.1, explains how adaptive filters work in general, and how a cerebellar-like architecture underlies adaptive filtering and phenomena such as negative **aftereffects**.

The remainder of this chapter depends on the idea that the cerebellar mechanisms of Pavlovian learning provide some hints about how both the cerebellum and basal ganglia might function in learning to reach and point. A fundamental aspect of Pavlovian learning involves the prediction of incoming signals: the sensory inputs hardwired by evolution to produce particular reflexive outputs. Both the cerebellum and basal ganglia play a role in certain forms of Pavlovian learning, and despite vast differences in their architecture and organization, surprisingly similar mechanisms appear to operate in these two structures.

This chapter presents the idea that both the cerebellum and the basal ganglia learn to predict neural signals, and that in both the validity of these predictions is evaluated at a **comparator**. The result of this test then adjusts the predictions computed in the future. It appears that the cerebellum predicts neural signals that are important in producing and adjusting reflexes and other motor commands. Some of these neural signals reflect stimuli that might cause harm, and thus have an *aversive* character. The basal ganglia predicts inputs of a different kind: those that provide positive biological value, often in the form of nutrients or inputs that satisfy other biological needs. (Note, however, that the basal ganglia also plays a role in learning about aversive stimuli, as in fear conditioning.) Recall that chapter 17 also dealt with the prediction of neural signals, in part based on **efference copy**. It introduced the idea that **forward models** predict the sensory feedback that should occur, in a given state, for a given motor command. Although that presentation focused on the PPC and this chapter deals in large part with the cerebellum, keep in mind the fact that the cerebellum and the PPC have a close functional relationship.

In terms of the control structure depicted in figure 19.4, the presentation in this chapter and the following one offers some speculation on the role of the cerebellum in learning the alignments of the location, displacement, and dynamics maps (networks 1, 4, and 5 in figure 19.4). In contrast, the basal ganglia may be a structure that contributes in part to the learning of control policies, i.e., contributing to a next-state planner (network 6 in figure 19.4). Recall that a next-state planner directs actions for the purpose of attaining a goal at the conclusion of these actions (section 19.3). In this framework, learning to perform a movement resembles learning the parameters of a control policy.

23.1 Cancellation of Predicted Signals by Adaptive Filtering

Your CNS can predict some sensory signals because they result from your own motor commands. The term **reafference** refers to the sensory inputs caused by your own movements, and your CNS has mechanisms both for predicting those signals and for canceling them to find other information hidden in such self-generated "noise."

A great deal of reafference involves proprioceptive input, but the sight of your hand moving in front of you provides another example. A computation that estimates how much sensory feedback results from reafference involves a forward model, a form of internal model discussed in section 17.3.3. Your CNS can use efference copy, delayed sensory feedback, and an internal model to predict the sensory consequences of its motor commands at some future time. For example, a forward model can predict the proprioceptive signals that return from your arm muscles as a consequence of motor commands that direct your biceps to contract. Physiological findings in many vertebrate species suggest that **projection neurons** in cerebellar-like structures predict reafference and remove the predictable component of sensory inflow. They do so through *adaptive filtering*. The term "cerebellar-like" refers to the fact that among cerebellar-like structures, a number can be found in the brainstem rather than in the cerebellum per se.

The cerebellar-like structures in the brainstem of electric fish provide a good example of the prediction of reafference and adaptive filtering. *Mormyrid* fish from Africa and *Gymnotid* fish from South America both produce weak electrical discharges to communicate and navigate. These two fishes independently evolved an *electric organ* that generates an electric field. Within the electric organ, modified muscle cells called *electrocytes* produce an electric pulse when they discharge synchronously. Their small membrane voltages add like a set of batteries in series. You can think of the electric-organ discharge as a large EMG signal broadcast to the world. The fish controls its electric field with the same kind of motor commands used for skeletal movements and uses same kind of cholinergic motor neurons to generate electricity. A signal from a nucleus located in the caudal medulla controls the motor neurons that drive the electric organ.

Electric fish have specialized receptors that detect electric fields and currents in their environment. By sensing minor distortions caused by objects in the electric field, the fish can detect objects and other features in its environment. However, electric fish have a problem: Their electroreceptors also pick up the electric signals caused by the fish's movements, including respiratory EMG and their own electric-organ discharges. Indeed, the electric-organ discharge dwarfs in magnitude the small distortions that provide the fish with important information. The fish needs to "know" what part of the *electrosensory* signal results from its own motor commands before it can extract the information it needs.

The sensory nerves from electroreceptors terminate in parts of the brainstem that have an architecture resembling that of the cerebellar cortex. In these cerebellar-like systems, granule cells give rise to parallel fibers that contact projection neurons resembling Purkinje cells[1] (the principal cell layer in figure 23.1A). The signals received by the granule cell–parallel fiber system include proprioceptive inputs indicating the configuration of the fish's body and fins, and an efference copy of the signals that drive the electric organ's motor neurons. The projection neurons also receive input from the electroreceptors, much like other brainstem sensory nuclei. These cerebellar-like nuclei differ from other sensory-relay nuclei in that they use efference copy to predict some of the input that should come from the electroreceptors and to filter that signal in order to detect other signals.

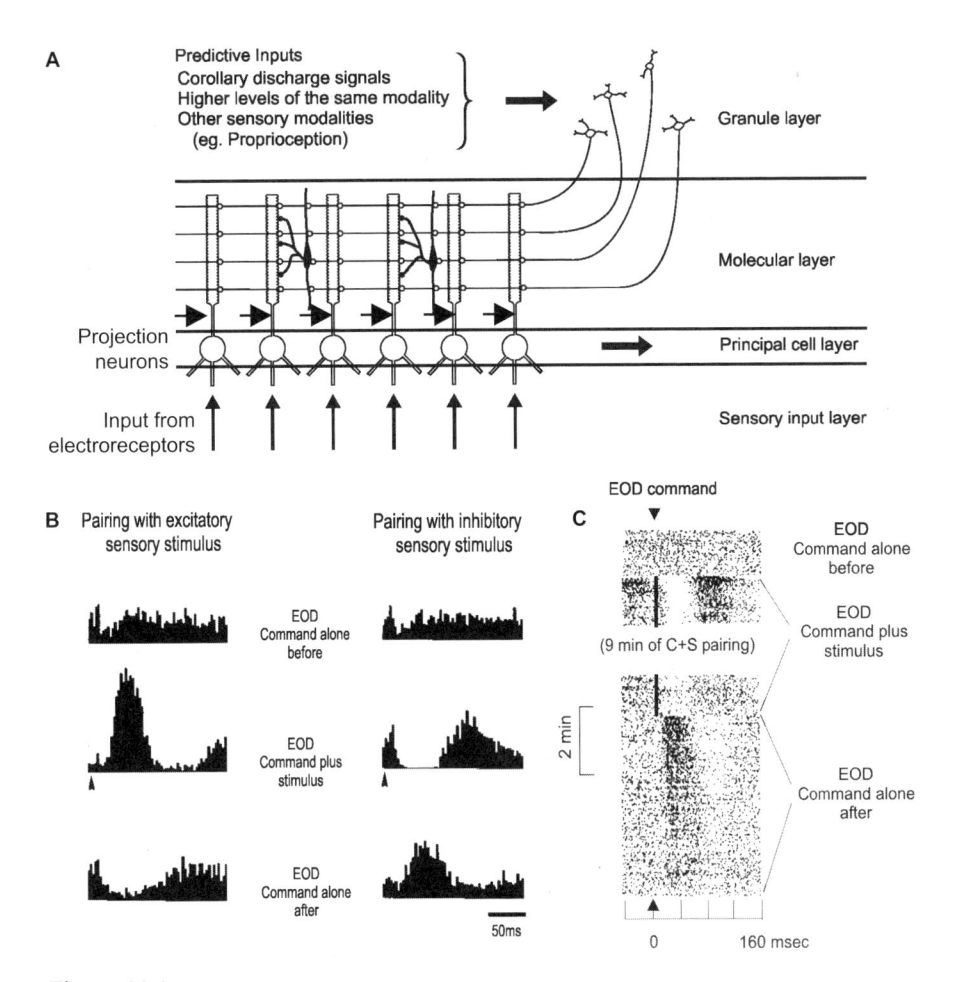

23.1. Cancellation of Predicted Signals by Adaptive Filtering

Figure 23.1

A cerebellar-like adaptive filter in electric fish. (*A*) Architecture of cerebellum-like structures in the brainstem of electric fish. The architecture of these sensory nuclei resembles that of the cerebellar cortex, except that these cerebellar-like structures do not have climbing-fiber inputs. (*B*) Discharge of projection neurons during adaptation. Adaptation to an excitatory (left) or an inhibitory (right) electrical stimulus that was time-locked to the motor command, corresponding to efference copy. Abbreviation: EOD, electric organ discharge. (*C*) Time course of adaptation to an inhibitory electrical stimulus. After adaptation, the EOD command alone produced a negative image of the sensory stimulus. Abbreviations: C (motor) command; S, stimulus. (From Bell[1] with permission.)

The kind of adaptive filtering used by electric fish requires learning. You might think that the fish could learn the currents caused by its own electric discharges once and then maintain that knowledge forever. However, the sensory signals received from electric-organ currents vary with the salinity and conductivity of water and other aspects of the environment. For example, the rainy season changes the water's conductivity. The electrical environment also changes when the fish swim near a riverbank or rest in small burrows. Accordingly, their predictions about the sensory signals received as a consequence of their own electromotor commands must adapt.

The sensory signals also depend on the curvature of the fish's body. For example, in order to better investigate an object, electric fish tend to bend their body around it as they produce electric discharges, which influences the sensory signal. Thus, the fish's CNS needs to integrate efference copy with proprioception in order to learn to predict the signals that result from its own motor commands. After predicting these self-generated signals, the fish's CNS can subtract them from the actual sensory signals and detect the components that stand apart. The projection neurons in the cerebellar-like structures play a prominent role in this computation.

Curtis Bell, David Bodznick, and their colleagues[2,3] performed experiments in which they used curare to block the effects of acetylcholine, thus preventing electric-organ discharge. Therefore, when the fish generated a motor command, it did not cause the usual sensory consequences. Instead, the experimenter stimulated the fish's electroreceptors with artificial currents that occurred at the same time as the fish's motor commands. In theory, the fish's CNS should have learned to filter this artificial electric signal, as if it resulted from the fish's motor commands.

At first, when the experimenters presented no artificial currents (termed the *sensory stimulus* in figure 23.1B), the fish's motor command (i.e., the efference copy) produced little response in the projection neurons (figure 23.1B, top left). Later, the experimenters presented an artificial excitatory sensory stimulus that was synchronized with the electric-organ discharge that corresponded to the efference copy. In that condition, the projection neurons responded vigorously (figure 23.1B, middle left). However, over the course of a few minutes, as the sensory stimulus continued to occur in conjunction with the efference copy, the projection neurons adapted so that their discharge returned to near baseline. After adaptation, the experimenters eliminated the sensory stimulus. As a result, the projection neurons' response had the opposite sign of the initial response to the artificial stimulus (figure 23.1B, bottom left). The projection neurons had formed a "negative image" of the sensory inputs that they consistently encountered, which corresponds to a negative aftereffect.[4]

A similar process took place when the experimenters paired artificial inhibitory currents with efference copy. After adaptation, when the experimenters withheld this stimulus, projection neurons showed an increase in their activity (bottom, right column of figure 23.1B). Figure 23.1C shows the time course of this adaptation. The sensory stimulus initially inhibited the discharge of the projection neurons, but then after 9 min of pairing (C + S pairing in figure 23.1C), the discharge adapted. Removal of the sensory stimulus resulted in a negative neural image, which returned to baseline after about 4 min of efference copy without the sensory stimulus.

The mechanism for this adaptation depends on a change in the synapses from the parallel fibers onto the projection neurons.[5–7] Recall that the projection neurons in these sensory nuclei resemble Purkinje cells of the cerebellar cortex and, like them, receive parallel-fiber inputs. The parallel fiber–Purkinje cell synapses transform the efference-copy signal into a negative image of the predicted sensory feedback, and thus act like weights of a forward model. As described in section 17.3.3, a forward model uses efference copy and proprioceptive inputs to compute predictions about future feedback. In these cerebellar-like nuclei, the parallel fiber–Purkinje cell synapses appear to perform this computation. The projection neuron compares sensory input from electroreceptors against the parallel-fiber's negative image, and the neuron's output reflects the difference between these two neural signals. These neurons appear to use an anti-Hebbian rule to govern the changes in synaptic weights that generate the negative image.[8] In anti-Hebbian learning, a coincidence of presynaptic impulses and postsynaptic depolarization leads to a reduction in the strength of the synapse.

In summary, these neurons appear to use efference copy and a forward model to adaptively filter the sensory signal, and pass through only the non-self-generated component. John Montgomery and his colleagues described the function of this system as follows:

Experiments show that the projection neurons learn to cancel an external stimulus that is paired with the animal's own activity; and a defining feature of these adaptive filters is the negative image, or cancellation signal, that is revealed when the stimulus is withheld. One way of describing cerebellar-like function is that the ongoing activity of the projection neurons themselves is, in effect, the error that selects the parallel fiber composite that forms the required negative sensory image. (Montgomery et al.,[9] pp. 238–239)

The experiments on adaptive filtering in electric fish have led to the model illustrated in figure 23.2, taken from David Bodznick and his colleagues.[2] The data suggest that the fish cancels self-generated "noise" that results from its own EMG activity or its own electric-organ discharge. Purkinje cell-like projection neurons adapt to the self-generated signals in order to "cancel" them and thereby reveal unexpected inputs. In figure 23.2, W represents the weight matrix needed for parallel fiber–Purkinje cell synapses to represent the self-generated noise N_{sg} or some other signal that the CNS needs to cancel. This representation depends on motor commands $MC_{1...n}$, in the form of efference copy; proprioceptive inputs $P_{1...n}$; EMG activity related to ventilation $V_{1...n}$; and other relevant signals. The latter two signals, of course, represent reafference (RA). Afferent inputs include both this self-generated noise and the sensory signal of interest. The figure depicts the changes in the synaptic weight matrix that occur when the signal occurs along with the efference copy. As a consequence

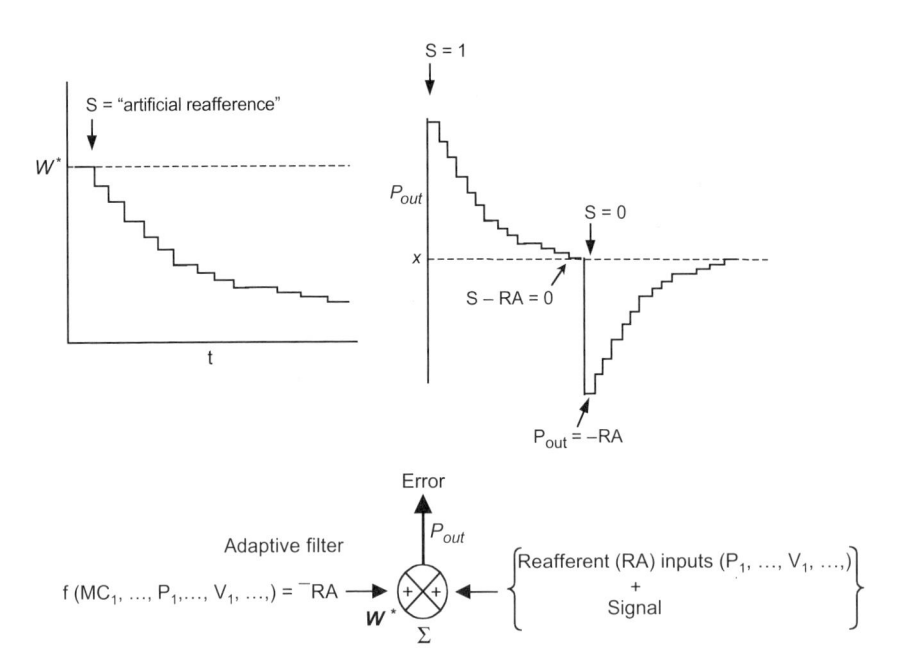

Figure 23.2
A cerebellar-like adaptive filter in electric fish. Abbreviations: RA, reafferent signals; MC, motor command signal; P, proprioceptive signal; P_{out}, output of projection neuron; S, stimulus; V, ventilation signal. W* signifies the weight of synapses at the site of plasticity, the parallel fiber–Purkinje cell synapses. (Based on the model of Bodznick et al.[2])

of the long-term depression (LTD) in synaptic strength that results from anti-Hebbian synaptic-plasticity rules, the projection-neuron output P_{out} declines. If the canceled signal goes away, a negative aftereffect remains.

The lessons taught by the sensory-relay nuclei of electric fish have important implications for motor adaptation. The way your CNS adapts often requires it to extract an "unexpected" part of a complex incoming signal from a large amount of predictable input, much of which results from your own motor commands. A mechanism like adaptive filtering shows not only how the CNS can use its motor commands to predict expected inputs, but also how it can produce a negative image of those inputs to subtract them from the incoming data stream. The negative afterimage bears an intriguing similarity to the negative aftereffects seen when perturbations, such as prismatic distortions of the visual field (see section 15.5) or imposed force fields (see section 20.1), go away.

23.2 Predicting and Responding to a Stimulus

Synapses on the projection neurons in cerebellar-like structures change so that efference copy produces a negative image of the self-generated feedback and therefore cancels that feedback in order to pass through novel sensory signals. In this case, local, anti-Hebbian rules guide learning. Signal prediction, however, plays a more general role in motor learning: Learning may be guided not by the prediction of a self-generated signal but by predictions of other neural signals, based on experience. In the projection neurons of the cerebellar cortex and the basal ganglia, this kind of learning appears to be guided by a neural "critic" that compares predictions with feedback and provides a training signal. The critic's feedback may be available only after a long string of outputs has resulted in either success or failure. The critic can change, and thus is sometimes called an **adaptive critic**.

Consider, for example, an air puff to the eye, electrical shocks to your face, or mechanical taps to your forehead, all of which activate a reflex that closes your eyelids. Section 4.2.4 introduced the concept of linking initially neutral stimuli with sensory inputs that—through innate mechanisms—trigger protective reflexes. If your CNS could predict these stimuli, then it could generate motor commands in anticipation of them. For example, if there was some signal that preceded the air puff, your CNS could predict occurrence of the air puff and avoid any of its unwanted effects by closing your eyelids in advance. A major function of the cerebellum involves not only predicting such sensory signals but also producing motor commands appropriate for them. The basal ganglia performs a similar function, but for signals that predict reward and other inputs of positive or negative biological value.

23.2.1 A Training Signal for the Cerebellum from the Inferior Olive

If a neural critic compares predicted signals against those received, then unexpected inputs should excite it much more than predicted ones. For the cerebellum, many experts believe that neurons of the inferior olivary nucleus (ION) in the brainstem serve as the critic. Recall that the ION sends climbing fibers to the Purkinje cells of the cerebellar cortex and to the deep cerebellar nuclei, and that these inputs generate complex spikes (see chapter 5). Jim Houk and his colleagues[10] reported some properties for ION neurons in cats. They observed that when cats rested, their ION neurons showed exquisite sensitivity to tactile stimuli. Furthermore, when an experimenter passively moved a cat's limb, the ION neurons fired vigorously. However, when the animal produced the same movement itself, the cells discharged little, if at all. If the cat's paw accidentally hit a barrier during the cat's own movements, however, these cells discharged vigorously. According to the ideas presented here, this response occurred because the sensory inputs that followed contact with the barrier were unpredicted (see also section 24.1.1).

The finding that ION neurons respond with excitation to unexpected stimuli was—pun intended—exciting, because a large body of animal learning theory points to the importance of "surprising" stimuli in associative learning, generally, including Pavlovian and instrumental learning

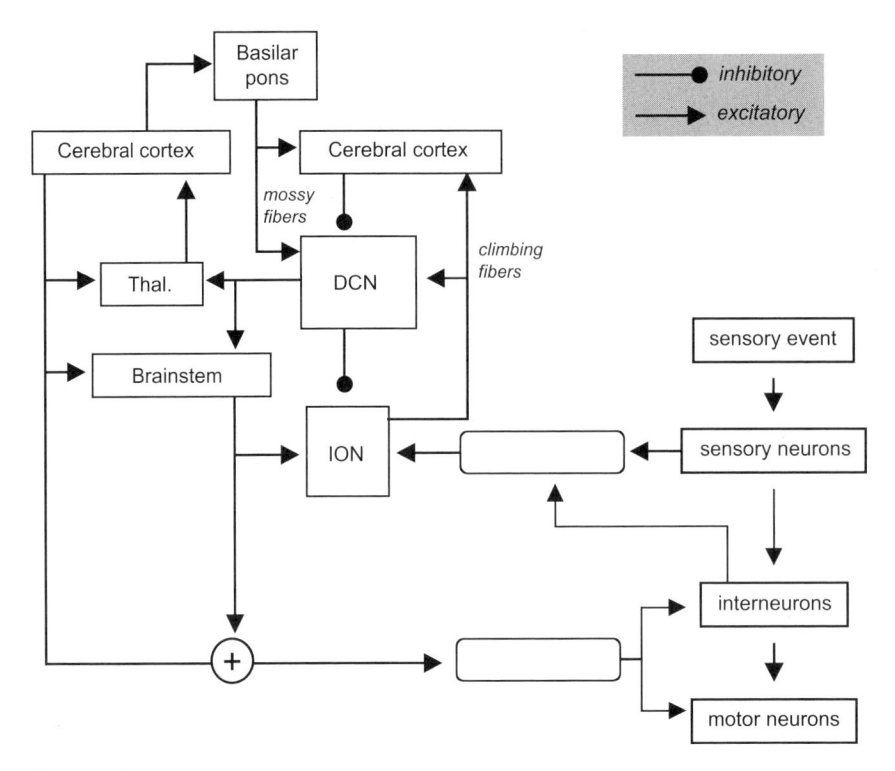

456   Chapter 23. Predicting Inputs and Correcting Errors I: Filtering and Teaching

Figure 23.3
Schematic of the cerebellar learning system. Arrowheads indicate excitatory connections. Filled circles represent inhibitory synapses. Abbreviations: DCN, deep cerebellar nuclei; ION, inferior olivary nuclei; Thal, thalamus. (Adapted from DeZeeuw et al.[13] with permission.)

(see sections 4.2.4 and 4.4.2). Figure 23.3 shows a basic block diagram of selected aspects of the cerebellar network and its interaction with brainstem and cortex, and figure 23.4 shows a sketch of the cerebellar circuits involved in Pavlovian learning. These figures primarily reflect models of cerebellar function in Pavlovian learning put forward by Mike Mauk and his colleagues Javier Medina and William Nores,[11] and by Mark Gluck, Todd Allen, Catherine Myers, and Richard Thompson.[12] The central idea of both these and other models is that the output of the ION neurons serves as a critic for teaching the cerebellar cortex.

Also central to these models is evidence showing that inhibitory input from a deep cerebellar nucleus (specifically, the interpositus nucleus) to the ION plays a necessary role both for *extinction*[11] and for a phenomenon called the *blocking effect*.[15] These findings provide a key to understanding the cerebellar mechanism of Pavlovian learning. Of course, to understand them, you need to know what these terms mean.

• *Extinction.* In Pavlovian learning, the CNS associates an originally neutral stimulus, called the conditioned stimulus (CS), with a second stim-

23.2. Predicting and Responding to a Stimulus

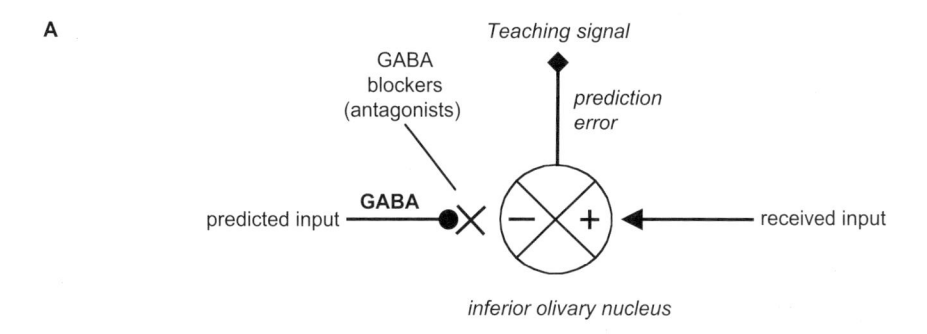

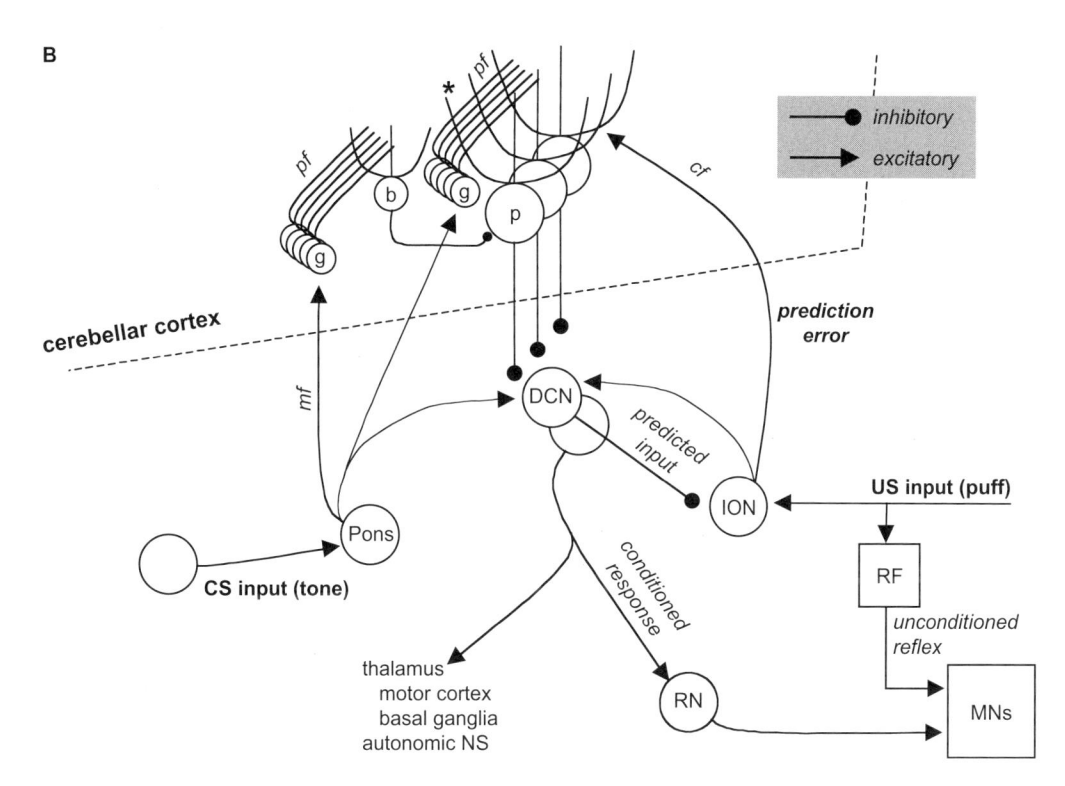

Figure 23.4
Schematic model of Pavlovian learning. Arrowheads indicate excitatory connections, and filled circles, inhibitory ones. (A) A comparator model of inferior olivary neurons. (B) Schematic model of cerebellar learning. Abbreviations: b, basket cell; cf, climbing fiber; CS, conditioned stimulus; DCN, deep cerebellar nucleus; g, granule cell; ION, inferior olivary nuclei; mf, mossy fiber; MNs, motor neurons; p, Purkinje cell; pf, parallel fiber; RF, reticular formation; RN, reticular nucleus; US, unconditioned stimulus. The conditioned response occurs when the Purkinje cells decrease their suppression of cells in the interpositus nucleus (DCN). The tone input from the mossy-fiber pathway suffices to trigger the reflex when the suppression decreases at the appropriate (predicted) time interval after tone onset. The asterisk indicates a site of synaptic plasticity. (B adapted from Steinmetz[14] with permission.)

ulus, called the unconditioned stimulus (US). Innate mechanisms program the US to elicit a reflex response, known as the unconditioned reflex (UR). Presenting the CS at a fixed time prior to the US causes the induction of a conditioned reflex or response (CR). That is, the animal responds to the CS with a CR, which resembles the UR originally caused by the US. In *extinction*, the US no longer occurs at the predicted time following the CS, and the CR accordingly declines (in effect, the CNS unlearns the association between the CS and the US).

- *The blocking effect*. A number of prominent theories of learning stress the key role that prediction plays in learning. Leon Kamin discovered that the mere occurrence of a potential CS in association with a US does not induce Pavlovian learning. Instead, learning depends upon whether the CS is unexpected or "surprising." In his classic experiment, Kamin studied two groups of rats. One group (the experienced group) heard a hissing noise (the CS) from a speaker, and a mild foot shock occurred at a fixed time later (the US), whereas a second group of rats (the naïve group) had no such experience. Both groups then received an equal amount of experience with a compound stimulus composed of the same noise and a light. Later, presentation of the light stimulus alone elicited a CR in naïve rats, but rats that had experienced the noise–shock pairing showed no response to the light. Thus, prior experience with the noise *blocked* the light as a potential CS. Kamin reasoned that this blocking effect occurred because the light was "redundant" (with the noise stimulus) in the experienced group. The ability to learn the light–shock pairing appeared to be "blocked" by virtue of this redundancy.

Jeansok Kim, David Krupa, and Richard Thompson[15] demonstrated that blocking inhibitory inputs to the ION rendered the redundant stimulus learnable (i.e., this manipulation undid the blocking effect). Mauk et al.[11] showed that a similar manipulation blocked extinction. How could the same manipulation both improve learning for redundant stimuli, which suggests an increase in neural plasticity, and prevent unlearning through extinction, which suggests the opposite?

One key to understanding this neuronal mechanism lies in appreciating the role of interpositus-nucleus neurons in both stimulus prediction and the release of motor-pattern generators. Another key involves recognizing that ION neurons act as critics that provide an error signal to the cerebellum, which reflects the difference between a prediction and what actually occurs (figure 23.4). At rest, ION neurons discharge at about 1 impulse/sec (1 Hz, for short). If only the expected happens, then they continue to discharge at 1 Hz, which, as it turns out, maintains the strength of the synapses between parallel fibers and Purkinje cells. When the ION neurons decrease their discharge, the parallel fibers–Purkinje cell synapses strengthen; when ION neurons increase their discharge, those synapses weaken. The former mechanism is called long-term potentiation (LTP), and the latter is called long-term depression (LTD). What do these mechanisms produce?

23.2.2 Eyeblink Conditioning

Imagine that a tone precisely predicts an air puff to your eye at some fixed time later. In this case, the tone is the CS and the air puff is the US. Through brainstem mechanisms involving the trigeminal nuclei, interneurons in the reticular formation, and motor neurons in the facial nucleus, evolution has hardwired a reflex response that meliorates the potentially damaging effects of the air puff. The US (the air puff) leads to a UR that closes the eyelid. The same general pathways mediate similar reflex responses to shocks, taps, or startling sounds. Eventually, after Pavlovian learning has occurred, an initially neutral stimulus such as a tone will generate a CR. The tone excites neurons in the pons, which send excitatory inputs to the interpositus neurons and to the granule cells of the cerebellar cortex (figure 23.4B). Through direct inputs to the interpositus nucleus, indirect inputs to the interpositus via the cerebellar cortex, and a series of premotor networks in the brainstem, the tone causes the generation of the CR, an eyeblink. But before that happens, the CNS must learn to associate the initially neutral stimulus with the US. The cerebellum plays a central role in this learning, but how? The answer to that question lies, in part, in predicting the US.

When the tone first occurs, the CNS does not, of course, predict the occurrence of the subsequent air puff. When the air puff unexpectedly occurs, excitatory inputs drive ION neurons to discharge at a firing rate between 3 Hz and 8 Hz. Recall that ION neurons, which give rise to the climbing-fiber inputs to the cerebellum, respond to unexpected inputs. They rest at 1 Hz, which causes no change in synaptic strength at the parallel fiber–Purkinje cell synapses. A discharge rate of 3–8 Hz causes the discharge of Purkinje cells and weakens the parallel fiber–Purkinje cell synapses through LTD. Purkinje-cell activity thus will decrease for future inputs from these parallel fibers. The next time the tone arrives, the parallel fibers will drive the Purkinje cells less, and therefore the Purkinje cells will inhibit interpositus neurons less. This relative lack of inhibitory inputs *releases* the neurons of the interpositus nucleus (at an appropriate interval from the tone's onset) and, with several repetitions of this cycle, the tone produces a blink. That is, the cerebellum learns to *predict* that the tone will be followed by an air puff, and at the appropriate time generates a response that precedes and substitutes for the UR. (Note the resemblance between this LTD mechanism and that described above for the parallel fiber–Purkinje cell synapses in sensory nuclei of electric fish in section 23.1.)

The results of Kim et al.[15] on the blocking effect can be understood in the context of this model. According to this model, the ION acts as a critic that provides an error signal to the cerebellum. The ION computes this error signal as the difference between an excitatory, sensory input reflecting the US and an inhibitory input from the interpositus nucleus that reflects a prediction of the US, as computed by the cerebellum (figure 23.4A). The ION thus acts as a comparator. **GABAergic**, inhibitory neurons transmit the "prediction" to the ION from the interpositus nucleus. Now consider again the blocking effect. Once the CNS has learned a tone–puff association, learning of a light–puff association does not take place (when the light and tone occur at the same time). According to current thinking, the *blocking effect* occurs because, after the learning of the tone–puff association has taken place, the tone already predicts the air puff completely. Therefore, in the presence of a compound stimulus—a light in addition to the tone—no error signal flows from the ION to the cerebellum: the tone generates a "prediction" signal from the interpositus to the ION that balances the excitatory inputs caused by the air puff. With no (positive) error signal from ION to the cerebellum, the light–puff association cannot be learned. The critic has, in a sense, informed the cerebellum that there is nothing to learn, which results in the blocking effect.

The model also accounts for the finding that GABA blockade of the ION enhances the learnability of the redundant stimulus (the light)—it undoes the blocking effect. If GABA blockers prevent the prediction signal from reaching the ION (figure 23.4A), the air puff should generate a positive error signal in the ION cells, and the cerebellum can then learn the light–puff association, even in animals previously trained on the tone–puff association. Kim et al.[15] confirmed this prediction. GABA antagonists injected into the ION block the inhibitory prediction signal coming from the interpositus nucleus. Accordingly, the puff-related sensory input to the ION is no longer balanced by the inhibition coming from the interpositus nucleus. The puff again becomes "surprising" (i.e., unpredicted) through this experimental manipulation, and therefore subject to association with a new CS. Blocking is reversed and the learnability of the redundant CS is restored. You can remember this finding and the model by using a mnemonic device: "GABA blockers block blocking."

Research on developing rats supports this model further. Daniel Nicholson and John Freeman[16] studied the development of activity in Purkinje cells and synaptogenesis in the ION. The emergence of eye-blink conditioning between the 17th and 24th days of a rat's life corresponds to the initial appearance of inhibitory input from the Purkinje cells to the ION.

The same model explains the lack of *extinction* when GABA antagonists block inhibition in ION neurons.[11] When the tone occurs but the air puff does not, the "predicted input" signal (the output of the interpositus neurons) inhibits ION, but an equal excitation from the air puff does not match this inhibition. Therefore, ION activity falls to nothing. In the absence of climbing-fiber activity, the synapses between the parallel fibers and Purkinje cells strengthen through LTP. This causes the tone to drive the Purkinje cells a little more strongly the next time it occurs, which suppresses the activity of the interpositus neurons. This suppression amounts to a weakening of the prediction that an air puff will occur. With repeated failures in its predictive value, the system will learn to predict nothing and, accordingly, the CR stops. Extinction has occurred. Pharmacological blockade of the GABAergic signal from the interpositus nucleus to the ION blocks the prediction signal, the ION activity does not fall to 0, and unlearning does not occur.

The molecular mechanisms by which learning and extinction occur have some interesting features. Recall that initial learning of the association depends on LTD, whereas extinction relies on LTP. The strengthening of the parallel fiber–Purkinje cell synapse through LTP appears to occur presynaptically (i.e., in mechanisms involving the parallel fiber's side of the synapse), whereas the weakening of the synapse through LTD occurs postsynaptically (i.e., on the Purkinje cell's side of the synapse). LTP extinguishes the response without undoing the LTD that underlies learning, at least for some time. Relearning therefore can occur more quickly than initial learning, through a speedy removal of the presynaptic LTP to unmask the effects of the still-present LTD. (Figure 23.6B shows a distinction between learning and unlearning for reward prediction in the basal ganglia.)

Of course, you should expect significant differences in the mechanisms of Pavlovian learning and those involved in other forms of motor learning, such as those involved in reaching and pointing. It seems likely, however, that the mechanisms of Pavlovian learning offer important clues to cerebellar mechanisms in motor learning generally.[17,18] Chapter 24 takes up this discussion.

23.2.3 Further Evidence That the Cerebellum Plays a Role in Predicting Neural Signals

In a brain-imaging study on Pavlovian learning, Narender Ramnani and his colleagues[19] showed that as human participants learned the tone–air puff association, local rates of blood flow changed in the ipsilateral cerebellar cortex in correlation with the learning of the response. When the predicted air puff did not occur (i.e., on catch trials), there was an increased blood flow in the contralateral cerebellum. It has been shown that the cerebellum controls eye-blink reflexes, so this finding provides some support for the idea that the human cerebellar cortex receives an error signal of the sort postulated in the model.

In a different experiment, monkeys with cerebellar lesions showed impairment on a serial reaction-time task, in which they learned to press a series of targets in a complex sequence (see also section 25.4.1), and similar observations have been made in patients with cerebellar lesions. A study by Phil Nixon and Dick Passingham[20,21] showed that this deficit was not specific for sequences of movements, but instead reflected a more general deficit in using sensory inputs to predict other sensory inputs. In their experiment, they trained monkeys to make rapid reaching movements to a single target on a video monitor. Sometimes the target could appear at any of three possible locations, and in that sense had spatial unpredictability. At other times, the target had complete spatial predictability—it appeared in the same location for a long sequence of consecutive movements—but it had temporal unpredictability. At still other times, the target appeared with both spatial and temporal predictability.

Normally, spatial and temporal predictability both result in faster reaction times, compared to movements made in the absence of such predictability. Following lesions that included the dentate and interpositus nuclei, these reaction-time savings disappeared. In other words, monkeys with lesions in those deep cerebellar nuclei had deficits in predicting sensory inputs, perhaps in "learning to prepare responses to predictable sensory events" (Nixon and Passingham[20]). This deficit in monkeys helps explain why cerebellar damage disrupts both sequential reaching movements and Pavlovian learning: Both depend on the prediction of sensory inputs.

In a related recent study, Richard Ivry and his colleagues[22] showed that the deficits incurred after cerebellar damage apply only to discontinuous movements. Cerebellar patients appear normal in their ability to make continuous, rhythmic movements. These investigators concluded that the cerebellum becomes necessary when the movement requires predicting events. To put this result in perspective, think about a continuous movement in terms of a control policy used by a next-state planner (see section 19.3.2). Recall that a *control policy* (or *feedback control policy*), in this sense, corresponds to a set of parameters for a next-state planner, and that these parameters remain constant during a movement. For discrete movements, the parameters change from one movement to the next. The results of Ivry et al. suggest that you need your cerebellum for predicting *when* to change those parameters. As long as you do not need to change the parameters, or the basis for such a change has no temporal predictability, you do not need your cerebellum for planning that movement.

23.3 Similar Learning Mechanisms in Basal Ganglia and Cerebellum

The previous section outlined a model of one kind of Pavlovian learning. It focused on cerebellar mechanisms and depended on the prediction of neural inputs, specifically the prediction of reflex-triggering inputs. However, Pavlovian learning is not the exclusive province of the cerebellum: Basal ganglia mechanisms also contribute to certain kinds of Pavlovian learning.

The basal ganglia and cerebellum have long been thought to play an important role in motor learning, including associative learning of the kind typified by Pavlovian conditioning, and both probably play a more general role in learning and memory as well. On the surface, the basal ganglia and cerebellum seem to have few anatomical or physiological features in common, and you might therefore assume that their learning mechanisms would differ dramatically.

As suggested in figures 6.2 and 23.5A, however, the basal ganglia and cerebellum have parallels in their functional architecture. For instance, both the striatum and the cerebellar cortex have GABAergic projection neurons that probably act as pattern recognizers.[23] In the cerebellum, these projection neurons are the Purkinje cells of the cerebellar cortex; in the basal ganglia, they are the medium spiny cells of the striatum. Projection neurons in both the cerebellar cortex and the striatum respond to a pattern of synaptic inputs from a large number of input elements according

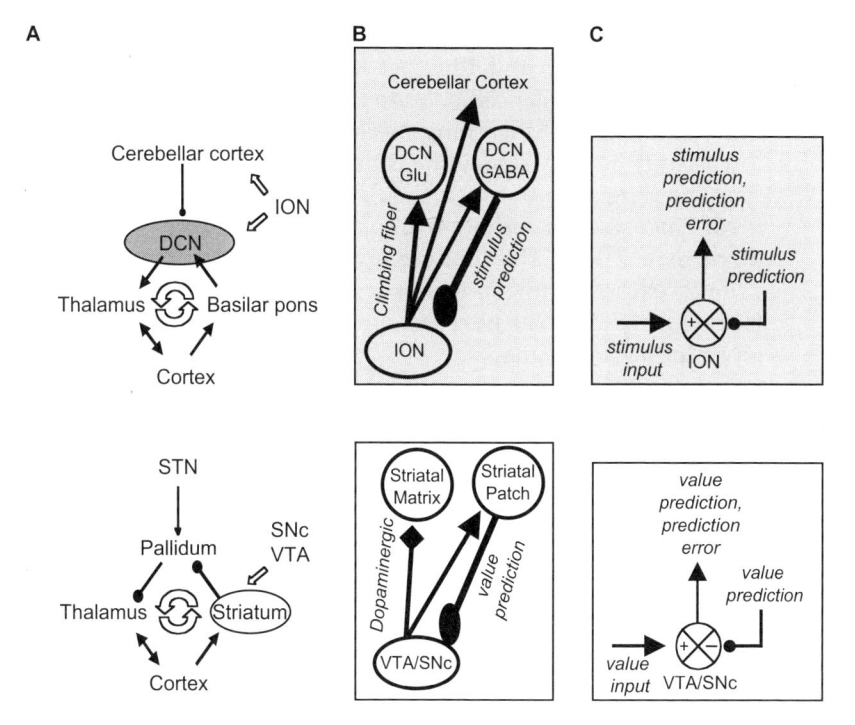

23.3. Similar Learning Mechanisms in Basal Ganglia and Cerebellum

Figure 23.5
Schematic of the cerebellar and basal ganglia learning system. (*A*) Recurrent modules, or "loops," for both the cerebellum and the basal ganglia, as described in section 6.2.1. Abbreviations: DCN, deep cerebellar nucleus; ION, inferior olivary nuclei; SNc, substantia nigra, compact part; VTA, ventral tegmental area. (*B*) Teaching signals and their inputs from predictors (i.e., their "students"). Arrowheads indicate excitatory synapses, circles indicate inhibitory ones, and diamonds indicate other kinds. Abbreviation: glu, glutamatergic neurons. (*C*) Comparator model of teachers.

to prior adjustments of their synaptic weights. In both systems, synaptic weights are adjusted through the influence of signals from a critic. And in both systems, the training signals from that critic are thought to reflect the difference—an error—between received inputs and predicted inputs (figure 23.5B and 23.5C). For the basal ganglia, the input from dopaminergic (DA) neurons provides the error signal;[24] for the cerebellum, it arises from neurons in the ION. Both systems also have a strong "return" projection to the critic (figure 23.5B). In the basal ganglia, this return projection comes to DA neurons from small groups of striatal medium spiny neurons in the *patches* or *striosomes*[25] (see chapter 6). In the cerebellum, this "return" projection comes to ION neurons directly from the deep cerebellar nuclei or indirectly via the small-cell part of the red nucleus[14] (see chapter 5).

In both systems, the "return projection" mentioned above is thought to carry a prediction signal, and DA and ION neurons both compute the error between that prediction and the inputs that actually occur (figures 23.4A and 23.5C), and thus act as comparators. GABAergic inhibitory neurons convey this prediction to the critic in both systems. Another common feature of their learning mechanism (see section 23.2.2) involves the low, steady activity rate of the comparator neurons, with a dynamic range (i.e., the difference between their highest and lowest discharge rate) that favors increases in activity (up to 8 Hz) over decreases in activity (rarely more than 2 Hz). This finding accords with the fact that learning rates usually exceed extinction rates in Pavlovian and instrumental learning (see figure 23.6B). The next section presents some of the evidence for these ideas for basal ganglia.

23.4 A Training Signal for the Basal Ganglia

Most expert opinion holds that the basal ganglia functions in the selection and initiation of movements, the regulation of continuous movements, the control of movement sequences and of other automated actions, and the scaling of movement parameters. Micrographia in Parkinson's disease serves as a prototypical example for the scaling concept. In these patients, the writing movements have approximately the correct form, but the patients' script becomes very small as the disease progresses. One model[26] discussed in chapter 19 posits that the basal ganglia provides a scaling ("go") signal to the motor cortex. Complementing these traditional views of basal ganglia function, other proposed functions involve different concepts: reward prediction, chunking of event or movement sequences, reducing the number of dimensions of inputs to the basal ganglia, and a host of competing ideas.

Another, admittedly speculative, way to view these ideas is to posit that the basal ganglia makes an essential contribution to the function of the next-state planner (section 19.3) by breaking down long-range goals into plans that describe the desired state of the limb in the near future. By learning to compute this transformation, the basal ganglia learns to achieve ultimate goals through intermediate steps and to predict the outcome of a series of such steps at the outset. The presentation here focuses on the role of the basal ganglia and the DA system in predicting outcomes and evaluating the accuracy of those predictions.

23.4.1 Dopamine Cells Signal Predicted Reward and Errors in Reward Predictions

The earliest studies of neuronal activity in the dopaminergic (DA) cells of the midbrain led to few published reports. The investigators conducting those experiments often reported frustration over the fact that the neurons never seemed to "do anything." For example, the DA cells discharged at about 2–4 Hz, but this rate never seemed to change. Although they did not recognize it at the time, the people who made these observations had actually stumbled upon a crucial and important property of DA cells. The earliest recordings were made in monkeys that performed highly **overlearned** movements—a simple flexion and extension of the arm—in a highly stereotyped and repetitive manner. The monkeys performed the task perfectly and learned nothing during the recording sessions.

When Wolfram Schultz and his colleagues[29] reexamined the activity of midbrain DA cells years later, they placed their monkeys in a situation involving learning, and the cells showed significant modulation in their activity level. They found that these cells carried a signal that conveyed an error in the prediction of *reward* (i.e., feedback about goal achievement and outcome in biological terms). The monkeys in the early studies never made errors and never received an unexpected reward; hence the cells never registered any error in reward prediction. As figure 23.6A shows, when an unexpected reward occurs, the DA cells increase their discharge rates. After repeated experience, the cells become unresponsive to rewards per se, but discharge after signals that predict reward. The omission of expected rewards leads to a decrease in the firing rate of DA neurons.[30] Thus, DA cells seem to encode both errors in predicted reward and events that predict reward. They also carry a signal that reflects the receipt of a reward under certain circumstances.

These basic results have recently been confirmed in experiments with rats. The experimenters measured the amount of dopamine released in the nucleus accumbens, a part of the striatum.[31] Brief puffs of dopamine were released just before rats approached a lever that would, when pressed, produce a reward. As the rats physically approached the lever, dopamine levels continued to increase, and peaked within seconds of the lever press. In rats trained to associate an initially neutral cue with the reward, those cues alone caused an increase in dopamine release, much as the cues come to evoke activity in dopaminergic cells when they become associated with reward (figure 23.6A).

23.4.2 Caudate Neurons Have a Reward-Prediction Signal

Okihide Hikosaka and his colleagues[28,32] studied the prediction of reward by striatal neurons (figure 23.6B). In their experiment, the monkey had to make either a left or a right saccadic eye movement, cued by a light spot, in order to keep the trials coming. However, for a block of consecutive trials, only movements to the left target would produce a reward. Then the experimenters reversed that *contingency*, and after the reversal, only movements to the right target produced reward. The activity of striatal neurons related to leftward saccades increased after just one trial when a leftward target became associated with reward. In conjunction with that increase in activity, the monkeys responded faster to that cue. Then, after the experimenters switched the reward contingency so that rightward, not leftward, movements produced rewards, the cell slowly returned to its original, lower level of activity, as the monkey responded more slowly for the leftward target.

Now examine the activity of a caudate cell when leftward saccades first produced a reward (left arrow in figure 23.6B), as opposed to when

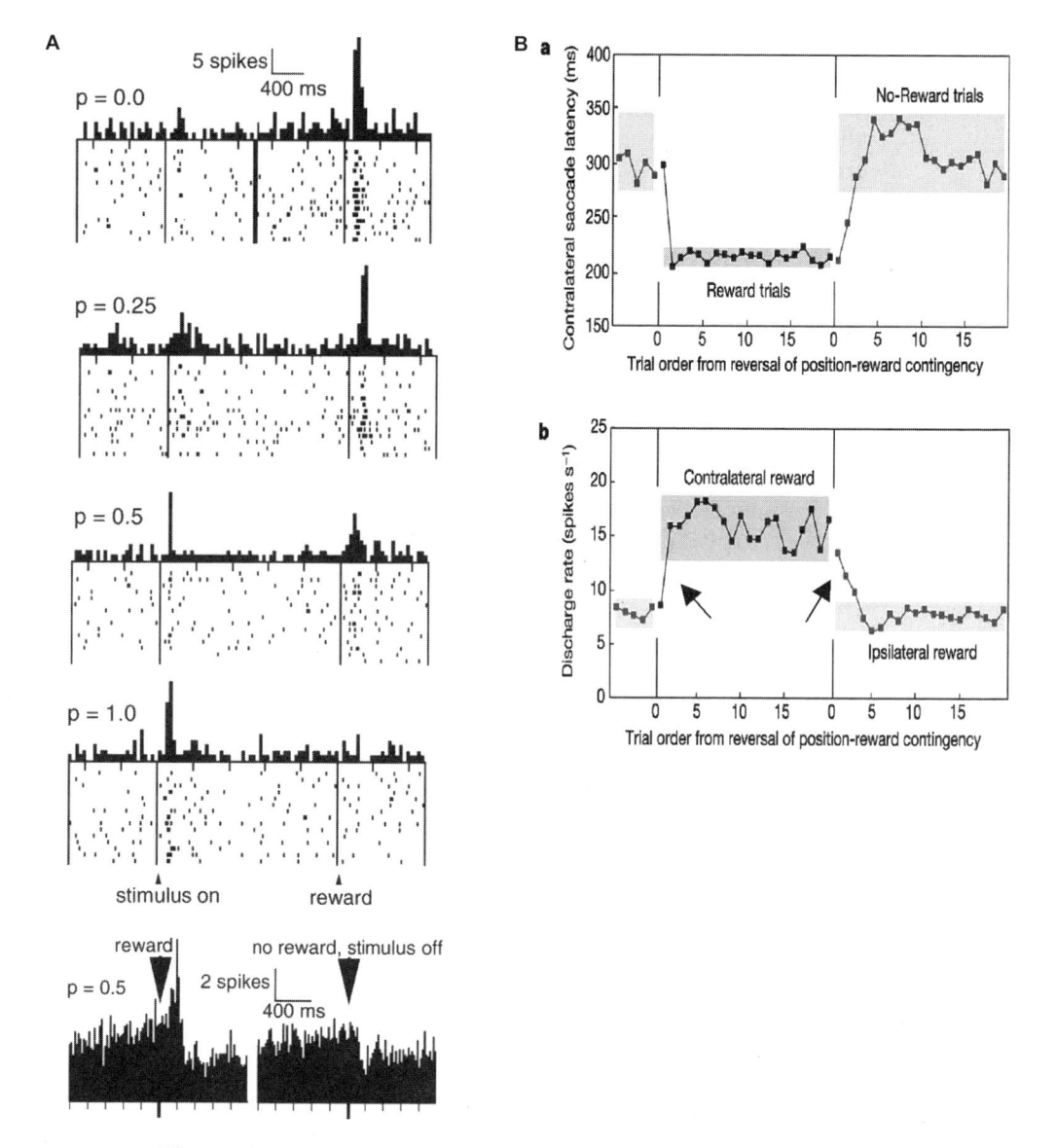

Figure 23.6

Reward prediction. (*A*) A dopamine neuron that initially (top) responded to unexpected rewards that occurred in association with a visual cue, but gradually adapted to this predicted reward, responding instead to the visual cue, which predicted reinforcement. At the bottom, the cell shows excitation when rewards exceeded expectations, and inhibition when predicted rewards did not occur. Abbreviation: p, probability of reward. (*B*) Activity of a neuron in the caudate nucleus as a monkey made saccades to a leftward target. At the beginning of the plot, the leftward saccade did not produce reward, but a rightward saccade did. The monkey reacted to the onset of the left target with a response latency of approximately 300 msec, and the cell discharged at about 8 impulses per sec. On the trial following the left vertical line (the first 0 on the *x*-axis), the leftward saccades produced a reward. After the monkey experienced this contingency once, its reaction time quickened by nearly 100 msec and the cell's discharge rate nearly doubled. When the contingency reversed again (at the second 0 on the *x*-axis), it took longer for the cell's discharge rate to "de-adapt" than it took to adapt. (*A* from Fiorillo et al.,[27] *B* from Lauwereyns et al.[28] with permission.)

the same movement (to the same target) first failed to produce an expected reward. Note that for both caudate cell activity and the monkey's reaction time, learning the association between leftward saccades and rewards had a faster effect than learning that they were not associated (for the time being)—that is, extinction. The changes for learning occurred in just one trial; extinction took at least three, and probably four, trials. As noted above, the baseline discharge of DA cells was so low that the training signal could increase for positive errors more than it could decrease for negative errors. The different rates of learning and unlearning probably resulted from the fact that the DA neurons could reduce their discharge to 0 from a resting level only 2–4 Hz, but could increase their activity much more.

23.4.3 Basal Ganglia During Pavlovian Learning

Section 23.2 presented a model of Pavlovian learning in the cerebellum, but not all such learning depends on the cerebellum. Recall from section 4.2.4 that Pavlovian learning comes in several types. Some types involve responses of a part of the body, such as an eyelid or a limb, to aversive stimuli. Those types of Pavlovian learning depend on the cerebellum. Other types of Pavlovian learning depend on other kinds of innate reflexes, such as **autonomic** ones, to both aversive and appetitive stimuli. The behavior referred to in section 4.2.4 as "Pavlovian approach" depends on basal ganglia and involves reflexes triggered by appetitive stimuli. Behaviors such as fear conditioning involve aversive stimuli and depend on the basal ganglia and amygdala.

A number of studies have demonstrated that striatal cells show learning-related activity during Pavlovian conditioning. For example, a specific population of neurons within the striatum, known as tonically active neurons, or TANs, have activity related to Pavlovian eye-blink conditioning. At first glance, this result seems odd: Pavlovian learning of the type involving protective reflexes, such as eye blinks, does not require the basal ganglia. This kind of Pavlovian learning depends on the cerebellum.[14] But put that notion aside for the time being. TANs are believed by many to correspond to the cholinergic interneurons that constitute ~5% of the striatal cell population.[33] They have been shown to respond to conditioned stimuli that are associated with either aversive stimuli[34] or appetitive ones,[35] and to appetitive stimuli per se.[34,36]

Ann Graybiel and her colleagues, Pablo Blazquez, Jun Kojima, and Noataka Fujii,[37] recorded from striatal neurons in monkeys during either appetitive or aversive Pavlovian-learning tasks. In addition to finding responses to aversive stimuli (air puffs) and appetitive ones (water), they noted that as monkeys learned each association (CS–air puff or CS–water), more TANs responded to the CS. Further analysis of the population responses of TANs revealed that their activity correlated with the probability of occurrence of the conditioned response. Graybiel and her colleagues concluded that TANs encode the probability that a given stimulus will elicit a behavioral response.

If eye-blink conditioning depends on the cerebellum rather than the striatum, why would cells in the striatum, interneurons or not, reflect the probability of a response? The most likely possibility, according to Joseph Steinmetz,[14] is that the basal ganglia use information about the performance of these protective reflexes in order to incorporate them into ongoing sequences of behavior. As is always the case with neurophysiological data, a cell's activity may be "related" to a behavior for many reasons, only one of which involves acting as causal agents in mediating that behavior. Section 25.4.1 takes up the role of the basal ganglia and cerebellum in sequential movements.

23.4.4 Basal Ganglia, Pavlovian Learning, and the Blocking Effect

Wolfram Schultz and his colleagues[38] predicted that because DA neurons reflect errors in reward expectancy, they should show properties corresponding to the blocking effect, much as described above for studies of the cerebellum and its ION critic (see section 23.2.2). That is, blocking of CS–US learning for redundant stimuli should be reflected in DA neurons. As predicted, many striatal neurons responded to a nonredundant stimulus, but not to a redundant stimulus, and no neuron showed the opposite result. These cells predicted reward in the same way that the monkeys predicted reward, consistent with the adaptive-critic theory: DA cells encode reward-prediction errors (along with other aspects of reward prediction and uncertainty[27]).

23.5 Why Does Huntington's Disease Result in Disorders in Reaching?

In section 19.5, you saw that patients with Huntington's disease (HD) and asymptomatic people carrying the gene for that disease have jerky movements and fail to adjust properly for errors in their movement trajectory. However, if their movements begin accurately, their trajectories closely resemble the trajectories of people without either the disease or the gene. How can those findings be understood in the context of the ideas presented in this chapter? (We can only offer speculation, of course.)

Before the speculation begins, however, a couple of facts and an inference are worth noting. Early in HD, degeneration begins in the striatal "patches" (see section 19.5), and patch neurons inhibit DA neurons (figure 23.5B, bottom). It is likely that asymptomatic gene carriers have a dysfunction mainly in the striatal patches or in the corticostriatal inputs to patch neurons.

According to the adaptive-critic theory, the "return" projection from the striatum to the DA neurons serves as the pathway conveying a "prediction" signal computed by the striatum (figure 23.5B and 23.5C). We do not know the nature of the prediction signal, but it could involve something about the biological value of the current limb state (e.g., its location and velocity). For example, this signal may convey information about the likelihood that an ongoing movement will succeed in reaching a goal. According to this idea, the striatal patch cells may convey a real-time measure of the likelihood of success as the movement unfolds. In this framework, the loss of a proportion of the striatal patch cells early in HD would result in a pathological reduction in the magnitude of the "prediction" signal to the DA neurons. In other words, as the movement progresses, its likelihood of success will be underestimated.

A real-time measure of the likelihood of success could be important because it might signal whether the next-state planner (figure 19.4 and section 19.3) should continue with the current *feedback control policy* or should abandon it and try something else. This change takes time, of course, and 300 msec or so seems reasonable for such a complex, feedback-dependent computation. For example, if a movement starts by going in a slightly wrong direction, the likelihood of success will drop as the movement progresses, especially if success requires reaching the target within a relatively short time. If the expected success for the current control policy is low, it is reasonable to change that policy in the middle of a movement. A change in control policy would be likely to result in increased jerkiness in the movement. However, when a movement progresses accurately, the likelihood-of-success signal could be large enough to maintain the current control policy.

According to this idea, the striatum detects the context for altering an ongoing feedback control policy in order to increase the likelihood of success. In the absence of disease, the original control policy would remain in effect for small errors, and the next-state planner would smoothly guide the end effector to its goal. In healthy people, larger movement errors would lead both to a new control policy and a signal from the critic that promotes learning. In early HD, and especially in asymptomatic gene carriers, an underestimate of the chances of success would lead to control policies that change too often, for errors that are too small, thus generating an excessive number of corrective submovements. Kevin Novak, Lee Miller, and Jim Houk[39,40] have also suggested that the basal ganglia play a central role in these kinds of corrective submovements.

In summary, this chapter has emphasized the similarities between the basal ganglia and the cerebellum in learning mechanisms, but there is also a division of labor between them. Chapter 25 returns to the idea that the basal ganglia plays a large role in context-dependent goal selection (decisions). Chapter 24 deals with the idea that the cerebellum learns to correct motor errors made subsequent to a decision. The disorders of Huntington's disease, at first glance, do not appear to accord with these ideas. However, if you view a control policy—moving fast or slow, straight or zigzag, and so forth—as a goal, then the effects of basal ganglia disease seem more sensible.

References

1. Bell, CC (2002) Evolution of cerebellum-like structures. Brain Behav Evol 59, 312–326.

2. Bodznick, D, Montgomery, JC, and Carey, M (1999) Adaptive mechanisms in the elasmobranch hindbrain. J Exp Biol 202, 1357–1364.

3. Bell, CC (2001) Memory-based expectations in electrosensory systems. Curr Opin Neurobiol 11, 481–487.

4. Bell, CC, and Grant, K (1992) Sensory processing and corollary discharge effects in mormyromast regions of mormyrid electrosensory lobe. II. Cell types and corollary discharge plasticity. J Neurophysiol 68, 859–875.

5. Appollonio, IM, Grafman, J, Schwartz, V, et al. (1993) Memory in patients with cerebellar degeneration. Neurology 43, 1536–1544.

6. Bell, CC, Han, VZ, Sugawara, Y, and Grant, K (1997) Synaptic plasticity in a cerebellum-like structure depends on temporal order. Nature 387, 278–281.

7. Bastian, J (1999) Plasticity of feedback inputs in the apteronotid electrosensory system. J Exp Biol 202, 1327–1337.

8. Bell, CC, Caputi, A, Grant, K, and Serrier, J (1993) Storage of a sensory pattern by anti-Hebbian synaptic plasticity in an electric fish. Proc Natl Acad Sci USA 90, 4650–4654.

9. Montgomery, J, Carton, G, and Bodznick, D (2002) Error-driven motor learning in fish. Biol Bull 203, 238–239.

10. Gellman, R, Gibson, AR, and Houk, JC (1985) Inferior olivary neurons in the awake cat: Detection of contact and passive body displacement. J Neurophysiol 54, 40–60.

11. Medina, JF, Nores, WL, and Mauk, MD (2002) Inhibition of climbing fibres is a signal for the extinction of conditioned eyelid responses. Nature 416, 330–333.

12. Gluck, MA, Allen, MT, Myers, CE, and Thompson, RF (2001) Cerebellar substrates for error correction in motor conditioning. Neurobiol Learn Memory 76, 314–341.

13. De Zeeuw, CI, Hoogenraad, CC, Koekkoek, SKE, et al. (1998) Microcircuitry and function of the inferior olive. Trends Neurosci 21, 391–400.

14. Steinmetz, JE (2000) Brain substrates of classical eyeblink conditioning: A highly localized but also distributed system. Behav Brain Res 110, 13–24.

15. Kim, JJ, Krupa, DJ, and Thompson, RF (1998) Inhibitory cerebello-olivary projections and blocking effect in classical conditioning. Science 279, 570–573.

16. Nicholson, DA, and Freeman, JH (2003) Addition of inhibition in the olivocerebellar system and the ontogeny of a motor memory. Nat Neurosci 6, 532–537.

17. Bracha, V, Kolb, FP, Irwin, KB, and Bloedel, JR (1999) Inactivation of interposed nuclei in the cat: Classically conditioned withdrawal reflexes, voluntary limb movements and the action primitive hypothesis. Exp Brain Res 126, 77–92.

18. Maschke, M, Erichsen, M, Drepper, J, et al. (2002) Limb flexion reflex-related areas in human cerebellum. NeuroReport 13, 2325–2330.

19. Ramnani, N, Toni, I, Josephs, O, et al. (2000) Learning- and expectation-related changes in the human brain during motor learning. J Neurophysiol 84, 3026–3035.

20. Nixon, PD, and Passingham, RE (2001) Predicting sensory events: The role of the cerebellum in motor learning. Exp Brain Res 138, 251–257.

21. Nixon, PD, and Passingham, RE (2000) The cerebellum and cognition: Cerebellar lesions impair sequence learning but not conditional visuomotor learning in monkeys. Neuropsychologia 38, 1054–1072.

22. Spencer, RM, Zelaznik, HN, Diedrichsen, J, and Ivry, RB (2003) Disrupted timing of discontinuous but not continuous movements by cerebellar lesions. Science 300, 1437–1439.
23. Houk, JC, and Wise, SP (1995) Distributed modular architectures linking basal ganglia, cerebellum, and cerebral cortex: Their role in planning and controlling action. Cerebral Cortex 5, 95–110.
24. Suri, RE, Bargas, J, and Arbib, MA (2001) Modeling functions of striatal dopamine modulation in learning and planning. Neuroscience 103, 65–85.
25. Gerfen, CR, and Wilson, CJ (1996). In Handbook of Chemical Neuroanatomy, Vol 12, Integrated systems of the CNS, Part III, eds Björklund, A, Hökfelt, T, and Swanson, LW (Elsevier Science, Amsterdam), pp 369–466.
26. Bullock, D, and Grossberg, S (1988). In Dynamic Patterns in Complex Systems, eds Kelso, JAS, Mandell, AJ, and Shlesinger, MF (World Scientific Publishers, Singapore).
27. Fiorillo, CD, Tobler, PN, and Schultz, W (2003) Discrete coding of reward probability and uncertainty by dopamine neurons. Science 299, 1898–1902.
28. Lauwereyns, J, Watanabe, K, Coe, B, and Hikosaka, O (2002) A neural correlate of response bias in monkey caudate nucleus. Nature 418, 413–417.
29. Schultz, W, and Dickinson, A (2000) Neuronal coding of prediction errors. Annu Rev Neurosci 23, 473–500.
30. Schultz, W (1998) Predictive reward signal of dopamine neurons. J Neurophysiol 80, 1–27.
31. Phillips, PE, Stuber, GD, Heien, ML, et al. (2003) Subsecond dopamine release promotes cocaine seeking. Nature 422, 614–618.
32. Itoh, H, Nakahara, H, Hikosaka, O, et al. (2003) Correlation of primate caudate neural activity and saccade parameters in reward-oriented behavior. J Neurophysiol 89, 1774–1783.
33. Aosaki, T, Tsubokawa, H, Ishida, A, et al. (1994) Responses of tonically active neurons in the primate's striatum undergo systematic changes during behavioral sensorimotor conditioning. J Neurosci 14, 3969–3984.
34. Ravel, S, Legallet, E, and Apicella, P (1999) Tonically active neurons in the monkey striatum do not preferentially respond to appetitive stimuli. Exp Brain Res 128, 531–534.
35. Kimura, M (1986) The role of primate putamen neurons in the association of sensory stimuli with movement. Neurosci Res 3, 436–443.
36. Apicella, P, Legallet, E, and Trouche, E (1997) Responses of tonically discharging neurons in the monkey striatum to primary rewards delivered during different behavioral states. Exp Brain Res 116, 456–466.
37. Blazquez, PM, Fujii, N, Kojima, J, and Graybiel, AM (2002) A network representation of response probability in the striatum. Neuron 33, 973–982.
38. Waelti, P, Dickinson, A, and Schultz, W (2001) Dopamine responses comply with basic assumptions of formal learning theory. Nature 412, 43–48.
39. Novak, KE, Miller, LE, and Houk, JC (2003) Features of motor performance that drive adaptation in rapid hand movements. Exp Brain Res 148, 388–400.
40. Novak, KE, Miller, LE, and Houk, JC (2002) The use of overlapping submovements in the control of rapid hand movements. Exp Brain Res 144, 351–364.

41. Blakemore, SJ, Frith, CD, and Wolpert, DM (2001) The cerebellum is involved in predicting the sensory consequences of action. NeuroReport 12, 1879–1884.

Reading List

Suri et al.[24] present a very interesting model of appetitive learning. Blakemore et al.[41] have explored the role of the cerebellum in predicting neural signals based on motor commands.

24 Predicting Inputs and Correcting Errors II: Learning from Reflexes

*Overview: When stimuli engage your reflexes, your CNS generates signals that guide learning in the cerebellum. In some cases, these stimuli are externally generated (e.g., an air puff to the eye, which produces the eye-blink reflex discussed in chapter 23). In many cases, however, the inputs result from your own actions. For example, motor commands for moving your forearm around the elbow produce **torques** that, due to **inertial** properties of the arm, also move your upper arm around the shoulder. If your goal is to move only your forearm, the movements of your upper arm are motor errors. The cerebellum learns to predict these errors and to produce motor commands that compensate for them. The cerebellum plays an essential role in learning internal models of dynamics.*

Reflexes are **primitive**, often inflexible mechanisms for responding to certain stimuli. For example, the **short-loop stretch reflex** is a spinal circuit that provides a fast neural pathway for your CNS to respond to unexpected stretching of a muscle (section 8.8). However, reflexes not only generate fast responses to unexpected stimuli; they also act as a sort of "wakeup call" for the cerebellum. According to the view presented here, the cerebellum learns the patterns of neural input that predict reflex-eliciting stimuli, especially when those reflexes compensate for undesired movements. By recognizing those inputs—much as it does for the CSs that predict USs in Pavlovian learning (section 23.2)—the cerebellum learns to anticipate undesirable neural inputs and to generate the motor commands that prevent their occurrence.

For example, imagine that you are holding a handle and it suddenly moves. If that movement stretches your biceps, stretch reflexes activate that muscle to resist the stretch. However, if a tone consistently predicts the movement of the handle, then you would learn to activate your biceps in advance of the anticipated stretch. Now extend this idea beyond maintaining a steady posture, as in stretch reflexes, and generating protective reflex responses, as in certain forms of Pavlovian learning. Consider making a movement with some novel and weighty object in your hand. As you reach with the object, your CNS cannot accurately estimate the object's dynamics, and therefore your hand moves in an unexpected way. These unexpected movements generate stretch-related reflexes. The stretch-related reflexes occur because the weight has introduced errors; muscles changed lengths in an unexpected way. Recall from chapter 7 that when movements progress as planned, coactivation of inputs to muscles and muscle-spindle afferents leads to a null signal from primary muscle-spindle afferents and a predictable muscle-length signal from secondary muscle-spindle afferents.

This chapter presents the idea that these error signals, which indicate a deviation from the desired feedback, guide learning in your cerebellum. Responding to its critic in the ION, the cerebellum learns to *augment* the existing motor commands, such as those generated by motor cortex or brainstem nuclei. This additional motor command corrects for unexpected movements.

In the examples presented in chapter 23, external stimuli such as lights and tones predicted the neural signals that generated reflexes. More generally, however, **efference copy** and other neural signals, including proprioceptive ones, can lead to important predictions. Your skeleton has a complex series of linkages that produce coupled dynamics. For example, consider the act of rapidly flexing your elbow. The flexion torques on the elbow joint accelerate your forearm. That is the desired effect of your motor command. But because of the forearm's inertia, its acceleration produces torques on the shoulder. These **interaction torques** have the undesired effect of accelerating your upper arm and shoulder. If you do not want your upper arm to move, stretch reflexes on the shoulder muscles are activated, which respond to the undesired consequences of your own motor commands. But efference copy of the motor commands can predict that such reflexes will occur, and your CNS can learn to compensate for them in advance. Figure 20.6 shows an example of such predictive compensation.

As a second example, consider something that moves your head as you fixate an object. Head movements carry your eyes along and could cause images on your retina to slip. Fortunately, you have a reflex developed by your most remote vertebrate ancestors (see box 11.1), called the vestibulo-ocular reflex (VOR, for short). The VOR rotates the eyes in the direction opposite to a head movement in order to maintain image stability.

Note some similarities between these two kinds of reflexes. Interaction torques engage stretch reflexes that activate muscles opposing the torques. Head movements engage reflexes that activate muscles opposing those movements. In each case, an error of some kind plays an important role in learning to regulate the reflex. The reflex not only corrects the error, but the reflex's motor commands could, in theory, teach the cerebellum how to prevent the error. If the cerebellum could learn to recognize the signals that predict the error, it might also learn to produce motor commands that compensate for and preempt these errors. This chapter presents the idea that the cerebellum learns to supplement existing motor commands, adding to them or subtracting from them, so that the net effect is a correct internal model of dynamics (i.e., network 5 in figure 19.4). Although, on this view, the cerebellum is not necessary for generating a movement, it is necessary for making a coordinated movement.

Unfortunately, the role of the cerebellum in the learning of reaching and pointing movements remains less well understood than its role in certain other forms of motor learning. Accordingly, the following presentation is based on withdrawal reflexes and VOR adaptation as models for exploring the role of the cerebellum in motor learning.

24.1 Climbing Fibers Encode a Signal That Represents Motor Error

As discussed in chapter 23, the ION sends an input to the cerebellum that reflects unexpected or "surprising" inputs. In the case of Pavlovian learning, these unexpected inputs were sensory signals that, after learning, generated a reflexlike response. Recall that through the mediation of the cerebellum, the CS predicts the occurrence of the US, and ION activity reflects any error in that prediction. Olov Oscarsson,[1] Mitsuo Kawato,[2] and Martin Garwicz[3] have suggested that the ION functions more generally. According to their ideas, ION activity sometimes reflects errors in the motor commands necessary to respond to an *undesired* sensory input. In other words, if some motor command should have prevented the undesired sensory signal, then the existence of that signal reflects an error in the motor command which needs to be corrected.

24.1.1 Withdrawal Reflexes

One specialized type of undesired sensory input involves aversive mechanical stimuli, which trigger withdrawal reflexes. To consider these reflexes in more detail, return to figure 23.3, which presents a schematic model of cerebellar inputs and outputs. It shows some inputs to the ION from interneurons that contribute to a reflex pathway. By having activity that leads to a reflex, these interneurons have, in a sense, already computed a motor response to some sensory input. Reflex pathways in the spinal cord often depend on such interneurons, and they provide the fastest routes for producing motor commands in response to unexpected stimuli. If the cerebellum has access to information that predicts the aversive input and to signals from the reflex pathway's interneurons, then perhaps the cerebellum produces a motor command ahead of time to avoid the aversive stimulation. In that case, the absence of input from the reflex pathway's interneurons would signal success.

To explore this idea, Martin Garwicz and his colleagues[4] used mechanical stimulation of a cat's paw to activate withdrawal reflexes and simultaneously recorded from the anterior part of the interpositus nucleus (NIA). By recording *complex spikes*, they identified NIA cells that responded to dorsal paw stimulation. They knew that these complex spikes were due to activity in the ION's climbing fibers. Electrical stimulation of these NIA neurons activated wrist and elbow flexors, withdrawing the limb from the site of the skin stimulation. Other NIA neurons responded to stimulation of the ventral skin. Stimulation of these NIA neurons tended to activate extensor muscles, again withdrawing the limb from the stimulation site. In a separate experiment, Garwicz and his colleagues[4] found that climbing fibers responded not only to skin stimulation but also to a stretch of the muscles that produced the withdrawal reflexes. Therefore, they suggested that the information contained in the climbing-fiber signal reflected the output of interneurons that form a reflex pathway in the spinal cord. Kris Horn, Milton Pong, and Alan Gibson[5] directly recorded from ION neurons while a cat reached to a target (figure 24.1). ION cells had very little discharge during undisturbed reaches and during reaches in which target was unexpectedly changed. However, ION cells discharged when the limb was tapped during stance. These cells also tended to discharge if during the reach, the limb was mechanically perturbed. This finding suggests that climbing-fiber activity is likely when spinal reflex pathways relay an unexpected event. Climbing-fiber activity is unlikely when other kinds of errors, such as target jumps, affect reach (see also section 23.2.1).

The spinal interneurons have converted an aversive sensory stimulus into the motor commands that will pull the limb away from the site of stimulation. Perhaps this motor command is part of the signal conveyed by the climbing fibers to the cerebellum.

You can view this interneuronal signal as a kind of motor error. In a perfect world, the CNS could predict the occurrence of the noxious stimulus and move the limb away in advance. Thus, the need for the withdrawal reflex, and the interneuronal activity that drives it, represents a failure of the CNS to predict and avoid the unwanted input. As another way to think about this idea, consider the possibility that an undesired outcome, such as intense stimulation of the skin, has two aspects: (1) it involves the undesired sensory input per se and (2) it generates a reflex response mediated by interneurons in the spinal cord. You can view the interneurons' activity as a motor error because it would not have occurred if the aversive stimulus had been predicted and avoided, which would preempt the reflex response.

Why is an anticipatory, predictive response preferable to a reflex-driven one? Because even the fastest reflexes take many tens of milliseconds. Reflexes alone cannot prevent the consequences of the stimulus; they can only meliorate them. To entirely cancel the effect of the unwanted stimulus, the CNS needs to learn to predict it and act ahead of time. The cerebellum appears to play a central role in this kind of learning.

Now extend the concept of *undesired* neural signals beyond inputs such as shock, noxious heat, and other aversive stimuli. In the example of withdrawal reflexes, painful stimuli prompt the withdrawal of the limb, and no one would doubt their undesirability. But stimuli may be undesired in a different sense. Inputs that signal an unwanted displacement of a limb—an unwanted lengthening or shortening of a muscle—can be equally disadvantageous.

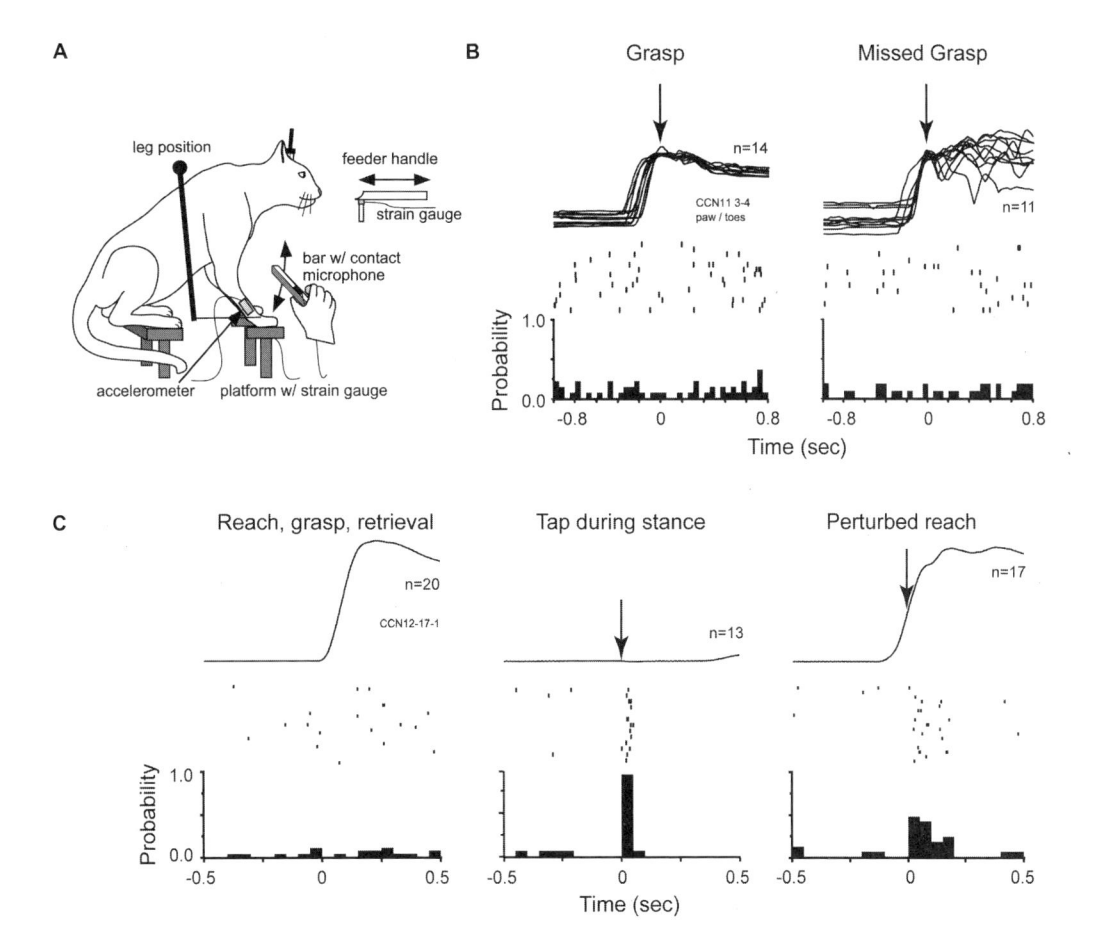

24.1. Climbing Fibers Encode a Signal That Represents Motor Error

Figure 24.1
Discharge of inferior olive cells during reaching errors and perturbations. (*A*) Cats were trained to reach and grasp a target (feeder handle). Perturbations to the limb were introduced by holding a bar in the path of the limb. A contact microphone attached to the bar noted time of perturbation. Perturbations to the target were introduced by moving the feeder handle. A lever arm attached to the cat's wrist measured leg position. (*B*) Discharge of an inferior olive cell during unperturbed reaches and during reaches in which the target was moved. Note the multiple attempts to grasp the handle. Top traces show limb position records, middle traces show neuronal activity rasters, and bottom histograms show the discharge probability, 50-ms bins. (*C*) Discharge of another inferior olive cell. The cell had little response during unperturbed reaches, but responded vigorously when the limb was tapped during stance or the limb was perturbed during reach. (From Horn et al.[5] with permission.)

24.1.2 Rate of Complex Spikes During Adaptation

Peter Gilbert and Tom Thach[6] examined how the discharge of Purkinje cells changed as a monkey learned to anticipate force pulses applied to the arm. Such perturbations cause just the sort of undesired lengthening and shortening of muscles mentioned above. Gilbert and Thach recorded Purkinje-cell activity as a monkey moved a handle attached to a motor. As the monkey held the handle steady at a central target, the motor produced a 300-g *flexor* load, which displaced the handle and stretched the biceps (figure 24.2A, left). In this experiment, the load remained "on," but the monkey nevertheless rapidly brought the handle back to the target zone, working against the load. After a variable period of a few seconds, the load reversed to a 300-g *extensor* load. The monkey again brought the handle back to the center (figure 24.2A, right). The magnitudes of the flexor and extensor loads remained constant during 2 mo of training. As the monkey gained experience with the loads, it learned to activate its arm muscles so that its hand came smoothly back to the target zone. However, because the monkey could not predict the timing of the load change, stretch reflexes serving the biceps or triceps—for extensor or flexor loads, respectively—were at least partially engaged, especially at the onset of the load change.

However, one day, following a few trials with the familiar 300-g loads, the experimenters increased the extensor load to 450 g (arrow in figure 24.2A) but kept the flexor load as before. The increased extensor load took its hand farther than the monkey had previously experienced, and the motor commands that previously had restored the initial hand location no longer did so. Accordingly, the monkey's hand came back to the target more slowly than usual. In the next couple of extensor trials, the monkey pushed more strongly against the load and caused its arm to return toward the target too fast, overshooting it. In all of these trials with the novel 450-g load, the motor commands did not accomplish the intended movement, and so spinal reflexes contributed to correcting hand location. It took the monkey 40–80 trials to adapt to the change from a 300-g load to a 450-g one.

Figure 24.2B shows the discharge in a Purkinje cell during trials with the initially novel load. The small dots show simple-spike discharges, and large dots show when complex spikes occurred. Notice that when the monkey moved the handle against the familiar loads, few complex spikes were observed (top traces of figure 24.2B), and they closely followed the onset of the perturbation (i.e., the time when the load changed direction and engaged stretch reflexes). When the extensor load increased to a novel level, the probability of complex spikes increased, particularly in the period 50–150 msec after the perturbation onset (figure 24.2C). As performance improved in the extensor trials, the total number of complex spikes per trial decreased.

Figure 24.2C sketches the frequency of complex and simple spikes as a function of trial number, as the monkey responded to novel loads. In the extensor trials, the number of complex spikes reached a peak around trial

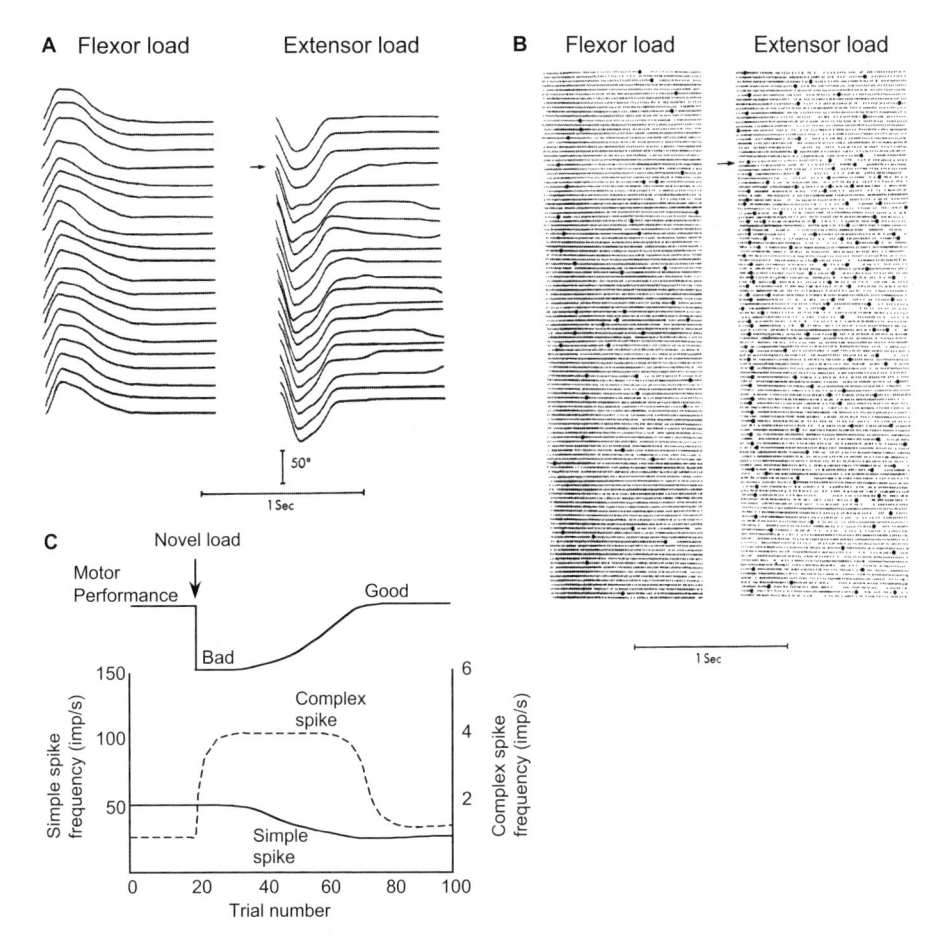

24.1. Climbing Fibers Encode a Signal That Represents Motor Error

Figure 24.2
Changes in simple and complex spikes during adaptation of a limb movement. The monkey held a handle attached to a torque motor. (*A*) The traces indicate the location of a handle. Trials run from top to bottom. At the start of each trial, the handle applied a flexor or an extensor load to the arm. The monkey brought the handle back to the center position against that load. At the trial marked by an arrow, the experimenters applied a larger-than-usual extensor load. (*B*) Simple spikes (small dots) and complex spikes (large dots) recorded from a Purkinje cell. (*C*) Average number of simple and complex spikes after load application as a function of trial number. The downward-pointing arrow indicates the trial on which the extensor load changed. (From Gilbert and Thach[6] with permission.)

20, and then decreased to baseline levels by trial 80. The increase in complex spikes coincided with a decrease in simple spikes. As the complex spikes decreased, the frequency of simple spikes rebounded somewhat. This observation is consistent with the idea that complex spikes reduce the parallel fiber–Purkinje cell synaptic weights. However, the reduction in Purkinje-cell discharge might have been due simply to changes in mossy-fiber input (i.e., some change in the proprioceptive feedback from the arm). This interpretation seems unlikely, though, because Masao Ito and his colleagues[7] showed a cause-and-effect relationship between reduced simple spikes and increased complex spikes when climbing and mossy fibers were simultaneously stimulated.

Notice, however, that as the frequency of simple spikes decreased for the extensor load, it appeared to increase slightly for the flexor load. This observation implies that the error that was experienced for one direction of the load affected the discharge of the cell not just for that direction, but also for another direction (in this case the opposite direction). This result provides an example of generalization of motor learning at the neurophysiological level, in accord with the psychophysical data presented in section 21.3.

24.2 Predictively Correcting Motor Commands

You might expect that only external, unexpected stimuli—such as the torque pulses used by Gilbert and Thach and the noxious inputs used by Garwicz and his colleagues—engage your reflexes. However, your own motor commands can also have undesirable sensory consequences. Consider two further generalizations of the concept of undesired inputs. First, rapid limb movements that nominally produce a torque on one joint result in the motion of limb segments around other joints. The undesired input consists of the excessive shortening or lengthening of muscles that span those other joints. Second, head movements also move the retina. In that case, the undesired input consists of a signal indicating the slippage of the image on your retina. In both conditions, the cerebellum plays a prominent role in predicting the motor commands necessary to minimize these undesirable sensory consequences.

24.2.1 Predicting and Compensating for Interaction Torques

In the equations of motion that describe the relationship between acceleration of your arm and torques on your joints, some terms predict torques that occur on one joint because of acceleration of limb segments around another joint. For example, a rapid forearm flexion will produce an extension torque on the shoulder, called an interaction torque. The dynamics of multijoint limbs often causes such torques. In the example above, the undesired shoulder-extension torque will generally cause extension movements, which in turn engage stretch reflexes that activate shoulder flexors.

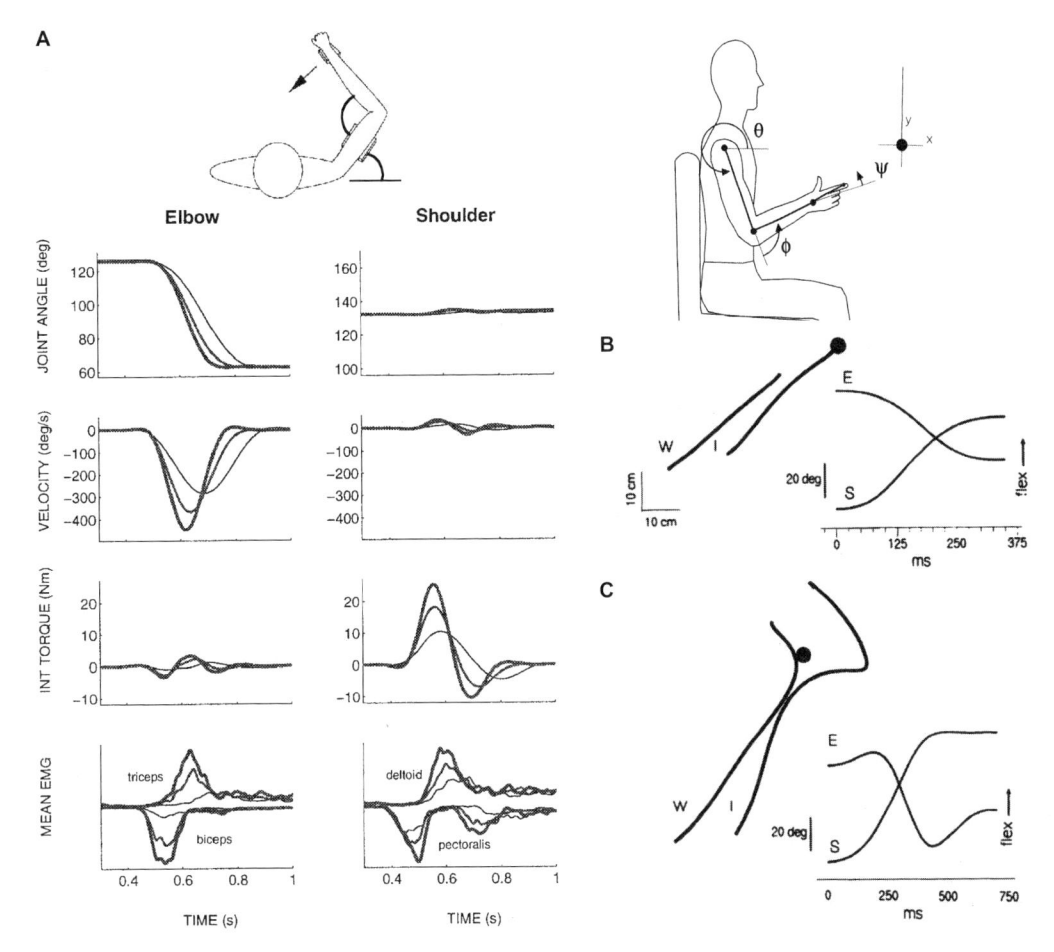

24.2. Predictively Correcting Motor Commands

Figure 24.3
Compensation for interaction torques in a healthy person and in a patient with cerebellar damage. (A) Muscle activity during rapid elbow flexion. The figure plots the joint angles of the elbow and the shoulder along with integrated, rectified activity in elbow and shoulder muscles for three different speeds of movement. Abbreviations: Int torque, interaction torque; Nm, newton-meters. The increased EMG activity in biceps and pectoral muscles appears on an inverted scale. (B) A typical reach in a healthy participant. Abbreviations: E, elbow angle; I, index finger location; S, shoulder angle; W, wrist location; (C) Abnormal control of interaction torques in a cerebellar patient. Format as in B. (A from Gribble and Ostry,[8] B and C from Bastian et al.[9] with permission.)

Figure 24.3A shows an example of this idea. Paul Gribble and David Ostry[8] asked participants to flex their elbow rapidly without moving their shoulder. From the inertial dynamics of the arm, they predicted that flexion torques on the elbow should result not only in acceleration of the elbow (flexion) but also in acceleration of the shoulder (extension). Therefore, in order to maintain a constant shoulder angle, the participants would have to activate shoulder flexors at about the same time as elbow flexors. Figure 24.3A shows that acceleration of the elbow produced an interaction torque on the shoulder. However, the participants predicted these torques and proactively compensated for them by activating the pectoral muscle, a shoulder flexor. Presumably, their CNS had learned that in order to make a rapid elbow flexion, it needs to activate both elbow and shoulder flexors. It had learned, in effect, an internal model of the **inverse dynamics** of the limb.

24.2.2 Cerebellar Damage Affects Prediction of Interaction Torques

Amy Bastian, Tom Thach, and their colleagues[9] studied interaction torques in both patients with cerebellar lesions and control participants (figure 24.3B and C). From a time series of joint-angle changes and estimates of limb mass, they used the equations of motion (inverse dynamics) to estimate the torques at each joint during each reaching movement. They concluded that cerebellar patients had specific deficits in their predictive compensation for the interaction torques. For example, figure 24.3C shows the accurate reach of a control participant. The elbow and shoulder joints rotated together: Flexion around the shoulder coupled with extension around the elbow. Accordingly, the participant's index finger reached the target (a ball hanging from the ceiling) without overshooting it. In a cerebellar patient, the shoulder's flexion produced extension torques on the elbow, which caused larger-than-needed extension around the elbow. Accordingly, the patient's index finger missed the target. Apparently, the acceleration of the shoulder caused uncompensated interaction torques acting on the elbow.

24.2.3 Encoding Head and Eye Movements in the Cerebellum

Turn now to another example of undesired sensory input. Although the present understanding of reaching and pointing movements does not permit a thorough understanding of how the cerebellum might prevent unwanted signals for those movements, results from an eye-movement paradigm provide some hints about how it might do so. As explained in box 11.1, eye movements evolved to hold the eyes still with respect to the outside world. The retina works slowly, so the image must not slip across the retina, or else vision is blurred. One neural mechanism that keeps the eyes still with respect to the world is the *optokinetic reflex* (see box 11.1). It is generated by a pathway that responds to retinal slip (i.e., the motion of an image on your retina). Retinal slip activates vestibular-nucleus neurons

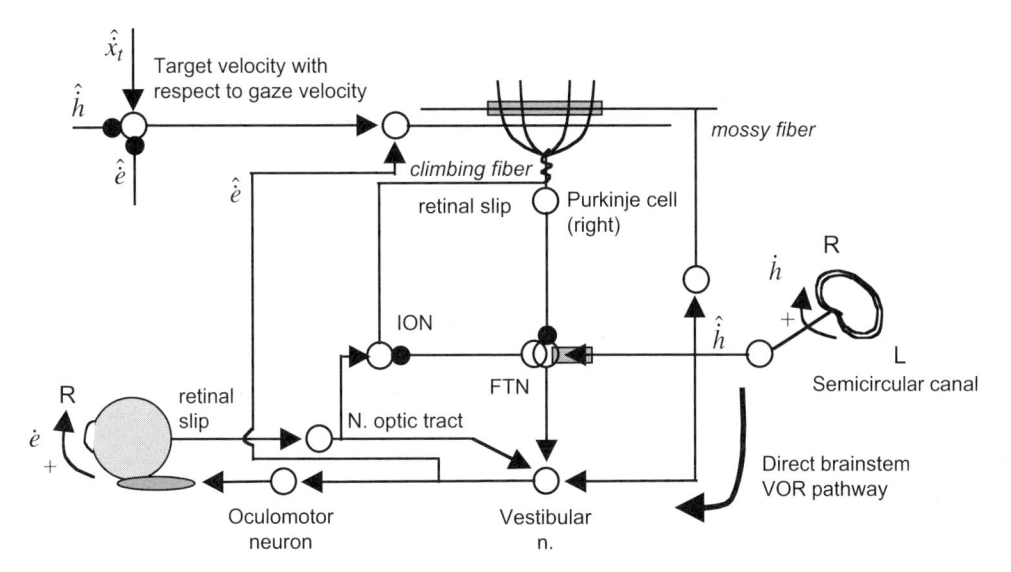

Figure 24.4
A model of the VOR. The afferent limb of the optokinetic reflex senses retinal slip for an object moving to the left. This reflex generates movements in the direction of the retinal slip and also provides a signal to the inferior olivary nuclei (ION), which in turn provide an error signal for teaching the cerebellum about errors in the VOR. Arrows indicate excitatory synapses; filled circles, inhibitory ones. Positive eye and head velocities are indicated. Gray regions on the flocculus target neuron (FTN) and Purkinje cell indicate sites of plasticity during modification of VOR, according to Lisberger's model.[13] Abbreviations: e, eye orientation; h, head orientation; L, left; N and n, nucleus; R, right; x_t, target location. (Modified from Lisberger[13] with permission.)

that, in turn, activate **oculomotor** neurons to compensate for the motion of the object on the retina. However, retinal slip results not only from the movement of objects in the world but also from movement of your head. As your head moves, unless your eyes move in the opposite direction at precisely the same speed, the image on your retina will slip. The VOR accomplishes this objective, and the cerebellum plays an important role in adjusting the motor commands needed to maintain the accuracy of that reflex response.[10,11]

To study the VOR, Steve Lisberger and Al Fuchs[12] recorded from Purkinje cells in the flocculus of a head-fixed monkey. Recall that the flocculus is a small (and phylogenetically ancient) part of the cerebellar cortex and, unlike most of the cerebellar cortex, its Purkinje cells project directly to a brainstem vestibular nucleus (specifically, to cells called flocculus target neurons, FTNs, in figure 24.4), rather than to one of the deep cerebellar nuclei. Lisberger and Fuchs studied Purkinje-cell activity in several conditions: (1) as a monkey fixated a moving light with its head still (called a smooth-pursuit eye movement), (2) as a monkey used its VOR to maintain fixation of a stationary light as its head moved (passively), and

(3) as a monkey suppressed its VOR to maintain fixation of a light that moved with the monkey's head.

During suppression of the VOR, when the monkey's eyes remained stationary in the monkey's head and the head rotated, the Purkinje-cell discharge reflected head velocity. Generally, when the head moved in a direction ipsilateral to the recording site, Purkinje-cell discharge increased. In figure 24.4, this corresponds to a rightward head movement for Purkinje cells in the right flocculus. Therefore, if during baseline conditions (no eye, head, or target movements), the discharge of the Purkinje cell p was p_0, then when only the head moved (as occurs during suppression of the VOR), $p(t) \approx p_0 + k_1 \hat{\dot{h}}$, where $\hat{\dot{h}}$ represents an estimate of head velocity (positive to the right) and k represents a gain factor.

When only the monkey's eyes moved (as occurs during smooth pursuit when the head is fixed), Lisberger and Fuchs found that Purkinje-cell discharge increased for eye movements in a direction ipsilateral to the recording site. Thus, $p(t) \approx p_0 + k_2 \hat{\dot{e}}$, where $\hat{\dot{e}}$ represents an estimate of eye velocity (positive to the right). When the eyes and the head moved simultaneously (as occurs during normal VOR), Purkinje-cell discharge equaled the sum of these two factors: $p(t) \approx p_0 + k_1 \hat{\dot{h}} + k_2 \hat{\dot{e}}$. If the gains k_1 and k_2 differed, then during normal VOR—when the head moved in one direction and the eyes moved with precisely equal and opposite velocity—the Purkinje-cell discharge modulated to the extent that those gains differed. For most cells, however, the gains were similar. In this case, Purkinje-cell discharge during the VOR did not change very much even though both the monkey's eyes and its head moved (figure 24.5). That is, when the head moved (to the right, for example) but the eyes moved with equal velocity in the opposite direction (to the left, yielding negative values of $\hat{\dot{e}}$), the discharge of Purkinje cells in the flocculus changed little, if at all.

Notice what happened when the monkey's head moved but the eyes remained stationary in the orbits, as occurs during suppression of the VOR. In that condition, if $k_1 \approx k_2$, then the output of the Purkinje cell changed by an amount that reflected a negative image of eye velocity in the normal VOR. This results from the fact that in the normal VOR, $\dot{h} \approx -\dot{e}$. This observation implies that during VOR suppression, when only the head moves, the Purkinje cell predicts the *negative image* of eye velocity (i.e., the oculomotor commands necessary to minimize retinal slip due to motion of the head). This is precisely the signal that the CNS needs to anticipate in order to prevent motion of one body part when another body part moves, as occurs for interaction torques. Note also that the existence of a negative image of $\dot{e}$ resembles the signal needed for adaptive filtering, as discussed in section 23.1. It seems likely that the prediction computed by Purkinje cells results from parallel-fiber inputs, also like the adaptive-filter mechanism of electric fish. Not only does the parallel-fiber input reflect the negative composite image of another input, as in electric fish, but the combined effect of that prediction and the expected input results in signal cancellation at the level of the projection neuron.

24.2. Predictively Correcting Motor Commands

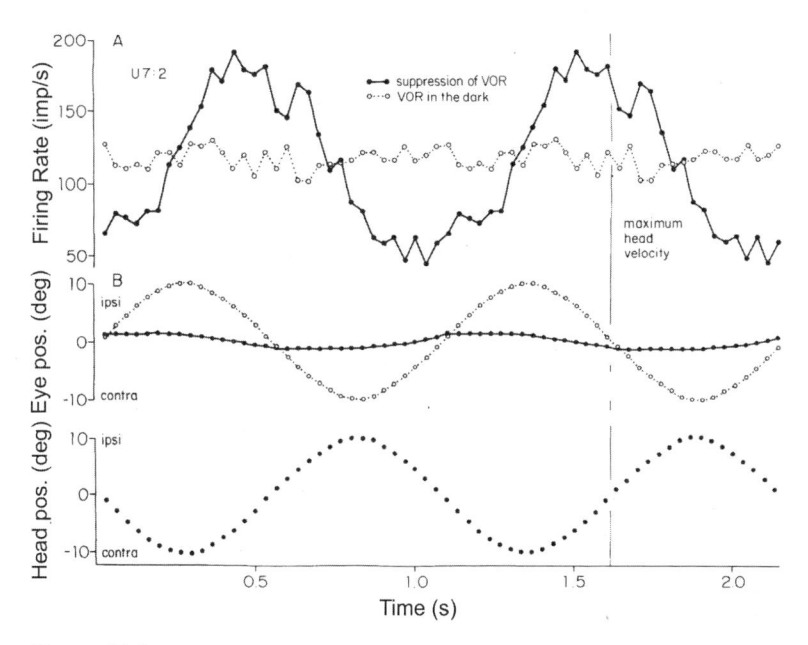

Figure 24.5
Average discharge of a Purkinje cell during head rotation under conditions of VOR in the dark and VOR suppression. Most Purkinje cells in the flocculus and paraflocculus did not modulate their discharge much during VOR in the dark. (From Lisberger and Fuchs[12] with permission.)

As a second consequence of $k_1 \approx k_2$, motor commands that move the eyes during normal VOR cannot depend principally on a pathway that includes the Purkinje cells. Because of signal cancellation, Purkinje cells do not usually provide much of an input to the sensorimotor pathways that produce the reflex (figure 24.4). Rather, under well-adapted conditions, head movements produce eye movements through two pathways: a pathway from the semicircular canals via the flocculus target neurons, and a pathway from the semicircular canals via other vestibular-nucleus neurons. The "side-loop" pathway through the cerebellar cortex, in this case the flocculus, contributes something occasional or facultative. Consistent with this idea, the VOR remains relatively unaffected after bilateral ablation of the flocculus. Importantly, however, although normal VOR goes on as usual, adaptations to changed conditions depend on the flocculus and paraflocculus. Damage to those parts of the cerebellar cortex prevents adaptation of the VOR to changed conditions, but not its performance in steady conditions.[14] This finding indicates that the ability to predicatively compensate for such movements ultimately must reside in memory in other parts of the CNS, and that the cerebellum contributes signals that increase or decrease these reflex pathways. Somehow, the cerebellum plays a central role in the mechanism that adjusts the gains of this reflex when circumstances change and the current state of adaptation no longer suffices to prevent errors. But how? The answer to that question has generated controversy for decades, and the following presentation outlines one line of thinking.

24.2.4 Adapting the VOR

In the laboratory, experimenters can induce a change in the VOR by having monkeys wear glasses that either magnify or miniaturize the visual scene.[15] After a few hours of experience, the gain of the VOR, defined as eye speed divided by head speed in darkness, increases or decreases, respectively. In both conditions, the gain of the reflex changes so that a given head movement produces an eye movement that minimizes retinal slip. For example, in the case of VOR gain of less than 1, the eyes move at a slower speed with respect to the head.

Masao Ito,[10] and Hiroaki Gomi and Mitsuo Kawato[16] hypothesized that the change in the VOR gain begins with adaptation that occurs in the flocculus (the gray region of Purkinje cells in the top part of figure 24.4). They suggested that climbing fibers guide this adaptation. If the gain of the VOR is too low, a given head movement produces an eye movement that is too slow. This relatively low eye velocity produces retinal slip, which in turn produces complex spikes and alters the synaptic weights that carry head-velocity information to the Purkinje cells.

Ito predicted that adaptation of the VOR should produce changes in the discharge of the Purkinje cells, depending on whether the gain was increased or decreased. A gain greater than 1 should produce a reduction in Purkinje-cell discharge. A gain less than 1 should produce an increase. Experiments later confirmed this prediction.[17,18]

For example, figure 24.6 shows Purkinje-cell discharge during gain changes in the VOR. With the normal VOR gain of 1, there was little or no change in Purkinje-cell discharge when the head rotated in the dark (not shown). When the monkey adapted to a low VOR gain, a given head velocity was accompanied by only a small eye velocity (in the opposite direction), and Purkinje-cell discharge increased. Because $p(t) \approx p_0 + k_1\hat{h} + k_2\hat{e}$, the increased discharge in the low-gain condition suggested that the eye and/or head velocity gains had changed. However, because changes in gain of the VOR did not alter Purkinje-cell discharge during head-fixed, smooth-pursuit eye movements,[19] Steve Lisberger[13] suggested that the increase in Purkinje-cell response resulted from a small *decrease* in k_1. Similarly, his model suggested that at high VOR gains, the small decrease in Purkinje-cell response resulted from a small *increase* in k_1.

However, Lisberger's gaze–velocity model also suggested that these small changes in Purkinje cells' sensitivity to head velocity could not account for the changes in motion of the eyes during altered VOR gains. For example, FTN discharge increased dramatically under the high-gain condition, whereas the Purkinje-cell discharge decreased only slightly under the same conditions. Lisberger suggested that this large change in FTN discharge resulted from adaptation not only at the Purkinje cell but also at the synapses of the vestibular afferent neurons projecting to the FTNs (figure 24.4). According to Lisberger, with high VOR gains, the sensitivity

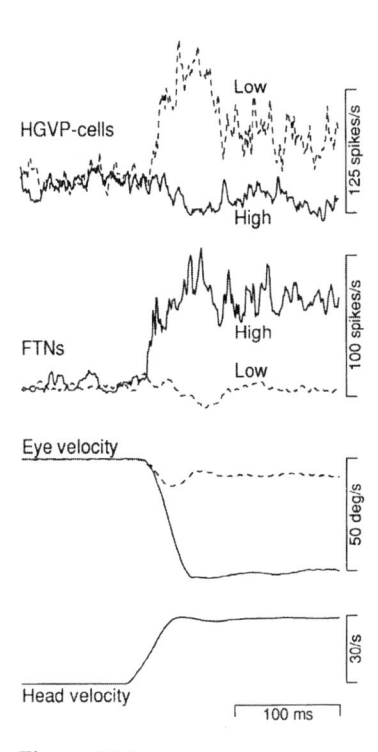

Figure 24.6
Examples of the effect of changes in VOR gain. Response of horizontal gaze-velocity Purkinje cells (HGVP cells), flocculus target neurons (FTNs), and eye velocity during rapid changes in head velocity. Traces for firing rates were taken from different cells. (From Lisberger[13] with permission.)

of FTNs to head velocity increased. With low VOR gains, this sensitivity decreased.

In summary, one view of VOR adaptation holds that it depends on synaptic plasticity of inputs to both FTNs and Purkinje cells (figure 24.4). It appears that changes at the Purkinje-cell level may be required to allow changes to occur at the FTN level,[20] but this issue remains poorly understood. For example, an alternative view has been put forward by Mitsuo Kawato and his colleagues.[21] The have argued against the two-site theory of VOR plasticity. To fairly represent their views, we quote extensively from their recent paper:

> The model proposed by Lisberger ... assumes that the learning that occurs in both the cerebellar cortex and the vestibular nucleus is necessary for VOR adaptation.... An alternative hypothesis ... is equivalent to assuming that there are two parallel neural pathways for controlling VOR and smooth pursuit.... First, we theoretically demonstrate that this parallel control-pathway theory can reproduce the various firing patterns of horizontal gaze velocity Purkinje cells ... at least equally as well as the gaze velocity theory, which is the basic framework of Lisberger's model. Second, computer simulations based on our hypothesis can stably reproduce neural firing data as well as behavioral data obtained in smooth

pursuit, VOR cancellation, and VOR adaptation, even if only plasticity in the cerebellar cortex is assumed.... Our results indicate that different assumptions about the site of pursuit driving command maintenance computationally lead to different conclusions about where the learning for VOR adaptation occurs. (Tabata et al.,[21] p. 2176)

Despite the persistent disagreements, these models and findings have enough in common to allow some consideration of how the cerebellum functions in adapting to changes in reaching and pointing movements, including both **dynamics** and kinematics. It seems clear that in contributing to VOR adaptation, the cerebellum learns to supplement the motor commands in brainstem pathways. This part of the system may be analogous to a computation that predicts some head velocity and generates oculomotor commands that partially compensate for it. It also seems clear that an error signal, in this case one reflecting retinal slip, drives adaptation in the Purkinje cells, which in turn results in adaptation of the VOR. Error signals probably drive cerebellar learning in general. However, despite extensive investigation of the neural basis of the VOR, the question of what Purkinje cells compute remains unresolved. Nevertheless, the hints these studies provide form a foundation for theorizing about the role of the cerebellum in both eye- and arm-movement control.

24.2.5 Computing a Model of Inverse Dynamics

Mitsuo Kawato and his colleagues[22,23] have argued that at least part of the output of the Purkinje cells represents an internal model of inverse dynamics. According to their theory, a region of cerebellar cortex computes an internal model of inverse dynamics of the arm or the eyes. For example, they noted that during head-fixed, smooth-pursuit eye movements, Purkinje-cell discharge leads eye velocity. If Purkinje cells encode a model of inverse dynamics, then the exact time course of firing rates should be a function of future eye orientation, velocity, and acceleration. Kawato et al. found this to be the case. In their experiment, they trained monkeys to look at a random-dot display and track it with their eyes. Then they fitted Purkinje-cell discharge to eye orientation, velocity, and acceleration (figure 24.7) and found coefficients for the following expression: $p(t) = a\ddot{e}(t + \Delta) + b\dot{e}(t + \Delta) + ce(t + \Delta) + p_0$. A 10-msec lead time Δ best fit the data, and the coefficients a and b roughly equaled those calculated for oculomotor neurons. This finding suggests that Purkinje cells may transform a desired eye velocity and acceleration signal into motorlike commands. (We say motorlike because Purkinje cells are inhibitory neurons that do not drive motor neurons, but influence downstream pathways that eventually produce motor commands.)

However, the cells did not appear to encode the entire inverse-dynamics function because the constant c differed dramatically from the value calculated for oculomotor neurons. Nevertheless, the findings of Kawato and his colleagues suggest that Purkinje-cell output predictively compensates for a component of dynamics that might otherwise result in retinal slip during smooth pursuit. According to their theory, the cere-

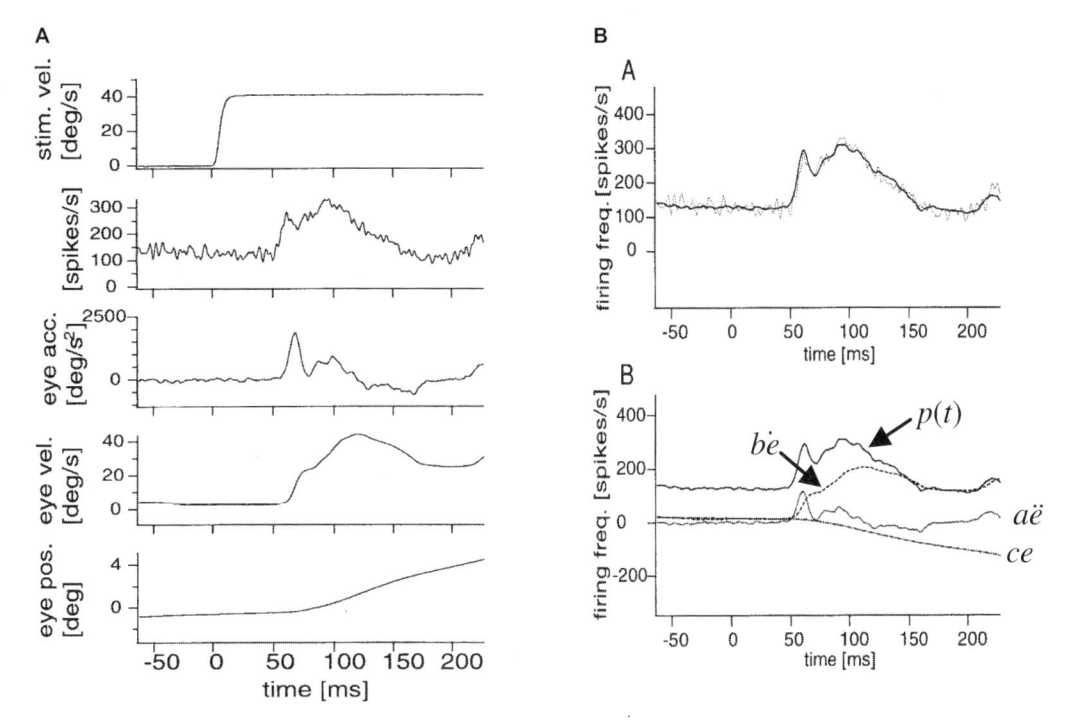

Figure 24.7
Example of Purkinje-cell discharge during smooth-pursuit eye movements. (A) Discharge for a visual stimulus that moved at constant speed and the accompanying eye movement. (B) Fit of the discharge function to a linear sum of eye orientation, velocity, and acceleration. (From Shidara et al.[22,23] with permission.)

bellum receives a readout of the motor command error (dashed line in figure 24.8B) via climbing fibers (dashed line in figure 24.8A), uses that signal to modify its computation of the inverse model, and provides a feedforward motor-command signal that sums with the output of a feedback controller.

The fact that during VOR, Purkinje cells do not modulate their activity much (figure 24.5) suggests that in a well-adapted VOR, FTNs and brainstem pathways compute much of the inverse dynamics of the desired eye movement. Nicholas Schweighofer, Mike Arbib, and Mitsuo Kawato[24,25] have argued that, in general, the cerebellar cortex may compute motor commands that represent a *partial* internal model of inverse dynamics of the task. These commands supplement motor commands generated elsewhere so that the sum results in actions that minimize undesired sensory consequences. The teaching signal for this internal model could reflect the motor error generated by reflexes either in the spinal cord (in the case of limb movements) or in the brainstem (in the case of eye movements). Perhaps the ION transmits the teaching signals to the cerebellum via its climbing fibers.

These notions must await further experimental and theoretical work for evaluation. Nevertheless, both the cerebellar cortex and the deep

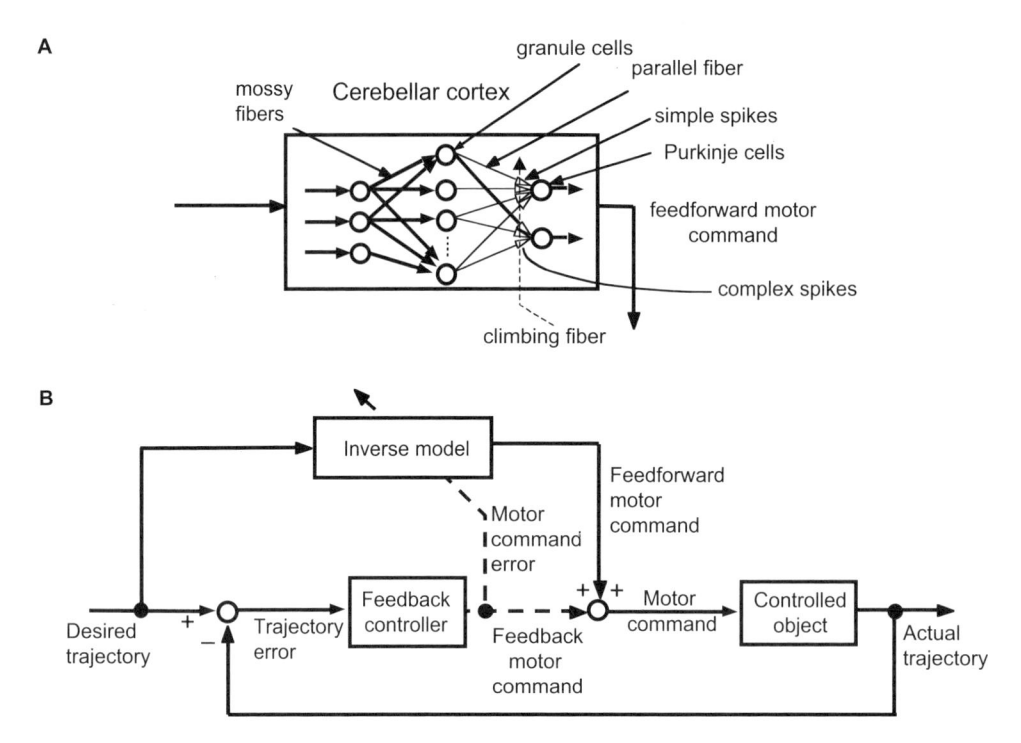

Figure 24.8
A cerebellar-feedback error-learning model. According to this model, the cerebellar cortex acquires an internal model of the inverse dynamics of the effector, this internal model predictively compensates for the undesired consequences of other motor commands, and the teaching signal for the cerebellar cortex comes, in part, from reflexes. Also according to this model, simple spikes represent motor-like commands, complex spikes represent errors in motor-command coordinates, parallel-fiber inputs represent desired trajectory, and the cerebellar cortex computes a model of inverse dynamics. (Modified from Wolpert et al.[28] with permission.)

cerebellar nuclei adapt during learning, and the idea that cerebellar output provides something to brainstem and spinal pathways that adds to or subtracts from other motor commands has wide support. According to Jennifer Raymond, Steve Lisberger, and Mike Mauk,[26] output from the cerebellar Purkinje cells guides learning in the targets of these projection neurons, so that, eventually, some or all of the requisite motor memory transfers there. According to Mike Mauk and Nelson Donegan,[27] adaptation in Pavlovian learning occurs both in the cerebellar cortex at the parallel fiber–Purkinje cell synapse and in collaterals of mossy-fiber inputs to the deep cerebellar nuclei. Eventually, the former become less important as the latter take over.

Taken together, the available evidence indicates that the cerebellum plays an important role in signal prediction generally, and in contributing to outputs that anticipate and counter the consequences of undesired neural signals. Further (although more controversial), it appears that the cerebellar cortex plays a larger role in learning conditions, and the deep cerebellar nuclei, along with brainstem pathways, play a larger role in well-adapted conditions. Aversive inputs number among the undesired neural signals, of course, and the role of the cerebellum in some kinds of Pavlovian learning and withdrawal reflexes exemplifies that. But so do motor errors, such as those associated with unexpected interaction torques, which exemplify the way reflex-mediating interneurons could provide the cerebellum with teaching inputs—via the ION—that guide motor adaptation and the development of internal models of inverse dynamics. As discussed in chapters 20 and 21, an accurate internal model of inverse dynamics leads to movements with minimal motor errors. In effect, they predict the forces and torques needed to follow the control policy adopted by the next-state planner, typically a straight, smooth movement in fixation-centered coordinates (see chapters 18 and 19).

References

1. Oscarsson, O (1973). In Handbook of Sensory Physiology: Somatosensory System, ed Iggo, A (Springer-Verlag, New York), pp 339–380.
2. Kawato, M (1989) Adaptation and learning in control of voluntary movement by the central nervous system. Adv Robotics 3, 229–249.
3. Garwicz, M (2002) Spinal reflexes provide motor error signals to cerebellar modules—relevance for motor coordination. Brain Res Rev 40, 152–165.
4. Ekerot, CF, Jorntell, H, and Garwicz, M (1995) Functional relation between corticonuclear input and movements evoked on microstimulation in cerebellar nucleus interpositus anterior in the cat. Exp Brain Res 106, 365–376.
5. Horn, KM, Pong, M, and Gibson, AR (2004) Discharge of inferior olive cells during reaching errors and perturbations. Brain Res 996, 148–158.
6. Gilbert, PFC, and Thach, WT (1977) Purkinje cell activity during motor learning. Brain Res 128, 309–328.
7. Ito, M, Sakurai, M, and Tongroach, P (1982) Climbing fibre induced depression of both mossy fibre responsiveness and glutamate sensitivity of cerebellar Purkinje cells. J Physiol (London) 324, 113–134.
8. Gribble, PL, and Ostry, DJ (1999) Compensation for interaction torques during single- and multijoint limb movement. J Neurophysiol 82, 2310–2326.
9. Bastian, AJ, Martin, TA, Keating, JG, and Thach, WT (1996) Cerebellar ataxia: Abnormal control of interaction torques across multiple joints. J Neurophysiol 76, 492–509.
10. Ito, M (1972) Neural design of the cerebellar control system. Brain Res 40, 80–82.
11. Kawato, M, and Gomi, H (1992) A computational model of four regions of the cerebellum based on feedback-error learning. Biol Cybern 68, 95–103.
12. Lisberger, SG, and Fuchs, AF (1978) Role of primate flocculus during rapid behavioral modification of vestibuloocular reflex. I. Purkinje cell activity during visually guided horizontal smooth-pursuit eye movements and passive head rotation. J Neurophysiol 41, 733–763.

13. Lisberger, S. G. (1994) Neural basis for motor learning in the vestibuloocular reflex of primates. III. Computational and behavioral analysis of the sites of learning. J Neurophysiol 72, 974–998.
14. Lisberger, SG, Miles, FA, and Zee, DS (1984) Signals used to compute errors in monkey vestibulo-ocular reflex: Possible role of flocculus. J Neurophysiol 52, 1140–1153.
15. Miles, FA, and Fuller, JH (1974) Adaptive plasticity in the vestibulo-ocular responses of the rhesus monkey. Brain Res 80, 512–516.
16. Gomi, H, and Kawato, M (1992) The cerebellum and VOR/OKR learning models. Trends Neurosci 15, 445–453.
17. Lisberger, SG, Pavelko, TA, Bronte-Stewart, HM, and Stone, LS (1994) Neural basis for motor learning in the vestibuloocular reflex of primates. II. Changes in the responses of horizontal gaze velocity purkinje cells in the cerebellar flocculus and ventral paraflocculus. J Neurophysiol 72, 954–973.
18. Watanabe, E (1984) Neuronal events correlated with long-term adaptation of the horizontal vestibulo-ocular reflex in the primate flocculus. Brain Res 297, 169–174.
19. Miles, FA, Braitman, DJ, and Dow, BM (1980) Long-term adaptive changes in primate vestibuloocular reflex. IV. Electrophysiological observations in flocculus of adapted monkeys. J Neurophysiol 43, 1477–1493.
20. Nagao, S, and Kitazawa, H (2003) Effects of reversible shutdown of the monkey flocculus on the retention of adaptation of the horizontal vestibulo-ocular reflex. Neuroscience 118, 563–570.
21. Tabata, H, Yamamoto, K, and Kawato, M (2002) Computational study on monkey VOR adaptation and smooth pursuit based on the parallel control-pathway theory. J Neurophysiol 87, 2176–2189.
22. Shidara, M, Kawano, K, Gomi, H, and Kawato, M (1993) Inverse dynamics model eye movement control by Purkinje cells in the cerebellum. Nature 365, 50–52.
23. Gomi, H, and Osu, R (1998) Task-dependent viscoelasticity of human multijoint arm and its spatial characteristics for interaction with environments. J Neurosci 18, 8965–8978.
24. Schweighofer, N, Arbib, MA, and Kawato, M (1998) Role of the cerebellum in reaching movements in humans. I. Distributed inverse dynamics control. Eur J Neurosci 10, 86–94.
25. Schweighofer, N, Spoelstra, J, Arbib, MA, and Kawato, M (1998) Role of the cerebellum in reaching movements in humans. II. A neural model of the intermediate cerebellum. Eur J Neurosci 10, 95–105.
26. Raymond, JL, Lisberger, SG, and Mauk, MD (1996) The cerebellum: A neuronal learning machine? Science 272, 1126–1131.
27. Mauk, MD, and Donegan, NH (1997) A model of Pavlovian eyelid conditioning based on the synaptic organization of the cerebellum. Learn Mem 4, 130–158.
28. Wolpert, DM, Miall, RC, and Kawato, M (1998) Internal models in the cerebellum. Trends Cog Sci 2, 338–347.
29. Liu, XG, Robertson, E, and Miall, RC (2003) Neuronal activity related to the visual representation of arm movements in the lateral cerebellar cortex. J Neurophysiol 89, 1223–1237.

30. Imamizu, H, Miyauchi S, Tamada T, et al. (2000) Human cerebellar activity reflecting an acquired internal model of a new tool. Nature 403, 192–195.

Reading List

The Cerebellum and Adaptive Control, by J.S. Barlow (Cambridge University Press, New York, 2002) provides a broad overview of the topics covered in this chapter. For more on the possible role of the cerebellum in correcting sensory errors, as opposed to the motor errors stressed here, see the recent paper by Liu et al.[29] They have recently provided evidence that more neurons in the cerebellum reflect visual information, such as moving cursors, than reflect the direction of hand movement, when these directions differ. For more on neuroimaging of internal models, you might begin with the paper by Kawato and his colleagues.[30]

25 Deciding Flexibly on Goals, Actions, and Sequences

Overview: There is more to life than reaching and pointing directly to stimuli, one at a time. Your reaching and pointing movements require decisions and choices. You must compare and contrast alternative targets and control policies (goals), and you must evaluate potential goals—or a sequence of goals—among several possibilities. You must decide, choose, and then act, all the while suppressing rejected alternatives and those held in abeyance. In addition, your reaching and pointing movements need not be aimed directly at the stimuli that instruct and guide your action. Your CNS can guide those and other movements by external cues, by internal cues, and by combinations of both. And you can learn to reach and point to places because of rules, strategies, and abstract goals.

According to the model of reaching and pointing presented in figure 19.4, a series of neural networks computes a difference vector, which extends to a current target location from a current end-effector location, each represented in fixation-centered coordinates. Then, a next-state planner implements a *control policy* that determines the trajectory that the end effector follows to the target. The previous chapters mainly discuss reaching and pointing movements aimed directly at a single target, made one at a time, in one "task" at a time. However, you know that your behavior is not like that at all, not even when restricted to reaching and pointing. You can simultaneously juggle multiple tasks, perform sequences of movements, and *choose* them from a multitude of options. You can reach to and point at targets directly or away from them by some chosen amount. In terms of a control policy, the previous chapters have rarely deviated—pun intended—from a smoothly accelerating, smoothly decelerating, straight path to a stationary target. In the motor control literature, you can find many references to *invariant* characteristics of movements that—a moment's reflection will reveal—you can vary any time you want. As your hand or some tool progresses to a chosen target, you can make it go fast, go slow, go straight, curve, stop along the way, or just give up and do something else. This chapter addresses some of these capabilities, and more. Recall that chapter 4 laid out three kinds of motor learning (figure 4.1):

- Learning across generations, including reflexes and other innate behaviors. This kind of learning leads to the ability to link other stimuli

with reflexes, and therefore enables predictive learning. Chapters 23 and 24 present some ideas about that kind of motor learning.
- Learning that promotes stability and control, involving skill acquisition and motor adaptation. Chapters 8–22 dealt with those mechanisms in considerable detail.
- Deciding where, whether, and when to move.

This chapter briefly addresses some of the issues pertaining to the third kind of motor learning listed above. It takes up two topics: (1) accumulator models of decision and choice and (2) choices among the different kinds of cues that can guide movements and movement sequences, as well as the way your CNS can use those cues to guide movements flexibly.

25.1 Deciding on a Target

Decision theory,[1,2] in part, attempts to explain the way that you and other animals make choices about what to do and when to do it, given conflicting demands. The theory depends on the idea that your CNS accumulates evidence and combines this evidence with an estimate about the likelihood of various sensory inputs. The CNS also computes the efficacy of potential actions in the context of its decisions about that evidence.

Typically, such models break decisions and choices into phases, with the first phase dealing primarily with the evaluation of sensory evidence and later phases, usually thought to overlap in time with the earlier ones, devoted to deciding among some set of potential actions, choosing one of them, and implementing that choice. According to these models, your CNS first represents the state of the world and, second, selects the best action given a current state. In other terms, a sensory-categorization process, involving the accumulation of evidence, precedes motor planning and preparation. The CNS makes decisions at both levels: the level of deciding what is out there and the level of deciding what is to be done about it.

Because the first stage involves accumulating information, models of decision have been termed *accumulator models* (there are several). When combined with the idea that competing accumulators generate different decisions, they are also called *racetrack models*. The idea is that the first accumulator to reach a threshold wins. (Alternatively, the accumulator that gets "far enough" ahead of the others wins.) In racetrack models, the two or more accumulators compete independently of each other, until one has crossed its threshold or gets "far enough" ahead of the others. Then that accumulator suppresses the activity in the competing ones. The relationship between accumulators is therefore strongly nonlinear: Up to their thresholds, they influence each other very little, but above their thresholds, they suppress each other markedly. An alternative to the racetrack model is called the *diffusion model*. Such a model has only one decision variable, which reflects the ratio of evidence for and against a particular choice. Thus, the main difference between diffusion and racetrack models lies in the amount and form of interaction among accumulators.

25.1. Deciding on a Target

Most of these models assume that you should choose the movement that will lead to the most favorable outcome, based on your current state, and that your CNS predicts the potential benefits and costs (called *cost functions*) associated with potential actions.

Neurophysiological evidence has begun to support these models. Neurons in the PPC[3] and the prefrontal cortex,[4] among other areas, appear to accumulate evidence about what is out there. Most of the research on accumulator models focuses on perceptual decisions, but similar ideas apply to decisions about motor planning and preparation. Accumulator networks thus perform several related functions. They discriminate objects that might serve as movement targets,[5–9] and they assign a **salience**—a combined measure of both biological importance and noticeability—to a representational map of potential movement targets. Often called a *salience map*, this representation reflects what you can expect to gain, based on experience, from choosing a potential target or targets. Motor networks select, as the target(s) of movement, a stimulus or a set of stimuli with a relatively high salience.[6,10–12] Many parts of your CNS reflect relative value, including at least LIP,[1] the prefrontal cortex,[13] the superior colliculus,[14] and the striatum.[15] Such signals may be ubiquitous, or at least widely distributed. Your CNS thus guides decisions by calculating the expected biological value to be obtained by choosing some targets as opposed to others. The results of Okihide Hikosaka and his colleagues, described in section 23.4.2, provide a good example of this idea at the neurophysiological level.

25.1.1 Accumulator Models

One way to think about the decision-making process is that the choice of a target arises from competition among neural networks, arranged largely in parallel. According to accumulator models, as these networks accumulate information about stimuli and potential movement targets, their activity increases and they approach a threshold for decision, first about what is out there and then about what to do. At the single-cell level, the neuronal activity appears to ramp up as the CNS accumulates evidence. Figure 25.1 shows an example of such activity for movement-related cells in the frontal eye field (FEF) (see figure 6.3) and saccadic eye movements. Jeff Schall and his colleagues averaged the activity of FEF neurons separately for fast reaction times and slow ones. Movements made with the fastest reaction times were associated with the fastest ramp-up in activity, and it appeared that the saccade occurred when the neuron, presumably representative of many neurons in the network, reached a threshold.

In racetrack models,[16] objects in the environment compete to be classified as response targets. As evidence concerning various targets accumulates in the context of a given task instruction, the decision involves a "race" among the competing, alternative targets. The "runner" that finishes first wins, or some set of early finishers wins collectively. These models assume that accumulators work something like leaky integrators, accumulating evidence independently as an exponential function of time.

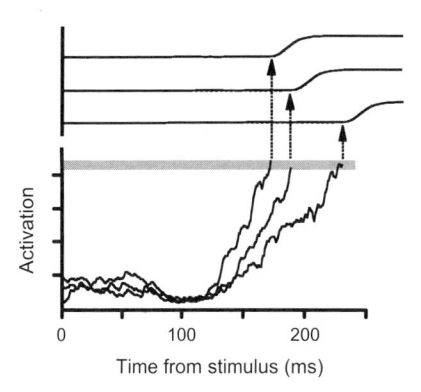

Figure 25.1
A ramp-to-threshold for reaction time. Top: Three traces show the horizontal eye orientation for saccades made with the fastest, middle, and slowest reactions times. Bottom: The activation level of an FEF neuron for three sets of trials, each corresponding to one of the reaction times shown at the top. For the fastest reaction time, the activity rises with the highest slope and reaches the threshold for action (shaded horizontal line) fastest. (From Schall and Thompson[12])

The rate parameter $v(i)$—v for velocity—characterizes this accumulation for target i, and the probability of choosing target i [$P(i)$] is given by the ratio of its rate parameter to the sum of all possible targets j in a response set **R** (i.e., all responses $j \in \mathbf{R}$):

$$P(i) = \frac{v(i)}{\sum_{j \in R} v(j)}.$$

As described up to this point, accumulator models may seem to reflect only passive integration of ascending sensory information, in what you might think of as a *bottom-up* manner. However, accumulator models incorporate *top-down* sensory information processing as well. The CNS uses memory, in the form of representations of event probabilities, to bias sensory analysis. Experience (not to mention the Bayes' theorem) dictates that not all sensory inputs have equal likelihood; indeed, they rarely do. Instead, memory-based interpretations of inputs provide top-down biases that represent hypotheses about the world and what happens in it.[17] For example, consider the case in which proprioception and vision both provide information about location of the hand, but in one case the visual information comes from directly viewing your hand, while in the other case it comes from viewing in a virtual reality setup in which you can see only some animated version of your hand. If there is a discrepancy between proprioception and vision information, the alignment of the two will be influenced by the likelihood that you assign to each kind of information. The likelihood will bias your estimate to rely more on one sensory modality (see section 15.9).

In terms of selecting actions from among a set, these biases may take the form of an altered threshold for one movement as opposed to alternatives or scalars that change the accumulation rates of neural integrators (see box 2.1). Accumulator models posit a steady increase in network activity as the network's threshold is approached.

Accumulator models also accord well with classical reaction-time theories[18]—as extended to include neurophysiological data and concepts—in which variation in both perceptual accuracy and reaction time depends upon stochastic variability in a neural integrator. This similarity raises an important point: The rate of activity growth in the networks representing the "correct" decision depends on both the strength of the signal and the time available for the integrator to accumulate information.[8]

25.1.2 Neuronal Correlates of Decision

Two independent series of experiments, by Mike Shadlen[8] and Jeff Schall[12] and their colleagues, examined activity as a monkey's CNS accumulates sensory information. Schall's laboratory exploited a well-known perceptual phenomenon known as *pop out*. This visual effect occurs when there are several objects of given type in the visual field (e.g., red squares), along with one object of a different type (e.g., a green square). Perceptually, the green square "pops out" of the field of red squares. Schall and his colleagues trained a monkey to make a saccadic eye movement to fixate the stimulus that "popped out." They found that the representation of a movement target emerges gradually as neural signals progress through the visual system, from occipital to frontal cortex. Ultimately, the network for one potential target, the "popped-out" stimulus, ramps up to its threshold.

According to the accumulator–racetrack models, this amounts to a decision to target that stimulus and this information eventually leads to the selection of and preparation of a movement. The accumulator that ramps to threshold first wins the race.[12] (Note, however, that these models usually make some provision for noise by requiring some degree of difference between competing accumulators.) As illustrated in figure 25.1 for a movement-related cell in FEF, it appears that these perceptual and motor networks reach a threshold when cells reach some level of activity.[19] This idea gains further support from the effects of cortical stimulation on reaction time in humans.[20,21]

In related studies of decision making, Bill Newsome, Mike Shadlen, and their colleagues[3] trained monkeys to detect the net movement of a screen full of moving light spots. The spots did not all move in the same direction, but some proportion of them did, a property called *coherent motion*. If hundreds of light spots move in the same direction at the same speed, people and monkeys perceive coherent motion. If only a proportion of the inputs move coherently, however, and the remainder move randomly, people and monkeys may still perceive the motion if enough light spots move together. In addition, the longer they can accumulate information about the inputs, especially for low levels of coherence, the better they can detect the motion. Coherent-motion experiments with monkeys require that they *report* whether the dots move to the left or the right with some movement, typically a saccadic eye movement. Two areas appear to compute the visual information needed to make the required perceptual decision: the medial temporal cortex (MT) and a nearby area called the medial superior temporal cortex (MST) (see figure 6.3). These areas pass information to other cortical and subcortical structures involved in selecting the movement dictated by the perceptual decision. These structures include LIP and the prefrontal cortex, to some extent, but most experts believe that cortical areas such as the FEF and brainstem structures such as the superior colliculus play the largest role in controlling movement execution, at least for eye movements.

In related work from Mike Shadlen and his colleagues,[8] the experimenters presented monkeys with stimuli that had such low coherence that they were near the threshold for detecting the motion. This level of input is often called a *perceptual threshold*, which turns out to be approximately 6%–7% coherence. In one experiment, the monkeys could observe the stimuli as long as necessary to make a decision, and they then initiated their report without further sensory input. Near the perceptual threshold, the monkeys took up to a second to respond, but with as much as 35% coherence, they responded in 300–400 msec. In accord with accumulator models, strong stimulus coherence led to a rapid increase in neuronal activity, whereas with weak coherence, the activity increase took longer. Figure 25.2 shows some data from Jamie Roitman and Mike Shadlen[22] for area LIP neurons. In many ways, these data resemble those Schall and his colleagues obtained from FEF. The LIP cell activity illustrated in figure 25.2B appeared to ramp up to some sort of threshold. This ramp-up takes longer for weaker signals, and so the monkey took longer to respond. Further, if the experimenters restricted the amount of time during which the monkey's CNS could accumulate evidence about coherent motion, the monkey's response accuracy depended on signal strength, in accord with accumulator models (figure 25.2B).

Similar results have been obtained by comparing neuronal activity during *antisaccades* with that during *prosaccades*. The term *prosaccade* refers to a saccadic eye movements made to fixate a visible target. An *antisaccade* is the same eye movement, but is made directly away from a visible target. (Section 25.5.1 places this kind of behavior in the context of a concept called *transformational mapping*, of which antisaccades are a special case.) In both the superior colliculus[23] and the FEF,[9,24] prosaccades are associated with a more rapid increase in activity compared to antisaccades, as well as with faster reaction times.

25.1.3 Deciding to Veto a Planned Movement

The results summarized above deal with overt movements. However, experiments can also detect aspects of the decision-making process before the decision is final. For example, intracortical microstimulation in the FEF evokes eye movements that deviate systematically in the direction of a decision in the making.[8] Further, these nascent decisions can be coun-

25.1. Deciding on a Target

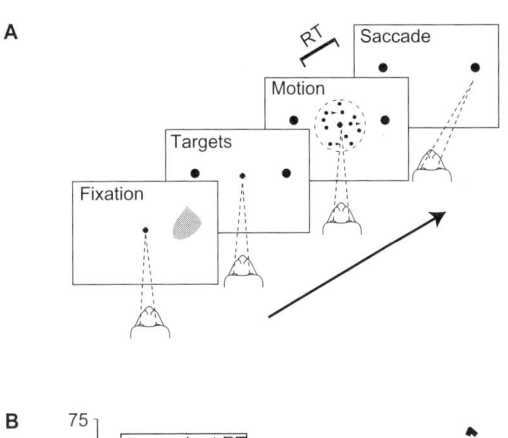

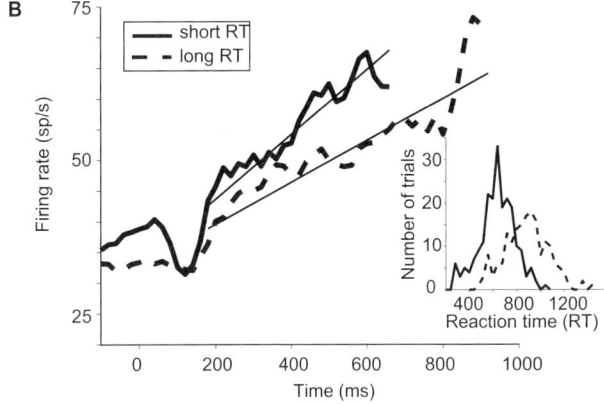

Figure 25.2

Decisions about random dot motion in LIP. (*A*) The experimenters trained a monkey to fixate a light spot at the center of the screen. After fixation, two target choices appeared in the periphery, one to the left and one to the right. One of the targets falls within the activity field of the neuron (gray shading). Dots appeared within a 5° circle around the fixation point and moved around in a complex way. Some proportion of the dots moved either to the left or to the right. The monkey made a saccade to the left or right target to indicate the net direction of random dot motion. In one condition, the monkey made its choice as soon as it could. In another condition, the monkey had to decide after a fixed, 1-sec viewing period and a subsequent delay. Disappearance of the fixation spot triggered the response in that condition. (*B*) Relationship between LIP activity and reaction time (RT) for ~6% image coherence, which was near the monkey's detection threshold. The main figure plots the population average for neuronal activity as a function of time from onset of motion. The experimenters divided the trials from each experiment into short and long reaction times (inset). The solid line shows neuronal activity for short RT, correctly executed trials; the dashed line shows activity for the same cells during long RT, correctly executed trials. Note that when the monkey reacted more rapidly, the average activity accumulated faster. The rate of activity accumulation also increased with increasing image coherence (not shown). (From Roitman and Shadlen[22] with permission.)

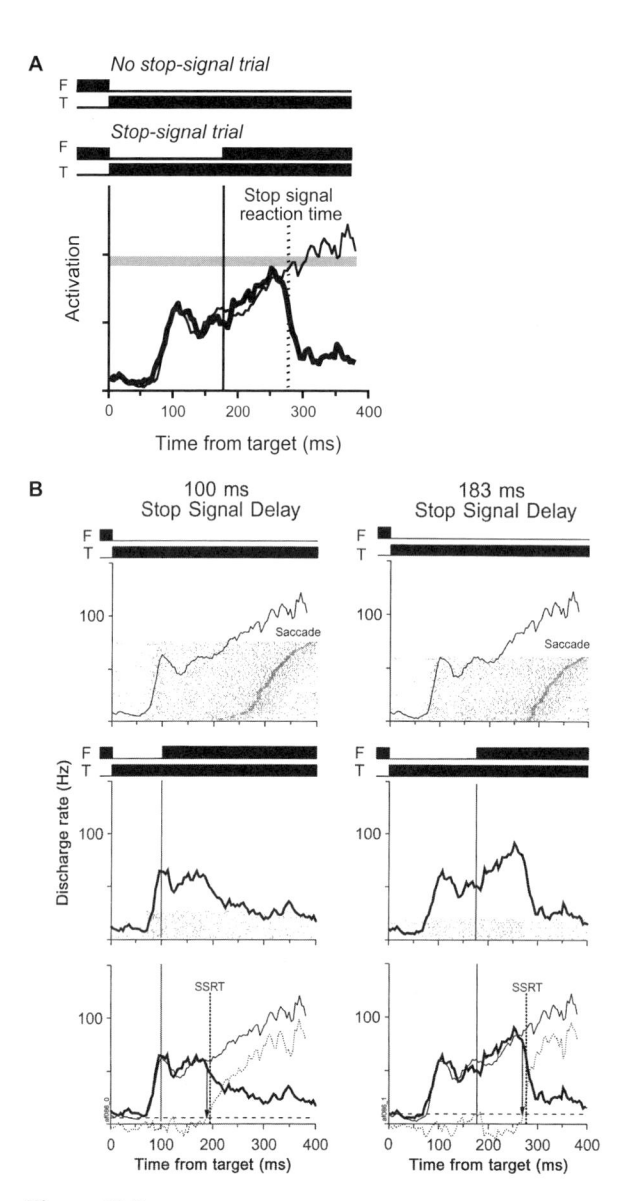

Figure 25.3

Activity for vetoed saccades in FEF. (*A*) Activity of an FEF neuron. After the instruction stimulus appeared at time 0, the activity of the cell increased. The solid vertical line (at 175 msec) shows when the stop signal appeared (if it did). The dotted vertical line shows the time that it took for the monkey to withhold half of the saccades, termed the stop-signal reaction time (SSRT). Note that the signals for regular trials (thin solid line) and for stop-signal trials (thick solid line) diverge shortly before the stop-signal reaction time. The gray horizontal bar shows the approximate threshold for action. Abbreviations: F, fixation point; T, target. The thick bars show when the fixation point and target were on; thin bars, when they were off. (*B*) Activity of a representative FEF neuron with activity aligned on the termanded, or vetoed, by subsequent ones, but the neural activity remains observable. In an eye-movement task, Jeff Schall and his colleagues[25] presented a monkey with a visual target for a saccade (figure 24.3A). In this experiment, the fixation point went out when the target appeared. At some variable time later, the fixation point might reappear. This later event, called a *stop signal*, instructed the monkey to withhold, or veto, the saccade. If the stop signal appeared soon after the target had appeared, the monkey could withhold the saccade on nearly every trial. If the stop signal came at about the same time as the saccade, the monkey could not stop the saccade on any trials. At a certain delay between the target and the stop signal, called the *stop-signal reaction time*, the monkey withheld the movement on half of the trials.

Figure 25.3A and 25.3B shows that when the monkey successfully withheld movement and the saccade did not occur, neuronal activity in the FEF decreased just before reaching a level that appeared to correspond to a decision threshold for making the movement. According to the race-track–accumulator models, your CNS accumulates evidence for both the target and the stop signal. The one that wins the race to threshold determines whether you can successfully veto a planned movement. It appears that the "winning" accumulators suppress the activity of the losing ones, at least after the winner has reached a certain threshold level of activity.

Decisions and choices thus depend on the accumulation of evidence, but what determines the accumulation rate? Evidence from the experiments presented in this section shows that the strength of the signal, the time for gathering the evidence, and the activity of competing accumulators all affect this process. But what determines the choice among potential targets when they do not differ in any of these ways?

25.2 Choosing Among Multiple Potential Targets of Movement

25.2.1 Deciding on Actions Based on Predictions of Reward

Mike Platt and Paul Glimcher[1] explored decisions among competing targets. They compared activity in LIP as a monkey made saccadic eye movements under various expectations of receiving a reward. In one

Figure 25.3 (continued)
time of target presentation. The displays show data for stop-signal delays of 100 msec (left) and 183 msec (right). Note that the SSRT occurred later in the trial when the stop signal appeared later. Top: Activity for trials without a stop signal, matched for reaction time with the trials shown below. The horizontal tick marks show the time of saccade initiation. Middle: Activity for canceled trials. Bottom: Activity for the top and middle, with the dotted lines showing the difference between them. The downward-pointing arrow shows when the differential activity became significant (*A* adapted from Schall and Thompson,[12] *B* from Hanes et al.[25] with permission.)

block of trials, the experimenters rewarded the monkey 80% of the time for fixating a given target; in another block, 20% or some other percentage. Platt and Glimcher observed that LIP cells showed activity that reflected the predicted value of the stimulus in their **activity field**. The more reward the monkey expected for saccades to fixate a particular stimulus location, the more the cell discharged after that stimulus appeared. Neurons in the prefrontal cortex and the striatum have similar properties,[26,27] and some reflect preferred reward. They show more activity when, for example, an apple is the preferred choice of two potential rewards, but not when the apple is the less preferred one.[27,28]

25.2.2 Deciding on Actions Based on Instructions

Platt and Glimcher studied decisions based on one aspect of salience, the predicted reward. Another aspect of salience involves the amount of attention devoted to a stimulus. Indeed, in the Platt–Glimcher experiment, you might reasonably assume that the monkey devoted more attention to locations that would produce reward. To study this aspect of neuronal activity, Mickey Goldberg and his colleagues[29] presented monkeys with an array of eight potential targets for saccadic eye movements. Each target had a particular set of colors and shapes, and the monkeys could always see the array, which remained stationary. An additional visual cue, called a *sample*, appeared, and it matched one of the cues on the screen. The animal used the sample to choose the saccade target. By rule, the monkey had to make a saccade to the stimulus that matched the sample's color and shape. LIP cells did not discharge just because a stimulus appeared in their activity field. The cells discharged sometime after the appearance of the sample cue, but only if a *matching* (i.e., target) stimulus happened to be in their activity field. Presumably, cells in other parts of PPC reflect similar decisions for reaching and pointing movements.

Paul Cisek and John Kalaska[30] extended this work to PMd and reaching movements. They presented two potential movement targets, one red and one blue. The target of movement remained uncertain until a subsequent cue designated one of those stimuli as the target of reach. The later cue, either red or blue, instructed the monkey to choose the target of the same color. Cisek and Kalaska observed two classes of neurons in PMd. *Potential-response* cells showed high levels of activity whenever any potential target fell in their activity field. *Selected-response* cells showed high levels only after a target in their activity field had been designated as *the* specific target for that trial. Rostral PMd tended to have potential-response cells; caudal PMd and M1 had more selected-response cells.

25.3 Deciding on Multiple Movements

The potential-response cells resemble the area 5d neurons illustrated in figure 13.4. Recall that in area 5d, cells showed delay-period activity for stimuli in the cells' activity field (termed the preferred direction in those studies), for both "go" and "no-go" stimuli. PMd cells, however, showed little delay-period activity for "no-go" stimuli. Thus, once a given stimulus cannot be the target of the next reaching movement, PMd discards its representation. In the go–no-go task, this process turns off delay-period activity for "no-go" cues (figure 13.4) and suppresses activity once a *potential* target in its activity field has been designated a nontarget. Area 5d, however, does not discard the nontarget data until some event presents an alternative target.

Taken together with the evidence presented in section 25.2, the evidence shows that more than sensory signal strength and accumulation time affects decisions about movement targets. Accumulator networks also reflect the biological value of potential targets, the attention allocated to them (perhaps because they "grab" attention, perhaps because of rules), and other inputs that designate stimuli and locations as targets or nontargets. But beyond those factors, decisions also depend on a balance between the use of internal cues—such as the state of the motor system and the body—versus external ones.

25.4 Action Selection Based on Estimates of State

The framework presented in figure 19.4 indicates that both proprioceptive and **extrinsic** (predominantly visual) inputs contribute to the computation of a difference vector, which corresponds to the high-level plan for a reaching or pointing movement. Recall that your estimate of current end-effector location results from an alignment that takes into account both the vision of your hand and proprioceptive inputs that, through internal models of **forward kinematics** and **forward dynamics**, predict end-effector location. If, somehow, you "knew" where the target was relative to your hand, you could dispense with visual information altogether. Reaching in the dark, or when you hand is under an opaque table, requires you to do this. It appears that as you gain experience with moving in a certain context, your CNS learns about both the extrinsic and the proprioceptive cues that could guide the movements. But after you have gained a great deal of experience with a certain context, proprioceptive cues and learned motor sequences become more important than external cues and learned event sequences (i.e., the patterns of sensory events). It is as if your CNS no longer cares as much about "what is out there," at least as a series of ongoing actions unfolds. Instead, some initial context triggers the entire series. It is also as if your CNS cares less and less about the consequences of those actions (outcome, in the terminology presented in chapter 4).

To make these ideas concrete, imagine that, new in town, you decide to drive down an unfamiliar road, looking for a particular intersection, at which you plan to turn right. When you find it, you make a mental note of some nearby landmark. As it happens, you end up living in an apartment down the road that follows the right-hand turn. The next few times you travel toward the intersection, the landmark serves as the cue that prompts your decision to turn right. After several weeks of repeating this procedure, the road becomes familiar, and you no longer need the landmark to locate the intersection. Driving down the road and turning right has become so routine that the one time that you need to turn left, you forget and turn right. What neural processes are responsible for this behavior?

25.4.1 Reliance on Internal vs. External Cues to Select Actions

Mark Packard and James McGaugh[31] approached this question by putting rats in the arm of the maze illustrated as a "training trial" in figure 25.4 (right inset). The experimenters blocked entry into the far arm of the maze, so that when the rats reached the junction, they had to turn either left or right. The rats obtained food for only one of those choices, the right

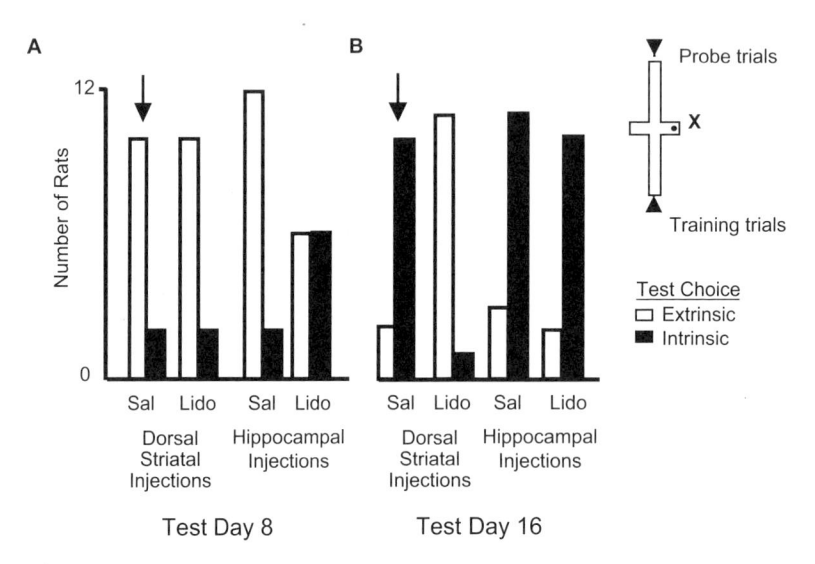

Figure 25.4

Intrinsic versus extrinsic guidance strategies. Right: Rats were trained on the maze illustrated by placing them at the triangle labeled "training trials." To test whether the rats used intrinsic or extrinsic cues, the experimenters placed the rats at the origin labeled "probe trials." X marks extramaze cues, and the black circle indicates the location of a food pellet. Early in training (day 8) most rats chose responses based on extrinsic cues (downward arrow in A). After more experience (day 16), most rats chose a response based on intrinsic cues (downward arrow in B). (*A*) Early in training (day 8), inactivation of the dorsal striatum by injections of lidocaine (lido) had no effect, but inactivation of the hippocampus disrupted the use of extrinsic cues. Injections of saline (sal) had no effect at either site. (*B*) After more experience (day 16), inactivation of the hippocampus had no effect, but inactivation of the dorsal striatum caused the rats to cease relying on intrinsic cues and to rely instead on extrinsic ones, as they had earlier in training. (Redrawn from Packard and McGaugh[31] with permission.)

turn in the illustration. Visual cues outside the maze, called extramaze cues (e.g., objects on the walls in the room where the maze was located), could guide the rats' choice, and the salience of those cues affected their use. The rats' alternative to using these extrinsic, extramaze cues was to use internal, intrinsic ones. For example, internal cues might have included an estimate of how far they had run since the start of the maze. Once they had run a certain distance, they could have decided to turn right to get food. In the conditions used by Packard and McGaugh, most rats used the extramaze cues at first. The experimenters tested which kind of cues the rats used by placing them at the "top" of the maze, labeled "probe trials" in figure 25.4. In probe trials, rats had to run to the junction from the "top" and turn either left or right. If they relied on the extramaze cues (depicted collectively by the X in the figure), they would turn toward these cues, making a left turn. If the rats instead relied on intrinsic cues, they would turn right.

Figure 25.4 (arrows) shows that most rats used the extrinsic cues after 8 days of training, but by day 16 of training, most rats used intrinsic cues. To explore the brain structure underlying the differential cue use, the experimenters injected a local anesthetic, lidocaine (lido), into either the dorsal striatum or the hippocampus, at both stages of training (8 days and 16 days). (Unfortunately, many researchers who study rats call the entire dorsal striatum the "caudate" or the "caudatoputamen." The latter term captures the idea that the part of the striatum studied by these researchers consists of homologues (see appendix A) to both the caudate nucleus and the putamen in primates. Collectively, the caudate and the putamen compose the dorsal striatum.)

Early in training, when most rats used extrinsic cues to guide movement, inactivation of the hippocampus, but not of the dorsal striatum, disrupted their performance (figure 25.4A). Later in training, when most rats used intrinsic cues to guide their movements, the experimenters obtained the opposite result: inactivation of the dorsal striatum, but not of the hippocampus, disrupted the rats' performance (figure 25.4B). It appears that after inactivation of the dorsal striatum, the rats ceased using intrinsic cues and resumed their earlier strategy, which depended on extrinsic cues. This suggests that by day 16, the animals had at their disposal the ability to make a decision based on either extrinsic or intrinsic cues. In rats with a fully functioning CNS, the length of training had resulted in a bias toward intrinsic cues. With the dorsal striatum inactivated, that bias disappeared, and the rats' decisions again depended on extrinsic cues.

In the prevailing interpretation of this result,[32] the rats are said to have shifted from **declarative** to **procedural knowledge** from day 8 to day 16, with the hippocampus underlying the former and the basal ganglia the latter. (Chapter 4, however, noted the difficulty in determining whether nonhuman animals use declarative knowledge at all.)

A different, and more parsimonious, explanation involves the nature of the cues that the rats used to make a decision. If we assume that the process of decision making in this case depends on estimating the rats' states (i.e., their current location relative to a target), then the results suggest that during the initial phase of training, the estimates of state depend on external cues. With training, the rats learn to predict their current state from their previous actions rather than from extrinsic, visual inputs. The rats' decisions continue to rely on an estimate of state, but those state estimations become biased toward internal predictions that rely more on **efference copy** and proprioception than on evidence accumulated by the sensory system.

On this view, with increasing experience, the accumulator networks that made decisions based those decisions on internal cues, while suppressing and "defeating" those based on extrinsic cues. Like the rats, when you drive down the street mentioned above, and turn right when you intended to turn left, your CNS has based its choice of movements on the series of efference copy and proprioceptive signals that occured in a well-learned sequence. Although this example does not specifically involve reaching or pointing movements, the occurrence of events in a familiar sequence lies at the heart of those behaviors as well.

25.4.2 A Sequence of Reaching Movements

Okihide Hikosaka and his colleagues[33,34] have extensively studied the problem of learning to make a sequence of reaching movements in monkeys. They trained monkeys to reach to a sequence of 10 targets. As shown in figure 25.5A, the monkey faced a panel with 16 potential targets, in a 4 × 4 array. As each monkey began the trial, cues appeared at two of the 16 locations. Call them cues 1 and 2. The monkey touched one of the cues, which the experimenters had arbitrarily designated as "first" or "second." If the monkey "mistakenly" touched the "second" cue first, both cues disappeared and the trial ended without the monkey receiving any reward. If the monkey touched the "first" cue first, it could then touch the other one, which caused the delivery of reward. In that case, the first two cues disappeared, and two additional cues appeared. Call these new stimuli cues 3 and 4. The monkey again had to decide between an arbitrarily designated "first" and "second" cue for that pair. Again, if the monkey decided wrongly, both cues disappeared and the trial began again with cues 1 and 2. If the monkey chose correctly, then cues 3 and 4 disappeared and a third pair appeared, until the monkey learned to decide correctly for five consecutive pairs.

A series of experiments from Hikosaka's laboratory, using this design—and a series of related studies in Jun Tanji's laboratory,[35] using a different experimental design—explored the neural basis of sequential movements. Both groups of investigators concluded that medial parts of the motor cortex, specifically the preSMA and the SMA (see figure 6.3), participate in the control of sequential movements, but in different ways.

Hikosaka and his colleagues[36] described changes in neuronal activity as a monkey learned novel sequences. Their data suggest that the preSMA contributes more to learning about the correct order of sequential targets, whereas the SMA (or SMA proper) underlies the speeding up of sequence

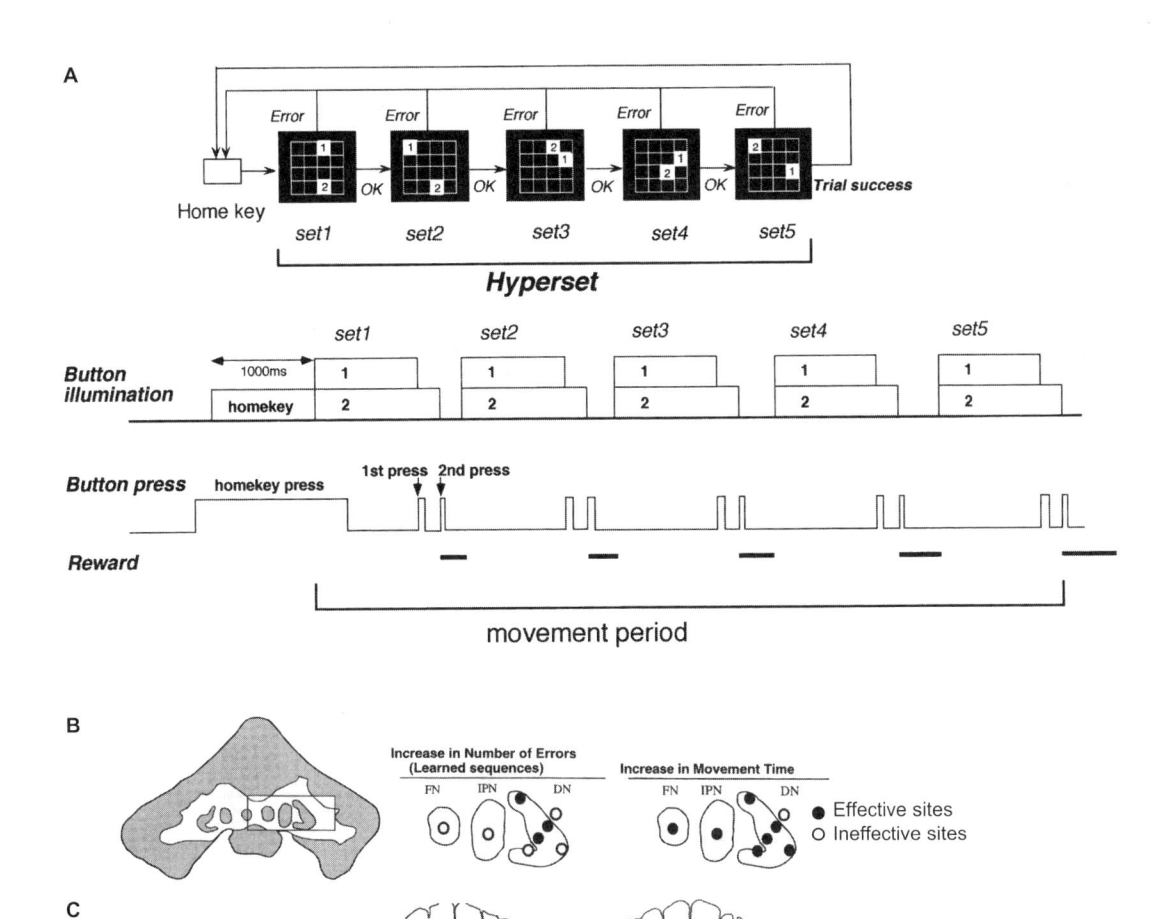

25.4. Action Selection Based on Estimates of State

Figure 25.5
Learning to reach to a sequence of spatial targets. (A) The task devised by Hikosaka and his colleagues for studying sequence learning. Horizontal bars show reward presentation. (B) Injection sites for the GABA agonist muscimol, which inactivated the injected regions. Inactivation of the dentate nucleus (DN) of the cerebellum disrupted the performance of learned sequences. Injections in all three deep cerebellar nuclei, including the fastigial (FN) and interpositus nuclei (IPN) caused a slowing of movement. (C) Injections in the caudate nucleus and rostral parts of the putamen disrupted the learning of new movement sequences. These injections also affected the performance of well-learned movement sequences, but to a lesser extent. Inactivation of a different part of the putamen, a middle part along the rostral-to-caudal axis, did not affect the learning of new movement sequences, but disrupted the performance of well-learned ones. (A from Rand et al.,[33] B and C from Hikosaka et al.[34] with permission.)

execution with experience. Importantly, like PMd neurons, those in preSMA show similar activity for reaching movements with either hand, whereas SMA cells often show more activity modulation for reaches with the contralateral hand. Not only do cells in the preSMA and the SMA change activity during learning,[36] but inactivation of those areas disrupts the performance of movement sequences.[35]

Although the preSMA and SMA play an important role in movement sequences, you should not get the idea that sequential movement depends entirely on the cortex. As suggested by the work of Packard and McGaugh,[31] you might expect the dorsal basal ganglia to play a role in movement sequences and, based on the ideas presented in chapter 16 for prism adaptation and in chapter 20 for force adaptation, you might also expect a contribution by the cerebellum.

Hikosaka and his colleagues obtained evidence that both basal ganglia and the cerebellum contribute to the learning of movement sequences. These investigators injected muscimol, a GABA agonist, to inactivate various parts of the dorsal striatum and deep cerebellar nuclei. Figure 25.5B and 25.5C shows the sites effective for disrupting the learning or performance of movement sequences, using the experimental design depicted in figure 25.5A. Inactivation of the anterior striatum, including the anterior parts of the putamen and caudate nucleus, causes an increase in the number of errors monkeys make in *learning* a new sequence.

Note that, to a first approximation, these anterior parts of the striatum receive most of their inputs from the prefrontal cortex (PF in figure 6.3) and from some sensory cortical areas, such as the inferotemporal cortex (IT), as well. The same injections cause some disruption, in the form of erroneous reaches, of well-learned reach sequences, but to a lesser extent than for novel sequences. Inactivation of the middle parts of the putamen—the parts that receive inputs from frontal motor areas, somatosensory areas, and the PPC—has no effect on the rate of learning novel sequences. It does, however, cause the monkey to make more errors while performing well-learned sequences. As with those more posterior striatal injections, inactivation of the dentate nucleus causes an increase in reaching errors for well-learned sequences. Inactivation of all deep cerebellar nuclei causes a slowing of movements, and none of these inactivations affect the rate of learning new sequences. The effects are generally specific for the hand ipsilateral to the inactivated nucleus.

According to Hikosaka and his colleagues, their data indicate a shift in the structures controlling sequences of reaching movements, as acquisition of a given sequence skill gives way to routine performance. Early in learning, control by the prefrontal cortex (including preSMA[37,38]) and anterior basal ganglia plays the most important role in skill acquisition. Later, more caudally located structures, involving premotor areas (including the SMA), the PPC, and the middle parts of the striatum, control the routine performance of sequences along with the cerebellum. Brain-imaging studies of sequence learning yield a similar picture.[39,40] Hikosaka and his colleagues interpreted their data in terms of declarative and procedural knowledge. They suggest that preSMA and other parts of prefrontal cortex underlie the declarative knowledge of the sequence and that the SMA, middle basal ganglia, and cerebellum underlie the procedural knowledge of the sequence. Note the similarity of their interpretation to those of Packard and McGaugh, except that those experimenters attributed declarative knowledge to the hippocampus.

Notice, however, that the same difficulty arises for both interpretations: How can you evaluate whether rats, monkeys, or other nonhuman animals have declarative knowledge (see chapter 4)? Experts have engaged in seemingly endless arguments about what, if anything, constitutes declarative knowledge in nonhuman animals, and the issue remains unresolved. Perhaps the distinction made above, between the use of extrinsic and intrinsic cues to compute the system's state (with respect to a target), provides a way forward. In the previous section, this framework was used to account for the rat data. We argued that rather than thinking of a transition from declarative to procedural knowledge as the training and testing situation became more familiar, you could think about a transition from using external cues to compute the rat's state to using internal cues, such as efference copy and proprioception, as illustrated in figure 19.4. Can this framework account for the results of Hikosaka and his colleagues?

We think that it can. You do not have to assume that the monkeys have declarative knowledge of the sequence of spatial targets $x_{t1}, \ldots, x_{tn}$ any more than you must assume that they have declarative knowledge of a sequence of joint rotations $\Delta\theta_1, \ldots, \Delta\theta_n$. Both extrinsic guidance in terms of the spatial targets, and intrinsic guidance in terms of a sequence of joint rotations, proprioceptive cues, and motor commands, play an important part in sequence learning. It only stands to reason that as a monkey gains experience with a sequence of targets, it begins by using extrinsic (e.g., visuospatial) cues and gradually makes a transition to intrinsic (e.g., joint-centered) ones. This idea makes certain predictions. For example, the extrinsic, visuospatial guidance of a movement sequence should lead to high levels of arm-to-arm generalization. Hikosaka and his colleagues indeed found that early in learning a sequence, the hand used by the monkey does not matter very much. After extensive experience with a given sequence, however, the monkey can perform the sequence well only with the trained hand. Physiological studies support the idea that the SMA,[41,42] the caudate nucleus,[43] the globus pallidus,[44] and the dentate nucleus of the cerebellum[45] show sequence-related neural activity. Examples of serial-order information have also been obtained for M1.[46]

To put these ideas on a more concrete footing, once again imagine the series of movements needed to tie your shoelaces. Acquisition of this skill began a long time ago, with trial-and-error practice in which you probably used spatial targets about what you should do (e.g., loop one lace under the other, etc.). This strategy gradually gave way to a sequence of joint rotations. That is, you made a transition from using extrinsic cues to using intrinsic cues. As you acquired the shoelace-tying skill, you learned the sequence of joint rotations and forces needed to do the job, and your SMA, M1, and cerebellum became more important, along with the parts of the dorsal striatum receiving inputs from those cortical areas, such as the PPC. Now you can tie your shoes without thinking about the movements or any spatial targets. When you view sequence learning this way, you do not need to speculate about whether monkeys have declarative knowledge. You, of course, might plan your spatial sequences with such knowledge, but you can also learn movement sequences without being aware of the sequence at all, as shown by the *serial reaction-time task* (SRTT), taken up next.

25.4.3 The Serial Reaction-Time Task

SRTT measures the time between movements for a fixed, repeating sequence. For a random sequence of targets, the speed of executing the sequence never improves. However, for a sequence of sufficient length and complexity—one you cannot recognize easily as targets appear one by one—the time between each movement gradually decreases before you become aware of the sequence. Studies by Alvaro Pascual-Leone and his colleagues[47] have shown that the excitability of M1 cortex increases as you implicitly learn the spatial sequence, but decreases once the sequence becomes explicit. Dan Willingham[48] has shown that patients with Huntington's disease fail to show normal learning in an SRTT.

If you substitute the declarative–procedural distinction for the extrinsic–intrinsic one, you can interpret the data of Hikosaka and his colleagues as indicating that sequence learning involves a transition from extrinsic to intrinsic cues, and that the cerebellum and basal ganglia participate in different aspects of sequence learning because of that transition. More realistically, as you gain the ability to use intrinsic cues, you lose the *need* to use extrinsic cues, but not the *ability* to do so; the two are learned in parallel.[49] That is, for a sequence of visually guided movements, there is both a sequence of stimuli in visual space and a sequence of motor commands.

Ann Graybiel[50] referred to the process of learning a sequence of events or motor commands as "chunking." Her idea was that with extensive practice, the basal ganglia encodes long chains of stimulus sequences or sequences of joint rotations, depending on the part of the basal ganglia involved. The benefits of chunking come in the form of smoothly and efficiently executed sequences of movement. Note that early learning involves more use of individual sensory cues. For a visually presented sequence of targets, at first you must attend to each of the visual cues in the sequence. After extensive experience, the first cue triggers the entire sequence. This process might remind you of the functions of the next-state planner in figure 19.4, which computes the next end-effector displacement that should occur to make progress toward a goal, even though the ultimate goal lies many "steps" into the future.

In terms of internal models, as the sequence becomes routine and you learn a task well, the next-state planner (section 19.3) relies less and less on estimates of state derived from external cues, and more and more on estimates of state derived from internal models (e.g., forward models).

Planning actions requires knowing your current state and the state of your environment. For a sequence of actions, in early aspects of training, each action relies strongly on the sensory feedback available immediately preceding the decision about what to do next. As you practice the sequence of actions, you begin to learn to predict the sensory consequences of your actions, and instead of relying on the immediate feedback, you rely on the predicted sensory consequences. In this sense, you switch action selection from a process that relies primarily on external cues that dictate your estimate of state to a process that relies primarily on internally generated cues that estimate that state. The action-selection process tilts toward the internal models as they become better at predicting things. After you have achieved reliable success with the internal models, you become essentially blind to the actual consequences of your actions. You have developed a **habit**, according to the concepts presented in chapter 4.

The cerebellum also contributes to both extrinsically and intrinsically guided movements, but in a different way. It encodes information and makes predictions about the timing of events, including both motor and sensory signals, as outlined in chapters 23 and 24.

25.5 Moving to Places Other Than a Stimulus: Standard Mapping vs. Nonstandard Mapping

Up to this point, the model in figure 19.4 has dealt exclusively with reaching directly to a stimulus or a series of stimuli. In those reaching and pointing movements, your CNS brings the end effector to a stimulus location, x_t. However, you can interact with stimuli in much more flexible ways.[51] Consider, one last time, the imaginary robot in figure 15.1A. Recall that the robot has a gripper as its end effector and a camera. In previous chapters, you imagined programming the robot to use a visible target or a sound source as the target for the gripper. That kind of sensorially guided movement can be called *standard sensorimotor mapping*. You could also program the robot to use a remembered target location without much further development of the robot's control system. The tricky part involves using the information coming from the camera in ways that are more flexible. Collectively, these more flexible relations between inputs and targets can be called *nonstandard sensorimotor mappings*.[51] Figure 25.6A shows an example of one kind of nonstandard mapping, called arbitrary mapping, but first consider a different kind, called transformational mapping.

25.5.1 Transformational Mapping

You might want the robot to have the capability to move its gripper in relation to a visible target, but not directly to it. For example, you might program the control system so that when a stimulus appears at a particular location, the gripper will move to a location 90° clockwise with respect to that location. Imagine that a clockwise-spinning wheel, located above

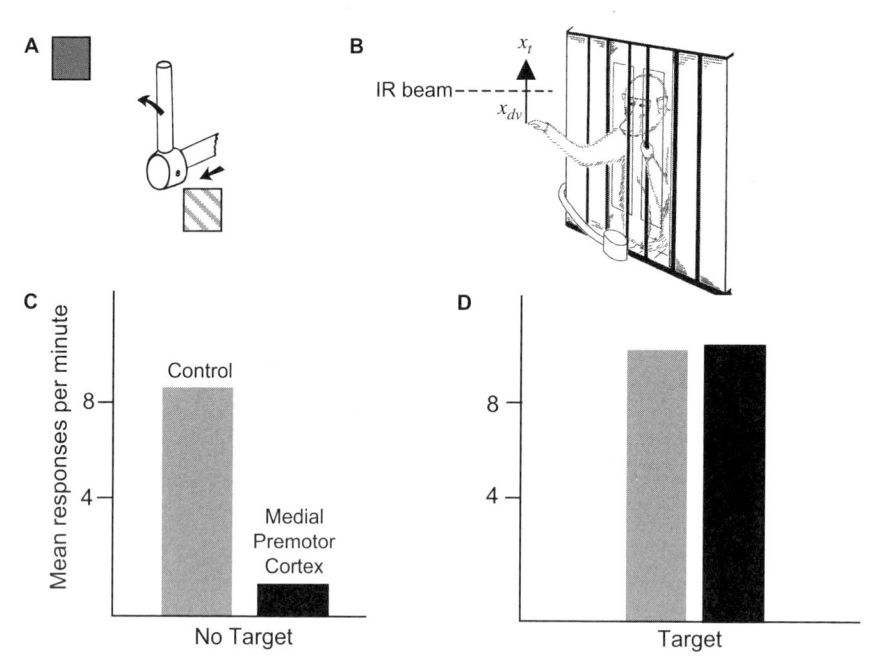

Figure 25.6
Nonstandard sensorimotor mapping. (*A*) Arbitrary visual cues (squares) instruct the monkey either to twist or to pull a handle. Damage to PMd disrupted this behavior. (*B*) Damage to SMA and preSMA caused deficits in arm raising in total darkness. Abbreviation: IR, infrared beam (dashed line). (*C*) Data for task illustrated in B, for total darkness. The bars show the mean number of arm lifts per minute in a control group versus monkeys with lesions of medial premotor areas. (*D*) As in (*C*), but for a visible target. (Drawings and data from Passingham and his colleagues[68,69] with permission.)

the robot, spits out a ball that the gripper must catch by moving to its predicted landing point. Assume that, given the momentum involved, this amounts to reaching at a 90° "lead" in radial coordinates. You can call that kind of relationship between sensation and movement *transformational sensorimotor mapping* because the target of movement depends on cue location, but that location is not the target of movement. Note that this situation *superficially* resembles the "rotation experiments" described in chapters 15 and 16. The two differ, however, in a crucial way: In rotation experiments, the end effector moves as it does in standard mapping, from an initial location to a target. The joint-angle changes needed to make that movement change, but the end effector still moves to the sensory cue. In transformational mapping, the end effector does not move to the visual cue, at least not to the cue that instructed the movement, but moves instead to a target that is a function of the instructing cue's location.

Figure 25.7 develops the model presented in figures 14.1 and 19.4 to incorporate nonstandard sensorimotor mappings. In the fine print, the figure also indicates some of the brain structures that might participate

25.5. Moving to Places Other Than a Stimulus

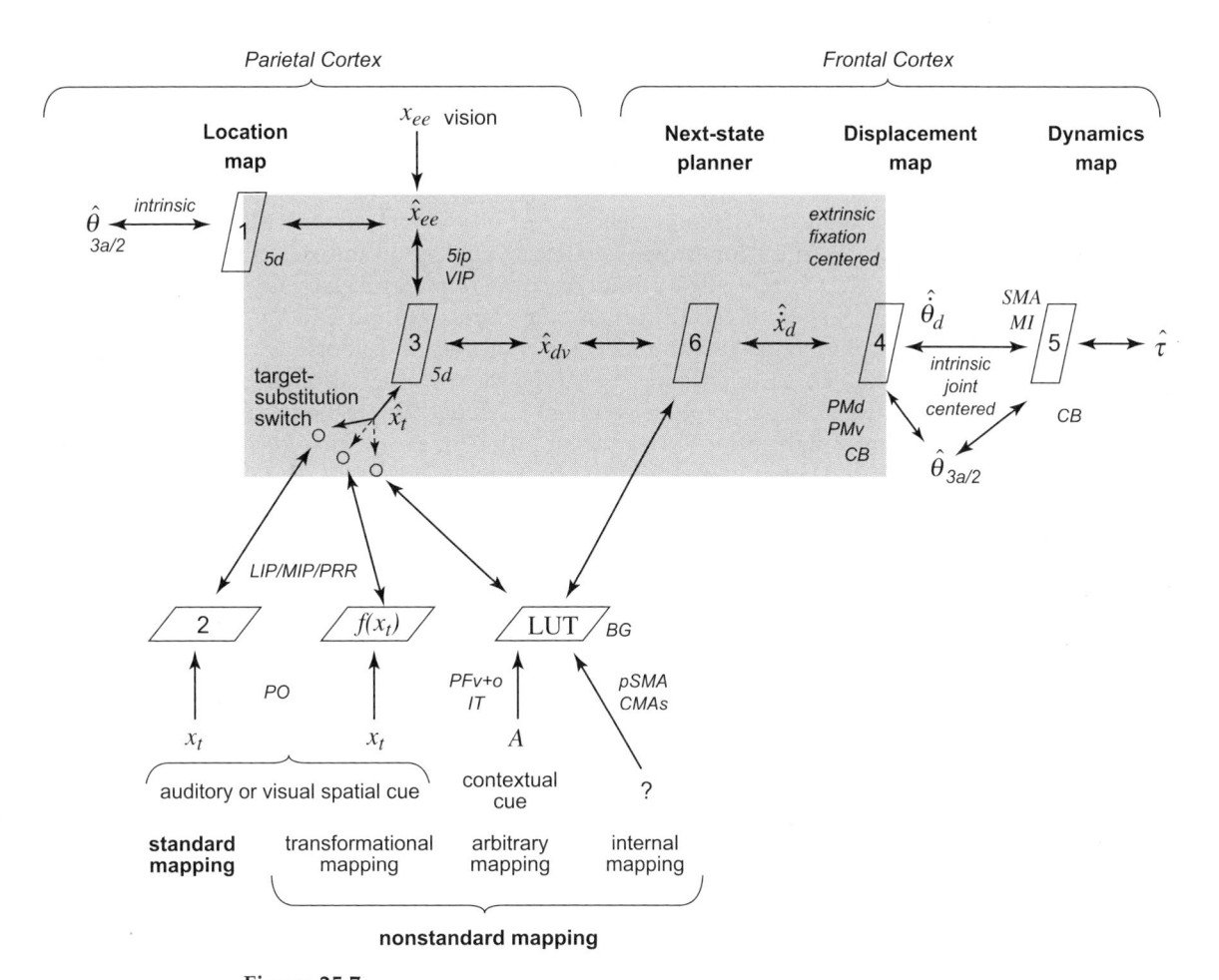

Figure 25.7
Summary. Format as in figure 14.1. Area designations as in the legend to figure 6.3. Note the switch, depicted as a linear array of unfilled circles. Abbreviation: LUT, lookup table. (Drawn by George Nichols.)

most directly in the various mappings and internal models. (We asked the publisher to put that part in disappearing ink, but that technology costs too much.)

For transformational mapping, imagine that a neural network transforms x_t to a target location $\hat{x}_t$ through the population coding of the map $\hat{x}_t = f(x_t)$. In section 21.2, you saw how neural networks can compute these kinds of functions, as well as nonlinear ones. You can think of this kind of mapping as a target-substitution mechanism. The "antisaccade" task mentioned in section 25.1.2 and its analogues for reaching movements (section 13.1.6), represent a special case of transformational mapping, one that involves a 180° angle between the cue and the target of movement (and no change in movement amplitude). Although this task can be described in terms of a "rotation" of the difference vector, x_{dv}, the evidence that the motor system implements transformational mapping through an analogical, algorithmic transform[52] remains controversial. Several models suggest that a more parsimonious, target-substitution mechanism mediates transformational mappings.[53–55] Work on monkeys shows that they can switch between standard and transformational mapping on a trial-by-trial basis.[52] Thus, no learning or other form of adaptation need be invoked to account for transformational mapping; your CNS can switch goals based on target substitution or some related mechanism.

Some results of certain prism-adaptation experiments might become clearer in the context of transformational mapping. For example, Dottie Clower and Driss Boussaoud[56] studied prism adaptation in two conditions. In the first, participants received feedback about the prism-distorted reach from a light attached to the pointing finger, which became illuminated for 200 msec at the end of the reach. Otherwise, the participants received no visual feedback from their arm or hand. In the second condition, the participants reached similarly, but received feedback in the form of a cursor $\hat{x}_{ee}$ on a video screen in relation to a target $\hat{x}_t$ on the same screen. The participants adapted their reaching movement at the same rate for both conditions, but aftereffects occurred only in the first. The model depicted in figure 25.7 accounts for this finding if you assume that in responding to cursors, participants ignored the distortions in visual feedback caused by the prisms. Having recognized from the context that they were now playing a video game, they might have used their proprioceptive mappings $\theta \leftrightarrow \hat{x}_{ee}$ to compute current hand location. Then, based on their errors, participants guessed where the target must be located and reached to that cognitively substituted $\hat{x}_t$. None of the mappings in figure 25.7 needed to adapt in this situation, and therefore no aftereffect occurred. Another study similarly showed that the mode of feedback delivered to participants affects the magnitude of aftereffects, and can be interpreted similarly.[57]

25.5.2 Arbitrary Mapping

For further flexibility, you might want to program the virtual robot to respond to a symbol detected by the camera (such as an object shaped like the letter A). For example, A could serve as an instruction for the robot to move rightward for a given distance, regardless of the location of that symbol in the camera's field of view. You can call that kind of stimulus–response association an *arbitrary mapping*.

Dick Passingham and Mike Petrides, working separately, showed that PMd plays an important role in this kind of nonstandard mapping. Figure 25.6 shows the setup for one experiment. The monkey saw a color cue, call it color A, which instructed it to twist a handle. A cue of color B served as an instruction to pull the handle. Now imagine implementing this kind of flexibility in the virtual robot. Color A provides an instruction to move the robot's gripper to a target in one place, and color B instructs the robot to move it somewhere else.

25.5. Moving to Places Other Than a Stimulus

Figure 25.7 depicts the neural network that converts the symbolic cue into a movement target x_t, a difference vector x_{dv}, or a pattern of joint rotations $\Delta\theta$. You could call the first approach a visuo*spatial* mapping because the arbitrary cue becomes associated with a spatial target $\hat{x}_t$ or a displacement in visual terms x_{dv}. You could call the second computational approach a visuo*motor* mapping because A or B becomes associated with a pattern of joint rotations $\Delta\theta$. Because no systematic relationship exists between the symbolic cue and the target location or any of the motor parameters, the neural network must learn the transform by trial and error, or you could program a lookup table (LUT) into the controller. Then, when the cue appears at some location in the camera's coordinate system, you need some way to decode the pattern and colors of the activated pixels, recognize the pattern, and use the lookup table to choose a target or movement. Note that although the visual cue appears at a location x_t, its location is irrelevant. Indeed, for your CNS, direct reaching to x_t needs to be suppressed.

It would seem reasonable to use a target-substitution mechanism $A \to \hat{x}_t$ rather than the alternatives. As attractive as this idea is, there are problems with it. Matthew Rushworth, Phil Nixon, and Dick Passingham[58] tested monkeys with ablations of areas 5d, MIP, and 7b before and after surgery. Figure 25.6A illustrates the task. Removal of those parts of the PPC did not cause a deficit. Accordingly, on the assumption that network 3 in figure 25.7 depends on the PPC, these findings suggest that arbitrary mappings need not depend on a target-substitution mechanism, at least not unless one exists independently of the PPC. However, the specifics of the arbitrary mapping task used by Rushworth and his colleagues weakens this conclusion. Twisting or pulling a given handle differs from choosing a spatial target or reach direction based on an arbitrary context. Consider this an open question.

Considerable evidence from monkeys suggests that networks involving PMd, prefrontal cortex, and parts of the basal ganglia play important roles in arbitrary mapping. The inferior temporal cortex (IT in figure 6.3) processes inputs that convey nonspatial visual information, such as color and form. Interactions between inferior temporal cortex and prefrontal cortex are necessary for normal rates of learning arbitrary mappings.[59-61] The medial parts of the temporal lobe also assist arbitrary sensorimotor learning by speeding the rate of learning,[62,63] but these areas are not necessary for learning the mappings more slowly.

Note that figure 25.7 does not show an interaction between the lookup table and either forces or torques. In the present model, the final stage of the transformations involves mappings that align joint rotations with muscle forces and torques. For example, in order to lift a full soda can, your CNS sends motor commands to muscles that produce larger forces than for an empty can (see chapter 20). Can you associate nonspatial visual cues, such as color, with these internal models of dynamics?

Francesca Gandolfo, Sandro Mussa-Ivaldi, and Emilio Bizzi[64] studied force-field adaptation for two force patterns, each associated with a specific color of light in the testing room. Participants never learned to use the color to predict the pattern of forces, despite hundreds of trials. Ashwini Rao and Shadmehr[65] did a similar experiment, for movements in only one direction. On any given trial, the target was either red or green, with each color indicating a different force field. Participants practiced for over 2000 trials, spread over three days. They could not learn this skill at all. (Note, however, that John Kalaska and his colleagues[66] showed such learning in a very well practiced monkey.) In contrast, with spatial cues (a little to one side or the other of the target), the participants could learn the task in 2000 trials.[65] Remarkably, once they could associate the spatial cues with the patterns of force, they quickly learned to associate colors with the same forces, as could people given randomly varied colors.[67]

These results suggest that the map of dynamics (network 5 in figure 25.7) has somewhat limited access to color information and requires a spatial intermediary. This conclusion is consistent with situating this network partly in M1 (and probably SMA, as well) and with the idea that M1 cortex receives only highly indirect inputs from cortical areas that encode nonspatial visual information. Likewise, the cerebellum seems to receive only sparse inputs from cortical areas that encode nonspatial information, such as color.

25.5.3 Internal versus External Mapping

Little is known about the vast aggregate of processes grouped in figure 25.7 as internal mappings, so we leave it to you to imagine programming the virtual robot to do such things. The gradual substitution of intrinsic for extrinsic cues relies on such internal mapping. In monkeys, according to Dick Passingham and his colleagues,[68,69] the medial parts of the motor cortex, including the SMA and the cingulate motor areas, subserve internally selected reaching movements. Figure 25.6B illustrates the experiment that led to this conclusion. In a totally dark room, the monkey had to lift its hand to break an infrared beam (in order to receive some reward). Thus, the monkey could have imagined some virtual target $\hat{x}_t$, a pattern of joint rotations $\Delta\theta$, or a displacement x_{dv}, in which the hand's trajectory brought it past the infrared beam. Ablation of SMA, preSMA, and some cingulate motor areas caused a deficit in a monkey's ability to make such a reaching movement in total darkness (figure 25.6C). However, if a visible target appeared, the monkey made the same number of reaching movements as controls did (figure 25.6D). Ablations of PMd had no such effect.

And, of course, a really smart android could do a lot more than that. It could, for example, make reaching and pointing movements based on abstractions such as rules and strategies. The prefrontal cortex plays an important role in such decisions. For example, bilateral damage to prefrontal cortex prevented monkeys from choosing a reach target on the basis of two particular strategies.[60] These findings underline the fact that the targets for reaching and pointing arise from rules, strategies, and abstract goals, as well as from standard and nonstandard sensorimotor mappings.

25.6 Summary

Reaching and pointing movements rely on computations that transform proprioceptive and visual information into motor commands. These transformations depend on internal models that (1) align proprioception with vision; (2) compute target location; (3) calculate a difference vector; (6) use a next-state planner to compute next end-effector location according to a control policy that determines its trajectory; (4) align this vector with the joint rotations needed to reach the target; and (5) convert these joint rotations into force and torque commands. (The numbers are out of sequence because they correspond to the networks numbered 1–6 in figure 25.7.)

Because of the predominance of vision, your CNS plans movements in terms of visual coordinates: It uses a fixation-centered reference frame, even for movements to auditory targets. Given the choice between a visually straight and an actually straight movement, your CNS chooses a visually straight one. The dependence on fixation-centered coordinates requires your CNS to update these mappings when your eyes change orientation. It also does so when either the target or your end effector moves, and it can do so predictively (i.e., in advance of the movement). Both of these updating mechanisms depend on efference copy and the predictions of forward models (figure 17.10). Although the neural basis of each internal model remains incompletely understood, some neurophysiological evidence indicates that mappings 1, 2, 3, and 6 appear to depend on parts of the PPC, whereas mappings 4 and 5 appear to involve frontal motor areas, functioning both in reciprocal parietofrontal networks and in recurrent modules involving the cerebellum and basal ganglia (see section 6.2.1).

All of these mappings adapt. Altering vision with prisms that displace the visual world, for example, changes the alignment between proprioception and end-effector location (mapping 1) as well as that between the output of the next-state planner and joint rotations (mapping 4). Altering the weight of the arm or imposing forces on it during movement changes the mapping of joint rotations onto force and torque commands (mapping 5). With repeated adaptation, often consuming many weeks and hundreds of movements, new internal models form that can be recalled with contextual cues. For visually guided reaching in primates, motor learning begins with the adaptation of existing internal models and ends in the formation of new ones. The system has acquired a new skill. The nature of the representations that your CNS uses to encode each variable in these maps affects the way it generalizes from the precise conditions encountered during motor learning to other situations. The tuning curves and other encoding mechanisms used by the elements computing these internal models also affect the ease or difficulty of adapting to particular kinds of perturbations and acquiring new skills.

The internal models for reaching to sensory targets underlie *standard sensorimotor mapping*. They also support the flexibility that characterizes reaching and pointing using *nonstandard sensorimotor mappings* (figure 25.7). The prefrontal and premotor cortex play the most important roles in nonstandard mappings, acting in concert with the basal ganglia and both the medial and the lateral parts of the temporal cortex. These capacities include the ability to reach—based on the locations of sensory cues—to other locations (*transformational mappings*) and to guide reaching through the nonspatial features of symbolic stimuli, such as their color or shape (*arbitrary mappings*). Finally, reaching may be based on nonsensory neural signals (*internal mappings*), including intrinsic cues such as proprioception and the efference copy of motor commands. In addition, and perhaps most important, your reaching and pointing movements can also arise from rules, strategies, and goals of unlimited abstraction.

References

1. Platt, ML, and Glimcher, PW (1999) Neural correlates of decision variables in parietal cortex. Nature 400, 233–238.
2. Gold, JI, and Shadlen, MN (2000) Representation of a perceptual decision in developing oculomotor commands. Nature 404, 390–394.
3. Shadlen, MN, and Newsome, WT (2001) Neural basis of a perceptual decision in the parietal cortex (Area LIP) of the rhesus monkey. J Neurophysiol 86, 1916–1936.
4. Kim, JN, and Shadlen, MN (1999) Neural correlates of a decision in the dorsolateral prefrontal cortex of the macaque. Nat Neurosci 2, 176–185.
5. Thompson, KG, Bichot, NP, and Schall, JD (1997) Dissociation of visual discrimination from saccade programming in macaque frontal eye field. J Neurophysiol 77, 1046–1050.
6. Thompson, KG, Hanes, DP, Bichot, NP, and Schall, JD (1996) Perceptual and motor processing stages identified in the activity of macaque frontal eye field neurons during visual search. J Neurophysiol 76, 4040–4055.
7. Thompson, KG, and Schall, JD (1999) The detection of visual signals by macaque frontal eye field during masking. Nat Neurosci 2, 283–288.
8. Gold, JI, and Shadlen, MN (2000) Representation of a perceptual decision in developing oculomotor commands. Nature 404, 390–394.
9. Everling, S, and Munoz, DP (2000) Neuronal correlates for preparatory set associated with pro-saccades and anti-saccades in the primate frontal eye field. J Neurosci 20, 387–400.
10. Schall, JD (2000) From sensory evidence to a motor command. Curr Biol 10, R404–R406.
11. Bichot, NP, and Schall, JD (1999) Effects of similarity and history on neural mechanisms of visual selection. Nat Neurosci 2, 549–554.
12. Schall, JD, and Thompson, KG (1999) Neural selection and control of visually guided eye movements. Annu Rev Neurosci 22, 241–259.
13. Leon, MI, and Shadlen, MN (1999) Effect of expected reward magnitude on the response of neurons in the dorsolateral prefrontal cortex of the macaque. Neuron 24, 415–425.

14. Dorris, MC, and Munoz, DP (1998) Saccadic probability influences motor preparation signals and time to saccadic initiation. J Neurosci 18, 7015–7026.

15. Kawagoe, R, Takikawa, Y, and Hikosaka, O (1998) Expectation of reward modulates cognitive signals in the basal ganglia. Nat Neurosci 1, 411–416.

16. Logan, GD, and Gordon, RD (2001) Executive control of visual attention in dual-task situations. Psychol Rev 108, 393–434.

17. MacKay, DM, and Gardiner, MF (1972) Two strategies of information processing. Neurosci Res Prog Bull 10, 77–78.

18. Luce, RD (1986) Response Times: Their Role in Inferring Elementary Mental Organization (Oxford University Press, New York).

19. Hanes, DP, and Schall, JD (1996) Neural control of voluntary movement initiation. Science 274, 427–430.

20. Pascual-Leone, A, Brasil-Neto, JP, Valls-Sole, J, et al. (1992) Simple reaction time to focal trandcranial magnetic stimulation. Brain 115, 109–122.

21. Pascual-Leone, A, Valls-Sole, J, Wassermann, EM, et al. (1992) Effects of focal transcranial magnetic stimulation on simple reaction time to acoustic, visual and somatosenory stimuli. Brain 115, 1045–1059.

22. Roitman, JD, and Shadlen, MN (2002) Response of neurons in the lateral intraparietal area during a combined visual discrimination reaction time task. J Neurosci 22, 9475–9489.

23. Everling, S, Dorris, MC, Klein, RM, and Munoz, DP (1999) Role of primate superior colliculus in preparation and execution of anti-saccades and pro-saccades. J Neurosci 19, 2740–2754.

24. Sato, TR, and Schall, JD (2003) Effects of stimulus-response compatibility on neural selection in frontal eye field. Neuron 38, 637–648.

25. Hanes, DP, Patterson, WF, and Schall, JD (1998) Role of frontal eye fields in countermanding saccades: Visual, movement, and fixation activity. J Neurophysiol 79, 817–834.

26. Schultz, W, Tremblay, L, and Hollerman, JR (2000) Reward processing in primate orbitofrontal cortex and basal ganglia. Cerebral Cortex 10, 272–283.

27. Hikosaka, K, and Watanabe, M (2000) Delay activity of orbital and lateral prefrontal neurons of the monkey varying with different rewards. Cerebral Cortex 10, 263–271.

28. Tremblay, L, and Schultz, W (2000) Reward-related neuronal activity during go-nogo task performance in primate orbitofrontal cortex. J Neurophysiol 83, 1864–1876.

29. Gottlieb, JP, Kusunoki, M, and Goldberg, ME (1998) The representation of visual salience in monkey parietal cortex. Nature 391, 481–484.

30. Cisek, P, and Kalaska, JF (2002) Simultaneous encoding of multiple potential reach directions in dorsal premotor cortex. J Neurophysiol 87, 1149–1154.

31. Packard, MG, and McGaugh, JL (1996) Inactivation of hippocampus or caudate nucleus with lidocaine differentially affects expression of place and response learning. Neurobiol Learn Memory 65, 65–72.

32. Packard, MG, and Knowlton, BJ (2002) Learning and memory functions of the basal ganglia. Annu Rev Neurosci 25, 563–593.

33. Rand, MK, Hikosaka, O, Miyachi, S, Lu, X, and Miyashita, K (1998) Characteristics of a long-term procedural skill in the monkey. Exp Brain Res 118, 293–297.

34. Hikosaka, O, Sakai, K, Nakahara, H, et al. (2000). In The New Cognitive Neurosciences, ed Gazzaniga, MS (MIT Press, Cambridge, MA), pp 553–572.

35. Shima, K, and Tanji, J (1998) Both supplementary and presupplementary motor areas are crucial for the temporal organization of multiple movements. J Neurophysiol 80, 3247–3260.

36. Nakamura, K, Sakai, K, and Hikosaka, O (1998) Neuronal activity in medial frontal cortex during learning of sequential procedures. J Neurophysiol 80, 2671–2687.

37. Picard, N, and Strick, PL (2001) Imaging the premotor areas. Curr Opin Neurobiol 11, 663–672.

38. Luppino, G, Matelli, M, Camarda, R, and Rizzolatti, G (1993) Corticocortical connections of area F3 (SMA-proper) and area F6 (pre-SMA) in the macaque monkey. J Comp Neurol 338, 114–140.

39. Jueptner, M, Stephan, KM, Frith, CD, et al. (1997) Anatomy of motor learning. I. Frontal cortex and attention to action. J Neurophysiol 77, 1313–1324.

40. Hikosaka, O, Sakai, K, Miyauchi, S, et al. (1996) Activation of human pre-supplementary motor area in learning of sequential procedures: A functional MRI study. J Neurophysiol 76, 617–621.

41. Mushiake, H, Inase, M, and Tanji, J (1991) Neuronal activity in the primate premotor, supplementary, and precentral motor cortex during visually guided and internally determined sequential movements. J Neurophysiol 66, 705–718.

42. Clower, WT, and Alexander, GE (1998) Movement sequence-related activity reflecting numerical order of components in supplementary and presupplementary motor areas. J Neurophysiol 80, 1562–1566.

43. Kermadi, I, and Joseph, JP (1995) Activity in the caudate nucleus of monkey during spatial sequencing. J Neurophysiol 74, 911–933.

44. Mushiake, H, and Strick, PL (1995) Pallidal neuron activity during sequential arm movements. J Neurophysiol 74, 2754–2758.

45. Mushiake, H, and Strick, PL (1993) Preferential activity of denate neurons during limb movements guided by vision. J Neurophysiol 70, 2660–2664.

46. Carpenter, AF, Georgopoulos, AP, and Pellizzer, G (1999) Motor cortical encoding of serial order in a context-recall task. Science 283, 1752–1757.

47. Pascual-Leone, A, Grafman, J, and Hallett, M (1994) Modulation of cortical motor output maps during development of implicit and explicit knowledge. Science 263, 1287–1289.

48. Willingham, DB, and Koroshetz, WJ (1993) Evidence for dissociable motor skills in Huntington's disease patients. Psychobiology 21, 173–182.

49. Willingham, DB, and Goedert-Eschmann, K (1999) The relation between implicit and explicit learning: Evidence for parallel development. Psychol Sci 10, 531–534.

50. Graybiel, AM (1998) The basal ganglia and chunking of action repertoires. Neurobiol Learn Memory 70, 119–136.

51. Wise, SP, di Pellegrino, G, and Boussaoud, D (1996) The premotor cortex and nonstandard sensorimotor mapping. Can J Physiol Pharmacol 74, 469–482.

52. Lurito, JT, Georgakopoulos, T, and Georgopoulos, AP (1991) Cognitive spatial-motor processes. 7. The making of movements at an angle from a stimulus direction: studies of motor cortical activity at the single cell and population levels. Exp Brain Res 87, 562–580.

53. Whitney, CS, Reggia, J, and Cho, S (1997) Does rotation of neuronal population vectors equal mental rotation? Connect Sci 9, 253–268.
54. Cisek, P, and Scott, SH (1999) An alternative interpretation of population vector rotation in macaque motor cortex. Neurosci Lett 272, 1–4.
55. Moody, SL, and Wise, SP (2001) Connectionist contributions to population coding in the motor cortex. Prog Brain Res 130, 245–266.
56. Clower, DM, and Boussaoud, D (2000) Selective use of perceptual recalibration versus visuomotor skill acquisition. J Neurophysiol 84, 2703–2708.
57. Norris, SA, Greger, BE, Martin, TA, and Thach, WT (2001) Prism adaptation of reaching is dependent on the type of visual feedback of hand and target position. Brain Res 905, 207–219.
58. Rushworth, MFS, Nixon, PD, and Passingham, RE (1997) Parietal cortex and movement. I. Movement selection and reaching. Exp Brain Res 117, 292–310.
59. Eacott, MJ, and Gaffan, D (1992) Inferotemporal-frontal disconnection: The uncinate fascicle and visual associative learning in monkeys. Eur J Neurosci 4, 1320–1332.
60. Bussey, TJ, Wise, SP, and Murray, EA (2001) The role of ventral and orbital prefrontal cortex in conditional visuomotor learning and strategy use in rhesus monkeys. Behav Neurosci 115, 971–982.
61. Bussey, TJ, Wise, SP, and Murray, EA (2002) Interaction of ventral and orbital prefrontal cortex with inferotemporal cortex in conditional visuomotor learning. Behav Neurosci 116, 703–715.
62. Murray, EA, and Wise, SP (1996) Role of the hippocampus plus subjacent cortex but not amygdala in visuomotor conditional learning in rhesus monkeys. Behav Neurosci 110, 1261–1270.
63. Brasted, PJ, Bussey, TJ, Murray, EA, and Wise, SP (2002) Fornix transection impairs conditional visuomotor learning in tasks involving nonspatially differentiated responses. J Neurophysiol 87, 631–633.
64. Gandolfo, F, Mussa-Ivaldi, FA, and Bizzi, E (1996) Motor learning by field approximation. Proc Natl Acad Sci USA 93, 3843–3846.
65. Rao, AK, and Shadmehr, R (2001) Contentual cues facilitate learning of multiple models of arm dynamics. Soc Neurosci Abstr 27, 3024.
66. Krouchev, NI, and Kalaska, JF (2003) Context-dependent anticipation of different task dynamics: Rapid recall of appropriate motor skills using visual cues. J Neurophysiol 89, 1165–1175.
67. Osu, R, Horai, S, Yoshioka, T, and Kawato, M (2004) Random presentation enables subjects to adapt to two opposing forces on the hand. Nat Neurosci 7, 111–112.
68. Chen, Y-C, Thaler, D, Nixon, PD, et al. (1995) The functions of the medial premotor cortex (SMA) II. The timing and selection of learned movements. Exp Brain Res 102, 461–473.
69. Thaler, D, Chen, Y-C, Nixon, PD, et al. (1995) The functions of the medial premotor cortex (SMA) I. Simple learned movements. Exp Brain Res 102, 445–460.
70. Schall, JD (2001) Neural basis of deciding, choosing and acting. Nat Rev Neurosci 2, 33–42.
71. Lebedev, MA, and Wise, SP (2002) Insights into seeing and grasping: Distinguishing the neural correlates of perception and action. Behav Cog Neurosci Rev 1, 108–129.

72. Wise, SP (1996). In Acquisition of Motor Behavior in Vertebrates, eds Bloedel, JR, Ebner, TJ, and Wise, SP (MIT Press, Cambridge, MA), pp 261–286.

73. Wise, SP, and Murray, EA (1999) Role of the hippocampal system in conditional motor learning: mapping antecedents to action. Hippocampus 9, 101–117.

74. Wise, SP, and Murray, EA (2000) Arbitrary associations between antecedents and actions. Trends Neurosci 23, 271–276.

75. Murray, EA, Bussey, TJ, and Wise, SP (2000) Role of prefrontal cortex in a network for arbitrary visuomotor mapping. Exp Brain Res 133, 114–129.

76. Murray, EA, Brasted, PJ, and Wise, SP (2002). In The Neurobiology of Leaning and Memory, eds Squire, LR, and Schacter, D (Guilford, New York), pp. 339–348.

Reading List

Gordon Logan and Robert Gordon,[16] Paul Glimcher in *Decisions, Uncertainty, and the Brain: The Science of Neuroeconomics* (MIT Press, Cambridge, MA, 2002), and Jeff Schall[70] have discussed the mechanisms and concepts underlying decisions and choice. You might consult Wise et al.[51] and Lebedev and Wise[71] for a more detailed discussion of section 25.5. Arbitrary sensorimotor mapping has been the subject of numerous review articles, which have focused on the changes in cortical activity that accompany the learning of these mappings,[72] the role of the medial parts of the temporal cortex[73,74] and that of the prefrontal cortex,[75] and the relevance of this kind of learning to the life of monkeys and people.[76]

V Glossary and Appendixes

Glossary

absolute coordinates, absolute reference frame   reference frame that remains unchanged with movements of the body or parts of the body

activity field   region of space for which a neuron is responsive; in sensory physiology, activity fields are most commonly called receptive fields, and in motor physiology they are called motor fields

actuator   a biological or mechanical device that produces force

adaptation   motor learning that results in changes in existing attractors in the neural networks that align, map, or control action; typically, the rate of adaptation is similar to the rate of de-adaptation

adaptive critic   teaching signal that changes to reduce errors

advanced   in evolutionary biology, an evolutionary development that represents a change from the **primitive**, ancestral condition

aftereffect   inaccurate motor performance that follows **adaptation** when the perturbation causing the adaptation is removed

agonist   a muscle that generates movement toward a goal, contrasts with antagonist; in pharmacology, a chemical agent that activates a receptor

angular velocity   rate of change in an angle, such as a joint angle; see appendix D

antagonist   a muscle that generates force in a direction opposite to that of an agonist muscle

appendage   an outgrowth of the body wall, such as a fin or a limb

appendicular   relating to an **appendage**

attention   allocation of information-processing resources

autonomic nervous system   nervous system output operating autonomously (i.e., without voluntary control); controls activities of body organs and tissues, such as heart rate

ballistic movement   a movement so fast that **feedback** from the periphery does not have time to influence the trajectory

basis function   a function that encodes a subspace; through a collection of basis functions, one can **represent** a space

Bayesian theory   using the knowledge of prior events and probabilities to predict the probability of future or current events

caudal   toward the tail; often, but not always, synonymous with posterior

central pattern generator, CPG   neural network in the CNS that produces rhythmic motor commands

coactivation   simultaneous activation of **agonist** and **antagonist** muscles

cognitive relating to knowledge, often implying conscious awareness of that knowledge

comparator device or neural network that subtracts one signal from another to compute their difference

conditional probability the probability of something happening if some other thing happens

consolidation change in **memory** from a labile form to a more stable form

controller neural or machine mechanism that controls movements in order to achieve a goal

corollary discharge synonymous with **efference copy**

corticospinal neurons neurons projecting from the cerebral cortex directly to the spinal cord

critic see **adaptive critic**

deafferented referring to the loss of somatosensory inputs to the CNS

declarative knowledge information available to consciousness; contrasts with procedural knowledge. See also **explicit, implicit**

degrees of freedom number of independent dimensions of movement allowed by a joint or a set of joints

direction of action direction in which muscles generate force and **torque**

distributed modules neural circuits involving two or more distinct anatomical structures, often recurrent

dynamics in motor-systems research, the generation of force or pertaining to force generation; contrasts with **kinematics**

efference copy motor command signal sent somewhere in the CNS

end effector body part, object, or stimulus being controlled

endocrine see **neuroendocrine**

endoskeleton limb architecture in which hard elements are on the inside, such as vertebrate limbs; contrasts with **exoskeleton**

epigenesis unfolding of a genetic program to create biological structures and support their functions

exoskeleton limb architecture in which hard elements are on the outside, such as insect limbs; contrasts with **endoskeleton**

explicit tasks performed with conscious awareness, thought to depend upon the declarative memory system; contrasts with **implicit**

extrafusal muscle fibers that generate force; contrasts with **intrafusal**

extrinsic coordinates, extrinsic reference frame reference frame based on an origin and axes outside the body

feedback, feedback control control of motor output by the consequence of that output; termed negative when output diminishes further output, and positive when output enhances further output

fitness, inclusive fitness net reproductive advantage and traits maximized or optimized in association with gene survival across generations

forward dynamics movements that result when **actuators** generate a given pattern of force

forward engineering intentional design of an artifact to achieve a goal

forward kinematics computation of limb locations for a given pattern of joint rotations

forward model computation of a future state based on current feedback and motor commands

fovea part of the retina with the highest resolution (and color vision), which serves central vision

fractionate, fractionation separate control of elements of a skeletomuscular system that usually work together

GABA, GABAergic neurons using the neurotransmitter γ-aminobutyric acid (GABA), which is usually inhibitory, especially in adults

gain field neuronal response modulated by some multiplicative factor, which depends on some other variable

ganglion, ganglia knot or knots of cells, usually in the peripheral nervous system; the basal ganglia is a group of telencephalic nuclei, not ganglia in this sense

gaze orientation of the eyes, retina, and **fovea**

genome the entirety of an animal's genes and its apparatus for replication

glia nonneuronal cells in the nervous system, thought to function in support and development

habit in *biology*, an innate form of behavior; in *psychology*, an automatic action learned on the basis of associations among stimuli, actions, and trial-and-error outcome

homeostasis state of a biological system operating in a normal range, in equilibrium between factors tending to increase and to decrease a controlled variable (e.g., body temperature, blood pressure, heart rate, et cetera), typically a feedback control system

homunculus picture of the human body, distorted to reflect neural representations, often used for other species (by analogy)

implicit tasks performed without conscious awareness, thought to depend upon the **procedural** memory system; contrasts with **explicit**

inertia, inertial the tendency of a body with mass to remain in motion

insertion attachment of muscle tendon to bone

instrumental learning, instrumental conditioning changes in performance based on an animal's history with stimuli, responses, and outcomes

interaction torque forces generated by the movement of one body part on another, usually unwanted

internal model a CNS **representation** of the **kinematics** and **dynamics** of a motor task

intrafusal muscle fibers that receive innervation by muscle-spindle afferents; contrasts with **extrafusal**

intrinsic coordinates, intrinsic reference frame reference frame based on the body or parts of the body

inverse dynamics computation of muscle activations and forces needed to reach a goal

inverse kinematics computation of joint rotations or movements needed to reach a goal

isometric generation of force against an immovable object or an equal force

joint coordinates a description of the configuration and movements (velocity, acceleration, etc.) in terms of joint angles

kinematics in motor-systems research, movement or the generation of movement; contrasts with **dynamics**

learning acquisition of information for storage in **memory**

ligand in pharmacology, a chemical agent that activates a receptor

lineage evolutionary group that involves an ancestral species and descendants
long-loop reflex reflexes involving the cerebral cortex
manipulandum object grasped by a participant in an experiment, often a handle or joystick
memory stored information; in psychology, sometimes used synonymously with declarative memory (archaic)
moment arm length of a lever transmitting **torque** to a rotating object
monosynaptic a neuron-to-neuron connection that involves a direct projection and only one synapse
motor command neural or machine signal that moves an **end effector**
motor neuron cell that directly, synaptically contacts muscle fibers
motor pool nucleus of **motor neurons** projecting to a single muscle
motor primitive a fundamental **motor command**, usually taken to occur in limited numbers, that can be summed in various combinations to produce movements
motor program series of **motor commands**
muscle-spindle afferent nerve innervating the force and length sensory transducer in the muscle
myelin, myelinate, myelination the membranous wrapping around axons, which functions to speed conduction; originates from **glia**
myoblasts immature precursors to striated muscle fibers
myotome segmental muscle mass found primarily in fish
neglect persistent failure to perceive and respond to stimuli in one part of space
neural crest a group of cells that appear early in development and lie between the neural tube, which is the precursor of the CNS, and the developing skin
neuroendocrine neural system for controlling internal secretions, typically into the circulation; endocrine refers to a gland with secretions that act generally and not just locally
oculomotor regarding eye movements
olfaction sense of smell
overcomplete system a system with more **degrees of freedom** than needed for a task
overlearned a behavior that is very frequently performed and has long since been learned to maximal proficiency
placodes skin thickenings that appear early in development and give rise to sensory **transducers**
population vector an average directional signal derived from a population of neurons, each with a **preferred direction**
posture fixed positioning of **end effectors**
preadaptation also known as an exaptation, an evolutionary development that permits future, different developments
preferred direction direction associated with the highest level of neuronal or EMG activity
prime mover muscles that provide the most force or **torque** for a given movement
primitive having evolved earlier or resembling what has evolved earlier; not necessarily simple or simpler than its antonym, **advanced**
procedural knowledge information stored in the CNS that is not available to conscious awareness; contrasts with **declarative knowledge**

projection neurons neurons that send signals from one neural structure to another, as opposed to local circuit neurons, which convey signals within a structure

pronation for an outstretched forearm, a rotation such that palm is moved to face downward; contrasts with **supination**

proprioceptive system, neurons motor interneurons (i.e., neurons that do not project outside the CNS)

proprioceptor sensory receptor transducing mechanical information

propriospinal system of axonal projections connecting different segments of the spinal cord

psychophysics the study of behavior through measurement

reafference the sensory consequences of a system's own movements

reflex a sensory-to-motor circuit that involves relatively direct pathways in which the motor output (M) can readily be described as a function of sensory input (S), such that $M = f(S)$

relative coordinates, relative reference frame a coordinate frame based on the moving part of the body; contrasts with **absolute reference frame**

representation information about the world or the body in the CNS

reverse engineering puzzling out how an existing system works

rostral literally, beakward, toward the front; often, but not always, synonymous with anterior

saccade a **ballistic** eye movement for reorienting the **fovea** and thereby changing gaze orientation

salience, salience map a computation that assigns a biological significance to stimuli

serial homologue similar structure in several segments of an animal's body

short-loop reflex reflex circuit confined to the spinal cord (and its serial homologues in the brainstem)

skeletomuscular, skeletomotor system the system of bones and muscles

skill motor **learning** that results in formation of new attractors in the neural networks that align, map, or control action; enhances the system's capability

somatotopy, somatotopic organization a map of the body laid out in a brain structure

stretch reflex motor output that is a function of muscle stretch

supervised learning learning in which the correct output is provided

supination for an outstretched forearm, a rotation such that the palm is moved to face upward, contrasts with **pronation**

synaptic weight efficacy of the influence of one neuron on another

synergist, synergistic muscles that work together to produce a given movement

tension force

torque angular force; see appendix D

transcranial magnetic stimulation, TMS induction of electrical current in the CNS with electromagnetic coils held close to the head

transduce to convert one form of information into another

vestibular system sensory system receiving inputs regarding head movements and gravity

virtual work work (see appendix D) done in an infinitely short period of time

work force times distance, in units of energy (see appendix D)

Appendix A: Biology Refresher

Nothing makes sense in biology except in the light of evolution, and one theme of this book is that motor **learning** overcomes the problems posed by your evolutionary history. If you do not have much of a background in biology, a few basic principles will help you understand the biological aspects of this book. Remember that the philosophy of biology differs from the philosophy of physics. Less concerned with inviolate laws of nature, biology concerns the history of life on Earth and the principles of living systems, as often as not illuminated by the exceptions that "prove the rule."

The three basic facets of biological organization are history, structure, and function—in other words, evolution, anatomy, and physiology—and their interactions. As multicellular animals evolved, they required developmental programs to elaborate their form from a single cell. A complex series of regulatory genes, transcription factors, chemical attractants and repellants, receptors, secondary messengers, and other molecular mechanisms made all of this happen according to a developmental program encoded into the **genome**. Thus, the evolution of organisms involves the evolution of their developmental programs.

Homology and Analogy

Although evolutionary relationships reflect descent and inheritance, in practice they have traditionally been revealed through a comparison among traits in existing species and in the fossils of extinct ones. More recently, the methods of molecular biology and genetics, especially regarding transcription factors and other gene regulators, have come to play an increasingly important role in understanding evolutionary relationships.

Comparative anatomy emphasizes two fundamental kinds of similarity between structures: homology and analogy. A *homologous* relationship implies that a trait is similar in two species because they inherited it from a common ancestor. *Analogy* implies only that a trait serves a similar function, without reference to an evolutionary relationship. Thus, the wings of birds and bats are usually said to be analogous but not homologous. Those structures provide lift for flying, but no ancestor common to bats and birds had wings. Wings evolved independently several times in evolution, as did paired eyes and many other structures. Note, however, that homology can depend on hierarchy. As structures subserving flight, the wings of birds and bats are analogous because no vertebrate ancestor common to bats and birds had wings. As forelimbs, however, the wings of birds and bats are homologous because they were inherited from a common reptilian ancestor.

Shared Traits

Traits include behaviors, functions, and structures. Many traits are shared among members of a phylogenetically related group. Understanding two classes of these shared traits helps you to understand evolutionary history. One class of shared traits, called shared derived traits (technically, *synapomorphies*) are those that evolved in a stem species. All mammals have hair. Mammals share hair because stem mammals evolved that structure and not because any more distant ancestor did so. All descendants of stem mammals (i.e., all mammals) have hair by virtue of inheritance. Similarly, all birds share feathers because stem birds evolved that trait. Sometimes, biologists have an odd way of talking about shared derived traits. Take, for example, the trait of having four legs and feet, common to so many land animals (technically, *tetrapods*). The first land animals evolved this trait, and all descendants of stem tetrapods belong to the group of four-footed animals. In this sense, snakes are said to be "four-footed" animals even though they come up about four feet short. They are four-footed in a historical sense.

A second class of shared traits, termed shared primitive traits (technically, *plesiomorphies*), reflects a different history. All mammals and birds have vertebrae, but not because a mammal was an ancestor of any bird or vice versa. Rather, all mammals and birds have this trait because they inherited vertebrae from ancestors more distant than the stem groups that gave rise to either. Distant vertebrates "invented" vertebrae, and all vertebrates share this trait by virtue of that inheritance.

Causation in Biology

In biology, causality takes two forms: ultimate causation and proximate causation. Ultimate causation involves the evolutionary causes for a biological structure or function. Proximate causation involves the physiological mechanisms of forming that structure or performing that function. Take the question: What causes a muscle to produce force? The proximate causes of force, as described in section 7.2.1, involve the Ca^{++}-mediated interaction of actin and myosin molecules within muscle, the mechanisms for initiating contraction, et cetera. These causes account for *how* a muscle produces force. Other causal factors, the stuff of ultimate causation, explain *why* muscles attained the relevant properties in the first place. These factors include the adaptive pressures that caused less-differentiated cells to become specialized for the production of force in multicellular animals and the genetic, evolutionary mechanisms for selecting and replicating those properties.

Brutality in Biology

Life on Earth is hardly benign. The history of life has been punctuated by occasional catastrophes, and what goes on between those events is no picnic either. In the evolutionary history of vertebrates, two major catastrophes destroyed most of the animals on land. Large numbers of **lineages** died out in these two events. One, about 250 million years ago, is called the *Permian* extinction. It took out most of the original group of animals, called *tetrapods*, that populated the land. This event opened opportunities for one of their descendant groups, called *amniotes* (because they have eggs with three membrane layers, one of which is called the amniotic sac). Amniotes in the form of reptiles, including the dinosaurs, "ruled the land"—as the saying goes—until the next catastrophe, called the *Cretaceous* extinction, about 65 million years ago. An unassuming group of furry amniotes—mammals, of course—filled many of the niches left vacant by the extinction, and birds filled many others.

Appendix B: Anatomy Refresher

The presentation of neuroanatomy in this book avoids traditional Latin nomenclature to the extent possible. As many a student has wisely requested, *Heu, modo itera omnia quae mihi nunc nuper narravisti, sed nunc Anglice?* (Listen, would you repeat everything you just told me, only this time in English?[1]) There was a time when nearly every well-educated person knew Latin and at least a little Greek. It seems pointless, however, to pretend that this remains so. Nevertheless, an anatomical name such as *substantia nigra pars compacta* does not benefit from translation into "dense dark stuff." Where translation is not an option, the text often mentions what some of the terms mean in English as a mnemonic aid or uses an acronym, such as SNc for substantia nigra pars compacta.

Standard anatomical nomenclature, unsurprisingly, uses body-centered coordinates. Imagine a dog standing on all fours, nose directly to the front, tail straight out to the rear. "Dorsal" refers to up (i.e., toward the vertebrae and spinal cord; "ventral," to down (toward the belly). "Rostral" (beakward) refers to a direction toward the mouth. "Caudal" means toward the tail. A near-synonym for "rostral" is "anterior," and in many cases the synonym of "caudal" is "posterior." The midsagittal plane, also known as the midline, intersects the spinal column perpendicular to the ground and extends from the middle of the nose at its front tip to the tip of the tail, with one foreleg and one hindleg on each side.

This reference system works reasonably well for most vertebrates, which have the good sense to keep their heads forward and their vertebrae oriented parallel with the ground or seafloor. Translating this coordinate scheme to humans generates some confusion, however. The problem stems from your habit of upright posture. "Anterior," the word used for "toward the front," now takes on a different meaning than in most other vertebrates. "Anterior" no longer corresponds to "rostral," but instead it means the same thing as "ventral." Thus, you can find, especially in older texts, the ventral horn of the spinal cord referred to as the anterior horn. Similarly, at least to an approximation, "posterior" loses its correlation with "caudal" and gains one with "dorsal."

Reference

1. Henry Beard, *Latin for All Occasions* (Pastimes Press, London, 1991).

Appendix C: Mathematics Refresher

Differentials and Notation

An object's change in *location* (x) from an initial condition to an end condition is denoted Δx. For the present purposes, *position* has the same meaning as location. A common way of expressing the rate of change in location involves dots over the variable:

$$\dot{x} = \frac{dx}{dt},$$

which indicates that velocity is the first derivative of location. The first term is pronounced "x dot."

Supplementary and Complementary Angles

A supplementary angle is the angle that needs to be added to another angle to reach $180°$; a complementary angle is the angle needed to reach $90°$.

Laws of Sines and Cosines

For the triangle with sides of length, $a, b,$ and c and angles opposite those sides of $A, B,$ and C, respectively, the law of sines states that

$$\frac{a}{\sin(A)} = \frac{b}{\sin(B)} = \frac{c}{\sin(C)} = 2r = d,$$

where r is the radius of the circle containing the triangle and d is its diameter.

The law of cosines states that $c^2 = a^2 + b^2 - 2ab\cos(C)$.

Degrees and Radians

Most angular measurements are described in units of degrees, a measure that dates to the ancient Babylonians and Egyptians. A relic of this past can be found in a hieroglyphic still in use, the degree sign °, which

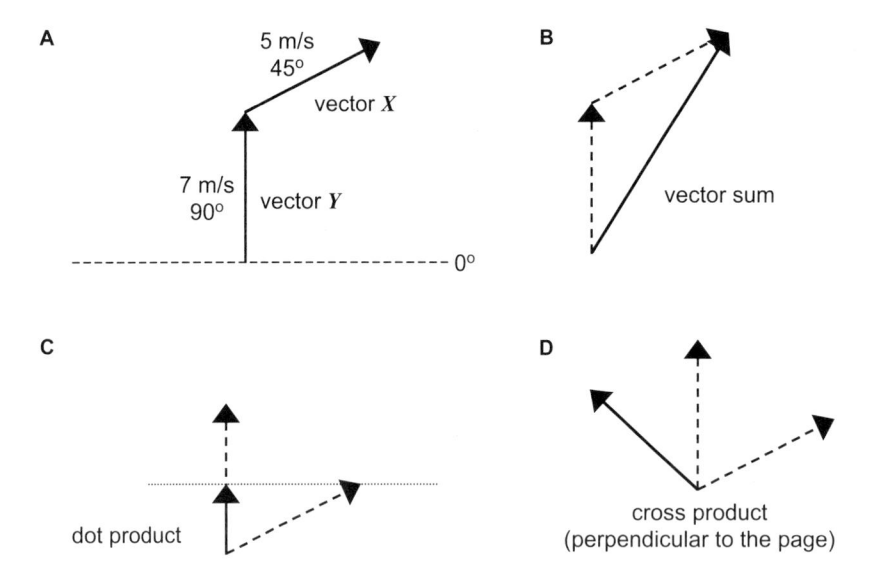

Figure C.1
Vector sums and products. Each vector has a direction, denoted by the arrow's orientation, and a magnitude, indicated by the arrow's length. In this convention, 0° is to the right. (*A*) Two vectors placed tip to tail. (*B*) The component vectors (dashed lines) sum to the resultant vector (solid line). (*C*) The dot product (solid line) of two vectors (dashed lines). (*D*) The cross product (solid line) of the other two vectors (dashed lines) should be understood as being perpendicular to the page (i.e., straight up if the page is horizontal).

represents the sun. Another useful measure of angle is the *radian*. One radian is the angle needed to subtend an arc that is the same length as the radius (r) of the circle. Because the circumference of a circle is $2\pi r$, the number of radians in a circle is approximately 6.28. Accordingly, one radian equals ~57.3°.

Vectors and Matrices

A vector is a multidimensional mathematical descriptor essential for describing movements. Nearly all movements have more than one dimension (e.g., direction and speed).

Figure C.1 shows two vectors, their sums, their dot products, and their cross products. In mathematical notation, you can write these two vectors as a single matrix

$$\begin{bmatrix} 5 & 7 \\ 45 & 90 \end{bmatrix},$$

where the left column corresponds to vector **X** in figure C.1 (upper left) and the right column corresponds to vector **Y**. Matrices have determinants (det)

$$\det \begin{bmatrix} a & b \\ c & d \end{bmatrix} \equiv \begin{vmatrix} a & b \\ c & d \end{vmatrix} \equiv ad - bc$$

and transpositions

$$\mathbf{A} = \begin{bmatrix} a & b \\ c & d \end{bmatrix} \quad \mathbf{A}^T = \begin{bmatrix} a & c \\ b & d \end{bmatrix},$$

the latter of which has the following identity:

$$\frac{1}{\mathbf{A}^T} = (\mathbf{A}^{-1})^T.$$

In the example, above,

$$\det \begin{bmatrix} 5 & 7 \\ 45 & 90 \end{bmatrix} = 450 - 315 = 135.$$

Note that the determinant of vector A equals that of its transpose A^T.

The dot product of two vectors, such as **X** and **Y** in the figure, can be viewed as the projection of vector **X** onto vector **Y** (or, in other words, the amount that vector contributes to the vector sum in the direction of vector **Y**).

$\mathbf{X} \cdot \mathbf{Y} = |X||Y| \cos(\theta)$, where θ is the angle between **X** and **Y**.

The cross product of two vectors is

$\mathbf{X} \times \mathbf{Y} = |X||Y| \sin(\theta).$

Thus it is another vector that is perpendicular to the plane containing the first two. Its direction is given by the right-hand rule, which is that if **X** is crossed with **Y**, as in the illustration above, then the fingers of the right hand point to **Y** and the thumb points away from the page. The cross product $\mathbf{Y} \times \mathbf{X}$ is a vector pointing directly away from you, into the book (figure C.1).

Appendix D: Physics Refresher

In classical, Newtonian mechanics, the study of systems in which momentum does not change is called statics, whereas dynamics involves the study of changes in momentum and the forces, including torques, that cause momentum to change. Kinematics is the term used to describe changes in the location of objects, independent of the forces that cause those movements. Movements from one location to another are called translational movements; those around an angle are called rotational movements. In this book, however, "dynamics" refers to forces and "kinematics" refers to movements (e.g., changes in joint angle or end-effector location).

Translational Physics

Mass is a property of matter: it has weight and takes up space (i.e., it has a **location**).
Velocity is a description of the motion of objects (i.e., the change in location). Velocity is a vector having a size, a speed, and a direction.
Momentum (P) is mass times velocity. Objects in motion have momentum, which is a vector having a size and a direction. In the equation below, the arrow over a symbol indicates that it is a vector. The size of that momentum is the mass of the object (m) times the object's velocity (v). The direction of the momentum is the same as the direction of the object's velocity.

$$\vec{P} = m\vec{v}$$

Impulse is the change in momentum (i.e., $mv_{\text{final}} - mv_{\text{intial}}$).
Force is mass times acceleration, Newton's second law. The force (F) required to accelerate an object of 5 kg by 1 meter per second (m/sec), every second, is 5 kg-m/sec^2, which equals 5 newtons (N).
Weight is the force of the Earth's gravitational acceleration (g) on an object and has the same units as force.
Work (W) is defined as the transfer of energy; work is done on an object when you transfer energy to that object: $W = F \cdot x$, where x is the change in location of the object. Applying a force of 5 newtons to move an object over a distance of 10 m = 50 N-m = 50 joules (J), a unit of energy.

Energy is the ability to do work.

Kinetic energy, $E_K = mv^2/2$, is how much energy is in the motion of an object, independent of its location.

Power (P) is the rate of energy use. The relationship between work and power is important in motor physiology. Imagine moving a mass of 5 N from point A to point B, a distance of 10 m. The amount of energy required to do that work does not vary as a function of the time taken to move the mass. Clearly, however, more of something is required to move the mass in a minute than in over an hour. That something is power. For reaching movements of the arm, this means that a given hand movement requires more power, depending on its speed. Thus, as demonstrated in the equation below, power is the product of force times velocity:

$$P = \frac{dW}{dt} = \frac{d(F \cdot x)}{dt} = F\frac{dx}{dt} = Fv.$$

A unit of power is a watt, which equals one joule of energy expended per sec:

$$1 \; watt = \frac{1 \, J}{s} = \frac{1 \, N \cdot m}{s} = 1 \, N \cdot \frac{m}{s} = 1\frac{kg \cdot m}{s^2} \cdot \frac{m}{s}.$$

Rotational Physics

All of the properties and measures listed above for translational physics have angular analogues, based on the force applied to a **moment arm**, which you can imagine as a long stick.

- Force is analogous to torque
- Mass is analogous to the moment of inertia
- Momentum is analogous to angular momentum.

Moment of inertia (I) is the rotational analogue to mass. Inertia is the tendency of objects in motion to remain in motion; rotational inertia, the moment of inertia, is the tendency of objects that are spinning to continue spinning: $I = mr^2$.

Angular velocity (ω) is the rate of rotation around a joint. In physiology and kinematics, angular velocity is often termed joint velocity, although the joint does not move. Its units are degrees per second (°/sec):

$$\omega = \frac{d\theta}{dt}.$$

Angular momentum, $(L) = I\omega$, is a vector like linear momentum.

Angular acceleration (α) is the tangential acceleration of a moment arm times its length, and also, of course, the derivative of angular velocity and the second derivative of angle:

$$\alpha = \frac{d\omega}{dt} = \frac{d^2\theta}{dt^2}.$$

Torque is a force applied to a moment arm. Torque (τ) equals the force times the length of the moment arm. Thus, its units are newtons times meters, abbreviated N-m, Nm, or N · m:

$$\tau = I\omega.$$

Angular work (W) = torque (τ) times the change in angle (θ), analogous to force times distance. $W = \tau\theta$ and, like linear work, its units are energy. **Angular power** (P) is, like power generally, the rate of energy use, and also is the product of torque and angular velocity:

$$P = \frac{dW}{dt} = \frac{d(\tau \cdot \theta)}{dt} = \tau\frac{d\theta}{dt} = \tau \cdot \omega.$$

This parameter has been called joint angular power, and its units are N-m-deg/sec.

Spring Loads

Spring loads and forces. $F = -k \cdot x$, where x is the distance that a spring is stretched from its resting length, the equilibrium condition. Force is proportional to a spring constant k. The negative sign comes from the fact that the spring force is always directed toward the equilibrium point. The concept of a spring is related to that of elasticity. A spring, however, is a highly linear system in which the spring constant, k, is the same during stretching and relaxing, and no energy is lost in cyclic loading. Elasticity, by contrast, may have constants that differ for stretching and relaxing, and may have other nonlinearities as well.

Viscosity

Viscosity is a property of matter, usually a fluid, in which the force required to move through the fluid is proportional to the velocity of movement. For example, to move a stick through oil at a very slow speed requires a certain amount of energy to do that work. As viscosity increases, the same movement requires more force, and therefore more energy, and the amount of that force increment increases with the velocity of movement.

Technically, muscles have the properties of both viscosity and elasticity, and hence are termed viscoelastic. To a good approximation, the elastic element can be modeled as a spring, notwithstanding the difference between springy and elastic elements.

Appendix E: Neurophysiology Refresher

Action potentials transmit information from one cell to another through synaptic transmission. The current caused by an action potential is large enough that a thin metal wire, called a microelectrode, can detect a single cell's action potential from outside the cell. The waveform of an extracellularly monitored action potential lasts ~1 msec. Careful examination of the waveform makes it possible, often enough, to distinguish the activity of one cell from that of neighboring ones, and this information is usually expressed in terms of discharge rate, usually impulses (or "spikes") per sec. This measure is often referred to simply as a cell's *activity*, but is also called firing rate, activity rate, or impulse rate. A single neuron is often called a *unit*, because the potential recorded from the cell has an all-or-none, unitary property that contrasts with other, more graded and slowly changing potentials picked up by electrodes in and near the nervous system. Excitatory input to a neuron is often called "drive."

The terminology used in this book depends on two signals common in neurophysiological experiments: an *instruction stimulus* that guides the selection of a movement, including its target and its goal, and a *trigger stimulus* that signals the time when the movement should be launched, often called a "go" stimulus. The time between these events is an *instructed delay period* (or, simply, a *delay period*).

In many experiments, the instruction and trigger stimuli are combined into a single input. Tasks in which the instruction and trigger occur simultaneously are often called choice reaction-time tasks because the participant is unaware of which target to choose until the time comes to react. In other experiments, the instruction and trigger stimuli are distinct, which allows activity related to the selection and preparation of movement to be separated in time from that involved in the execution and monitoring of movement.

An activity increase immediately after an instruction stimulus is called *signal-* or *cue-related activity*. A sustained activity increase that continues for up to several seconds after an instruction stimulus and fills most of the time between the instruction stimulus and the trigger stimulus is termed either *delay-period activity* or *set-related activity*. An increase in activity immediately prior to (and during) movement is termed *movement-related activity*. It is common to divide movement-related activity into a phase after the trigger stimulus (TS) but prior to movement onset, called a "reaction-time" or RT period, and a phase during the movement, termed the "movement-time" or MT period.

The same terms apply to decreases in activity relative to a baseline or background level. Increases in discharge usually are called excitatory, and decreases are termed inhibitory. Those terms are misleading in a sense because the term "inhibitory" appears to imply that the cell inhibits other cells. Instead, the term means that the cell receives a net inhibitory influence, which could include a reduction in its excitatory drive or an enhancement of inhibitory inputs. Those terms are so common in the neurophysiological literature that the text uses them here and there. The absolute value of a difference in activity (e.g., delay-period activity), relative to a baseline or background level, is called *modulation*.

Index

Abbott, Larry, 198–201
Abeele, Sylvie, 314
Abstract mapping, 58
Accumulator models, 496–499
Acetylcholine, 452
Actin, 35, 94
Action potential, 547
Activity fields, 172–175, 184
  adaptation and, 295–296, 313–314
  Gaussian distribution and, 214–215
  gaze location and, 259
  multiple targets and, 503–504
  planning and, 239
  remapping and, 319–333
  target location and, 179–201
Actuators, 27, 245. *See also* Muscles
  biological vs. mechanical, 93–94
  land locomotion and, 31–32
  springy, 34–35
  target location and, 179–203
Adaptation, 18, 30, 403
  activity fields and, 295–296, 313–314
  aftereffects and, 48, 126, 276, 278, 286, 384, 386, 435, 448, 516
  altered dynamics and, 391–401
  arbitrary mapping and, 516–518
  basal ganglia and, 462–464
  cerebellum and, 301–302, 314–316, 447–448, 455–458
  complex spike rate and, 478–480
  consolidation and, 316, 435–444
  context and, 280–283
  decision making and, 51–58 (see also Decision making)
  EMG changes and, 393–395
  eyeblink conditioning and, 459–461
  eye rotation and, 275–278
  filtering and, 449–454
  Gaussian function and, 303–304, 313
  generalization functions and, 303–316, 420–431
  interaction torques and, 48
  ION and, 301–302
  Jacobian matrices and, 279, 307–308
  joint coordinates and, 308
  map realignment and, 305–316
  motor learning and, 47–48, 51–58
  newts and, 275–276
  planning and, 286–287
  PMv deactivation and, 295–296
  PPC and, 303
  pre-exposure and, 305
  preferred direction and, 313, 395–401
  primates and, 276–278
  prism, 276–286, 299–301, 516
  representations and, 304
  rotation and, 395–397
  signal cancellation and, 449–454
  sleep and, 441–444
  spatial neglect and, 296–301
  stiffness and, 423–424
  systems identification and, 423
  time-dependent change and, 436–441
  transformational mapping and, 513–516
  virtual robotics and, 273–275, 284–286
  VOR and, 486–488
Adaptive critic, 455

Aftereffects. *See* Adaptation
Agnathans, 12
Airplanes, xv–xvi
Ajemian, Robert, 266
Alexander, Gary, 252–254, 257, 259
Algorithms, 4
Allen, Todd, 456
Allocortex, 82
Amnesiacs, 435
Amphioxus, 15
Analogy, 533–534
Andersen, Richard, 324
 difference vectors and, 234, 241–242
 fixation-centered coordinates and, 211–212, 216
 target location and, 188, 190–191, 193–198
Angular acceleration, 120–121
Angular velocity, 102, 385–386
 defined, 544
 gain fields and, 426–429
 limb velocity and, 426–429
Antagonist muscles, 47
Anterior intraparietal cortex, 89
Anti-Hebbian learning, 452, 455
Appendages, 5, 27, 29–30, 39
Arbib, Michael, 354, 357–360, 489
Arbitrary mapping, 516–518
Ariff, Greg, 332
Arm configuration, 254–257
Artificial neural networks, 273–292
Ashe, James, 262
Attractor states, 220, 278
Auditory system, 159–162
Autonomic nervous system, 11, 18, 41

Balliene, Bernard, 51–52
Ballistic pinching, 440–441
Bannister, Roger, 40
Basal ganglia, 15, 62, 64, 447–448
 architecture of, 75
 context switching and, 78–79
 GABA and, 75–76
 Huntington's disease and, 76, 368–371
 inputs and, 76–77
 outputs and, 77–78
 Parkinson's disease and, 76–77, 79, 134–135
 Pavlovian learning and, 464–468
 population coding and, 412–413
 reflexes and, 134–135
 sequential movement and, 511
 skills and, 462–464
 spiny neurons and, 76–78
 striatum and, 76–77
 training signal for, 464
Basis functions, 413–414, 417, 423–424
Basket cells, 68–69
Bastian, Amy, 482
Batista, Aaron, 212, 234, 241–242, 324
Bauer, Joseph, 287
Bauswein, Erhard, 232
Bayesian theory, 292
Behavior. *See also* Motor learning
 crying, 65
 decision making and, 51–58 (*see also* Decision making)
 escape, 66
 habits and, 56–57
 hindbrain and, 65–68
 laughing, 65
 Pavlovian learning and, 41, 44–46, 52, 449, 455–468, 473–480
 periaqueductal gray and, 65
 spiny neurons and, 76
Bell, Curtis, 452
Bernard, Claude, 3
Bilateria, 13
Bizzi, Emilio, 517–518
 adaptation and, 283
 limb stability and, 127–127, 129, 136
 planning and, 364
 prediction and, 397–399, 404
Blazquez, Pablo, 467
Blindness, 348
Bloch, Jonathan, 51, 349
Blocking effect, 458–461, 468
Bock, Otmar, 208–209, 314
Bodznick, David, 452–453
Bones
 limb architecture and, 28–32 (*see also* Limbs)
 muscles and, 32–36
Boussaoud, Driss, 516
Boyer, Douglas, 51, 349
Brachiostoma, 15

Index

Brain, 15–16. *See also* Cortical areas; Specific part
  amnesia and, 435
  communication speed and, 37–38
  consciousness and, 54–55
  decision making and, 495–520 (*see also* Decision making)
  forebrain and, 75–89
  hindbrain and, 65–68
  M1 neuron sensitivity and, 163–165, 171–172
  medulla and, 450
  neocortex and, 22–23
  periaqueductal gray and, 65
  PPC neuron sensitivity and, 163–164, 171–172
  proprioceptive pathways and, 64–65
  Purkinje cells and, 68–70, 76, 404, 450, 453–462, 478–480, 483–491
  sexual behavior and, 64
  spiny neurons and, 76–78
  superior colliculus and, 73
  transcranial magnetic stimulation and, 133–134, 330–331, 440–441
Brakefield, Tiffany, 442, 444
Branchiostoma, 15
Brandt, Jason, 367, 435
Brashers-Krug, Thomas, 436–437
Brodmann, Korbinian, 83
Brodmann's areas, 83–84, 86–89
Buch, Ethan, 314
Bullock, Daniel, 266, 355, 357–358
Buneo, Christopher, 212, 214, 221, 324

Camera-centered coordinate system, 184–185
Caminiti, Roberto, 162, 171–172, 222
Carroll, Robert, 34
Cartesian space, 185, 265–268. *See also* Coordinate systems
Cartilage, 27
Caudate nucleus, 75, 465–467
Cell discharge rate, 193, 195, 198–199
  fixation-centered coordinates and, 212–216, 222–225
  movement direction and, 412
  target location and, 222–225
Center of gravity, 66–67

Central nervous system
  adaptation and, 273–292, 295–316, 449–454
  artificial stimulation and, 126–127
  basal ganglia and, 75–79
  center of gravity and, 66–67
  cerebellum and, 21–23, 68–71
  cerebral cortex and, 84–85 (*see also* Cortical areas)
  chordates and, 12
  coactivation and, 115
  communication speed and, 37–38
  components of, 10–11
  consolidation and, 435–444
  context switching and, 78–79
  control issues and, 19–20, 27–38
  crown group and, 12
  difference vectors and, 205–244 (*see also* Difference vectors)
  electricity and, 450–452
  end effectors and, 143–177 (*see also* End effectors)
  escape and, 66
  feedback and, 3–4
  fixation-centered coordinates and, 205–225
  functional understanding of, 3–4
  generalization functions and, 404–406, 414–431
  gnathostome and, 12
  Golgi tendon organs and, 111–112
  hindbrain and, 65–68
  hypothalamus and, 11
  inputs and, 69–70, 76–77, 84
  internal model and, 67–68, 379–401
  large-fiber sensory neuropathy and, 135
  M1 neuron sensitivity and, 163–165, 171–172
  mapping and, 58, 319–333 (*see also* Mapping)
  motor learning and, 1–2, 39–59
  motor system and, 9–24, 32–36, 93–94 (*see also* Motor system; Muscles)
  myoblasts and, 19–20
  neocortex and, 11, 22–23
  outputs and, 70–71, 77–78, 84–85
  perceptual threshold and, 500

Central nervous system (cont.)
  planning and, 205–225, 229–242, 353–354
  PPC neuron sensitivity and, 163–164, 171–175
  proprioception and, 273–292
  red nucleus and, 71–73
  redundancy and, 31
  reflexes and, 11, 20 (see also Reflexes)
  S1 nueron sensitivity and, 167–171
  sequential activation and, 127–129
  signal processing and, 449–454 (see also Signal processing)
  size principle and, 99
  spinal cord and, 61–65
  springy actuators and, 34–35
  state estimation and, 503–513
  stem group and, 12
  target location and, 179–201
  telencephalon and, 16–17
  thalamus and, 11, 80–81
  time-dependent change and, 436–440
  transcranial magnetic stimulation and, 133–134, 330–331, 440–441
Central pattern generators (CPGs), 11, 31, 62
  fractionation and, 65
  hindbrain and, 65–68
  myotomes and, 63–64
  sexual behavior and, 64
Cerebellum, 21
  adaptation and, 301–302, 314–316
  architecture of, 68–69
  augmentation by, 474
  basket cells and, 68–69
  climbing fibers and, 69
  damage to, 447–448, 482
  deep cerebellar nuclei and, 68–69
  depression and, 439
  dynamics and, 401
  eyeblink conditioning and, 459–461
  filtering and, 449–454
  GABA and, 70–71
  gnathostomes and, 21–22
  granule cells and, 68–69
  hemispheres of, 68
  inferior olive (ION) and, 69–71, 455–458
  inputs and, 69–70
  interaction torques and, 482
  ION and, 69–71, 455–458
  limb velocity and, 425–426
  mossy fibers and, 69
  neocortex and, 22–23
  outputs and, 69–70
  parallel fibers and, 68–69
  Pavlovian learning and, 41, 44–46, 52, 449, 455–468, 473–480
  population coding and, 412–413
  PPC and, 449
  predictions and, 455–464
  proprioception and, 401
  Purkinje cells and, 68–70, 404
  red nucleus and, 71–73
  reflexes and, 134–135
  sequential targets and, 511–512
  signal cancellation and, 449–454
  thalamus and, 80–81
  training signal for, 455–458
Chance, Frances, 201
Chordates, 12, 15, 18–19, 98
Cisek, Paul, 237, 239, 504
Clack, Jennifer, 31
Climbing fibers, 69, 475–480
Clower, Dottie, 303, 516
Coactivation, 384, 474
  artificial stimulation and, 126–127
  feedback and, 115
  motor learning and, 47–49
  stability and, 123
Coelem, 13
Coherent motion, 499–500
Colby, Carol, 322
Collinearity, 192
Commissurotomy, 84
Compliance matrix, 419
Conditional probability, 291
Conditt, Michael, 389–391
Connectin, 96–97
Consciousness, 54–55
Consolidation, 316
  amnesia and, 435
  cortical areas and, 440–441
  sleep and, 441–444
  time-dependent change and, 436–441

Context, 280–283, 447
  accumulator models and, 496–499
  swtiching and, 78–79
Contreras-Vidal, José, 314
Control, xvi, 49–50, 59. *See also* Signal processing
  ACGs and, 368
  adaptation and, 273–292, 295–316 (*see also* Adaptation)
  arbitrary mapping and, 516–518
  ejaculation and, 64
  feedback and, 381 (*see also* Feedback)
  homeostasis and, 3, 23, 43, 75
  Huntington's disease and, 76, 134–135, 368–371
  next-state planner and, 354–374
  oculomotor, 68, 86–87, 327–328, 482–486
  online correction and, 366–371
  Parkinson's disease and, 76–77, 79, 134–135
  planning and, 353–354 (*see also* Planning)
  redundancy and, 371–374
  remapping and, 319–333
  robotics and, 247 (*see also* Robotics)
  superior colliculus and, 73
  target location and, 179–201
  variable trade-offs and, 373–374
Coolen, Lique, 64
Coordinate systems, 144–146, 150–151
  adaptation and, 273–292, 307–316 (*see also* Adaptation)
  camera-centered, 184–185
  coordinate smoothness and, 341–352
  dynamics and, 403–410
  extrinsic, 404, 407
  fixation-centered, 205–225 (*see also* Fixation-centered coordinates)
  internal models and, 403–410
  intrinsic, 182, 405–407
  path alteration and, 344–347
  remapping and, 319–333
  retinotopic, 183–184
  shoulder-centered, 209–212
  visual smoothness and, 341–352
Cordo, Paul, 67

Coriolis forces, 385–386, 388
Corkin, Suzanne, 435
Corpus callosum, 84
Corrective movement, 128
Cortical areas
  5d, 230–235, 239, 257, 504–505
  7a, 175
  adaptation and, 273–292, 295–316
  anterior intraparietal cortex (AIP), 89
  arbitrary mapping and, 516–518
  basal ganglia and, 75–79
  Brodmann's areas and, 83–89
  caudal regions and, 162
  cerebellum and, 21–23 (*see also* Cerebellum)
  cerebral organization and, 81–85
  colonies and, 64
  columns and, 64
  consolidation and, 435–444
  crown group and, 12
  decision making and, 499–500 (*see also* Decision making)
  dorsal premotor cortex (PMd), 86, 230–231, 234, 237 (*see also* Dorsal premotor cortex [PMd])
  efference copy and, 322, 325–331
  frontal cortex, 85, 162
  gnathostome and, 21–22
  hand orientation and, 249–252
  hemispheres and, 81–82
  homunculus and, 85–86
  lateral intraparietal (LIP) area, 175, 192, 239, 322–324 (*see also* Lateral intraparietal (LIP) area)
  lesions and, 229
  limb velocity and, 425–426
  medial intraparietal area (MIP), 89
  motor cortex, 85–88 (*see also* Motor system, Primary motor cortex [M1])
  multiple targets and, 503–504
  myelin and, 83
  neglect and, 296–301
  neocortex and, 11, 22–23, 82
  optic ataxia and, 229
  perceptual threshold and, 500
  posterior parietal cortex (PPC), 81, 88–89, 162–164 (*see also* Posterior parietal cortex [PPC])

Cortical areas (cont.)
  primary motor cortex (M1), 163–165, 171–172, 229–234, 261 (*see also* Primary motor cortex [M1])
  primary somatosensory areas (S1), 167–171
  primary visual area (V1), 83
  Purkinje cells and, 68–70, 76, 404, 450, 455–462, 478–480, 483–491
  red nucleus and, 71–73
  remapping and, 319–333
  sequential movement and, 508–512
  striate cortex, 83 (*see also* Primary visual area [V1])
  supplementary eye field (SEF), 86, 88
  supplementary motor area (SMA), 86, 508–512
  thalamus and, 11, 80–81
  time-dependent change and, 440–441
  transcranial magnetic stimulation and, 133–134, 330–331, 440–441
  tuning function and, 229–231
  updating and, 331–336
  V3a, 192
  ventral premotor cortex (PMv), 86, 230–232, 234–235, 237 (*see also* Ventral premotor cortex [PMv])
  visual coordinate smoothness and, 341–352
Cortical fields, 85–89
Corticocortical projections, 84
Corticospinal projection, 23
Corticospinal tract, 86
Crackle, 351–352
Crammond, Donald, 230–232, 237, 239
Crawford, J. Douglas, 186, 297–298
Critic, 455
Crown group, 12
Cunningham, Helen, 311
Curl field, 393

Dance, 42
Danielli, James, 194
da Vinci, Leonardo, xv
Davson, Hugh, 194
Daw, Nigel, 280
Decision making, 495
  accumulator models and, 496–499
  coherent-motion experiments and, 499–500
  cues and, 504–505
  efference copy and, 508
  external cues and, 506–508
  flexibility and, 57–58
  habits and, 56–57
  instructions and, 504
  instrumental learning and, 56
  internal cues and, 506–508
  introspection and, 54–55
  mapping and, 513–519
  motor learning and, 51–58
  movement veto and, 500–503
  multiple movements and, 504–505
  neuronal correlates of, 499–500
  perceptual threshold and, 500
  preferred direction and, 504–505
  reflexes and, 51–52
  reward and, 503–504
  sequential reaching and, 508–512
  serial reaction-time task (SRTT) and, 512–513
  state estimation and, 505–513
  targets and, 496–504
Declarative knowledge, 507
Deep cerebellar nuclei, 68–69
Degeneracy, 16
Degrees of freedom, 28, 47, 172
  coordinate frames and, 144–146
  redundancy and, 372
Delay-period activity, 229–230, 547–548
  sequential movement and, 241–242
  target disappearance and, 231–232
Deneve, Sophie, 150, 217–219, 221–222, 274
Dentate nucleus, 70
Desmurget, Michel, 328, 330
Difference vectors, 147, 179, 347–348
  adaptation and, 273–292
  amplitude coding and, 252
  cue effects and, 257–261
  delay-period activity and, 229–231 (*See also* Delay-period activity)
  directional tuning and, 229–231
  error and, 205–209
  feedback and, 252

Index

fixation-centered coordinates and, 205–225
frontal lobe areas and, 249–261
Gaussian distribution and, 214–215
gaze effects and, 261
hand orientation and, 249–254
head orientation and, 205
M1 activity and, 261–268
mapping and, 247, 252–254, 319–333
minimization of, 343–344
planning and, 205–225, 229–242
posterior parietal cortex and, 229–231, 234–242
shoulder-centered coordinates and, 209–212
trajectories and, 354–357
varying arm configuration and, 254–257
varying gaze and, 249
Differentials, 539
Diffusion model, 496
di Pellegrino, Giuseppe, 257
Direction of action, 263, 265
Displacement map, 279
Distributed modules, 75, 80–81
Dizio, Paul, 130–131, 385
Donchin, Opher, 332, 417, 420–426
Donegan, Nelson, 490
Dopamine cells, 462–469
Dorsal horn, 62
Dorsal premotor cortex (PMd), 86, 230
   adaptation and, 295–296, 301–302
   deactivation of, 295–296
   gaze locations and, 259
   hand orientation and, 249–252
   movement planning and, 231, 234, 237
   population vectors and, 250–252
   visionless reaching and, 257
   visual coordinate smoothness and, 341–352
Dorsal spinocerebellar tract (DSCT), 62, 65
Dorsal striatum, 75
Dove prisms, 276–278
Duhamel, Rene, 322, 326
Dynamics, 49–50, 162. *See also* Kinematics
   adaptation and, 273–292, 295–316, 391–401
   arbitrary mapping and, 517–518
   attractor states and, 220, 278
   cerebellum and, 401
   coordinate system and, 403–410
   end effectors and, 144
   forward, 50, 336–338, 362–364, 381, 503–513
   generalization functions and, 404–406, 414–431
   internal models of, 379–401, 474–475
   inverse, 362–364, 381, 482, 488–491
   limb states and, 387–389
   mapping and, 247
   motor commands and, 382–384
   networks and, 245
   null-field condition and, 384
   planning and, 232–234
   preferred direction and, 395–401
   Purkinje cells and, 483–491
   robotic reaching and, 382–384
   rotation and, 384–387, 395–397
   state estimation and, 503–513
   target location and, 179–201
   time and, 389–391, 436–440
   torque and, 362–364
   tuning curves and, 414–416

Eating, 12–13
Ebner, Timothy, 167, 169, 252
Eckmiller, Rolf, 208–209
Efference copy, 322, 325–331, 333
   adaptive filtering and, 449–454
   decision making and, 508
   electric fish and, 450–452
   forward models and, 336–338
   reflexes and, 474
   trajectories and, 355
Ejaculation, 64
Elasticity, 99, 101, 113
Electric fish, 450–452
Electric organs, 450
Electrocytes, 450
Electromyographic (EMG) activity, 79, 134–135
   adaptive changes and, 393–395
   filtering and, 453–454
   preferred direction and, 395–401

End effectors, 245, 519
  adaptation and, 273–292, 295–316 (*see also* Adaptation)
  configuration coding and, 165–175
  coordinate frames and, 144–146
  current location estimation and, 205–208
  decision making and, 495 (*see also* Decision making)
  degrees of freedom and, 144–146
  difference vectors and, 205–244 (*see also* Difference vectors)
  displacement and, 247
  dynamics and, 144
  feedback and, 143–144
  fixation-centered coordinates and, 205–209, 221–225
  frontal neurophysiology and, 162–165
  Jacobian for, 355
  kinematics and, 144–157
  location prediction and, 147–157
  mapping and, 146–147, 252–254, 319–333
  movement of, 347–348
  planning and, 354–357 (*see also* Planning)
  parietal neurophysiology and, 162–165 (*see also* PPC and)
  PPC and, 171–175
  proprioception and, 148–162, 273–292 (*see also* Proprioception)
  reach planning and, 237
  robotics and, 148–151
  rotation matrices and, 151–154
  target location and, 179–201
  updating and, 331–336
  visual coordinates and, 147–148, 341–352
Endocrine functions, 41
Endoskeleton, 27
Energy, 544
Equations
  compliance matrix, 419
  generalization functions, 416, 419–420, 425, 429
  internal model, 388
  Jacobian, 279–280, 355
  smoothness, 350–351
  stiffness, 97, 136

Equilibrium-point hypothesis 129–131, 143–144
  limb stability and, 120–122, 126–127
Error, 47
  cerebellum and, 455–464
  climbing fibers and, 475–480
  difference vectors and, 205–209
  dopamine cells and, 464–465
  EMG changes and, 393–395
  eyeblink conditioning and, 459–461
  fixation-centered coordinates and, 205–212
  generalization functions and, 404–406, 414–431
  Huntington's disease and, 368–371
  motor command correction and, 480–491
  movement sequences and, 208–209
  online correction and, 366–371
  Pavlovian learning and, 41, 44–46, 52, 449, 455–468, 473–480
  reflexes and, 473–491
  target location and, 186–188
Essential tremor, 128
Eukaryotes, 36
Evarts, Edward, 229
Evolution, 3
  biology and, 533–535
  cerebellum and, 21–23
  cerebral cortex and, 82–84
  gnathostome and, 21–22
  land locomotion and, 31–32
  mammalian and, 22–23
  motor learning and, 39–59
  motor system and, 10, 12–24
  preadaptation and, 18, 30
  telencephalon and, 16–17
  tetrapod and, 21–22
  vertebrate brain and, 15–16
  vertebrate head and, 17–19
Explicit tasks, 52
Extension, 28
Extinction, 456, 458, 460, 535
Extrafusal fibers, 35–36, 98–99
Extrinsic coordinates, 404
Eye movements
  error and, 186–188
  gain fields and, 188–201
  modulation and, 188–201

planning saccades and, 234–236
remapping and, 319–333
superior colliculus and, 73
target location and, 186–188
updating and, 331–336
Eyes
  activity field and, 239
  camera-centered coordinates and, 184–185
  cerebellum and, 314–316, 482–486
  dominant eye and, 188
  dove prisms and, 276–278
  fixation-centered coordinates and, 188–192, 205–225
  fovea and, 88, 148, 182, 184, 188, 191
  frontal eye field and, 86, 234–236, 240–241, 497–503
  Gaussian distribution and, 214–215
  gaze location and, 259
  multiple targets and, 503–504
  optokinetic reflex and, 482–483
  Purkinje cells and, 68–70, 76, 404, 450, 453–462, 478–480, 483–491
  retinal ganglion cells and, 188–192
  retinal location and, 188–201
  retinotopic coordinates and, 183–184
  rotational adaptation and, 275–278
  supplementary eye field and, 86, 88
  target location and, 186–201
  vestibulo-ocular reflex and, 182

Feedback, 3–4, 27, 39
  adaptation and, 273–292, 295–316 (see also Adaptation)
  biological vs. mechanical actuators and, 93–94
  control and, 50, 462, 469
  corollary discharge and, 322, 325–331
  dove prisms and, 276–278
  efference copy and, 322, 325–331
  EMG changes and, 393–395
  end effectors and, 143–177 (see also End effectors)
  filtering and, 449–454
  fixation-centered coordinates and, 205–225
  forward models and, 336–339
  Golgi tendon organ and, 132
  internal dynamics models and, 379–401
  Jacobian and, 105–108
  large-fiber sensory neuropathy and, 135
  muscle afferents and, 108–117
  next-state planner and, 354–374
  path altering and, 344–347
  planning and, 353–354 (see also Planning)
  pointing and, 143–144
  reaching and, 123–126, 135, 143–144
  remapping and, 319–333
  serial reaction-time tasks (SRTT) and, 512–513
  signal noise and, 364–366, 369–371
  stability and, 123–126, 135
  torque conversion and, 102–108
  varying visual, 252
  vestibular, 325
Femur, 29
Feynman, Richard, 121
Fibula, 29
Fields of Forel, 20
Filaments, 35
Fins, 29–30
Fischer, Stefan, 441–442
Fish, 12–13, 29–30
  electric, 450–452
  land locomotion and, 31–32
  myotomes and, 63–64
Fitness, 41, 64
Fixation-centered coordinates, 184–186, 225, 247
  adaptation and, 273–292, 295–316
  cell discharge rates and, 212–216, 222–225
  decision making and, 500–503
  end-effector encoding and, 221–225
  error and, 205–212
  gain fields and, 211
  Gaussian distribution and, 214–215
  hand location and, 252–254
  neglect and, 296–301
  network analysis and, 211–212, 215–222
  planning and, 205–221 (see also Planning)

Fixation-centered coordinates (cont.)
  proprioception and, 273–292 (see also Proprioception)
  reciprocal connections and, 219
  remapping and, 319–333
  shoulder-centered coordinates and, 209–212
  stop signal and, 503
  trajectories and, 354–357, 360–364, 371–374
Fixation point, 220–221
  cell discharge rate and, 212–216
Flash, Tamar, 350
Flatworms, 13
Flexibility, 43, 57–58. See also Decision making
Flocculus, 483–486
Forces, 32, 36, 519. See also Dynamics
  adaptation and, 391–401 (see also Adaptation)
  angular velocity and, 102, 385–386
  arbitrary mapping and, 517
  arm configurations and, 404–410
  biological vs. mechanical actuators and, 93–94
  coordinate system and, 403–410
  Coriolis, 385–386, 388
  curl field and, 393
  defined, 543
  EMG changes and, 393–395
  end effectors and, 159–160 (see also End effectors)
  estimated trajectory, 414–416
  gain fields and, 426–429
  generalization functions and, 404–406, 414–431
  generation of, 94–96
  inertial, 50, 96, 120, 381–382, 420, 544
  internal dynamics models and, 379–401
  Jacobians and, 105–108, 119
  length-tension properties and, 97–98, 121–123
  limb states and, 387–389, 406–407
  M1 activity and, 261–263
  muscles afferents and, 108–117
  overcompensation and, 432
  preferred direction and, 231, 247, 313, 322, 395–401
  proteins and, 35
  restoring, 373
  rotation and, 384–387
  skill consolidation and, 435–444
  state maps and, 389–391
  stiffness and, 97, 136
  tension, 99
  torque and, 102–108 (see also Torque)
  tuning curves and, 414–416
Forebrain
  basal ganglia and, 75–80
  cortical organization and, 81–89
  thalamus and, 80–81
Forward models, 367, 449. See also Dynamics
  PPC damage and, 338–339
  Remapping and, 319, 336
  serial reaction-time tasks (SRTT) and, 512–513
Fossils, 9–10
Fovea, 88, 148, 182, 184, 188, 191
Fractionation, 65
Francis, Joseph, 422–423
Freeman, John, 460
Fromm, Christof, 232
Frontal eye field (FEF), 86
  difference vectors and, 234–236, 240–241
  motor commands and, 497–503
Frontal lobe
  coordinate frames and, 265–268
  difference vectors and, 249–261
  feedback variance and, 252
  force coding and, 262–263
  gaze effects and, 261
  hand location and, 249–252
  planning and, 245–268
  population coding and, 263–265
Fuchs, Albert, 483–484
Fujii, Noataka, 467
Function approximation, 217
Fuzzy categories, 40

Gaffan, David, 296–297, 299, 324
Gain fields, 186
  angular velocity and, 426–429

fixation-centered coordinates and, 211
head orientation and, 197–198
limb velocity and, 426–429
modulation and, 188–201
neural basis for, 198–201
γ-aminobutyric acid (GABA), 70–71
  adaptation and, 295–296
  basal ganglia and, 75–76
  eyeblink conditioning and, 459–461
  muscimol and, 295
  pallidum and, 75
  Pavlovian learning and, 459–464
  remapping and, 328
  sequential reaching and, 510
  thalamus and, 80
Gandolfo, Francesca, 389–391, 517–518
Garwicz, Martin, 476, 480
Gaussian distribution, 191, 194–195, 197-198, 214–215
  adaptation and, 303–304, 313
  basis functions and, 424
  generalization functions and, 420–421
  remapping and, 322
Gauthier, Gabriel, 325
Gaze, 249, 259, 261
Generalization functions
  adaptation and, 423–425
  arm configurations and, 404–406
  artificial systems and, 420–423
  bimodality and, 423–425, 429
  equations for, 416, 419–420, 425, 429
  gain fields and, 426–429
  learning and, 430–431
  measurement problems and, 416–417
  trial-to-trial changes and, 417–420
  tuning curves and, 414–416
  velocity encoding and, 425–426
Genetics, 39, 43, 533–535
Georgopoulos, Apostolos, 171–172, 230, 241, 257, 262–263
Ghahramani, Zoubin, 344, 346
Ghez, Claude, 311–312, 437
Gibson, Alan, 476
Gilbert, Peter, 478
Giszter, Simon, 126
Glickstein, Mitchell, 247
Glimcher, Paul, 503–504
Gluck, Mark, 456

Gnathostome, 12, 21–22
Goals, 41
Go function, 355, 357–360
Goldberg, Michael, 322, 504
Golgi tendon organs, 108–109, 111–112, 132
Gomi, Hiroaki, 486
Goodbody, Susan, 305
Graham, Robert, 1
Granule cells, 68–69
Graybiel, Ann, 467
Graziano, Michael, 127, 223, 259
Grea, Helene, 331
Gribble, Paul, 482
Grossberg, Steven, 266, 355, 357–358
Gymnotid fish, 450

Habits, 44, 54, 56–57
Hagfish, 12–13
Hallett, Mark, 301–302
Hand location
  adaptation and, 273–292
  cell discharge rates and, 212–216, 222–225
  directional mapping and, 252–254
  fixation-centered coordinates and, 205–216
  population vectors and, 250–252
  proprioception and, 222 (*see also* Proprioception)
  representations and, 217
  tuning function and, 215–217
  varying orientation and, 249–252
Harris, Christopher, 354, 365–366
Head orientation, 197–198, 205
  cerebellum encoding and, 482–486
  flocculus and, 483–486
  movement and, 324–325
Held, Richard, 280, 287
Helms-Tillery, Steven, 167, 169–170
Henriques, Denise, 186
Hepp-Reymond, Marie-Claude, 262
Hikosaka, Okihide, 465, 508–512
Hildreth, Ellen, 4
Hill, A. V., 99
Hindbrain
  CPGs and, 65–68
  escape behavior and, 66

Hindbrain (cont.)
  organization of, 65
  reaching and, 66–68
Hobson, Allan, 442, 444
Hoff, Bruce, 354, 357–360
Hoff-Arbib model, 357–360
Hoffman, Donna, 249
Hogan, Neville, 136, 350, 353, 372
Hollerbach, John, 4
Homeostasis, 3, 23, 43, 75
Homology, 12, 29, 533–534
Homunculus, 85–86
Horak, Fay, 79
Horn, Kris, 476
Hornak, Julia, 296–297, 299, 324
Hoshi, Eiji, 237, 283, 301
Houk, James, 80, 374, 455, 469
Humerus, 28, 29
Huntingtin gene, 368
Huntington's disease, 76, 134–135
  asymptomatic gene carriers (AGCs) and, 368
  cortical function and, 368–369
  noise adjustments and, 369–371
  reaching and, 468–469
  serial reaction-time task (SRTT) and, 512
  striatal cell death and, 368
Hwang, Eun-Jung, 426, 429
Hydrophilic substances, 37
Hydrophobic substances, 37
Hypothalamus, 11

Ijspeert, Auke Jan, 361
Implicit tasks, 52
Impulse, 543
Inertia, 50, 96, 120, 381–382, 420, 544
Inferior olivary nuclei (ION), 69–71, 301–302
  cerebral training signal and, 455–458
  eyeblink conditioning and, 459–461
  Pavlovian learning and, 455–468, 475–480
Insects, 27
Instincts, 41–46
Instruction stimulus, 229–232
Interaction torques, 48, 380, 480–482

Internal models, 50, 67–68, 79, 336, 379–381
  adaptation and, 391–401, 449–454 (see also Adaptation)
  amnesia and, 435
  consolidation and, 435–444
  coordinate system and, 403–410
  generalizations and, 416–431
  inverse dynamics and, 362–364
  kinematics and, 150
  limb states and, 387–389
  partial, 489
  planning and, 353–354 (see also Planning)
  population coding and, 410–416
  preferred direction and, 395–401
  reaching and, 382–387
  rotation and, 395–397
  signal cancellation and, 449–454
  sleep and, 441–444
  state estimation and, 503–513
  state maps and, 389–391
  synaptic weight matrix and, 245
  time and, 389–391, 436–441
Intrafusal fibers, 35–36, 98–99
Intrinsic coordinates, 185
Introspection, 54–55
Invertebrates, 15
Iriki, Atsushi, 172
Isometric contractions, 261, 429
Ito, Masao, 486
Ivry, Richard, 462

Jacobians
  adaptation and, 279, 307–308
  end-effector location and, 355
  joint velocity and, 406
  limb stability and, 119, 121
  torque and, 105–108, 121
Jarvis, Murray, 212
Jaws, 12–13, 29–30
Jerk, 350–351, 353
  difference-vector transformation and, 354–357
  Hoff-Arbib model and, 357–360
  planning and, 354–357, 360–364, 371–374
  redundancy and, 371–374

signal noise and, 369–371
Joint coordinates, 167, 279
  adaptation and, 308
  degrees of freedom and, 145
  difference vector minimization and, 343–344
  limb stability and, 139
  planning and, 245–248
Joint stiffness, 138
Joint velocity, 405–406, 410
Jordan, Michael, 344, 346, 373–374

Kakei, Shinji, 249–250
Kalaska, John, 171–172, 230–234, 237, 239–240, 252–257, 259, 262, 403, 504, 518
Kamin, Leon, 458
Kawato, Mitsuo, 302, 486–487, 489
Kim, Keansok, 458
Kinematics, 46, 404
  adaptation and, 273–292, 295–316
  alignment and, 147–149
  coordinate frames and, 144–146, 150–151
  degrees of freedom and, 144–146
  end effectors and, 144–157
  forward, 144–151, 159, 222, 265, 273
  fovea and, 148
  function approximation and, 150
  internal dynamics models and, 379–401
  mapping and, 146–147, 319–333
  muscle vs. movement, 162–165
  pattern-completion networks and, 150
  planning and, 232–234
  proprioception and, 273–292 (*see also* Proprioception)
  Purkinje cells and, 483–491
  state estimation and, 503–513
  target location and, 179–203
Kinesis, 43–44
Kinetic energy, 544
Kitazawa, Shegiru, 287
Kojima, Jun, 467
Kording, Konrad, 292
Krupa, David, 458
Kurata, Kiyoshi, 283, 301

Lackner, James, 130–131, 160, 384–385
Lacquaniti, Francisco, 172
Lamprey, 12–13
Lancelet, 15
Large-fiber sensory neuropathy, 135
Lateral intraparietal (LIP) area
  decision making and, 500
  multiple targets and, 503–504
  perceptual threshold and, 500
  remapping and, 322–324, 327
Latham, Peter, 150, 217
Law of sines, 105
Learning. *See* Motor learning
Lebedev, Mikhail, 58
Lee, Sohie, 316
Lesions, 72–73
  cerebellum and, 447–448, 482
  optic ataxia and, 229
Levine, Minna, 160
Li, Chiang-Shan Ray, 397
Ligands, 64
Lilienthal, Otto, xv
Limbs
  adaptation and, 30, 273–292, 295–316
  coding configuration of, 165–175, 414–410
  control and, 28–32 (*see also* Control)
  degrees of freedom and, 47, 144–146
  difference vectors and, 254–257 (*see also* Difference vectors)
  equilibrium-point hypothesis and, 129–131
  force states and, 387–389
  forward models and, 336–339
  hand location and, 205–217, 222–225, 249–254, 273–292
  inertia and, 50
  internal dynamics models and, 379–401
  intrinsic coordinates and, 406–407
  kinematics and, 144–146
  large-fiber sensory neuropathy and, 135
  M1 activity and, 171–172
  next-state planner and, 354–374
  passive properties and, 129–131
  PPC activity and, 171–175
  reflexes and add, 131–135

Limbs (cont.)
  remapping and, 319–333
  S1 activity and, 167–171
  stability and, 48–49, 119–139
  transcranial magnetic stimulation and, 133–134
  varying arm configuration and, 254–257
Lisberger, Stephen, 483–484, 486–488
Lobsters, 27
Location map, 205, 247
Loins, 61
Long-loop reflexes, 134–135, 368–369, 382
Long-term depression (LTD), 439, 453–454, 458
Long-term potentiaion (LTP), 458–461
Loops
  description of, 75, 80–81
  long, 134–135, 368–369, 382
  short, 132, 393, 473
Lumbar, 61

McGaugh, James, 506–507, 511
MacLean, Paul, 45–46
Malfait, Nicole, 410
Manipulandum, 123, 237, 288, 332–333, 343
Mapping, 58, 247
  accumulator models and, 496–499
  adaptation and, 273–292, 295–316 (see also Adaptation)
  arbitrary, 516–518
  decision making and, 513–519
  efference copy and, 322, 325–331
  end effectors and, 146–147
  hand location and, 205–217, 222–225, 249–254, 273–292
  internal models and, 389–391 (see also Internal models)
  internal vs. external, 518–520
  population coding and, 412–413
  remapping and, 319–333
  sensorimotor, 513–514, 520
  transformational, 500, 513–516
  updating and, 331–336
Marr, David, 3–4

Marsden, C. David, 133
Matrices
  Jacobian, 105–108, 119, 121, 279, 307–308, 355, 406
  rotation, 151–156
Mauk, Michael, 456, 490
Mauritz, Karl-Heinz, 232, 257
Maximum-likelihood estimation, 219
Medawar, Jean, 40, 43
Medawar, Peter, 40, 43
Medendorp, Pieter, 297–298
Medial intraparietal area (MIP), 89
Medial superior temporal area (MST), 500
Medina, Javier, 456
Medulla, 69–70, 450
Memory, 39–40, 43, 537
  adaptation and, 273–292
  amnesia and, 435
  consolidation and, 283, 435–444
  context and, 280–283
  declarative, 54
  habit and, 54
  internal dynamics models and, 389–391
  neglect and, 296–301
  procedural, 54
  rote learning and, 389–391
  sleep and, 441–444
  time-dependent change and, 436–441
  transcranial magnetic stimulation (TMS) and, 330–331, 440–441
Messier, Julie, 252
Midbrain locomotion region, 65
Midsagittal plane, 537
Miller, Lee, 374, 469
Minimal intervention principle, 373–374
Mnemonic aids, 537
Modulation, 548
Moment arm, 104–105, 380
  angular acceleration and, 120–121
  Jacobian and, 119
Momentum, 543
Monosynaptic connections, 86, 111
Montgomery, John, 453
Moran, Daniel, 250

Morasso, Pietro, 343–344, 372
Mormyrid fish, 450
Mossy fibers, 69
Motor commands, 3, 17, 24
  basal ganglia and, 76
  efference copy and, 76
  force generation and, 93
  force prediction and, 379–382
  hindbrain and, 68
  inverse dynamics and, 362–364
  planning and, 353–354 (see also Planning)
  pulses and, 36
  remapping and, 319–333
  serial reaction-time tasks (SRTT) and, 512–513
  signal noise and, 364–366
  skill acquisition and, 46
Motor cortex. See Cortical areas
Motor field, 184
Motor learning
  adaptation and, 47–58
  amnesia and, 435
  anti-Hebbian rules and, 452, 455
  appendages and, 39
  arm configurations and, 407–410
  cerebellum and, 68–71 (see also Cerebellum)
  conditioning and, 44–46, 56
  consolidation and, 435–444
  control and, 49–50
  decision making and, 495–520 (see also Decision making)
  defining, 39–41
  dynamics and, 382–384
  electricity and, 450–452
  escape and, 45
  flexibility and, 43
  forebrain and, 75–89
  fuzzy categories and, 40
  habits and, 56–57
  hindbrain and, 65–68
  instincts and, 41–46
  instrumental learning and, 56
  ION and, 455–458
  kinesis and, 43–44
  memory and, 39–40, 43
  motor system and, 1–2
  natural categories and, 40
  Pavlovian learning and, 41, 44–46, 52, 455–468, 475–480
  pointing and, 50–51
  predicting ease of, 430–431
  reaching and, 50–51, 508–512
  red nucleus and, 71–73
  reflexes and, 41–46, 473–491
  rote learning and, 389–391
  sensitization and, 44
  sequential and, 508–512
  skill acquisition and, 42, 46–47, 50–51 (see also Skills)
  sleep and, 441–444
  spinal cord and, 61–65
  stability and, 48–49
  superior colliculus and, 73
  taxis and, 43–44
  time-dependent change and, 436–441
  training signals and, 455–458, 464
Motor memories, 39–40, 43
Motor neurons, 98–99
  learning and, 49, 62, 86
  muscular force and, 94–96
  muscle afferents and, 115–117
Motor pool, 62, 98
Motor primitives, 126–127, 417
Motor system
  adaptation and, 273–292, 295–316
  arm configurations and, 404–410
  augmentation of, 474
  brain and, 15–16
  Brodmann's areas and, 83–84
  cerebellum and, 21–23
  chordates and, 12
  communication speed and, 37–38
  components of, 10–12
  control issues and, 27–38
  crown group and, 12
  decision making and, 495–520 (see also Decision making)
  difference vectors and, 249–261 (see also Difference vectors)
  efficiency of, 27
  evolution of, 9–24
  eyeblink conditioning and, 459–461
  forward models and, 336–339
  gnathostome and, 12

Motor system (cont.)
  hand location and, 205–217, 222–225, 249–254, 273–292
  head and, 17–19
  hypothalamus and, 11
  internal dynamics models and, 379–401
  limb velocity and, 425–426
  muscles and, 32–36 (see also Muscles)
  myoblasts and, 19
  neocortex and, 11, 22–23
  next-state planner and, 354–374
  perceptual threshold and, 500
  planning and, 205–225, 229–242, 353–354 (see also Planning)
  posture and, 20
  preparation and, 230–231
  reflexes and, 20, 473–491 (see also Reflexes)
  remapping and, 319–333
  signal cancellation and, 449–454
  telencephalon and, 16–17
  transcranial magnetic stimulation (TMS) and, 133–134, 330–331, 440–441
  trigger stimulus and, 230–231
  varying arm configuration and, 254–257
  vertebrates and, 4–5
  visual coordinate smoothness and, 341–352
Motor units, 98–99
Moussavi, Zahra, 405–407
Movement, 547–548. See also Motor system; Reaching; Pointing
  accumulator models and, 496–499
  adaptation and, 47–48, 51–58, 273–292, 295–316
  angular velocity and, 426–429
  cerebellum and, 455–464
  computing plan for, 229–242
  contraction and, 36
  corrective, 128
  decision making and, 495–520 (see also Decision making)
  delay-period activity and, 229–231
  difference vectors and, 205–244 (see also Difference vectors)
  direction and, 229–231, 412–413
  dynamics of, 46
  EMG changes and, 393–395
  eyeblink conditioning and, 459–461
  fixation-centered coordinates and, 205–225
  flocculus and, 483–486
  fovea and, 148
  generalization functions and, 404–406, 414–431
  go function and, 355, 357–360
  hand location and, 205–217, 222–225, 249–254, 273–292
  head-direction encoding and, 482–486
  Hoff-Arbib model and, 357–360
  instructions and, 504
  internal dynamics models and, 379–401
  Jacobians and, 105–108
  kinematics and, 46, 144–157 (see also Kinematics)
  kinesis and, 43–44
  land locomotion and, 31–32
  mapping and, 247, 252–254 (see also Mapping)
  muscles and, 93–94, 127–129, 162–165 (see also Muscles)
  myotomes and, 63–64
  Pavlovian learning and, 41, 44–46, 52, 449, 455–468, 473–480
  planning and, 353–357, 364 (see also Planning)
  population coding and, 410–416
  preparation and, 230
  Purkinje cells and, 68–70, 76, 404, 450, 453–462, 478–480, 483–491
  reflexes and, 41–46, 473–491 (see also Reflexes)
  saccades and, 234–236
  sensory cue effects and, 257–261
  sequential, 127–129, 241–242, 508–512
  serial reaction-time task (SRTT) and, 512–513
  shoulder-centered coordinates and, 209–212
  signal noise and, 364–366, 369–371
  S-shaped, 432

superior colliculus and, 73
target disappearance and, 231–232
target size and, 360
taxis and, 43–44
trajectory smoothness and, 350–352
tuning curves and, 414–416
visual impairment and, 348
zigzag paths and, 360–364
Muellbacher, Wolf, 440–441
Muscimol, 295
Muscles
actin and, 94
active mechanisms of, 94–96
as actuators, 34–35
agonist, 72, 128
angular acceleration and, 120–121
angular velocity and, 102
antagonist, 47, 72, 93, 120–121, 143–144
arm configurations and, 404–410
climbing fibers and, 475–480
coactivation and, 115, 123, 126, 384
communication speed and, 37–38
complex spikes and, 478–480
connectin and, 96–97
contraction of, 36, 96, 261, 429
control and, 49–50
degrees of freedom and, 47
dynamics and, 49–50, 379–401
elasticity and, 99, 101, 113
electrocytes and, 450
EMG and, 79, 134–135, 393–401, 453–454
end effectors and, 143–177 (*see also* End effectors)
equilibrium points and, 120–121, 126–131, 143–144
fibers and, 35–36, 62, 98–99, 135
filaments and, 35, 94, 96, 98
γ-motor neuron and, 115–117
Golgi tendon organs and, 111–112
inertia and, 96, 120
insertions and, 104
internal models and, 379–401
isometrics and, 96, 261, 429
Jacobians and, 105–108, 119, 121
joint velocity and, 405–406
land locomotion and, 31–32
large-fiber sensory neuropathy and, 135
length-tension properties and, 97–98, 121–123
limb architecture and, 32–34, 119–139 (*see also* Limbs)
M1 activity and, 163–165, 171–172
membrane voltage and, 37
moment arm and, 104–105, 119
motor units and, 98–99
movement and, 162–165 (*see also* Movement)
myelin and, 38, 62, 83
myoblasts and, 19, 35
myofibrils and, 35
myosin and, 35, 94
nuclear bag region and, 112–113
passive mechanisms of, 96–98
passive properties and, 129–131, 136–139
polarization of, 93–96
PPC activity and, 163–164, 171–172
preferred direction and, 231, 247, 313, 322, 395–401
pulling and, 119
reaction time and, 93–94
reflexes and, 131–135, 473–491 (*see also* Reflexes)
robotics and, 27–28
rotation and, 384–387, 395–397
S1 activity and, 167–171
sarcomeres and, 35, 96–98
sequential activation and, 127–129
signals and, 10–11
spindle afferents and, 36, 109–117, 133–134, 159–160
stability and, 48–49
stiffness and, 28, 96–97, 123, 136, 138, 372–373, 388, 423–424
torque and, 102–108, 121–123 (*see also* Forces; Torque)
transcranial magnetic stimulation (TMS) and, 133–134
vibration and, 159–162
viscosity and, 101, 388
Mushiake, Hajime, 252
Mussa-Ivaldi, Ferdinando, 126, 136–138, 372, 382, 384, 387–391, 517–518

Myelin, 38, 62, 83
Myers, Catherine, 456
Myoblasts, 19, 35
Myofibrils, 35
Myosin, 35, 94
Myotomes, 34, 63–64

Nakanishi, Jun, 361
Nanayakkara, Thrishantha, 332
Nashner, Lewis, 67
Neglect, 296–301
Neocortex, 11, 22–23, 82
Neoteny, 15
Networks, 200
 adaptation and, 273–292 (see also Adaptation)
 artificial neural, 282
 fixation-centered coordinates and, 205–222
 pattern-completion, 150
 synaptic weight matrix and, 245
Neural crest, 17–18, 38
Neural integrator, 20
Neural transmission, 37–38
Neuroendocrine functions, 11
Neuromuscular junctions, 94
Neurons, 10–11, 547–548. See also Feedback
 activity field and, 172–175, 184, 214–215, 239, 249
 adaptation and, 273–292, 295–316
 caudate, 465–467
 cerebellum and, 455–464
 climbing fibers and, 475–480
 complex spikes and, 478–480
 configuration coding and, 165–175
 decision making and, 499–500 (see also Decision making)
 dopaminergic, 462–469
 eyeblink conditioning and, 459–461
 FEF, 497–503
 fixation-centered coordinates and, 205–225
 γ-motor, 115–117
 M1 activity and, 163–165, 171–172
 motor, 450 (see also Motor system)
 oculomotor, 3, 68, 86–87, 327–328, 482–486 (see also Eyes)
 PPC, 162–164, 171–175, 188–201
 preferred direction and, 231, 247, 313, 322, 395–403
 projection, 77, 450, 452–453, 462–463, 484
 proprioception and, 295–303 (see also Proprioception)
 Purkinje cells and, 68–70, 76, 404, 450, 453–462, 478–480, 483–491
 remapping and, 319–333 (see also Mapping)
 S1 activity and, 167–171
 signal cancellation and, 449–454 (see also Signal processing)
 target location and, 179–203, 231–232
 tonically active, 467
Neurophysiology, 2, 547–548
Newtonian mechanics, 2
Newts, 275–276
Next-state planner
 difference vector and, 354–357
 Hoff-Arbib model and, 357–360
 Huntington's disease and, 366–371
 redundancy and, 371–374
 signal noise and, 364–366
 trajectories and, 354–357, 360–364
Nicholson, Daniel, 460
Nieuwenhuys, Rudolf, 19
Nixon, Philip, 175, 461–462, 517
Nonprimary motor cortex, 85
Nonstandard mapping, 58
Nores, William, 456
Northcutt, R. Glenn, 18
Notochord, 12
Novak, Kevin, 374, 469
Nuclear bag region, 112–113
Nucleus accumbens, 75
Null-field condition, 384

Ochiai, Tetsuji, 252
Oculomotor control, 3, 68, 86–87. See also Eyes
 neurons and, 482–486
 remapping and, 327–328
Olfaction, 9, 16–18
Optic ataxia, 229
Optokinetic reflex, 482–483

Ostry, David, 410, 482
Overcompleteness, 30–31
Overlearning, 57, 465

Packard, Mark, 506–507, 511
Padoa-Schioppa, Camillo, 397
Pallidum, 75
Parallel fibers, 68–69
Parietal reach region (PRR), 89, 239, 235, 322–324
Parkinson's disease, 76–77, 79, 134–135, 464
Pascual-Leone, Alvaro, 512
Passingham, Richard, 175, 461–462, 516–518
Pattern-completion networks, 150
Pavlovian learning, 44–46, 52, 449, 459–461, 473
  approach behavior and, 45
  basal ganglia and, 464–468
  cerebellum and, 455–464
  eyeblink and, 459–461
  ION and, 455–458, 475–480
  reflexes and, 41
Pectoral fins, 29
Pelvic fins, 29
Perceptual threshold, 500
Periaqueductal gray, 65
Perturbation, 334–336
Petride, Michael, 516
Phototaxis, 43–44
Physics, 1–3, 5, 533–535
  implicit knowledge and, 380
  rotational, 544–545
  spring loads, 545
  translational, 543–544
  viscosity, 545
Pinching, 440–441
Placodes, 17–18
Planning
  activity field and, 239
  adaptation and, 286–287
  amplitude coding and, 252
  controllers and, 247
  cue effects and, 257–261
  delay-period activity and, 229–231
  difference vectors and, 354–357 (*see also* Difference vectors)
  directional tuning and, 229–231
  direction coding and, 252
  end effectors and, 237
  feedback and, 252
  fixation-centered coordinates and, 205–225
  frontal eye field (FEF) and, 234–236, 240–241
  gaze effects and, 249, 261
  go function and, 355, 357–360
  hand orientation and, 249–254
  Hoff-Arbib model and, 357–360
  Huntington's disease and, 368–371
  kinematics and, 232–234
  M1 activity and, 261–268
  next-state, 354–374
  online correction and, 366–371
  problem of, 353–354
  redundancy and, 371–374
  remapping and, 319–333
  robotics and, 245–248
  saccades and, 234–236, 240–241
  sequential movement and, 241–242
  signal noise and, 364–366
  stretch reflex and, 247
  target disappearance and, 231–233
  torques and, 245–247
  trajectories and, 354–357, 360–364, 371–374
  trajectory smoothness and, 341–352
  varying arm configuration and, 254–257
  without execution, 237–241
Plasticity, 50
Platt, Michael, 503–504
Pointing, 1
  adaptation and, 47–48, 51–58
  decision making and, 495–520 (*see also* Decision making)
  difference vectors and, 205–244 (*see also* Difference vectors)
  feedback and, 143–144
  fovea and, 188
  land locomotion and, 31–32
  motor learning and, 50–51
  neural transmission and, 37–38
  regularity in, 343–350
  remapping and, 319–333

Pointing (cont.)
  trajectory smoothness and, 341–352, 350–352
  updating and, 331–336
Polit, Andreas, 123–126, 129
Pong, Milton, 476
Pop out, 499
Poppele, Richard, 165, 167, 170
Population vectors, 263–265
  internal models and, 410–416
  PMv and, 250–252
Posterior parietal cortex (PPC), 81, 88–89
  adaptation and, 296–299, 303
  arbitrary mapping and, 517
  cerebellum and, 449
  decision making and, 504
  difference vectors and, 221–222, 229–242
  end effectors and, 162–164, 171–175
  forward models and, 338–339
  motor cortex and, 247
  motor imagery and, 338–339
  planning and, 338–352
  prediction and, 449, 504, 510, 517
  proprioception and, 296–299, 303
  sequential reaching and, 510
  target location and, 186, 188–201
  trajectory smoothness and, 341–352
Posture, 66
  artificial stimulation and, 126–127
  force generation and, 93
  head orientation and, 197–198, 205
  internal model and, 67–68
  limb architecture and, 28
  stability and, 48–49
Pouget, Alexandre, 150, 197, 217, 413–414
Power, 32, 544–545
Preadaptation, 18, 30
Predictions
  basal ganglia and, 462–464
  consolidation and, 435–444
  coordinate system and, 403–410
  decision making and, 503–504 (see also Decision making)
  dopamine cells and, 464–465
  eyeblink conditioning and, 459–461
  filtering and, 449–454
  generalizations and, 416–431
  interaction torques and, 480–482
  internal models and, 410–416 (see also Internal models)
  motor command correction and, 480–491
  motor learning and, 430–431
  Purkinje cells and, 483–491
  reflexes and, 473–491
  reward and, 464–467, 503–504
  skill consolidation and, 435–444
  stimulus response and, 454–464
  training signal and, 455–458
Preferred direction, 231, 247, 313, 322
  arm configuration and, 404–406
  consolidation and, 439
  decision making and, 504–505
  gain field and, 426–429
  M1 activity and, 403–404
  Purkinje cells and, 68–69, 404
  rotation and, 395–401
Preparation, 230–231
Prepositus hypoglossi, 20
Primary motor cortex (M1)
  arbitrary mapping and, 518
  arm configuration and, 404–406
  consolidation and, 439–441
  coordinate frames and, 265–268
  decision making and, 504
  force coding and, 262–263
  hand orientation and, 249–254
  intrinsic coordinates and, 250
  limb velocity and, 426
  population coding and, 263–265
  preferred direction and, 247, 395–404
  trajectory smoothness and, 341–352
  transcranial magnetic stimulation (TMS) and, 440–441
  torque coding and, 262–263
  varying arm configuration and, 254–257
Primary sensory afferents, 61–63
Primary somatosensory areas (S1), 167–171
Primates, 9, 276–278

Prime mover, 34–35, 128
Prisms. *See* Adaptation
Procedural knowledge, 507
Projection neurons, 77, 450, 452–453, 462–463, 484
Proprioception, 62–65, 84, 125, 205, 448, 474
   adaptation and, 273–292, 295–316
   artificial stimulation and, 126–127, 159–162
   cerebellum and, 401
   end effectors and, 148–162
   fixation-centered coordinates and, 205–225
   forward models and, 336–339
   Huntington's disease and, 368–369
   kinematics and, 144
   remapping and, 319–333
   robotics and, 273–275
   rotation matrices and, 151–154
   state estimation and, 503–513
   target location and, 180–183
Proteins, 35
   actin and, 94
   myosin and, 94
   water and, 37
Protists, 36
Protostomes, 13, 15
Psychophysics, 5, 170, 183, 237
   complex spikes and, 478–480
   go function and, 357–360
   Hoff-Arbib model and, 357–360
   limb velocity and, 425–426
   remapping and, 319–333
   smoothness and, 341–352
Purkinje cells, 68–70, 76, 404, 462
   adaptive filtering and, 450, 453
   cerebral training signal and, 455–458
   complex spikes and, 478–480
   electricity and, 450
   eyeblink conditioning and, 459–461
   flocculus and, 483–486
   FTN discharge and, 486–487
   inverse dynamics and, 488–491
   training signal and, 455–458
   VOR and, 483–488
Putamen, 75

Pyramidal cells, 82
Pyramidal tract, 86

Raccoons, 50
Racetrack models, 496–499
Rao, Ashwini, 518
Reaching, 1
   adaptation and, 47–48, 51–58 (*see also* Adaptation)
   complex spikes and, 478–480
   context switching and, 78–79
   cortical fields and, 85–89
   decision making and, 495–520 (*see also* Decision making)
   difference vectors and, 205–244 (*see also* Difference vectors)
   end effectors and, 237, 331–336
   feedback and, 123–126, 135, 143–144
   gain fields and, 426–429
   hindbrain and, 66–68
   Huntington's disease and, 75, 134–135, 368–371, 468–469
   internal models and, 382–387
   kinematics and, 144–157
   land locomotion and, 31–32
   motor learning and, 50–51
   neural transmission and, 37–38
   Pavlovian learning and, 41, 44–46, 52, 449, 455–468, 473–480
   planning and, 229–242 (*see also* Planning)
   population coding and, 410–416
   reflexes and, 475–491 (*see also* Reflexes)
   regularity in, 343–350
   remapping and, 319–333
   robotics and, 382–384
   rotation and, 384–387
   saccades and, 234–236
   sequential, 508–512
   target location and, 179–201
   trajectory smoothness and, 341–352
   updating and, 331–336
   without vision, 257
Reaction time, 503–504, 512–513, 547–548
Reafference, 449
Receptive fields, 183–184

Reciprocal connections, 219
Reductionism, xvi
Redundancy, 31, 247
  control trade-offs and, 373–374
  degrees of freedom and, 372
  planning and, 371–374
  stiffness and, 372–373
  trajectories and, 371–374
Reference frames, 162–165
Reflexes, 11, 20
  autonomic, 467
  basal ganglia and, 134–135
  blocking effect and, 458
  cerebellum and, 134–135
  climbing fibers and, 475–480
  complex spike rate and, 478–480
  conditioning and, 44–46
  decision making and, 51–52, 495 (*see also* Decision making)
  delay period and, 229–231
  efference copy and, 474
  extinction and, 456, 458
  forward models and, 336–338
  interaction torques and, 480–482
  long-loop, 132, 368–369, 382
  motor command correction and, 480–491
  motor learning and, 41–46
  optokinetic, 482–483
  passive properties and, 136–139
  Pavlovian learning and, 41, 44–46, 52, 449, 455–468, 473–480
  primitive, 473
  redundancy and, 373
  short-loop, 132, 393, 473
  spinal pathways and, 420
  stability and, 48–49, 131–133
  stiffness and, 373
  stretch, 49, 111, 247, 473–474
  time delays and, 133–134
  unconditioned, 458
  vestibulo-ocular (VOR), 182
  withdrawal, 475–477
Reina, G. Anthony, 250, 265
Remapping
  efference copy and, 322, 325–331
  GABA and, 328
  Gaussian function and, 322
  head movement and, 324–325
  intervening eye movements and, 319–324
  LIP and, 322–324, 327
  oculomotor commands and, 327–328
  planned eye movement and, 326–327
  preferred direction and, 322
  transcranial magnetic stimulation (TMS) and, 330–331
  trunk movement and, 324–325
Representations, 165, 185. *See also* Mapping
  accumulator models and, 496–499
  adaptation and, 273–292, 295–316, 304
  adaptive filtering and, 453–454
  consolidation and, 435–444
  intrinsic coordinates and, 407
  planning and, 237
  somatotopic, 76, 85
  target location and, 217
Reticulospinal system, 65–68
Retinal location, 188–192
Retinotopic coordinates, 183–184
Reverse engineering, xv
Reward, 56, 464–467, 503–504
Reyes, Alexander, 201
Robotics, 1–2, 5, 93–94
  actuators and, 27–28
  adaptation and, 273–275, 284–286
  decision making and, 55
  endeffectors and, 148–151
  forces and, 245–248 (*see also* Forces)
  inertia and, 381–382
  internal dynamics models and, 379–401
  joint coordinates and, 139, 245–248
  manipulandum and, 123, 237, 288, 332–333, 343
  mapping and, 146–147
  planning and, 245–248
  preferred direction and, 395–401
  prism adaptation and, 284–286
  proprioception and, 273–275
  reaching and, 382–384 (*see also* Reaching)
  redundancy and, 247

target location and, 179–203
virtual, 148–151
visual coordinate smoothness and, 341–352
Roitman, Jamie, 500
Rossetti, Yves, 299
Rotation
eyes and, 275–278
internal models and, 384–387, 395–397
joint, 103, 410
matrices for, 151–156
physics and, 544–545
preferred direction and, 395–401
Rothwell, John, 135
Rushworth, Matthew, 175, 517

Saccades. *See also* Eye movement
accumulator models and, 496–499
activity field and, 239
antisaccades and, 500
double-jump tasks and, 328
multiple targets and, 503–504
planning and, 234–236, 240–241
prosaccades and, 500
reaching movement and, 234–236
remapping and, 319–333
reward-prediction and, 465–467
target location and, 184
transformational mapping and, 500
updating and, 331–336
Sacral designation, 61
Salinas, Emilio, 198–201
Sarcomeres, 35, 96–98
Sarcoplasm, 94
Sarcopterygians, 21
Scaling, 464
Schaal, Stefan, 361
Schall, Jeffrey, 240, 497, 499, 503
Schmidt, Richard, 40
Schultz, Wolfram, 465, 468
Schwartz, Andrew, 250, 257, 263, 265
Schweighofer, Nicholas, 489
Scott, Stephen, 254–257, 259, 265–266, 348
Sejnowski, Terrence, 197, 413–414
Sensitization, 44
Sensorimotor mapping. *See* Mapping

Sensory system, 547–548
adaptation and, 273–292, 295–316 (*see also* Adaptation)
basal ganglia and, 75–79
blocking effect and, 468
Brodmann's areas and, 86–89
conditioning and, 44–46
cue effects and, 257–261, 504–505
decision making and, 499–500 (*see also* Decision making)
efference copy and, 322, 325–331
electricity and, 450–452
escape and, 66
eyeblink conditioning and, 459–461
fixation-centered coordinates and, 205–225
forebrain and, 75–89
forward models and, 336–339
Gaussian distribution and, 214–215
hindbrain and, 65–68
instruction stimulus and, 229–231
kinesis and, 43–44
muscles and, 94–96 (*see also* Muscles)
neglect and, 296–301
perceptual threshold and, 500
preparation and, 230–231
Purkinje cells and, 68–70, 76, 404, 450, 453–462, 478–480, 483–491
reflexes and, 41–46, 472–491 (*see also* Reflexes)
remapping and, 319–333
sensitization and, 44
sequential targets and, 508–512
signal noise and, 364–366, 369–371
spatial neglect and, 296–301
stimulus prediction and, 454–464
stimulus response and, 454–464
target location and, 179–201
taxis and, 43–44
thalamus and, 80–81
trajectory smoothness and, 341–352
transcranial magnetic stimulation (TMS) and, 133–134, 330–331, 440–441
triggers and, 230–231
Sergio, Lauren, 254–257, 259, 262, 348, 403
Serial homologues, 29

Serial reaction-time task (SRTT), 512–513
Series elastic element, 99
Sessile animals, 15
Set-related activity, 547
Sexual behavior, 62, 64
Shadlen, Michael, 499–500
Shadmehr, Reza, 136, 332, 364, 367, 382–388, 393, 397, 405–407, 417, 421, 423, 426, 435–437, 518
Shared traits, 534
Shen, Liming, 252–254, 257, 259
Shenker, Barbara, 160
Sherrington, Charles, 111
Shiller, Douglas, 410
Short-loop reflexes, 132, 393, 473
Shoulders, 28–32
  elbow joint system and, 154–157
  fixation-centered coordinates and, 209–212
  interaction torques and, 480–482
  shoulder-centered coordinates, 209–212
Signal processing, 547–548
  adaptive critic and, 455
  adaptive filtering and, 449–454
  basal ganglia and, 462–464
  blocking effect and, 458–461, 468
  bottom-up, 498
  cerebellum and, 455–464
  complex spike rate and, 478–480
  cues and, 506–508
  decision making and, 495–520 (*see also* Decision making)
  dopamine cells and, 464–465
  electricity and, 450
  extinction and, 456, 458
  eyeblink conditioning and, 459–461
  feedback and, 447 (*see also* Feedback)
  Huntington's disease and, 369–371
  inferior olive (ION) and, 455–458
  noise and, 364–366, 369–371
  Pavlovian learning and, 41, 44–46, 52, 449, 455–468, 473–480
  Purkinje cells and, 68–70, 76, 404, 450, 453–462, 478–480, 483–491
  reflexes and, 473–491 (*see also* Reflexes)
  reward and, 464–467
  serial reaction-time tasks (SRTT) and, 512–513
  stimulus prediction and, 454–464
  stop signal, 503
  top-down, 498
  training signals and, 455–458, 464
Simple spikes, 69–70
Sines, 539
Sirigu, Angela, 338–339
Size principle, 99
Skills, 1–2, 42
  adaptation and, 47–48, 273–292, 295–316 (*see also* Adaptation)
  attractor states and, 220, 278
  basal ganglia and, 462–464
  benefits of, 50–51
  blocking effect and, 468
  cerebellum and, 455–464
  consolidation and, 435–444
  control and, 49–50 (*see also* Control)
  explicit, 83
  eyeblink conditioning and, 459–461
  implicit, 83
  Pavlovian learning and, 41, 44–46, 52, 449, 455–468, 473–480
  rote learning and, 389–391
  sleep and, 441–444
  stability and, 48–49
Sleep, 441–444
Smith, Maurice, 367, 426
Smoothness
  minimum-jerk trajectories and, 350–352
  visual coordinates and, 341–352
Snyder, Lawrence, 234, 239, 324
Somatosensory cortex, 84, 88
Somatotopic representations, 76, 85
Sommer, Marc, 327–328
Spatial neglect, 296–301
Sperry, Roger, 275–276, 280
Spinal cord, 10–11, 20. *See also* Vertebrates
  artificial stimulation and, 126–127
  caudal designation and, 61
  cervical segments and, 61–62
  configuration coding and, 165–167
  ejaculation and, 64

equilibrium points and, 126–127
escape behavior and, 66
feedback and, 123–126
ganglia and, 62
motor neurons and, 86
myotomes and, 63–64
organization of, 61–64
primary sensory afferents and, 61–63
proprioceptive pathways and, 62–65
reaching and, 66–68
red nucleus and, 71–73
rostral designation and, 61
stretch reflexes and, 247
Spiny neurons, 76–78
Spring loads, 545
Stability, xvi, 41, 48–49
  artificial stimulation and, 126–127
  coactivation and, 123
  feedback and, 123–126, 135
  length-tension properties and, 121–122
  moment arm and, 119–120
  passive properties and, 129–131, 136–139
  reflexes and, 48–49, 131–135
  sequential activation and, 127–129
State space, 391
Stein, John, 247
Stem group, 12
Stiffness, 28, 97, 136, 388
  adaptation and, 423–424
  coactivation and, 123
  joint, 138
  redundancy and, 372–373
  sarcomeres and, 96–97
Stikgold, Robert, 442, 444
Stimuli. *See* Sensory system
Stop signal, 503
Striatal cells, 368
Strick, Peter, 134, 249
Striosomes, 368, 463
Substantia innominata, 75
Sugita, Yoichi, 276
Superior colliculus, 73
Supination, 28–29
Supplementary eye field (SEF), 86, 88
Supplementary motor area (SMA), 86, 508–512

Swimming, 12–13, 32, 42, 63–64
Synapses. *See* Signal processing
Synaptic weights, 414, 439, 453, 486
Synergy, 164
Systems identification, 423

Tanji, Jun, 229, 237, 252
Target location, 41, 179, 245
  accumulator models and, 496–499
  adaptation and, 273–292, 295–316
  attractors and, 220
  cell discharge rates and, 212–216, 222–225
  common frame and, 180–183
  decision making and, 496–504
  difference vectors and, 205–244 (*see also* Difference vectors)
  efference copy and, 322, 325–331
  EMG changes and, 393–395
  error and, 186–188, 205–209
  fixation-centered coordinates and, 182–186, 205–225
  flocculus and, 483–486
  fovea and, 182, 184
  gain fields and, 188–201, 426–429
  Gaussian distribution and, 214–215
  hand orientation and, 249–252
  intrinsic coordinates, 182
  neglect and, 296–301
  network analysis and, 211–212, 215–218, 221–222
  path altering and, 344–347
  planning and, 354–357 (*see also* Planning)
  proprioception and, 180–183, 273–292 (*see also* Proprioception)
  Purkinje cells and, 483–491
  reciprocal connections and, 219
  remapping and, 319–333
  retinotopic coordinates and, 183–184
  saccades and, 184
  sequential movement and, 241–242, 508–512
  serial reaction-time task and, 512–513
  shoulder-centered coordinates and, 209–212
  signal noise and, 364–366
  target disappearance and, 231–233

Target location (cont.)
  trajectories and, 354–357, 360–364, 371–374
  transducers and, 180
  updating and, 331–336
  visual coordinate smoothness and, 341–352
  zigzag paths and, 360–364
Taxis, 43–44
Telencephalon, 16–17
Tension, 99
Tetrapods, 21–22, 534
Thach, W. Thomas, 280, 302, 478, 482
Thalamus, 11, 75–76, 80–81
Thick filaments, 94
Thigmotaxis, 44
Thin filaments, 94
Thompson, Richard, 456, 458
Thoroughman, Kurt, 393, 397, 417
Tibia, 29
Tinbergen, Niko, 42
Todorov, Emanuel, 364, 373–374
Tonically active neurons (TANs), 467
Torque, 48–49, 93, 179, 245
  angular acceleration and, 120–121
  angular velocity and, 102
  arbitrary mapping and, 517
  basal ganglia and, 134
  defined, 545
  dynamics and, 362–364
  force conversion and, 102–108
  interaction, 48, 380, 480–482
  inverse dynamics and, 362–364
  Jacobians and, 105–108, 119
  joint velocity and, 405–406
  length-tension properties and, 121–123
  negative, 119
  M1 activity and, 261–263
  moment arm and, 104–105
  positive, 119
  step, 123
  virtual work and, 102–104
Toughness, 28
Training signals, 455–458, 464
Trajectories, 432
  control trade-offs and, 373–374
  difference vector transformation and, 354–357
  Hoff-Arbib model and, 357–360
  planning and, 354–357, 360–364, 371–374
  redundancy and, 371–374
  signal noise and, 364–366
  smoothness and, 342–351
  S-shaped movements and, 432
  target size and, 360
  tuning curves and, 414–416
Transcranial magnetic stimulation (TMS), 133–134, 330–331, 440–441
Transducers, 27, 382
  difference vectors and, 250
  kinematics and, 144
  motor learning and, 44
  muscle afferents and, 108–117
  spinal cord and, 61–63
  target location and, 180
Transformational mapping, 58, 513–516
Truitt, William, 64
T-tubules, 94
Tunicates, 16
Tuning functions, 215–217, 229–231, 414–416

Ulna, 28–29
Unconditioned stimulus (US), 458–459, 473
Updating, 331–336
Up state, 77

van Beers, Robert, 170
Ventral horn, 62
Ventral premotor cortex (PMv), 86
  arbitrary mapping and, 517
  caudal, 504
  decision making and, 504–505
  hand location and, 252–254
  movement planning and, 230–232, 234–235, 237, 239–240, 242
  rostral, 504
  sensory cue effects and, 257–261
  sequential reaching and, 510
  varying feedback and, 252
  visionless reaching and, 257

Ventral striatum, 75
Vertebrates, 4–5, 10
  adaptation and, 275–276 (*see also* Adaptation)
  brain and, 15–16
  center of gravity and, 66–67
  head and, 17–19
  inertia and, 96
  limbs and, 28–32
  muscles and, 32–36 (*see also* Muscles)
  myoblasts and, 19
  sarcopterygians, 21
  spinal cord and, 61–65
  telencephalon and, 16–17
Vestibular afferents, 68
Vestibular nuclei, 20, 483
Vestibular receptors, 18
Vestibulo-ocular reflex (VOR), 182, 474–475
  adaptation and, 486–488
  cerebellum and, 483–486
  flocculus and, 483–486
  Purkinje cells and, 483–488
Vetter, Philipp, 305
Vibration, 159–162
Vindras, Philippe, 206–207
Virtual work, 104–108
Viscosity, 545
Vision, 9, 18, 21. *See also* Eyes
  activity field and, 172–175, 188–192, 239
  adaptation and, 273–292, 295–316
  artificial muscle stimulation and, 159–162
  coordinate smoothness and, 341–352
  end effectors and, 147–148
  fixation-centered coordinates and, 205–225
  flocculus and, 483–486
  gain fields and, 188–201
  gaze effects and, 261
  impairment and, 348–349
  modulation and, 188–201
  neglect and, 296–301
  optic ataxia and, 229
  proprioception and, 273–292 (*see also* Proprioception)
  reaching and, 257 (*see also* Reaching)
  remapping and, 319–333
  sensory cue effects and, 257–261
  spatial neglect and, 296–301
  superior colliculus and, 73
  V1 area and, 83
  varying feedback and, 252
  varying gaze location and, 249

Walker, Matthew, 442, 444
Walls, Gordon, 182
Weinrich, Michael, 230, 257, 261
Westphal, Carl, 111
Whitehead, Alfred North, 2
Wiesel, Torsten, 280
Willingham, Daniel, 512
Wise, Steven, 58, 230, 232, 257, 261, 316
Wolpert, Daniel, 292, 308, 344, 346, 354, 366
Work, 32
  defined, 543, 545
  joint rotation, 103
  torque conversion and, 102–108
Wright brothers, xvi–xvii
Wurtz, Robert, 327–328

Yin, Ping-Bo, 287
Young, Sereniti, 314

Zaccaria, Roberto, 372
Z disk, 96
Zigzag paths, 360–364
Zipser, David, 193–195, 197–198, 211, 216